THEORY OF ORDINARY DIFFERENTIAL EQUATIONS

Theory of Ordinary Differential Equations

EARL A. CODDINGTON

Assistant Professor of Mathematics
University of California, Los Angeles

NORMAN LEVINSON

Professor of Mathematics
Massachusetts Institute of Technology

KRIEGER PUBLISHING COMPANY
MALABAR, FLORIDA

Original Edition 1955
Reprint Edition 1984

Printed and Published by
**ROBERT E. KRIEGER PUBLISHING COMPANY, INC.
KRIEGER DRIVE
MALABAR, FLORIDA 32950**

Printed in the United States of America

Library of Congress Cataloging in Publication Data

Coddington, Earl A., 1920-
Theory of ordinary differential equations

Reprint. Originally published: New York : McGraw-Hill, 1955.
Bibliography: p.
Includes index.
1. Differential equations. I. Levinson, Norman, 1912-
II. Title.
QA 372.C6 1984 515.3'52 84-4438
ISBN 0-89874-755-4

12 13 14 15 16 17 18 19

To

FAGI AND SUE

PREFACE

This book has developed from courses given by the authors and probably contains more material than will ordinarily be covered in a one-year course. The selection of material is partly conditioned by the interests of the authors.

It is hoped that the book will be a useful text in the application of differential equations as well as for the pure mathematician. Prerequisite for this book is a knowledge of matrices and of the essentials of functions of a complex variable. The notion of the Lebesgue integral is used in Chaps. 2, 7, 9, and 10. However, Chap. 2 is needed only for certain parts of Chap. 15, which, so far as applications go, are adequately covered by Chap. 13. The Lebesgue integral can easily be avoided in Chap. 7, as is indicated there. However, a rigorous study of Chaps. 9 and 10 requires a mathematical sophistication that would certainly include the ability to understand the statements of the theorems required from integration theory. An alternative approach is to apply the theory of Chaps. 9 and 10 to a restricted class of functions as is done in the proof of Theorem 3.1 of Chap. 9. This approach requires a knowledge of the Riemann-Stieltjes integral only.

Chapters 3 through 12 are on linear equations. For linear theory, it is not necessary to cover the existence theory of Chap. 1. For Chap. 3, the necessary theorem is sketched in Prob. 1 at the end of that chapter. The discussion in Sec. 7 of Chap. 3 suffices for Chaps. 4 and 5. For Chaps. 7 through 12, Prob. 7 of Chap. 1 provides the additional existence theory needed.

Chapters 4, 5, and 6 are not needed for any later chapters. Chapter 8 is not required for any later chapter, nor are Chaps. 9 and 10. Chapter 8 does not depend on Chap. 7.

Chapter 12 requires only Chap. 7 and, for Sec. 5, also Chap. 11.

Chapters 1 and 3 only are required for Chaps. 13 and 14. Chapter 1 will suffice for most of Chap. 15 and for Chaps. 16 and 17.

No attempt has been made to give the historical origin of the theory, and only a limited number of references are given at the end of the book. In keeping with this approach, the authors make no mention in the text where they present new results.

The problems, in some cases, give additional material not considered in the text.

The preparation of this book was greatly facilitated by a grant from the Office of Naval Research.

The authors are indebted to a number of colleagues and students who read portions of the manuscript, in particular, F. G. Brauer, Prof. A. Horn, and Dr. J. J. Levin.

EARL A. CODDINGTON
NORMAN LEVINSON

CONTENTS

THEORY OF ORDINARY DIFFERENTIAL EQUATIONS

CHAPTER 1

EXISTENCE AND UNIQUENESS OF SOLUTIONS

1. Existence of Solutions

Let I denote an open interval on the real line $-\infty < t < \infty$, that is, the set of all real t satisfying $a < t < b$ for some real constants a and b. The set of all complex-valued functions having k continuous derivatives on I is denoted by $C^k(I)$. If f is a member of this set, one writes $f \in C^k(I)$, or $f \in C^k$ on I. The symbol $\in$ is to be read "is a member of" or "belongs to." It is convenient to extend the definition of C^k to intervals I which may not be open. The real intervals $a < t < b$, $a \leqq t \leqq b$, $a \leqq t < b$, and $a < t \leqq b$ will be denoted by (a,b), $[a,b]$, $[a,b)$, and $(a,b]$, respectively. If $f \in C^k$ on (a,b), and in addition the right-hand kth derivative of f exists at a and is continuous from the right at a, then f is said to be of class C^k on $[a,b)$. Similarly, if the kth derivative is continuous from the left at b, then $f \in C^k$ on $(a,b]$. If both these conditions hold, one says $f \in C^k$ on $[a,b]$.

If D is a *domain*, that is, an open connected set, in the real (t,x) plane, the set of all complex-valued functions f defined on D such that all kth-order partial derivatives $\partial^k f/\partial t^p \partial x^q$ $(p + q = k)$ exist and are continuous on D is denoted by $C^k(D)$, and one writes $f \in C^k(D)$, or $f \in C^k$ on D.

The sets $C^0(I)$ and $C^0(D)$, the continuous functions on I and D, will be denoted by $C(I)$ and $C(D)$, respectively.

Let D be a domain in the (t,x) plane and suppose f is a real-valued function such that $f \in C(D)$. Then the central problem of this chapter may be phrased as follows:

Problem. *To find a differentiable function φ defined on a real t interval I such that*

$$\text{(i)} \qquad (t,\varphi(t)) \in D \qquad (t \in I)$$

$$\text{(ii)} \qquad \varphi'(t) = f(t,\varphi(t)) \qquad \left(t \in I,\ ' = \frac{d}{dt}\right)$$

This problem is called an *ordinary differential equation of the first order*, and is denoted by

$$\text{(E)} \qquad x' = f(t,x) \qquad \left(' = \frac{d}{dt}\right)$$

If such an interval I and function φ exist, then φ is called a *solution of the differential equation* (E) *on* I. Clearly if φ is a solution of (E) on I, then $\varphi \in C^1$ on I, on account of (ii).

In geometrical language, (E) prescribes a slope $f(t,x)$ at each point of D. A solution φ on I is a function whose graph [the set of all points $(t,\varphi(t))$, $t \in I$] has the slope $f(t,\varphi(t))$ for each $t \in I$.

The problem (E) may have many solutions on an interval I. For example, the simple equation

$$x' = 1$$

has, for any given real constant c, the solution φ_c given by

$$\varphi_c(t) = t + c$$

on any t interval I. However, there exists only one solution passing through the point (1,1), say, and existing on an interval I containing $t = 1$, namely, φ_0. Therefore, in order to be able to talk about uniqueness of solutions of (E), one is led to the problem of finding a solution passing through a given point in the (t,x) plane.

Suppose (τ,ξ) is a given point in D. Then an *initial-value problem* associated with (E) and this point is defined in the following way:

Initial-value Problem. *To find an interval I containing τ and a solution φ of* (E) *on I satisfying*

$$\varphi(\tau) = \xi$$

This problem is denoted by

$$x' = f(t,x) \qquad x(\tau) = \xi$$

Suppose φ is such a solution which exists on an interval I. Then by integrating (ii) one obtains immediately the integral equation

$$\varphi(t) = \xi + \int_\tau^t f(s,\varphi(s))\,ds \qquad (t \in I)$$

Conversely, suppose $\varphi \in C$ is a function satisfying the above integral equation on I. Then clearly $\varphi(\tau) = \xi$, and by differentiating the equation it follows that φ is a solution of (E) on I. In other words, there is a correspondence between solutions φ of (E) on I satisfying $\varphi(\tau) = \xi$, and continuous functions φ satisfying the above integral relation on I. Thus the initial-value problem for (E) and (τ,ξ) on I is completely equivalent to the finding of all continuous φ on I satisfying the integral equation.

Given a continuous function f on a domain D as above, the first question to be answered is whether there exists a solution of the equation (E). The answer is yes, if I is properly prescribed. An indication of the limitation of any general existence theorem can be seen by considering the simple example

$$x' = x^2$$

It is clear that a solution of this equation which passes through the point $(1,-1)$ is given by $\varphi(t) = -t^{-1}$. However, this solution does not exist at $t = 0$, although $f(t,x) = x^2$ is continuous there. This shows that any general existence theorem will necessarily have to be of a *local* nature, and existence in the large can only be asserted under additional conditions on f.

The local existence proof proceeds by two stages. First, it is shown by an actual construction that there exists an "approximate" solution to (E), in a sense to be made precise below. Then one proves that there exists a sequence of these approximate solutions which tend to a solution of (E).

Let f be a real-valued continuous function on a domain D in the (t,x) plane. An *ϵ-approximate solution* of (E) on a t interval I is a function $\varphi \in C$ on I such that

(i) $(t,\varphi(t)) \in D \qquad (t \in I)$
(ii) $\varphi \in C^1$ on I, except possibly for a finite set of points S on I, where φ' may have simple discontinuities†
(iii) $|\varphi'(t) - f(t,\varphi(t))| \leqq \epsilon \qquad (t \in I - S)$

Any function $\varphi \in C$ satisfying property (ii) on I is said to have a *piecewise continuous derivative* on I, and this is denoted by $\varphi \in C_p^1(I)$.

If $f \in C$ on the rectangle

$$R: \quad |t - \tau| \leqq a \qquad |x - \xi| \leqq b \qquad (a,b > 0)$$

about the point (τ,ξ), it is bounded there. Let

$$M = \max |f(t,x)| \qquad ((t,x) \in R)$$

and

$$\alpha = \min\left(a, \frac{b}{M}\right)$$

Theorem 1.1. *Let $f \in C$ on the rectangle R. Given any $\epsilon > 0$, there exists an ϵ-approximate solution φ of* (E) *on $|t - \tau| \leqq \alpha$ such that $\varphi(\tau) = \xi$.*

Proof. Let $\epsilon > 0$ be given. An ϵ-approximate solution will be constructed for the interval $[\tau, \tau + \alpha]$; a similar construction will define it for $[\tau - \alpha, \tau]$. This approximate solution will consist of a polygonal path starting at (τ,ξ), that is, a finite number of straight-line segments joined end to end.

† A function g is said to have a simple discontinuity at a point c if the right and left limits of g at c exist but are not equal. In case $\epsilon = 0$, it will be understood that the set S is empty.

Since $f \varepsilon C$ on R, it is uniformly continuous there, and hence, for the given ϵ, there exists a $\delta_\epsilon > 0$ such that

$$|f(t,x) - f(\bar{t},\bar{x})| \leqq \epsilon \tag{1.1}$$

if

$$(t,x) \varepsilon R,\ (\bar{t},\bar{x}) \varepsilon R, \text{ and } |t - \bar{t}| \leqq \delta_\epsilon \qquad |x - \bar{x}| \leqq \delta_\epsilon$$

Now divide the interval $[\tau, \tau + \alpha]$ into n parts

$$\tau = t_0 < t_1 < \cdot \cdot \cdot < t_n = \tau + \alpha$$

in such a way that

$$\max |t_k - t_{k-1}| \leqq \min \left(\delta_\epsilon, \frac{\delta_\epsilon}{M} \right) \tag{1.2}$$

From (τ,ξ), construct a straight-line segment with slope $f(\tau,\xi)$ proceeding to the right of τ until it intersects the line $t = t_1$ at some point (t_1,x_1).

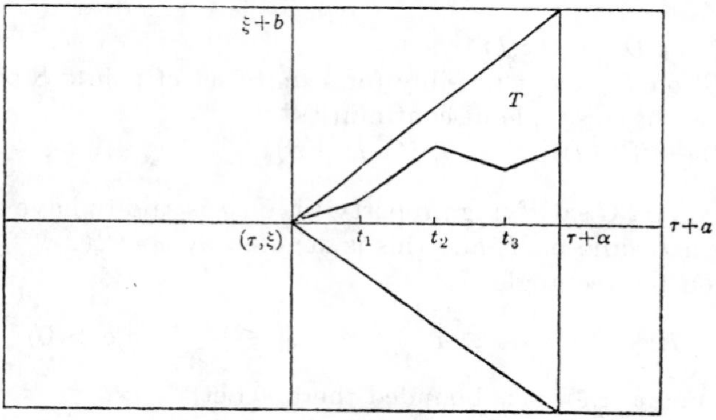

FIG. 1

This segment must lie inside the triangular region T bounded by the lines issuing from (τ,ξ) with slope M and $-M$, and the line $t = \tau + \alpha$ (see Fig. 1, where α is shown as b/M). This follows immediately from the definition of α and the fact that $|f(t,x)| \leqq M$. In particular, the constructed segment actually meets the line $t = t_1$ in T. At the point (t_1,x_1) construct to the right of t_1 a straight-line segment with slope $f(t_1,x_1)$ up to the intersection with $t = t_2$, say at (t_2,x_2). Continuing in this fashion, in a finite number of steps the resultant path φ will meet the line $t = \tau + \alpha$. Further, the path will lie completely within T.

This φ is the required ϵ-approximate solution. Analytically it may be expressed as

$$\begin{aligned} \varphi(\tau) &= \xi \\ \varphi(t) &= \varphi(t_{k-1}) + f(t_{k-1},\varphi(t_{k-1}))(t - t_{k-1}) \\ t_{k-1} &< t \leqq t_k \qquad k = 1, \ . \ . \ . \ , n \end{aligned} \tag{1.3}$$

From the construction of φ it is clear that $\varphi \in C_p^1$ on $[\tau, \tau + \alpha]$, and that

$$|\varphi(t) - \varphi(\tilde{t})| \leqq M|t - \tilde{t}| \qquad (t,\tilde{t} \text{ in } [\tau, \tau + \alpha]) \tag{1.4}$$

If t is such that $t_{k-1} < t < t_k$, then (1.4) together with (1.2) imply that $|\varphi(t) - \varphi(t_{k-1})| \leqq \delta_\epsilon$. But from (1.3) and (1.1),

$$|\varphi'(t) - f(t,\varphi(t))| = |f(t_{k-1},\varphi(t_{k-1})) - f(t,\varphi(t))| \leqq \epsilon$$

This shows that φ is an ϵ-approximate solution, as desired.

The construction of Theorem 1.1 is sometimes used as a practical means for finding an approximate solution. In fact, what has been found is really a set of points $(t_k,\varphi(t_k))$ and these are joined by line segments. The points, by (1.3), satisfy the difference equation

$$x_k - x_{k-1} = (t_k - t_{k-1})f(t_{k-1},x_{k-1})$$

This is a formulation that might be used on a digital computing machine, for example.

The existence of a solution of (E) will now be deduced. For the reader mainly interested in the applications, other existence proofs, under more restricted assumptions on f, are given in Theorems 2.3 and 3.1; the rest of this section can be omitted.

In order to prove the existence of a sequence of approximate solutions tending to a solution of (E), where the only hypothesis is $f \in C$ on R, the notion of an equicontinuous set of functions is required. A set of functions $F = \{f\}$ defined on a real interval I is said to be *equicontinuous* on I if, given any $\epsilon > 0$, there exists a $\delta_\epsilon > 0$, independent of $f \in F$ and also $t, \tilde{t} \in I$ such that

$$|f(t) - f(\tilde{t})| < \epsilon \qquad \text{whenever } |t - \tilde{t}| < \delta_\epsilon$$

The fundamental property of such sets of functions needed here is given in the following lemma:

Lemma (Ascoli). *On a bounded interval I, let $F = \{f\}$ be an infinite, uniformly bounded, equicontinuous set of functions. Then F contains a sequence $\{f_n\}$, $n = 1, 2, \ldots$, which is uniformly convergent on I.*

Proof. Let $\{r_k\}$, $k = 1, 2, \ldots$, be the rational numbers in I enumerated in some order. The set of numbers $\{f(r_1)\}$, $f \in F$, is bounded, and hence there exists a sequence of distinct functions $\{f_{n1}\}$, $f_{n1} \in F$, such that the sequence $\{f_{n1}(r_1)\}$ is convergent. Similarly, the set of numbers $\{f_{n1}(r_2)\}$ has a convergent subsequence $\{f_{n2}(r_2)\}$. Continuing in this way, an infinite set of functions $f_{nk} \in F$, n, $k = 1, 2, \ldots$, is obtained which have the property that $\{f_{nk}\}$ converges at $r_1, \ldots, r_k$. Define f_n to be the function f_{nn}. Then $\{f_n\}$ is the required sequence which is uniformly convergent on I.

Clearly $\{f_n\}$ converges at each of the rationals on I. Thus, given any

$\epsilon > 0$ and rational number $r_k \in I$, there exists an integer $N_\epsilon(r_k)$ such that

$$|f_n(r_k) - f_m(r_k)| < \epsilon \qquad (n,m > N_\epsilon(r_k))$$

For the given ϵ there exists a δ_ϵ, independent of $t,\bar{t}$ and $f \in F$ such that

$$|f(t) - f(\bar{t})| < \epsilon \qquad |t - \bar{t}| < \delta_\epsilon$$

Divide the interval I into a finite number of subintervals $I_1, \ldots, I_p$ such that the length of the largest subinterval is less than δ_ϵ. For each I_k choose a rational number $\tilde{r}_k \in I_k$. If $t \in I$, then t is in some I_k, and hence

$$|f_n(t) - f_m(t)| \leqq |f_n(t) - f_n(\tilde{r}_k)| + |f_n(\tilde{r}_k) - f_m(\tilde{r}_k)| + |f_m(\tilde{r}_k) - f_m(t)| < 3\epsilon$$

provided that $n,m > \max(N_\epsilon(\tilde{r}_1), \ldots, N_\epsilon(\tilde{r}_p))$. This proves the uniform convergence of the sequence $\{f_n\}$ on I.

Theorem 1.2 (**Cauchy-Peano Existence Theorem**). *If $f \in C$ on the rectangle R, then there exists a solution $\varphi \in C^1$ of* (E) *on $|t - \tau| \leqq \alpha$ for which $\varphi(\tau) = \xi$.*

Proof. Let $\{\epsilon_n\}$, $n = 1, 2, \ldots$, be a monotone decreasing sequence of positive real numbers tending to zero as $n \to \infty$. By Theorem 1.1, for each ϵ_n there exists an ϵ_n-approximate solution, φ_n, of (E) on $|t - \tau| \leqq \alpha$ such that $\varphi_n(\tau) = \xi$. Choose one such solution φ_n for each ϵ_n. From (1.4) it follows that

$$|\varphi_n(t) - \varphi_n(\bar{t})| \leqq M|t - \bar{t}| \tag{1.5}$$

Applying (1.5) to $\bar{t} = \tau$, one readily sees, since $|t - \tau| \leqq b/M$, that the sequence $\{\varphi_n\}$ is uniformly bounded by $|\xi| + b$. Moreover, (1.5) implies that $\{\varphi_n\}$ is an equicontinuous set. By the Ascoli lemma, there exists a subsequence $\{\varphi_{n_k}\}$, $k = 1, 2, \ldots$, of $\{\varphi_n\}$, converging uniformly on $[\tau - \alpha, \tau + \alpha]$ to a limit function φ, which must be continuous since each φ_n is continuous. [Indeed, it follows from (1.5) that $|\varphi(t) - \varphi(\bar{t})| \leqq M|t - \bar{t}|$.]

This limit function φ is a solution of (E) which meets the required specifications. To see this, one writes the relation defining φ_n as an ϵ_n-approximate solution in an integral form, as follows:

$$\varphi_n(t) = \xi + \int_\tau^t (f(s,\varphi_n(s)) + \Delta_n(s))\, ds \tag{1.6}$$

where $\Delta_n(t) = \varphi_n'(t) - f(t,\varphi_n(t))$ at those points where φ_n' exists, and $\Delta_n(t) = 0$ otherwise. Because φ_n is an ϵ_n-approximate solution, $|\Delta_n(t)| \leqq \epsilon_n$. Since f is uniformly continuous on R, and $\varphi_{n_k} \to \varphi$ uniformly on

$[\tau - \alpha, \tau + \alpha]$, as $k \to \infty$, it follows that $f(t,\varphi_{n_k}(t)) \to f(t,\varphi(t))$ uniformly on $[\tau - \alpha, \tau + \alpha]$, as $k \to \infty$. Replacing n by n_k in (1.6) one obtains, in letting $k \to \infty$,

$$\varphi(t) = \xi + \int_\tau^t f(s,\varphi(s))\, ds \tag{1.7}$$

But from (1.7), $\varphi(\tau) = \xi$, and, upon differentiation, $\varphi'(t) = f(t,\varphi(t))$, for $f(t,\varphi(t))$ is a continuous function. It is clear from this that φ is a solution of (E) on $|t - \tau| \leqq \alpha$ of class C^1.

In general, the choice of a subsequence of $\{\varphi_n\}$ in the above proof is necessary, for there exist polygonal paths $\{\varphi_n\}$ which diverge everywhere on a whole interval about $t = \tau$ as $\epsilon_n \to 0$; see Prob. 12.

If it is assumed that a solution of (E) through (τ,ξ) (if it exists) is unique, then *every* sequence of polygonal paths $\{\varphi_n\}$ for which $\epsilon_n \to 0$ must converge on $|t - \tau| \leqq \alpha$, and hence uniformly, to a solution, for $\{\varphi_n\}$ is an equicontinuous set on $|t - \tau| \leqq \alpha$. Suppose this were false. Then there would exist a sequence of polygonal paths $\{\varphi_n\}$ divergent at some point on $|t - \tau| \leqq \alpha$. This implies the existence of at least two subsequences of $\{\varphi_n\}$ tending to different limit functions. Both will be solutions, and this gives a contradiction. Therefore, if uniqueness is assured, the choice of a subsequence in Theorem 1.2 is unnecessary.

It can happen that the choice of a subsequence is unnecessary even though uniqueness is not satisfied. The example

$$x' = x^{\frac{1}{3}} \tag{1.8}$$

illustrates this. There are an *infinite* number of solutions starting at (0,0) which exist on [0,1]. For any c, $0 \leqq c \leqq 1$, the function φ_c defined by

$$\begin{aligned} \varphi_c(t) &= 0 \qquad (0 \leqq t \leqq c) \\ \varphi_c(t) &= \left(\frac{2(t-c)}{3}\right)^{\frac{3}{2}} \qquad (c < t \leqq 1) \end{aligned} \tag{1.9}$$

is a solution of (1.8) on [0,1]. If the construction of Theorem 1.1 is applied to Eq. (1.8), one finds that the only polygonal path starting at the point (0,0) is φ_1. This shows that this method cannot, in general, give *all* solutions of (E).

Theorem 1.3. *Let $f \in C$ on a domain D in the (t,x) plane, and suppose (τ,ξ) is any point in D. Then there exists a solution φ of* (E) *on some t interval containing τ in its interior.*

Proof. Since D is open, there exists an $r > 0$ such that all points, whose distance from (τ,ξ) is less than r, are contained in D. Let R be any closed rectangle containing (τ,ξ), and contained in this open circle of radius r. Then Theorem 1.2 applied to (E) on R gives the required result.

2. Uniqueness of Solutions

The example (1.8), with the solutions given in (1.9), shows that something more than the continuity of f in (E) is required in order to guarantee that a solution passing through a given point be unique. A simple condition which permits one to imply uniqueness is the Lipschitz condition. Suppose f is defined in a domain D of the (t,x) plane. If there exists a constant $k > 0$ such that for every (t,x_1) and (t,x_2) in D

$$|f(t,x_1) - f(t,x_2)| \leqq k|x_1 - x_2|$$

then f is said to satisfy a *Lipschitz condition* (with respect to x) in D, and this fact will be denoted by $f \in \text{Lip}$ in D. The constant k is called the *Lipschitz constant*. If, in addition $f \in C$ in D, one writes $f \in (C,\text{Lip})$ in D. If $f \in \text{Lip}$ in D, then f is uniformly continuous in x for each fixed t, although nothing is implied concerning the continuity of f with respect to t. If D is convex (that is, D contains the line segment connecting any two points in D), then an application of the mean-value theorem of differential calculus shows that the existence and boundedness of f_x $(= \partial f/\partial x)$ in D are sufficient for $f \in \text{Lip}$ in D.

Before proceeding to the uniqueness proof, an important inequality will be deduced. In the following, D is a domain in the (t,x) plane.

Theorem 2.1. *Suppose $f \in (C,\text{Lip})$ in D, with Lipschitz constant k. Let φ_1,φ_2 be ϵ_1- and ϵ_2-approximate solutions of* (E) *of class C_p^1 on some interval (a,b), satisfying for some $\tau, a < \tau < b$,*

$$|\varphi_1(\tau) - \varphi_2(\tau)| \leqq \delta \tag{2.1}$$

where δ is a nonnegative constant. If $\epsilon = \epsilon_1 + \epsilon_2$, then for all $t \in (a,b)$,

$$|\varphi_1(t) - \varphi_2(t)| \leqq \delta e^{k|t-\tau|} + \frac{\epsilon}{k}(e^{k|t-\tau|} - 1) \tag{2.2}$$

Theorem 2.1 is of practical as well as theoretical interest since in computational procedures it is always approximate solutions of a differential equation that are found.

Proof of Theorem 2.1. Consider the case where $\tau \leqq t < b$; a corresponding proof holds for $a < t \leqq \tau$. Since φ_1,φ_2 are ϵ_1- and ϵ_2-approximate solutions of (E),

$$|\varphi_i'(s) - f(s,\varphi_i(s))| \leqq \epsilon_i \qquad (i = 1, 2) \tag{2.3}$$

at all but a finite number of points on $\tau \leqq s < b$.

Integrating from τ to t, where $\tau \leqq t < b$, (2.3) yields

$$\left| \varphi_i(t) - \varphi_i(\tau) - \int_\tau^t f(s,\varphi_i(s))\, ds \right| \leqq \epsilon_i(t - \tau) \qquad (i = 1, 2)$$

Using the fact that $|\alpha - \beta| \leqq |\alpha| + |\beta|$, the above gives

$$\left| (\varphi_1(t) - \varphi_2(t)) - (\varphi_1(\tau) - \varphi_2(\tau)) - \int_\tau^t [f(s,\varphi_1(s)) - f(s,\varphi_2(s))]\, ds \right| \leqq \epsilon(t - \tau)$$

Let r be the function defined on $[\tau,b)$ by $r(t) = |\varphi_1(t) - \varphi_2(t)|$. Then the preceding inequality gives

$$r(t) \leqq r(\tau) + \int_\tau^t |f(s,\varphi_1(s)) - f(s,\varphi_2(s))|\, ds + \epsilon(t - \tau)$$

and using the fact that $f \in$ Lip in D, one gets

$$r(t) \leqq r(\tau) + k \int_\tau^t r(s)\, ds + \epsilon(t - \tau) \tag{2.4}$$

Define the function R by

$$R(t) = \int_\tau^t r(s)\, ds \qquad (\tau \leqq t < b)$$

In terms of R, (2.4) is

$$R'(t) - kR(t) \leqq \delta + \epsilon(t - \tau)$$

since by (2.1) $r(\tau) \leqq \delta$. Multiply both sides of this inequality by $e^{-k(t-\tau)}$ and integrate the resulting expression from τ to t, obtaining

$$e^{-k(t-\tau)}R(t) \leqq \frac{\delta}{k}(1 - e^{-k(t-\tau)}) - \frac{\epsilon}{k^2}e^{-k(t-\tau)}(1 + k(t - \tau)) + \frac{\epsilon}{k^2}$$

or

$$R(t) \leqq \frac{\delta}{k}(e^{k(t-\tau)} - 1) - \frac{\epsilon}{k^2}(1 + k(t - \tau)) + \frac{\epsilon}{k^2}e^{k(t-\tau)} \tag{2.5}$$

Combining (2.5) with (2.4), there results finally

$$r(t) \leqq \delta e^{k(t-\tau)} + \frac{\epsilon}{k}(e^{k(t-\tau)} - 1)$$

which is the desired result on $[\tau,b)$.

A particularly important case of Theorem 2.1 occurs when $\varphi_1 = \varphi$ is an actual solution of (E). The theorem then shows that as ϵ_2 and $\delta \to 0$ the approximate solution tends to the actual solution.

The inequality (2.2) is the best possible, in the sense that equality can be attained for nontrivial φ_1 and φ_2. For example, let k, ϵ_1, ϵ_2 be any real constants, and let P_1: $(0,\xi_1)$, P_2: $(0,\xi_2)$ be two points in the (t,x) plane. Let $\varphi_1(0) = \xi_1$, and $\varphi_2(0) = \xi_2$ and let φ_1 and φ_2 be solutions of the equations

$$x' = kx - \epsilon_1 \qquad x' = kx + \epsilon_2$$

respectively, on [0,1]. Then φ_1 and φ_2 are clearly ϵ_1- and ϵ_2-approximate solutions of

$$x' = kx$$

there. A simple calculation shows that for φ_1 and φ_2 the equality sign in (2.2) must hold if $\xi_2 \geqq \xi_1$.

Note that, roughly speaking, the inequality says that, if δ and ϵ are small, then so is $\varphi_1(t) - \varphi_2(t)$. In fact, if $\delta = \epsilon = 0$, then $\varphi_1 = \varphi_2$, and there is at most one solution of (E) going through any given point (τ,ξ) in D. This proves the following uniqueness result:

Theorem 2.2. *Let $f \in (C,\text{Lip})$ in D, and $(\tau,\xi) \in D$. If φ_1 and φ_2 are any two solutions of* (E) *on (a,b), $a < \tau < b$, such that $\varphi_1(\tau) = \varphi_2(\tau) = \xi$, then $\varphi_1 = \varphi_2$.*

Actually, in order to obtain uniqueness, it is not necessary to assume as strong a restriction on f as the Lipschitz condition. However, a more general discussion of the uniqueness problem will be deferred until Chap. 2, Sec. 2.

An existence proof can be based on the inequality (2.2) also.

Theorem 2.3. *Suppose $f \in (C,\text{Lip})$ on the rectangle*

$$R: \quad |t - \tau| \leqq a \qquad |x - \xi| \leqq b \qquad (a,b > 0)$$

and let

$$M = \max |f(t,x)| \qquad ((t,x) \in R)$$

and

$$\alpha = \min \left(a, \frac{b}{M}\right)$$

Then there exists a (unique) solution $\varphi \in C^1$ of (E) *on $|t - \tau| \leqq \alpha$ for which $\varphi(\tau) = \xi$.*

Proof. Let $\{\epsilon_n\}$ be a monotone decreasing sequence of positive real numbers tending to zero as $n \to \infty$. Choose for each ϵ_n an ϵ_n-approximate solution φ_n. These functions satisfy the relation

$$\varphi_n(t) = \xi + \int_\tau^t (f(s,\varphi_n(s)) + \Delta_n(s))\, ds \tag{2.6}$$

where $\Delta_n(t) = \varphi_n'(t) - f(t,\varphi_n(t))$ at those points where φ_n' exists, and $\Delta_n(t) = 0$ otherwise. Now $\Delta_n(t) \to 0$, as $n \to \infty$, uniformly on $|t - \tau| \leqq \alpha$, by the very definition of ϵ_n. From (2.2) applied to φ_n and φ_m one obtains for $|t - \tau| \leqq \alpha$,

$$|\varphi_n(t) - \varphi_m(t)| \leqq \frac{(\epsilon_n + \epsilon_m)(e^{k\alpha} - 1)}{k}$$

where k is the Lipschitz constant. Thus the sequence $\{\varphi_n\}$ is uniformly convergent on $|t - \tau| \leqq \alpha$, and therefore there exists a continuous limit function φ on this interval such that $\varphi_n(t) \to \varphi(t)$ as $n \to \infty$ uniformly on $|t - \tau| \leqq \alpha$. This fact, plus the uniform continuity of f on R, implies that

$$f(t,\varphi_n(t)) \to f(t,\varphi(t)) \qquad (n \to \infty)$$

uniformly on $|t - \tau| \leqq \alpha$. Hence

$$\lim_{n \to \infty} \int_\tau^t (f(s,\varphi_n(s)) + \Delta_n(s))\, ds = \int_\tau^t f(s,\varphi(s))\, ds$$

and from (2.6) one gets finally, by letting $n \to \infty$,

$$\varphi(t) = \xi + \int_\tau^t f(s,\varphi(s))\, ds$$

which proves the existence of a solution $\varphi \varepsilon C^1$ of (E) on $|t - \tau| \leqq \alpha$. It is unique by Theorem 2.2. Clearly

$$|\varphi(t) - \varphi_n(t)| \leqq \frac{\epsilon_n(e^{k|t-\tau|} - 1)}{k} \tag{2.7}$$

3. The Method of Successive Approximations

The existence proof given in Theorem 1.2 is unsatisfactory in one respect in that there is no constructive method given for obtaining a solution of (E). However, as was pointed out after that proof, if the solution through the given point is known to be unique, then the original polygonal approximate solutions can be used to obtain the solution; no subsequence need be chosen. In particular, if f satisfies a Lipschitz condition, the inequality (2.7) gives a bound for the error in using an ϵ_n-approximate solution in place of the actual solution. In the following a very useful method, known as the *method of successive approximations*, will be considered, and the existence of a solution will be deduced with its aid. Here again one can conveniently compute an upper bound on the error involved in stopping the process after a finite number of steps.

The results will be deduced for the case of the rectangle R defined by

$$R: \qquad |t - \tau| \leqq a \qquad |x - \xi| \leqq b$$

where (τ,ξ) is some point in the (t,x) plane, and $a > 0$, $b > 0$. It will be clear that the analogue of Theorem 1.3 also holds.

If $f \varepsilon C$ on R, then f is bounded there; let $\max |f| = M$ on R, and, as before, $\alpha = \min (a,b/M)$. It is clear that a solution φ of (E) on $|t - \tau| \leqq \alpha$ for which $\varphi(\tau) = \xi$ must satisfy the integral equation

$$\varphi(t) = \xi + \int_\tau^t f(s,\varphi(s))\, ds \qquad (|t - \tau| \leqq \alpha) \tag{3.1}$$

and conversely, if φ satisfies (3.1), it satisfies (E) and $\varphi(\tau) = \xi$. The *successive approximations* for (E) are defined to be the functions φ_0, φ_1, . . . , given recursively by the formulas

$$\varphi_0(t) = \xi$$
$$\varphi_{k+1}(t) = \xi + \int_\tau^t f(s,\varphi_k(s))\,ds \qquad (k = 0, 1, 2, \ldots ; |t - \tau| \leqq \alpha) \quad (3.2)$$

It is shown below that these functions actually exist on $|t - \tau| \leqq \alpha$.

Theorem 3.1 (Picard-Lindelöf). *If $f \in (C,\text{Lip})$ on R, then the successive approximations φ_k exist on $|t - \tau| \leqq \alpha$ as continuous functions, and converge uniformly on this interval to the unique solution φ of* (E) *such that $\varphi(\tau) = \xi$.*

Proof. Consider the interval $[\tau, \tau + \alpha]$; similar arguments hold for $[\tau - \alpha, \tau]$.

It will be shown that every φ_k exists on $[\tau, \tau + \alpha]$, $\varphi_k \in C^1$ there, and

$$|\varphi_k(t) - \xi| \leqq M(t - \tau) \qquad (t \in [\tau, \tau + \alpha]) \quad (3.3)$$

Obviously φ_0, being the constant ξ, satisfies these conditions. Assume φ_k does the same; then $f(t,\varphi_k(t))$ is defined and continuous on $[\tau, \tau + \alpha]$. From (3.2) this implies φ_{k+1} exists on $[\tau, \tau + \alpha]$, $\varphi_{k+1} \in C^1$ there, and $|\varphi_{k+1}(t) - \xi| \leqq M(t - \tau)$. Therefore these properties are shared by all the φ_k by induction. Geometrically, this means that all the φ_k start at (τ,ξ) and stay within a triangular region T between the lines

$$x - \xi = \pm M(t - \tau)$$

and $t = \tau + \alpha$.

It remains to prove the convergence of the φ_k. Let Δ_k be defined by

$$\Delta_k(t) = |\varphi_{k+1}(t) - \varphi_k(t)| \qquad (t \in [\tau, \tau + \alpha])$$

Then from (3.2) by subtraction and the fact that $f \in \text{Lip}$ on R with some constant $c > 0$,

$$\Delta_k(t) \leqq c \int_\tau^t \Delta_{k-1}(s)\,ds \quad (3.4)$$

But (3.3) gives for $k = 1$,

$$\Delta_0(t) = |\varphi_1(t) - \varphi_0(t)| \leqq M(t - \tau)$$

and an easy induction on (3.4) implies that

$$\Delta_k(t) \leqq \frac{M}{c}\,\frac{c^{k+1}(t - \tau)^{k+1}}{(k + 1)!} \qquad (t \in [\tau, \tau + \alpha])$$

This shows that the terms of the series $\sum_{k=0}^{\infty} \Delta_k(t)$ are majorized by those of the power series for $(M/c)e^{c\alpha}$, and therefore the series $\sum_{k=0}^{\infty} \Delta_k(t)$ is uniformly convergent on $[\tau, \tau + \alpha]$. Thus the series

$$\varphi_0(t) + \sum_{k=0}^{\infty} (\varphi_{k+1}(t) - \varphi_k(t))$$

is absolutely and uniformly convergent on $[\tau, \tau + \alpha]$; consequently the partial sum

$$\varphi_0(t) + \sum_{k=0}^{n-1} (\varphi_{k+1}(t) - \varphi_k(t)) = \varphi_n(t)$$

tends uniformly on $[\tau, \tau + \alpha]$ to a continuous limit function φ.

It will be shown that the function φ satisfies (3.1), and is hence a solution of (E) on $[\tau, \tau + \alpha]$ for which $\varphi(\tau) = \xi$. Since all the φ_k are within the region T, so is φ. Therefore $f(s,\varphi(s))$ exists for $s \in [\tau, \tau + \alpha]$. Clearly

$$\left| \int_\tau^t [f(s,\varphi(s)) - f(s,\varphi_k(s))] \, ds \right| \leqq \int_\tau^t |f(s,\varphi(s)) - f(s,\varphi_k(s))| \, ds$$
$$\leqq c \int_\tau^t |\varphi(s) - \varphi_k(s)| \, ds$$

the latter inequality being due to the fact that $f \in \text{Lip}$ on R. Now $|\varphi(s) - \varphi_k(s)| \to 0$ as $k \to \infty$ uniformly on $[\tau, \tau + \alpha]$, and thus the above inequalities show that (3.2) yields (3.1) as $k \to \infty$.

The solution φ is unique by Theorem 2.2, and this completes the proof.

An upper bound for the error in approximating the solution φ by the nth approximation φ_n is easily computed. It is given by

$$|\varphi(t) - \varphi_n(t)| \leqq \sum_{k=n}^{\infty} |\varphi_{k+1}(t) - \varphi_k(t)| \leqq \frac{M}{c} \sum_{k=n+1}^{\infty} \frac{c^k(t-\tau)^k}{k!}$$
$$\leqq \frac{M}{c} \sum_{k=n+1}^{\infty} \frac{(c\alpha)^k}{k!} < \frac{M}{c} \frac{(c\alpha)^{n+1}}{(n+1)!} \sum_{k=0}^{\infty} \frac{(c\alpha)^k}{k!} = \frac{M}{c} \frac{(c\alpha)^{n+1}}{(n+1)!} e^{c\alpha}$$

4. Continuation of Solutions

Suppose that $f \in C$ in some domain D of the (t,x) plane and that (E) has a solution φ which exists on a finite interval (a,b) and passes through some point $(\tau,\xi) \in D$, $a < \tau < b$. If $|f|$ is bounded by some constant $M < \infty$ on D, then it is easy to see that both the limits

$$\varphi(a+0) = \lim_{t \to a+0} \varphi(t) \qquad \varphi(b-0) = \lim_{t \to b-0} \varphi(t)$$

exist. This follows at once from the fact that

$$\varphi(t) = \xi + \int_\tau^t f(s,\varphi(s)) \, ds \qquad (t \in (a,b))$$

and therefore, if $a < t_1 < t_2 < b$,

$$|\varphi(t_1) - \varphi(t_2)| \leqq \int_{t_1}^{t_2} |f(s,\varphi(s))|\, ds \leqq M|t_2 - t_1|$$

Thus, as t_1 and t_2 tend to $a + 0$, $\varphi(t_1) - \varphi(t_2) \to 0$, which implies, by the Cauchy criterion for convergence, that $\varphi(a + 0)$ exists; similarly for $\varphi(b - 0)$.

Suppose that the point $(b,\varphi(b - 0))$ is in D. If $\tilde{\varphi}$ is the function defined by

$$\tilde{\varphi}(t) = \varphi(t) \qquad (t \in (a,b))$$
$$\tilde{\varphi}(t) = \varphi(b - 0) \qquad (t = b)$$

then $\tilde{\varphi}$ is a solution of (E) of class C^1 on $(a,b]$. Indeed,

$$\tilde{\varphi}(t) = \xi + \int_\tau^t f(s,\tilde{\varphi}(s))\, ds \qquad (t \in (a,b])$$

which implies that the left-hand derivative $\tilde{\varphi}'_-$ of $\tilde{\varphi}$ at b exists, and

$$\tilde{\varphi}'_-(b) = \tilde{\varphi}'(b - 0) = f(b,\tilde{\varphi}(b))$$

The function $\tilde{\varphi}$ is called a *continuation* of the solution φ to $(a,b]$.

The equation $x' = x^2$ has a solution $\varphi(t) = -t^{-1}$ through $(-1,1)$ which exists on $(-1,0)$ but *cannot* be continued to $(-1,0]$. Here φ does not stay within a region D, where $f(t,x) = x^2$ is bounded.

Actually the process can be carried further, for by Theorem 1.3 the equation (E) has a solution $\psi \in C^1$ passing through $(b,\varphi(b - 0))$ which exists on some interval $[b, b + \beta]$, $\beta > 0$. If now $\hat{\varphi}$ is defined by

$$\hat{\varphi}(t) = \tilde{\varphi}(t) \qquad (t \in (a,b])$$
$$\hat{\varphi}(t) = \psi(t) \qquad (t \in [b, b + \beta])$$

then $\hat{\varphi}$ is a solution of (E) of class C^1 on $(a, b + \beta]$, and $\hat{\varphi}(\tau) = \xi$. The only point to check is the existence and continuity of the derivative $\hat{\varphi}'$ at b. It will be shown that

$$\hat{\varphi}(t) = \xi + \int_\tau^t f(s,\hat{\varphi}(s))\, ds \qquad (t \in (a, b + \beta]) \tag{4.1}$$

This is obvious for $a < t \leqq b$. For $t > b$ it follows from the definition of $\hat{\varphi}$ that

$$\hat{\varphi}(t) = \varphi(b - 0) + \int_b^t f(s,\hat{\varphi}(s))\, ds$$

But

$$\varphi(b - 0) = \xi + \int_\tau^b f(s,\hat{\varphi}(s))\, ds$$

which proves (4.1) for $t > b$. The continuity of $\hat{\varphi}$ in (4.1) implies that of $f(s,\hat{\varphi}(s))$, and by differentiation of this integral equation for $\hat{\varphi}$, one obtains the fact that $\hat{\varphi}'(t) = f(t,\hat{\varphi}(t))$ for $t \in (a, b + \beta]$. Naturally, $\hat{\varphi}$ is called a *continuation* of the solution φ to $(a, b + \beta]$. There are just as many continuations of φ to $(a, b + \beta]$ as there are solutions of (E) issuing from $(b,\varphi(b - 0))$ which exist on $[b, b + \beta]$. If it is known that there exists at most one solution through $(b,\varphi(b - 0))$ (for example, if $f \in$ Lip on D), then one can speak of *the* continuation of φ to $(a, b + \beta]$. In general, if a continuation of a solution φ on (a,b) exists on some interval containing (a,b), then one says φ can be continued, or has a continuation.

The above remarks are summarized in the following theorem:

Theorem 4.1. *Let $f \in C$ in a domain D of the (t,x) plane, and suppose f is bounded on D. If φ is a solution of* (E) *on an interval (a,b) then the limits $\varphi(a + 0)$ and $\varphi(b - 0)$ exist. If $(a,\varphi(a + 0))$ [or $(b,\varphi(b - 0))$] is in D, then the solution φ may be continued to the left of a (or to the right of b).*

A more general analysis of the continuation problem appears in Chap. 2, Sec. 1.

5. Systems of Differential Equations

Suppose n is a positive integer and $f_1, \ldots, f_n$ are n real continuous functions defined on some domain D of the real $(t,x_1, \ldots, x_n)$ space. Just as in the case where $n = 1$, this is abbreviated $f_i \in C$ in D, $i = 1, \ldots, n$. One can then formulate the following problem:

Problem. *To find n differentiable functions $\varphi_1, \ldots, \varphi_n$ defined on a real t interval I such that*

(i) $$(t,\varphi_1(t), \ldots, \varphi_n(t)) \in D \qquad (t \in I)$$

(ii) $$\varphi_i'(t) = f_i(t,\varphi_1(t), \ldots, \varphi_n(t)) \qquad \left(t \in I;\ ' = \frac{d}{dt};\ i = 1, \ldots, n\right)$$

This problem is called a *system of n ordinary differential equations of the first order*, and is denoted by

(E) $$x_i' = f_i(t,x_1, \ldots, x_n) \qquad (i = 1, \ldots, n)$$

Correspondingly, if such an interval I and functions $(\varphi_1, \ldots, \varphi_n)$ exist, then the set of functions $(\varphi_1, \ldots, \varphi_n)$ is called a *solution of the system* (E) on I.

Let $(\tau,\xi_1, \ldots, \xi_n) \in D$. The *initial-value problem* consists of finding a solution $(\varphi_1, \ldots, \varphi_n)$ of (E) on an interval I containing τ such that $\varphi_i(\tau) = \xi_i$.

It turns out that the results so far obtained for the case $n = 1$ can be carried over successfully to the system (E). Let X denote the Euclidean n-dimensional space with points x having coordinates $(x_1, \ldots, x_n)$. Then the functions f_i defined on the $(t,x_1, \ldots, x_n)$ space give rise to functions $\tilde{f}_i$ on the (t,x) space defined by

$$\tilde{f}_i(t,x) = f_i(t,x_1, \ldots, x_n)$$

Also associated with any point x in the x space is the one-column matrix

$$\hat{x} = \begin{pmatrix} x_1 \\ \cdot \\ \cdot \\ \cdot \\ x_n \end{pmatrix}$$

called the *vector* associated with x; x_i is called the ith *component* of $\hat{x}$ Clearly $\tilde{f}_i$ defines a function $\hat{f}_i$ of t and the vector $\hat{x}$ by

$$\hat{f}_i(t,\hat{x}) = \tilde{f}_i(t,x)$$

Associated with the point (t,x) in the (t,x) space one has the vector

$$\tilde{f}(t,x) = \begin{pmatrix} \tilde{f}_1(t,x) \\ \cdot \\ \cdot \\ \cdot \\ \tilde{f}_n(t,x) \end{pmatrix}$$

and this gives rise to a vector $\hat{f}(t,\hat{x})$ defined by

$$\hat{f}(t,\hat{x}) = \begin{pmatrix} \hat{f}_1(t,\hat{x}) \\ \cdot \\ \cdot \\ \cdot \\ \hat{f}_n(t,\hat{x}) \end{pmatrix}$$

If the x_i are differentiable functions of t, then $\hat{x}'$, the *derivative* of $\hat{x}$, is defined by

$$\hat{x}' = \begin{pmatrix} x_1' \\ \cdot \\ \cdot \\ \cdot \\ x_n' \end{pmatrix}$$

Then the equation (**E**) may be written simply as

$$\hat{x}' = \hat{f}(t,\hat{x})$$

In actuality, there is seldom any chance of confusing the point x with the vector $\hat{x}$, and so the same notation will be used for both; the circumflex will be dropped. This has the effect of identifying the space X with the space $\hat{X}$ of all n-rowed one-column matrices, considered as a vector space. Similarly, no confusion results by identifying the functions f_i, $\tilde{f}_i$, and $\hat{f}_i$, and this will be done in all that follows. With this understanding, Eq. (E) may be written as

$$x' = f(t,x) \tag{E}$$

where f may be thought of as a vector function of real t and the point $x \in X$, or as a function of t and the one-column matrix

$$x = \begin{pmatrix} x_1 \\ \cdot \\ \cdot \\ \cdot \\ x_n \end{pmatrix}$$

Here

$$f(t,x) = \begin{pmatrix} f_1(t,x) \\ \cdot \\ \cdot \\ \cdot \\ f_n(t,x) \end{pmatrix}$$

A solution of (E) on an interval I becomes a vector function φ with components $\varphi_1, \ldots, \varphi_n$ defined on I satisfying

(i) $\quad (t,\varphi(t)) = (t,\varphi_1(t), \ldots, \varphi_n(t)) \in D \qquad (t \in I)$

(ii) $\quad \varphi'(t) = f(t,\varphi(t)) \qquad (t \in I)$

The *magnitude* (or *norm*) $|x|$ of a vector $x \in X$ with components $x_1, \ldots, x_n$ is defined by†

$$|x| = \sum_{i=1}^{n} |x_i|$$

The *Euclidean length* $\|x\|$ of a vector $x \in X$ is defined by

$$\|x\| = \left(\sum_{i=1}^{n} |x_i|^2 \right)^{\frac{1}{2}}$$

The *distance* between two vectors $x,y \in X$ is defined to be $|y - x|$. It is

† Other definitions for the magnitude of a vector can be used, such as

$$\boldsymbol{|} x \boldsymbol{|} = \max (|x_i|) \qquad (i = 1, \ldots, n)$$

or the Euclidean length $\|x\|$. As a matter of fact, the following inequalities relating the various norms are readily seen to hold:

$$\boldsymbol{|} x \boldsymbol{|} \leqq |x| \leqq n \boldsymbol{|} x \boldsymbol{|} \qquad \boldsymbol{|} x \boldsymbol{|} \leqq \|x\| \leqq n^{\frac{1}{2}} \boldsymbol{|} x \boldsymbol{|}$$

and

$$\|x\| \leqq |x| \leqq n^{\frac{1}{2}} \|x\| \qquad n^{-\frac{1}{2}} |x| \leqq \|x\| \leqq |x|$$

obvious that this distance satisfies the ordinary rules for a distance function

(a) $$|y - x| = |x - y|$$

(b) $$|y - x| \geqq 0 \quad \text{and } |y - x| = 0 \quad \text{if and only if } y = x$$

(c) $$|y - x| \leqq |y - z| + |z - x|$$

This distance function allows one to consider X as a metric space; a sequence of vectors $\{x_k\}$ is said to be convergent if it is convergent with respect to this distance function. Note that here the x_k are vectors and not components of a vector. Clearly $\{x_k\}$ is convergent if and only if each of the component sequences $\{x_{k_i}\}$ (x_k having components x_{k_i}, $i = 1, \ldots, n$) is convergent.

If g is a differentiable vector function on some t interval (a,b), that is, g' exists on (a,b), and $r = |g|$ is the function defined by

$$r(t) = |g(t)| \qquad (t \in (a,b))$$

then, if sgn $a = a/|a|$ for $a \neq 0$,

$$r(t) = \sum_{i=1}^{n} g_i(t) \operatorname{sgn} g_i(t)$$

If none of the components $g_i(t)$ of $g(t)$ vanishes, clearly

$$r'(t) = \sum_{i=1}^{n} g_i'(t) \operatorname{sgn} g_i(t)$$

exists and

$$|r'(t)| \leqq |g'(t)| \tag{5.1}$$

In any case [whether a component of $g(t)$ is zero or not], it is always true that if $g'(t)$ exists, then the right- and left-hand derivatives $r_+'(t)$ and $r_-'(t)$ exist† and satisfy

$$|r_\pm'(t)| \leqq |g'(t)| \tag{5.2}$$

For if t is an isolated zero of any component g_i of g, a straightforward calculation shows that the right- and left-hand derivatives of $|g_i|$ exist at t and do not exceed $|g_i'|$; and if t is a zero of g_i which is not isolated (that is, t is a limit point of zeros of g_i), then an approach to t through a sequence of these zeros shows that $g_i'(t) = 0$, and hence $|g_i|'(t) = 0$. In either case, (5.2) is valid.

† Notice that if g is absolutely continuous, then so is r, since

$$|r(t_2) - r(t_1)| \leqq |g(t_2) - g(t_1)|$$

If $\|g(t)\| \neq 0$, and $\|g\|$ is the function defined by $\|g\|(t) = \|g(t)\|$, then $\|g\|'(t)$ exists, and it is easily verified that

$$| \, \|g\|'(t) | \leqq \|g'(t)\| \tag{5.3}$$

For integration purposes the magnitude $|\ \ |$ is more useful than the length $\|\ \ \|$. If g and $r = |g|$ are integrable over an interval (a,b), then

$$\left| \int_a^b g(t)\, dt \right| \leqq \int_a^b r(t)\, dt \tag{5.4}$$

where by $\int_a^b g(t)\, dt$, of course, is meant the vector whose ith component is $\int_a^b g_i(t)\, dt$.

Suppose a vector f (which may have any finite number of components, not necessarily n) is defined on a domain D of the (t,x) space. If there exists a constant $k > 0$ such that for every (t,x) and $(t,\tilde{x})$ in D

$$|f(t,x) - f(t,\tilde{x})| \leqq k|x - \tilde{x}| \tag{5.5}$$

then the vector f is said to satisfy a *Lipschitz condition* (with respect to x) in D, and one writes $f \in \text{Lip}$ in D.

Suppose $f \in C$ on a domain D in the (t,x) space. An *ϵ-approximate solution* of (E) on an interval I is a vector function $\varphi \in C$ on I such that

(α)	$(t,\varphi(t)) \in D$	$(t \in I)$
(β)	$\varphi \in C^1$ on I	except for a finite set of points S on I
(γ)	$\lvert\varphi'(t) - f(t,\varphi(t))\rvert \leqq \epsilon$	$(t \in I - S)$

In terms of the definitions introduced above, all the theorems in Secs. 1–4 are valid for the vector equation (**E**) if, in their statements and proofs, x,f are replaced by the vectors x,f and the magnitude is understood in the sense defined above for vectors. (The Ascoli lemma is valid for vectors also.) Therefore it will be assumed from now on that these theorems have been proved for the more general vector equation (**E**).

A particularly interesting system is the *linear system*

$$x_i' = \sum_{j=1}^{n} a_{ij}(t)x_j \qquad (i = 1, \ldots, n) \tag{L}$$

where the a_{ij} are continuous functions on some closed bounded t interval $[a,b]$. If f is the vector with components f_i defined by

$$f_i(t,x) = \sum_{j=1}^{n} a_{ij}(t)x_j \qquad (i = 1, \ldots, n)$$

then clearly f satisfies a Lipschitz condition on the $(n+1)$-dimensional region

$$D: \quad a \leqq t \leqq b \quad |x| < \infty$$

(Here D is not a domain since it is not open.) In fact, if (t,x) and $(t,\tilde{x})$ are in D,

$$|f(t,x) - f(t,\tilde{x})| \leqq k|x - \tilde{x}|$$

where

$$k = \max \sum_{i=1}^{n} |a_{ij}(t)| \qquad (t \in [a,b];\, j = 1, \ldots, n)$$

Theorem 5.1. *For the linear system* (L), *where the functions* $a_{ij} \in C$ *on* $[a,b]$, *there exists one and only one solution* φ *of* (L) *on* $[a,b]$ *passing through any point* $(\tau,\xi) \in D$, *that is,* $\varphi(\tau) = \xi$.

Proof. Since the vector f satisfies a Lipschitz condition on D, the existence and uniqueness of a solution ψ through (τ,ξ) over some interval $[c,d] \subseteqq [a,b]$ are guaranteed. It remains to show that ψ can be continued to a unique solution φ on the whole interval $[a,b]$.

If $\tilde{\psi}$ is any solution of (L) through (τ,ξ) existing on any subinterval of $[a,b]$, then applying the inequality of Theorem 2.1 to $\varphi_1 = \tilde{\psi}$ and $\varphi_2 = 0$, one obtains

$$|\tilde{\psi}(t)| \leqq |\xi| e^{k(b-a)} \tag{5.6}$$

for t in the domain of definition of $\tilde{\psi}$. Now suppose ψ does not have a continuation to $[a,b]$, and for definiteness assume ψ has a continuation $\tilde{\psi}$ existing up to $\bar{t} < b$, but cannot be continued past $\bar{t}$. But, from (5.6), the path $(t,\tilde{\psi}(t))$ remains inside a closed bounded subset of D, where $f \in C$ and satisfies a Lipschitz condition. Therefore, by Theorem 4.1 (interpreted for systems), $\tilde{\psi}$ may be continued beyond $\bar{t}$. This results in a contradiction, thus proving a continuation φ of ψ exists on $[a,b]$. It is unique, for f satisfies a Lipschitz condition on D.

Theorem 5.2. *Let the functions* a_{ij} $(i,j = 1, \ldots, n)$ *be continuous on an open interval* I, *which may be unbounded. Then there exists on* I *one and only one solution* φ *of* (L) *satisfying*

$$\varphi(\tau) = \xi \qquad (\tau \in I,\ |\xi| < \infty) \tag{5.7}$$

Proof. By Theorem 5.1 there exists a solution of (L) satisfying (5.7) on every closed subinterval of I containing τ; using the same argument as in that proof, any such solution may be continued to the whole of I uniquely.

A more detailed treatment of the linear system will be the subject of Chap. 3.

6. The nth-order Equation

Suppose f is a real continuous function defined in a domain D of the real $(t,x_1, \ldots, x_n)$ space. Then the *nth-order equation associated with f*,

$$(\mathrm{E}_n) \qquad x^{(n)} = f(t,x,x', \ldots, x^{(n-1)}) \qquad \left({}^{(k)} = \frac{d^k}{dt^k}\right)$$

is defined to be the following problem:

Problem. *To find a function φ defined on a real t interval I possessing n derivatives there such that*

$$\text{(i)} \qquad (t,\varphi(t),\varphi'(t), \ldots, \varphi^{(n-1)}(t)) \in D \qquad (t \in I)$$
$$\text{(ii)} \qquad \varphi^{(n)}(t) = f(t,\varphi(t),\varphi'(t), \ldots, \varphi^{(n-1)}(t)) \qquad (t \in I)$$

If such an interval I and function φ exist, then φ is said to be a *solution of* (E_n) *on* I. If φ is a solution, clearly $\varphi \in C^n$ on I. Note that x, f, and φ are not vectors here.

Let $(\tau,\xi_1, \ldots, \xi_n) \in D$. Then the *initial-value problem* consists of finding a solution φ of (E_n) on an interval I containing t such that $\varphi(\tau) = \xi_1$, $\varphi'(\tau) = \xi_2, \ldots, \varphi^{(n-1)}(\tau) = \xi_n$.

The theory of the equation (E_n) can be reduced to the theory of a system of n first-order differential equations. Indeed, associated with Eq. (E_n) is the system of equations

$$(\tilde{\mathrm{E}}_n) \qquad \begin{aligned} x_1' &= x_2 \\ x_2' &= x_3 \\ &\cdot \\ &\cdot \\ &\cdot \\ x_{n-1}' &= x_n \\ x_n' &= f(t,x_1, \ldots, x_n) \end{aligned}$$

which is defined for $(t,x) = (t,x_1, \ldots, x_n) \in D$. If the vector $\tilde{\varphi}$, with components $(\varphi_1,\varphi_2, \ldots, \varphi_n)$, is a solution of $(\tilde{\mathrm{E}}_n)$ on I, then since $\varphi_2 = \varphi_1'$, $\varphi_3 = \varphi_2' = \varphi_1''$, $\ldots, \varphi_n = \varphi_1^{(n-1)}$,

$$f(t,\varphi_1(t), \ldots, \varphi_n(t)) = f(t,\varphi_1(t), \ldots, \varphi_1^{(n-1)}(t)) = \varphi_1^{(n)}(t)$$

and the first component φ_1 of $\tilde{\varphi}$ is a solution of (E_n) on I. Conversely, if φ_1 is a solution of (E_n) on I, then the vector $\tilde{\varphi}$ with components $\varphi_1,\varphi_1', \ldots, \varphi_1^{(n-1)}$ is a solution of the system $(\tilde{\mathrm{E}}_n)$ on I. The system $(\tilde{\mathrm{E}}_n)$ is called the *system* (or *vector equation*) *associated with the nth-order equation* (E_n). If $\varphi_1(\tau) = \xi_1, \ldots, \varphi_1^{(n-1)}(\tau) = \xi_n$, then the vector $\tilde{\varphi}$ satisfies $\tilde{\varphi}(\tau) = \xi$, where $\xi = (\xi_1, \ldots, \xi_n)$, and conversely.

It is thus clear that all statements proved about the system $(\tilde{\mathrm{E}}_n)$ carry over directly to statements about the nth-order equation (E_n). In particular, if $f \varepsilon C$ on a domain D of the $(t,x_1, \ldots, x_n)$ space, and P is a point of D, there exists a solution $\varphi \varepsilon C^n$ of (E_n) on some t interval and passing through P. If, in addition, $f \varepsilon$ Lip in D, that is, if

$$|f(t,x_1, \ldots, x_n) - f(t,\tilde{x}_1, \ldots, \tilde{x}_n)| \leqq k \sum_{i=1}^{n} |x_i - \tilde{x}_i|$$

for some constant $k > 0$, then the solution through P is unique.

7. Dependence of Solutions on Initial Conditions and Parameters

A solution of a differential equation on an interval I can be considered as a function, not only of $t \varepsilon I$, but of the coordinates of a point through which the solution passes. For example, the first-order equation in one dimension $x' = x$ has the solution $\varphi(t) = \xi e^{t-\tau}$ through the point (τ,ξ). This determines a function of (t,τ,ξ), which is also called† φ, given by $\varphi(t,\tau,\xi) = \xi e^{t-\tau}$. In the general situation, it is important to know how φ behaves as a function of (t,τ,ξ) together, and, in particular, under what circumstances φ is continuous in (t,τ,ξ). In the following the behavior of the solutions as functions of the initial conditions will be investigated for the general case of a system.

Let D be a domain in the $(n + 1)$-dimensional real (t,x) space and suppose $f \varepsilon (C,\text{Lip})$ in D. Let ψ be a solution of the equation

$$\textbf{(E)} \qquad x' = f(t,x)$$

on some interval I. Thus $(t,\psi(t)) \varepsilon D$ for $t \varepsilon I$. It follows from the existence theorem that (E) has a unique solution through any point (τ,ξ) close enough to the given solution. However, the existence theorem assures the existence of the solution only over some short t interval containing τ. Actually, it can be shown that the solution exists over the whole interval I, and is a continuous function of (t,τ,ξ). The following theorem will be proved.

Theorem 7.1. *Let $f \varepsilon (C,\text{Lip})$ in a domain D of the $(n + 1)$-dimensional (t,x) space, and suppose ψ is a solution of* (E) *on an interval I: $a \leqq t \leqq b$. There exists a $\delta > 0$ such that for any $(\tau,\xi) \varepsilon U$, where*

$$U: \qquad a < \tau < b \qquad |\xi - \psi(\tau)| < \delta$$

there exists a unique solution φ of (E) *on I with $\varphi(\tau,\tau,\xi) = \xi$. Moreover, $\varphi \varepsilon C$ on the $(n + 2)$-dimensional set*

$$V: \qquad a < t < b \qquad (\tau,\xi) \varepsilon U$$

† There will be little chance of confusing these two functions. If φ is thought of as a function of (t,τ,ξ), then φ' will always mean $\partial\varphi/\partial t$.

REMARKS: In many applications τ is fixed, and in this case U can be considered as the set of all ξ satisfying $|\xi - \psi(\tau)| < \delta$, and V the domain $a < t < b$, $\xi \varepsilon U$. The proof for this case is contained in the proof of Theorem 7.1. An important consequence of the proof in this case is that the mapping T_t which associates with each point (τ,ξ), $\xi \varepsilon U$, the point $(t,\varphi(t,\tau,\xi))$ for some fixed t, $a < t < b$, is a topological mapping.† The uniqueness of the solutions guarantees that T_t is one-to-one, and the continuity of φ in ξ implies T_t is continuous. Since ξ can be considered as the point $\xi = \varphi(\tau,t,\bar{\xi})$, where $\bar{\xi} = \varphi(t,t,\bar{\xi}) = \varphi(t,\tau,\xi)$, the continuity of φ again implies T_t^{-1} is continuous. Actually, the uniqueness of the solutions passing through (τ,ξ), $\xi \varepsilon U$, is sufficient for the continuity of φ in ξ; see Theorem 4.3, Chap. 2.

Often ψ can be continued outside of I, in which case U, V would include the end points a and b of I.

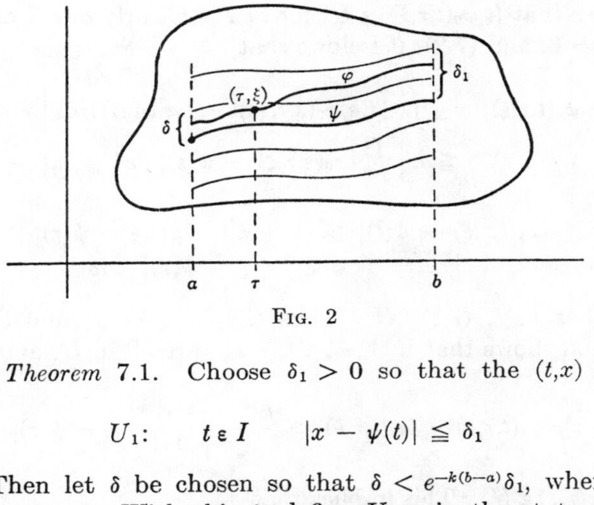

FIG. 2

Proof of Theorem 7.1. Choose $\delta_1 > 0$ so that the (t,x) region U_1, given by

$$U_1: \qquad t \varepsilon I \qquad |x - \psi(t)| \leqq \delta_1$$

is in D. Then let δ be chosen so that $\delta < e^{-k(b-a)}\delta_1$, where k is the Lipschitz constant. With this δ, define U as in the statement of the theorem; see Fig. 2 for the case $n = 1$. If $(\tau,\xi) \varepsilon U$, there exists a solution φ through (τ,ξ) locally, and this satisfies

$$\varphi(t,\tau,\xi) = \xi + \int_\tau^t f(s,\varphi(s,\tau,\xi))\, ds \tag{7.1}$$

as far as it exists. Moreover, for $t \varepsilon I$,

$$\psi(t) = \psi(\tau) + \int_\tau^t f(s,\psi(s))\, ds \tag{7.2}$$

† A topological mapping T of a set S onto a set $T(S)$ is a one-to-one mapping such that T and T^{-1} are continuous.

Thus, using the fundamental inequality (2.2) with $\epsilon = 0$, there results

$$|\varphi(t,\tau,\xi) - \psi(t)| \leqq |\xi - \psi(\tau)|e^{k|t-\tau|} < \delta_1$$

This proves φ cannot leave U_1, and can therefore, by Theorem 4.1, be continued to the whole interval I.

The continuity of φ on V will be proved by showing that φ is the uniform limit of continuous functions on V. Note that φ satisfies (7.1) on I. Define the successive approximations $\{\varphi_j\}$ for (7.1) by

$$\begin{aligned} \varphi_0(t,\tau,\xi) &= \psi(t) + \xi - \psi(\tau) \\ \varphi_{j+1}(t,\tau,\xi) &= \xi + \int_\tau^t f(s,\varphi_j(s,\tau,\xi))\,ds \qquad (j = 0, 1, 2, \ldots) \end{aligned} \tag{7.3}$$

Then for $(\tau,\xi) \in U$

$$|\varphi_0(t,\tau,\xi) - \psi(t)| = |\xi - \psi(\tau)| < \delta_1$$

which shows that $(t,\varphi_0(t,\tau,\xi)) \in U_1$ for $t \in I$. Clearly $\varphi_0 \in C$ on V. From (7.3) for $j = 0$, and (7.2), it follows that

$$\begin{aligned} |\varphi_1(t,\tau,\xi) - \varphi_0(t,\tau,\xi)| &= \left| \int_\tau^t \{f(s,\varphi_0(s,\tau,\xi)) - f(s,\psi(s))\}\,ds \right| \\ &\leqq k \left| \int_\tau^t |\varphi_0(s,\tau,\xi) - \psi(s)|\,ds \right| = k|\xi - \psi(\tau)|\,|t - \tau| \end{aligned}$$

and hence

$$\begin{aligned} |\varphi_1(t,\tau,\xi) - \psi(t)| &\leqq (1 + k|t - \tau|)|\xi - \psi(\tau)| \\ &< e^{k|t-\tau|}|\xi - \psi(\tau)| < \delta_1 \end{aligned}$$

provided that $t \in I$, $(\tau,\xi) \in U$. Thus $(t,\varphi_1(t,\tau,\xi)) \in U_1$ and $\varphi_1 \in C$ on V. An induction shows that if $\varphi_0,\varphi_1, \ldots, \varphi_j$ are all in U_1 and continuous on V, then

$$|\varphi_{j+1}(t,\tau,\xi) - \varphi_j(t,\tau,\xi)| \leqq \frac{k^{j+1}|t - \tau|^{j+1}}{(j+1)!}|\xi - \psi(\tau)| \tag{7.4}$$

if $t \in I$ and $(\tau,\xi) \in U$. This implies that

$$|\varphi_{j+1}(t,\tau,\xi) - \psi(t)| < e^{k|t-\tau|}|\xi - \psi(\tau)| < \delta_1$$

proving that $(t,\varphi_{j+1}(t,\tau,\xi)) \in U_1$. Also, from (7.3), $\varphi_{j+1} \in C$ on V. Thus by induction $(t,\varphi_j(t,\tau,\xi)) \in U_1$ and $\varphi_j \in C$ on V for all j.

Using (7.4), it follows that the φ_j converge uniformly on V to φ, which proves the continuity of φ on V. (Note that the uniform convergence of the φ_j also proves the existence of φ on I.)

Having established the existence and continuity of φ as a function of (t,τ,ξ), it is natural, and for purposes of application also important, to give reasonable sufficient conditions for the existence and continuity of the partial derivatives $\partial\varphi/\partial\tau$, $\partial\varphi/\partial\xi_j$ $(j = 1, \ldots, n)$, where the ξ_j are the components of ξ. Such a sufficient condition is the existence and continuity of the partial derivatives $\partial f/\partial x_j$ on D.

Let f_x denote the matrix (if it exists) with element $\partial f_i/\partial x_j$ in the ith row and jth column ($i,j = 1, \ldots, n$). Also let φ_ξ be the matrix (if it exists) with element $\partial \varphi_i/\partial \xi_j$ in the ith row and jth column ($i,j = 1, \ldots, n$). A matrix is said to be continuous if all its elements are. If $A = (a_{ij})$ is an n-by-n matrix, its determinant will be denoted by det A, and its trace, $\sum_{i=1}^{n} a_{ii}$, by tr A. The symbol exp u denotes e^u.

Theorem 7.2. *Let the hypothesis of Theorem 7.1 be satisfied, and suppose f_x exists and $f_x \in C$ on D. Then $\varphi \in C^1$ on V, and moreover*

$$\det \varphi_\xi(t,\tau,\xi) = \exp \int_\tau^t \operatorname{tr} f_x(s,\varphi(s,\tau,\xi))\, ds \tag{7.5}$$

REMARKS: The fact that $f_x \in C$ on D actually makes the explicit Lipschitz hypothesis for f superfluous.

Notice that det $\varphi_\xi(t,\tau,\xi)$ is just the Jacobian of the transformation, taking ξ into $\varphi(t,\tau,\xi)$, which was considered in the remarks following Theorem 7.1.

For the case where f is an analytic function, Theorem 7.2 is easily obtained from Theorem 7.1, as is shown in Sec. 8. The reader interested mainly in this important case can therefore omit Theorem 7.2.

Proof of Theorem 7.2. In order to prove the existence of φ_ξ, it is sufficient to consider the case of $\partial\varphi/\partial\xi_1$, where $\xi = (\xi_1, \ldots, \xi_n)$. Let $h = (h_1, 0, \ldots, 0)$, $\tilde{\xi} = \xi + h$, and let (τ,ξ) and $(\tau,\tilde{\xi})$ be in U. If χ is the function defined by

$$\chi(t,\tau,\xi,h) = \frac{\varphi(t,\tau,\tilde{\xi}) - \varphi(t,\tau,\xi)}{h_1}$$

for $(t,\tau,\xi) \in V$, then what has to be proved is that

$$\lim_{h\to 0} \chi(t,\tau,\xi,h) \tag{7.6}$$

exists. It will be shown that the limit in (7.6) exists uniformly on V and that the limit function is continuous on V. This will prove $\partial\varphi/\partial\xi_1$ exists and is continuous on V.

The motivation behind the proof is very simple: The solution φ satisfies (E), and so

$$\varphi'(t,\tau,\xi) = f(t,\varphi(t,\tau,\xi))$$

Thus, if φ and f are sufficiently differentiable,

$$\left(\frac{\partial\varphi}{\partial\xi_1}\right)'(t,\tau,\xi) = f_x(t,\varphi(t,\tau,\xi))\,\frac{\partial\varphi}{\partial\xi_1}(t,\tau,\xi)$$

where the latter product is an ordinary matrix product. Therefore $\partial\varphi/\partial\xi_1$ is a solution of a *linear* differential equation. All the following proof does is to justify this procedure.

Let

$$\theta(t,\tau,\xi,h) = \varphi(t,\tau,\tilde{\xi}) - \varphi(t,\tau,\xi)$$

Using the inequality (2.2), there results

$$|\theta(t,\tau,\xi,h)| \leqq |\theta(\tau,\tau,\xi,h)|e^{k|t-\tau|} \leqq |h_1|e^{k(b-a)} \tag{7.7}$$

Thus as $h_1 \to 0$, $\theta \to 0$ uniformly for $(t,\tau,\xi) \in V$.

Since φ is a solution of (E)

$$\theta'(t,\tau,\xi,h) = f(t,\varphi(t,\tau,\tilde{\xi})) - f(t,\varphi(t,\tau,\xi)) \tag{7.8}$$

Using the theorem of the mean on the right side of (7.8), and recalling that $f_x \in C$ on D, there exists a matrix $\Gamma = (\Gamma_{ij})$ such that

$$\theta'(t,\tau,\xi,h) = (f_x(t,\varphi(t,\tau,\xi)) + \Gamma)\theta(t,\tau,\xi,h) \tag{7.9}$$

where, given any $\epsilon_1 > 0$, there exists a δ_1, which depends on ϵ_1, such that $|\Gamma| = \sum_{i,j=1}^{n} |\Gamma_{ij}| < \epsilon_1$ if $|\theta| < \delta_1$ for $(t,\tau,\xi) \in V$.† By (7.7), then, $|\Gamma| \to 0$ as $h_1 \to 0$ uniformly for $(t,\tau,\xi) \in V$.

Since $\chi = \theta/h_1$, (7.9) yields

$$\chi'(t,\tau,\xi,h) = f_x(t,\varphi(t,\tau,\xi))\chi(t,\tau,\xi,h) + \gamma \tag{7.10}$$

where $\gamma = \Gamma\theta/h_1$ so that by (7.7)

$$|\gamma| \leqq |\Gamma|e^{k(b-a)}$$

Thus $\gamma \to 0$ as $h_1 \to 0$ uniformly on V. In particular, given any $\epsilon > 0$, there exists a $\delta_\epsilon > 0$ such that $|\gamma| < \epsilon$ if $|h_1| < \delta_\epsilon$. Thus (7.10) states that χ as a function of t is an ϵ-approximate solution of the linear equation

$$y' = f_x(t,\varphi(t,\tau,\xi))y \tag{7.11}$$

provided that $|h_1| < \delta_\epsilon$. The initial value $\chi(\tau,\tau,\xi,h)$ is e_1, where e_1 is the vector with components $(1, 0, \ldots, 0)$.

Consider now for fixed $(\tau,\xi) \in U$ the solution β of (7.11) which assumes the initial value e_1 at $t = \tau$. That this solution exists on I: $a \leqq t \leqq b$ follows from Theorem 5.1. The fact that χ is an ϵ-approximate solution of (7.11) for $|h_1| < \delta_\epsilon$ implies by Theorem 2.1 that

$$|\chi(t,\tau,\xi,h) - \beta(t,\tau,\xi)| \leqq \frac{\epsilon}{k}(e^{k(b-a)} - 1)$$

† Here use is made of the fact that for $(t,\tau,\xi) \in V$ the points $(t,\varphi(t,\tau,\xi)) \in U_1$, a closed bounded set. Thus f_x is uniformly continuous on U_1.

for (t,τ,ξ) on V. Clearly this implies that

$$\lim_{h\to 0} \chi(t,\tau,\xi,h) = \beta(t,\tau,\xi)$$

uniformly on V. This proves the existence of $\partial\varphi/\partial\xi_1$ and also proves that it is the solution of (7.11) which assumes the initial value e_1 at $t = \tau$. The uniformity of the convergence of χ as $h \to 0$, and the continuity of χ on V, imply the continuity of $\partial\varphi/\partial\xi_1$ on V.

An entirely similar argument proves the existence and continuity of $\partial\varphi/\partial\xi_j$, $j = 2, \ldots, n$, on V. Also if e_j is the vector with all components zero except the jth, which is 1,

$$\frac{\partial\varphi}{\partial\xi_j}(\tau,\tau,\xi) = e_j \qquad (j = 1, \ldots, n) \tag{7.12}$$

and $\partial\varphi/\partial\xi_j$ is a solution of (7.11). The columns of the matrix φ_ξ are precisely the vectors $\partial\varphi/\partial\xi_j$. Therefore the following matrix equation is valid:

$$\varphi_\xi'(t,\tau,\xi) = f_x(t,\varphi(t,\tau,\xi))\varphi_\xi(t,\tau,\xi) \tag{7.13}$$

where $\varphi_\xi' = \partial\varphi_\xi/\partial t$. The relation (7.12) may be written as

$$\varphi_\xi(\tau,\tau,\xi) = E \tag{7.14}$$

where E is the n-by-n unit matrix,

$$E = \begin{pmatrix} 1 & 0 & \cdots & 0 \\ 0 & 1 & \cdot & \cdot \\ \cdot & \cdot & \cdot & \cdot \\ \cdot & \cdot & \cdot & \cdot \\ \cdot & & \cdot & 0 \\ 0 & \cdots & 0 & 1 \end{pmatrix}$$

The relation (7.5) is a consequence of a general fact concerning matrix solutions of linear systems. Since this relation is of importance in itself, it will be proved in the next theorem. One obtains (7.5) from (7.18) below using (7.13) and (7.14) and the fact that det $E = 1$.

It is but a repetition of the previous arguments to show that $\partial\varphi/\partial\tau$ also satisfies the linear equation (7.11), once it is observed that it has the initial value given by

$$\frac{\partial\varphi}{\partial\tau}(\tau,\tau,\xi) = -f(\tau,\xi) \tag{7.15}$$

This is shown by a direct calculation as follows:

$$\begin{aligned}\varphi(\tau,\bar\tau,\xi) - \varphi(\tau,\tau,\xi) &= \varphi(\tau,\bar\tau,\xi) - \xi \\ &= \varphi(\tau,\bar\tau,\xi) - \varphi(\bar\tau,\bar\tau,\xi) \\ &= \int_{\bar\tau}^{\tau} f(s,\varphi(s,\bar\tau,\xi))\, ds\end{aligned}$$

Thus

$$\frac{\varphi(\tau,\bar{\tau},\xi) - \varphi(\tau,\tau,\xi)}{\bar{\tau} - \tau} = -\frac{1}{\bar{\tau} - \tau} \int_{\tau}^{\bar{\tau}} f(s,\varphi(s,\bar{\tau},\xi))\, ds$$

Since the integrand is continuous for $(s,\bar{\tau},\xi) \in V$, it follows that the limit as $\bar{\tau} \to \tau$ exists for $(\tau,\xi) \in U$ and gives (7.15).

Theorem 7.3. *Let A be an n-by-n matrix with continuous elements on an interval $I: a \leq t \leq b$, and suppose Φ is a matrix of functions on I satisfying*

$$\Phi'(t) = A(t)\Phi(t) \qquad (t \in I) \tag{7.16}$$

Then det Φ *satisfies on I the first-order equation*

$$(\det \Phi)' = (\operatorname{tr} A)(\det \Phi) \tag{7.17}$$

and thus for τ, $t \in I$

$$\det \Phi(t) = \det \Phi(\tau) \exp \int_{\tau}^{t} \operatorname{tr} A(s)\, ds \tag{7.18}$$

Proof. Let φ_{ij}, a_{ij} be the elements in the ith row and jth column of Φ and A, respectively. Then (7.16) says

$$\varphi_{ij}'(t) = \sum_{k=1}^{n} a_{ik}(t)\varphi_{kj}(t) \qquad (i,j = 1, \ldots, n) \tag{7.19}$$

The derivative of det Φ is a sum of n determinants

$$(\det \Phi)' = \begin{vmatrix} \varphi_{11}' & \varphi_{12}' & \cdots & \varphi_{1n}' \\ \varphi_{21} & \varphi_{22} & \cdots & \varphi_{2n} \\ \cdot & \cdot & \cdots & \cdot \\ \varphi_{n1} & \varphi_{n2} & \cdots & \varphi_{nn} \end{vmatrix} + \begin{vmatrix} \varphi_{11} & \varphi_{12} & \cdots & \varphi_{1n} \\ \varphi_{21}' & \varphi_{22}' & \cdots & \varphi_{2n}' \\ \cdot & \cdot & \cdots & \cdot \\ \varphi_{n1} & \varphi_{n2} & \cdots & \varphi_{nn} \end{vmatrix} + \cdots + \begin{vmatrix} \varphi_{11} & \varphi_{12} & \cdots & \varphi_{1n} \\ \varphi_{21} & \varphi_{22} & \cdots & \varphi_{2n} \\ \cdot & \cdot & \cdots & \cdot \\ \varphi_{n1}' & \varphi_{n2}' & \cdots & \varphi_{nn}' \end{vmatrix}$$

Using (7.19) in the first determinant on the right, one gets

$$\begin{vmatrix} \sum_k a_{1k}\varphi_{k1} & \sum_k a_{1k}\varphi_{k2} & \cdots & \sum_k a_{1k}\varphi_{kn} \\ \varphi_{21} & \varphi_{22} & \cdots & \varphi_{2n} \\ \cdot & \cdot & \cdots & \cdot \\ \varphi_{n1} & \varphi_{n2} & \cdots & \varphi_{nn} \end{vmatrix}$$

and this determinant is unchanged if one subtracts from the first row a_{12} times the second row plus a_{13} times the third row up to a_{1n} times the nth row. This gives

$$\begin{vmatrix} a_{11}\varphi_{11} & a_{11}\varphi_{12} & \cdots & a_{11}\varphi_{1n} \\ \varphi_{21} & \varphi_{22} & \cdots & \varphi_{2n} \\ \cdot & \cdot & \cdots & \cdot \\ \varphi_{n1} & \varphi_{n2} & \cdots & \varphi_{nn} \end{vmatrix}$$

which is just $a_{11} \det \Phi$. Carrying out a similar procedure with the remaining determinants, one obtains finally (7.17). The equation (7.17) is of the form $u' - \alpha(t)u = 0$ from which follows

$$u \exp\left[-\int_\tau^t \alpha(s)\,ds\right] = \text{constant}$$

which gives (7.18).

The case where the right member f of (**E**) contains a parameter vector μ can be readily dealt with. Suppose μ space has k real dimensions, and let I_μ be the domain of μ space, $|\mu - \mu_0| < c$, where μ_0 is fixed and $c > 0$. As above, D is a domain of (t,x) space. Let D_μ be the domain of (t,x,μ) space

$$D_\mu: \qquad (t,x) \in D \qquad \mu \in I_\mu$$

and let $f \in C$ on D_μ and satisfy a Lipschitz condition in x uniformly on D_μ. The differential equation

$$(\mathbf{E}_\mu) \qquad x' = f(t,x,\mu)$$

will be considered here. For a fixed given $\mu = \mu_0$, let ψ be a solution of $(\mathbf{E}_\mu)$ on an interval $a \leqq t \leqq b$. Then the following theorem, which includes Theorem 7.1 as a special case, will be proved:

Theorem 7.4. *Let ψ be the solution of $(\mathbf{E}_\mu)$ described above. There exists a $\delta > 0$ such that for any $(\tau,\xi,\mu) \in U_\mu$, where*

$$U_\mu: \qquad a < \tau < b \qquad |\xi - \psi(\tau)| + |\mu - \mu_0| < \delta$$

there exists a unique solution φ of $(\mathbf{E}_\mu)$ on $a \leqq t \leqq b$ satisfying

$$\varphi(\tau,\tau,\xi,\mu) = \xi$$

Moreover, $\varphi \in C$ on the $(n + k + 2)$-dimensional domain

$$V_\mu: \qquad a < t < b \qquad (\tau,\xi,\mu) \in U_\mu$$

REMARK: An alternative proof of this theorem under slightly more restrictive hypotheses is given in the course of proving Theorem 7.5 below.

Proof of Theorem 7.4. The proof is like that of Theorem 7.1. As remarked there, the successive-approximations procedure can be used to prove the whole theorem. Choose $\delta_1 > 0$ so that the (t,x,μ) region $U_{1\mu}$ given by

$$U_{1\mu}: \qquad a \leqq t \leqq b \qquad |x - \psi(t)| + |\mu - \mu_0| \leqq \delta_1$$

is in D_μ. Define the approximations $\{\varphi_j\}$ by

$$\varphi_0(t,\tau,\xi,\mu) = \psi(t) + \xi - \psi(\tau)$$
$$\varphi_{j+1}(t,\tau,\xi,\mu) = \xi + \int_\tau^t f(s,\varphi_j(s,\tau,\xi,\mu),\mu)\,ds$$

Clearly

$$|\varphi_0(t,\tau,\xi,\mu) - \psi(t)| = |\xi - \psi(\tau)|$$

and

$$|\varphi_1(t,\tau,\xi,\mu) - \varphi_0(t,\tau,\xi,\mu)| = \left| \int_\tau^t \{f(s,\varphi_0(s,\tau,\xi,\mu),\mu) - f(s,\psi(s),\mu_0)\}\,ds \right| \tag{7.20}$$

The uniform continuity of f in $U_{1\mu}$ implies that, given any $\epsilon > 0$, there exists a $\delta_\epsilon > 0$ such that

$$|f(s,\varphi_0(s,\tau,\xi,\mu),\mu) - f(s,\psi(s),\mu_0)| < \epsilon$$

provided that $a \leqq s \leqq b$, $(\tau,\xi,\mu) \in U_{1\mu}$, and

$$|\xi - \psi(\tau)| + |\mu - \mu_0| < \delta_\epsilon \tag{7.21}$$

Thus (7.20) implies

$$|\varphi_1(t,\tau,\xi,\mu) - \varphi_0(t,\tau,\xi,\mu)| < \epsilon|t - \tau|$$

provided (7.21) is valid. Proceeding as before, there now results

$$|\varphi_{j+1}(t,\tau,\xi,\mu) - \varphi_j(t,\tau,\xi,\mu)| \leqq \frac{\epsilon|t-\tau|^{j+1}k^j}{(j+1)!}$$

where k is the Lipschitz constant. Let ϵ be chosen so that

$$\frac{\epsilon}{k}\,(e^{k(b-a)} - 1) < \frac{\delta_1}{2}$$

and let $\delta = \delta_\epsilon < \delta_1/2$ be chosen as above for this ϵ. Then it follows easily by induction that, for all j, $(t,\varphi_j(t,\tau,\xi,\mu))$ is in the region $U_{1\mu}$ for all $(\tau,\xi,\mu) \in U_\mu$. The continuity and the uniform convergence of the φ_j on V_μ lead to the result of the theorem.

The generalization of Theorem 7.2 to $(\mathbf{E}_\mu)$ is valid. In fact, it follows directly from Theorem 7.2 itself.

Theorem 7.5. *Let the hypothesis of Theorem* 7.4 *be satisfied and suppose that $f_x \in C$, $f_\mu \in C$, on D_μ. Then the solution φ defined in Theorem* 7.4 *is of class C^1 on V_μ.*

Proof. Consider the $(n + k)$-dimensional u space consisting of points with coordinates

$$u_i = x_i \qquad (i = 1, \ldots, n)$$
$$u_{i+n} = \mu_i \qquad (i = 1, \ldots, k)$$

and define the vector function $F = (F_1, \ldots, F_{n+k})$ on D_μ by

$$F_i(t,u) = f_i(t,x,\mu) \qquad (i = 1, \ldots, n)$$
$$F_{i+n}(t,u) = 0 \qquad (i = 1, \ldots, k)$$

Then, by Theorem 7.1, the system of equations

$$u' = F(t,u) \tag{7.22}$$

has for a solution the vector $\chi = (\chi_1, \ldots, \chi_{n+k})$ given by

$$\begin{aligned} \chi_i(t) &= \varphi_i(t,\tau,\xi,\mu) \qquad (i = 1, \ldots, n) \\ \chi_{i+n}(t) &= \mu_i \qquad (i = 1, \ldots, k) \end{aligned}$$

since χ has the initial value given by

$$\begin{aligned} \chi_i(\tau) &= \xi_i \qquad (i = 1, \ldots, n) \\ \chi_{i+n}(\tau) &= \mu_i \qquad (i = 1, \ldots, k) \end{aligned}$$

Thus $\mu_1, \ldots, \mu_k$ may be thought of as forming part of the components of an initial-value vector for the system (7.22), and the F in (7.22) satisfies the conditions in Theorem 7.2. Therefore the first partial derivatives of χ with respect to τ, ξ_i, and μ_i exist and are continuous on V_μ, and from the definition of χ this implies the theorem.

From

$$\varphi(t,\tau,\xi,\mu) = \xi + \int_\tau^t f(s,\varphi(s,\tau,\xi,\mu),\mu)\, ds$$

it follows that

$$\frac{\partial \varphi}{\partial \mu_j}(t,\tau,\xi,\mu) = \int_\tau^t \left[f_x(s,\varphi(s,\tau,\xi,\mu),\mu) \frac{\partial \varphi}{\partial \mu_j}(s,\tau,\xi,\mu) + \frac{\partial f}{\partial \mu_j}(s,\varphi(s,\tau,\xi,\mu),\mu) \right] ds$$

This shows that $\partial\varphi/\partial\mu_j$ is the solution of the initial-value problem

$$y' = f_x(t,\varphi(t,\tau,\xi,\mu),\mu)y + \frac{\partial f}{\partial \mu_j}(t,\varphi(t,\tau,\xi,\mu),\mu) \qquad y(\tau) = 0$$

Hypotheses under which the existence of higher derivatives of φ with respect to τ, ξ_i, or μ_i can be shown to exist are readily ascertained from the fact that the first-order derivatives are solutions of a linear equation. For example, $\partial\varphi/\partial\xi_i$ is the solution β_i of

$$y' = f_x(t,\varphi(t,\tau,\xi,\mu),\mu)y \tag{7.23}$$

with the initial value e_i. Clearly $\partial^2\varphi/\partial\xi_j\partial\xi_i$ is $\partial\beta_i/\partial\xi_j$, if it exists. But ξ enters (7.23) as a parameter. If τ and μ are held fixed in (7.23), then ξ in (7.23) plays the role of μ in Theorem 7.5. Thus, if $f_x(t,\varphi(t,\tau,\xi,\mu),\mu)$ has a continuous derivative with respect to ξ_j, then $\partial\beta_i/\partial\xi_j$ exists. If f has continuous partial derivatives of the second order with respect to the components of x, then $f_x(t,\varphi(t,\tau,\xi,\mu),\mu)$ will have continuous first-order partial derivatives with respect to ξ_j.

In much the same way, if f has continuous partial derivatives of the second order with respect to the components of (x,μ), then $\partial^2\varphi/\partial\mu_i\partial\mu_j$

exists as do the mixed derivatives $\partial^2\varphi/\partial\mu_i\partial\xi_j$. The case where the partial derivatives of φ are taken with respect to the components of (τ,ξ,μ) is left to the reader as an exercise.

8. Complex Systems

So far it has been assumed in the equation (E) that t,x,f were all real. If f is a continuous complex-valued function on an open connected set D in the (t,w) space, where t is real and w is complex n-dimensional (real $2n$-dimensional), then the equation

$$(\mathbf{E}_1) \qquad w' = f(t,w)$$

is defined to be the problem of finding an interval I on the real t line and a (complex) differentiable function φ on I such that

$$\text{(i)} \qquad (t,\varphi(t)) \in D \qquad (t \in I)$$

$$\text{(ii)} \qquad \varphi'(t) = f(t,\varphi(t)) \qquad \left(t \in I,\ ' = \frac{d}{dt}\right)$$

It is an easy task to see that all the existence, uniqueness, continuation, and dependence theorems proved in Secs. 1 to 7 are valid for $(\mathbf{E}_1)$ as well, if one defines the norm $|w|$ of a complex vector $w = (w_1, \ldots, w_n)$ formally as before, namely,

$$|w| = \sum_{i=1}^{n} |w_i|$$

Here, of course, $|w_i| = ((\Re w_i)^2 + (\Im w_i)^2)^{\frac{1}{2}}$, where $\Re w_i$ and $\Im w_i$ are the real and imaginary parts of w_i. Moreover, the Theorems 7.4 and 7.5 concerning the equation

$$(\mathbf{E}_\mu) \qquad x' = f(t,x,\mu)$$

can be extended in an obvious way to the case where μ is a complex parameter vector, if f is defined for complex x and μ. Linear systems are an important case where the above remarks apply.

Usually a function defined on a set of complex numbers that occurs in a differential equation is analytic. Let F be a vector function defined on a domain (open connected set) D of the complex n-dimensional w space. Then F is said to be *analytic* at a point $\omega \in D$ if in some neighborhood $|w - \omega| < \rho$, $\rho > 0$, each component F_j of F is continuous in

$$w = (w_1, \ldots, w_n)$$

and is analytic in each w_k when all other w_l, $l \neq k$, are held fixed. An equivalent definition is that each F_j is representable by a convergent power series

$$F_j(w_1, \ldots, w_n) = \sum_{m_1=0}^{\infty} \cdots \sum_{m_n=0}^{\infty} A_{m_1 \ldots m_n}(w_1 - \omega_1)^{m_1} \cdots (w_n - \omega_n)^{m_n}$$

in some neighborhood $|w - \omega| < \rho$, $\rho > 0$. The $A_{m_1 \ldots m_n}$ are complex constants. A function F is said to be analytic in a domain D if it is analytic at each point of D.

It will be recalled that an analytic function in a domain D possesses derivatives of all orders on D. A basic property of analytic functions is that, if a sequence of analytic functions converges uniformly on a domain D, then the limit function is analytic in D.

It is evident that since an analytic function F in D is represented locally by a power series it is locally single-valued, that is, for every point $\omega \in D$ there is a $\rho > 0$ such that F is single-valued on $|w - \omega| < \rho$. However, in the large, it need not be single-valued. For example, the function F given by $F(w) = w^{\frac{1}{2}}$, where w has one complex dimension, is analytic in the ring $1 < |w| < 2$ but is double-valued there. If $w^{\frac{1}{2}}$ is taken as positive and real on the interval $1 < \Re w < 2$ and w is followed around a closed path ($|w| = \frac{3}{2}$, for example), then $w^{\frac{1}{2}}$ assumes negative real values when w again reaches the positive real axis. The function $F(w) = w^{\alpha}$, α real and irrational, assumes infinitely many values in the ring.

An important extension of the problem (**E**) is to the case where t may be complex. Suppose that f is an analytic complex-valued vector function defined on a domain D in the complex (z,w) space, where the z space has one complex dimension, and the w space is complex n-dimensional. Then the equation

$$(\mathbf{E}_2) \qquad w' = f(z,w)$$

is defined to be the problem of finding a domain H in the complex z plane and a (complex) differentiable locally single-valued function φ [a solution of ($\mathbf{E}_2$)] on H such that

$$\text{(i)} \qquad (z,\varphi(z)) \in D \qquad (z \in H)$$

$$\text{(ii)} \qquad \varphi'(z) = f(z,\varphi(z)) \qquad \left(z \in H,\ ' = \frac{d}{dz}\right)$$

The existence and uniqueness of solutions of ($\mathbf{E}_2$) can be inferred from the method of successive approximations. Indeed, suppose f has components $f_1, \ldots, f_n$, and $w = (w_1, \ldots, w_n)$, and f is analytic on the domain

$$R_2: \qquad |z - z_0| < a \qquad |w - w_0| < b \qquad (a,b > 0)$$

which will be called a rectangle, although it is $n + 1$ complex dimensional. Note that w_0 is a vector here and not a component.

Theorem 8.1. *Suppose f is analytic and bounded on the open rectangle R_2, and let*

$$M = \sup_{(z,w)\,\varepsilon\, R_2} |f(z,w)| \qquad \alpha = \min\left(a, \frac{b}{M}\right)$$

Then there exists on $|z - z_0| < \alpha$ a unique analytic function φ which is a solution of (E_2) *satisfying $\varphi(z_0) = w_0$.*

Proof. Since the matrix $f_w = (\partial f_i/\partial w_j)$ is bounded on any closed rectangle $\tilde{R}_2 \subset R_2$, it follows that f satisfies a Lipschitz condition on $\tilde{R}_2$. Therefore one can construct the successive approximations

$$\varphi_0(z) = w_0$$
$$\varphi_{k+1}(z) = w_0 + \int_{z_0}^{z} f(\zeta,\varphi_k(\zeta))\, d\zeta \qquad (k = 0, 1, 2, \ldots) \tag{8.1}$$

where the integrals can be taken along a straight line joining z_0 to z. Applying the argument in Theorem 3.1, one obtains the existence of a unique solution φ on the circle $|z - z_0| < \alpha$ which satisfies $\varphi(z_0) = w_0$. Clearly φ_0 is analytic in z on $|z - z_0| < \alpha$, and thus the function f_0 defined by $f_0(z) = f(z,\varphi_0(z))$, being an analytic function of an analytic function, is analytic on $|z - z_0| < \alpha$. From (8.1) it follows that φ_1 is analytic on $|z - z_0| < \alpha$, and an easy induction proves that all the approximations φ_k are analytic on $|z - z_0| < \alpha$. Since the solution φ is the uniform limit of the sequence $\{\varphi_k\}$ of analytic functions, it is itself analytic on $|z - z_0| < \alpha$. This completes the proof.

REMARK: Unless other restrictive assumptions are made on f, the circle of analyticity $|z - z_0| < \alpha$ cannot be improved. For $a \leqq b/M$, this is illustrated by the case where f is independent of w, and has singularities on the circle $|z - z_0| = a$. For $a > b/M$ the example

$$w' = f(w) = M\left[\frac{1}{2}\left(1 + \frac{w}{b}\right)\right]^{1/m}$$

where w is one dimensional, illustrates this. The solution φ of this equation for which $\varphi(0) = 0$ (here $z_0 = w_0 = 0$), is

$$\varphi(z) = b\left[\left(1 + \frac{z}{c_m}\right)^{m/(m-1)} - 1\right]$$

where

$$c_m = \left(\frac{m2^{1/m}}{m-1}\right)\frac{b}{M}$$

Clearly f is analytic and bounded in the circle $|w| < b$, and $\sup |f(w)| = M$ there. The solution φ has a singular point at $z = -c_m < -b/M$, and this tends to $z = -b/M$ as $m \to \infty$. Therefore, for any given $r > b/M$, the solution φ has a singularity in the region

$$\frac{b}{M} < |z| < r$$

if m is made large enough.

The analogue of Theorem 7.1 for the equation $(\mathbf{E}_2)$ is the following result:

Theorem 8.2. *Let f be analytic in a domain D of the (z,w) space, and suppose ψ is a solution of $(\mathbf{E}_2)$ on H, where H is a closed convex domain of the z plane. There exists a $\delta > 0$ such that for any $(\zeta,\omega) \in U$, where*

$$U: \qquad \zeta \in H \qquad |\omega - \psi(\zeta)| < \delta$$

there exists a unique solution $\varphi = \varphi(z,\zeta,\omega)$ of $(\mathbf{E}_2)$ on H with $\varphi(\zeta,\zeta,\omega) = \omega$. Moreover, φ is analytic on the $n + 2$ complex dimensional domain

$$V: \qquad z \in H \qquad (\zeta,\omega) \in U$$

REMARK: Actually H need not be convex. It is sufficient if H is simply connected and if there is some constant $c > 0$ such that any two points of H may be joined by a polygonal arc of length less than c.

Proof of Theorem 8.2. The proof follows that part of the proof of Theorem 7.1 that deals with the successive approximations. The path of integration from ζ to z in the successive approximations can be taken as a straight line if H is convex. In any case, the path can be taken as a polygonal path of length less than c. The argument of Theorem 7.1 carries over with the obvious modifications necessary to meet the requirements that the variables are complex. The approximations φ_j are all analytic on V. Thus the limit function, to which the approximations converge uniformly on V, must be analytic on V.

Since φ has all derivatives with respect to z,ζ,ω on V, the equation

$$\varphi'(z,\zeta,\omega) = f(z,\varphi(z,\zeta,\omega))$$

can be differentiated with respect to ω_j, giving

$$\frac{\partial\varphi'}{\partial\omega_j}(z,\zeta,\omega) = f_w(z,\varphi(z,\zeta,\omega))\frac{\partial\varphi}{\partial\omega_j}$$

Thus $\partial\varphi/\partial\omega_j$ is the solution of the linear equation

$$y' = f_w(z,\varphi(z,\zeta,\omega))y \tag{8.2}$$

with initial condition $(\partial\varphi/\partial\omega_j)(\zeta,\zeta,\omega) = e_j$. Thus the analogue of the main result of Theorem 7.2 is proved. The result analogous to (7.5) follows in much the same way as (7.5). The result here is

$$\det \varphi_\omega(z,\zeta,\omega) = \exp \int_\zeta^z \operatorname{tr} f_w(s,\varphi(s,\zeta,\omega))\, ds$$

where the path of integration of the integral is along an arc in H.

Since $\partial\varphi/\partial\zeta$ exists, it follows easily that it is the solution of (8.2) with initial value $-f(\zeta,\omega)$ at ζ.

The case

$$(\mathbf{E}_{2\mu}) \qquad w' = f(z,w,\mu)$$

where f is analytic in (z,w,μ) and μ is k complex dimensional can also be dealt with. Let I_μ be the domain of μ space given by $|\mu - \mu_0| < c$, where μ_0 is fixed and $c > 0$, and let D be a domain in the $n + 1$ complex dimensional (z,w) space. Let D_μ be the set of all (z,w,μ) such that $(z,w) \in D$ and $\mu \in I_\mu$. The analogue of Theorems 7.4 and 7.5 for $(\mathbf{E}_{2\mu})$ is the following result:

Theorem 8.3. *Let f be analytic in the domain D_μ and suppose ψ is a solution of $(\mathbf{E}_{2\mu})$ for $\mu = \mu_0$ which exists for $z \in H$, where H is a closed convex domain in the z plane. There exists a $\delta > 0$ such that for any $(\zeta,\omega,\mu) \in U_\mu$, where*

$$U_\mu: \qquad \zeta \in H \qquad |\omega - \psi(\zeta)| + |\mu - \mu_0| < \delta$$

there exists a unique solution $\varphi = \varphi(z,\zeta,\omega,\mu)$ of $(\mathbf{E}_{2\mu})$ on H with

$$\varphi(\zeta,\zeta,\omega,\mu) = \omega$$

Moreover, φ is analytic in the $n + k + 2$ complex dimensional domain

$$V_\mu: \qquad z \in H \qquad (\zeta,\omega,\mu) \in U_\mu$$

The proof can be obtained using either the method of proof of Theorem 7.4 or that of Theorem 7.5. The remark following Theorem 8.2 applies here also.

A theorem which is a mixture of the results of Secs. 7 and 8 is obtained when t is assumed to be real and w, μ, f complex. Let D be a domain of (t,w) space, where t is real and w complex n-dimensional. Let I_μ be the set of all μ satisfying $|\mu - \mu_0| < c$ for some $c > 0$, where μ is k complex dimensional. Finally let D_μ denote the set of all (t,w,μ) satisfying $(t,w) \in D$ and $\mu \in I_\mu$.

Theorem 8.4. *Let $f \in C$ on the domain D_μ, and for each fixed t suppose f is analytic in (w,μ). For $\mu = \mu_0$ let ψ be a solution of $w' = f(t,w,\mu_0)$ on some interval I: $a \leq t \leq b$ [thus $(t,\psi(t)) \in D$ for $t \in I$] satisfying $\psi(\tau) = \omega_0$, where $\tau \in I$. There exists a $\delta > 0$ such that for any $(\omega,\mu) \in U_\mu$ where*

$$U_\mu: \qquad |\omega - \omega_0| + |\mu - \mu_0| < \delta$$

there exists a unique solution $\varphi = \varphi(t,\omega,\mu)$ of $w' = f(t,w,\mu)$ on I with

$$\varphi(\tau,\omega,\mu) = \omega$$

Moreover, φ is continuous in (t,ω,μ) for $a \leqq t \leqq b$, $(\omega,\mu) \in U_\mu$, and for each fixed $t \in I$ is an analytic function of (ω,μ) for $(\omega,\mu) \in U_\mu$.

The proof is very much like that of Theorem 7.4 and is left to the reader. The uniform convergence of the successive approximations, each of which is analytic in (ω,μ) in U_μ, leads to the analyticity result for φ as a function of (ω,μ).

An important application of this result is to the case of a linear system involving a one-dimensional parameter μ linearly. For example, let A, B be continuous complex-valued n-by-n matrices defined on some open real t interval J, and consider the system

$$w' = (A(t) + \mu B(t))w$$

Then the vector f defined by $f(t,w,\mu) = (A(t) + \mu B(t))w$ is continuous on the domain D_μ given by

$$D_\mu: \qquad t \in J \qquad |w| + |\mu| < \infty$$

and for each fixed $t \in J$ it is analytic in (w,μ) for $|w| + |\mu| < \infty$. Applying Theorems 5.2 and 8.4, it follows that the solution φ passing through the point (τ,ω) $(\tau \in J,\ |\omega| < \infty)$ exists for all $t \in J$, is continuous in (t,ω,μ) for $t \in J$, $|\omega| + |\mu| < \infty$, and for each fixed $t \in J$ is analytic in (ω,μ) for $|\omega| + |\mu| < \infty$. In particular, for fixed (τ,ω) φ is entire in μ. (See also Prob. 7.)

A case of great importance where the above is applied occurs in the study of boundary-value problems involving a parameter; see Chaps. 7–12.

PROBLEMS

1. Let φ, ψ, χ be real-valued continuous (or piecewise continuous) functions on a real t interval $I: a \leqq t \leqq b$. Let $\chi(t) > 0$ on I, and suppose for $t \in I$ that

$$\varphi(t) \leqq \psi(t) + \int_a^t \chi(s)\varphi(s)\, ds$$

Prove that on I

$$\varphi(t) \leqq \psi(t) + \int_a^t \chi(s)\psi(s) \exp\left(\int_s^t \chi(u)\, du\right) ds$$

HINT: Let $R(t) = \int_a^t \chi(s)\varphi(s)\, ds$ and show that $R' - \chi R \leqq \chi\psi$.

2. A function f defined on a domain D of the real (t,x) space is said to be of class Lip (t) on D if there exists an integrable function k of t such that for all (t,x) and $(t,\bar{x})$ in D

$$|f(t,x) - f(t,\bar{x})| \leqq k(t)|x - \bar{x}|$$

Let $f \in$ Lip (t) on D. Let φ_1, φ_2 be two continuous functions on $I: a \leqq t \leqq b$ such that $(t,\varphi_i(t)) \in D$ for $t \in I$, and $f(t,\varphi_i(t))$ is integrable over I for $i = 1,2$. Let

$$\varphi_i(t) = \varphi_i(\tau) + \int_\tau^t f(s,\varphi_i(s))\,ds + E_i(t)$$

where $\tau \in I$, and suppose $|\varphi_1(\tau) - \varphi_2(\tau)| \leqq \delta$.

Prove, if $E(t) = |E_1(t)| + |E_2(t)|$, that for $\tau \leqq t \leqq b$

$$|\varphi_1(t) - \varphi_2(t)| \leqq \delta \exp\left[\int_\tau^t k(s)\,ds\right] + E(t) + \int_\tau^t E(s)k(s) \exp\left[\int_s^t k(u)\,du\right] ds$$

and a similar result for $a \leqq t \leqq \tau$.

HINT: Use Prob. 1.

3. Let the functions φ_1, φ_2 presented in Prob. 2 be of class C_p^1 on I, and in addition let $|\varphi_i'(t) - f(t,\varphi_i(t))| \leqq \epsilon_i(t)$, $\epsilon(t) = \epsilon_1(t) + \epsilon_2(t)$.

Prove that

$$|\varphi_1(t) - \varphi_2(t)| \leqq \delta \exp\left[\int_\tau^t k(s)\,ds\right] + \int_\tau^t \epsilon(s) \exp\left[\int_s^t k(u)\,du\right] ds$$

HINT: $E(t) \leqq \int_\tau^t \epsilon(s)\,ds$.

If $K = \int_a^b k(s)\,ds$, then the above inequality yields

$$|\varphi_1(t) - \varphi_2(t)| \leqq \left(\delta + \int_a^b \epsilon(s)\,ds\right) e^K$$

Clearly the above inequalities can be used to prove uniqueness of solutions of

$$x' = f(t,x)$$

if $f \in$ Lip (t) on D.

4. In the hypothesis of Theorem 3.1 let the condition (C,Lip) in R be replaced by

$$|f(t,\xi)| \leqq k(t)(1 + |\xi|)$$

and the condition Lip (t),

$$|f(t,x) - f(t,\bar{x})| \leqq k(t)|x - \bar{x}|$$

for (t,x) and $(t,\bar{x})$ in R. Assuming f is such that the integral of $f(t,\psi(t))$ is defined for any continuous function ψ, show that there exists an interval $\tau \leqq t \leqq \tau + \alpha_1$ $(\alpha_1 > 0)$ on which the successive approximations converge uniformly to a solution.

HINT: Let $K(t) = \int_\tau^t k(s)\,ds$. If $(1 + |\xi|)(e^{K(t_0)} - 1) = b$ for some t_0 in the interval $(\tau, \tau + a)$, let $t_0 - \tau = \alpha_1$. Otherwise let $\alpha_1 = a$. Show that all the successive approximations φ_j stay in $|x - \xi| \leqq (1 + |\xi|)(e^{K(t)} - 1)$ for $t \in [\tau, \tau + \alpha_1]$. Show

$$|\varphi_j(t) - \varphi_{j-1}(t)| \leqq \frac{(1 + |\xi|)(K(t))^j}{j!}$$

REMARK: If the above hypothesis on f is true for all x and $\bar{x}$ and for all $t \in [a,b]$, and if $\tau \in [a,b]$, then the successive approximations converge uniformly on $[a,b]$. This is the case when f is linear in x.

5. Let $f \in C^1$ on the (t,x,y) set given by $0 \leqq t \leqq 1$, and all x,y. Let φ be a solution of the second-order equation $x'' = f(t,x,x')$ on $[0,1]$, and let $\varphi(0) = a$, $\varphi(1) = b$. Suppose $\partial f/\partial x > 0$ for $t \in [0,1]$ and for all x,y. Prove that if β is near b then there exists a solution ψ of $x'' = f(t,x,x')$ such that $\psi(0) = a$, $\psi(1) = \beta$.

HINT: Consider the solution θ (as a function of (t,α)) with initial values $\theta(0,\alpha) = a$, $\theta'(0,\alpha) = \alpha$. Let $\varphi'(0) = \alpha_0$. Then for $|\alpha - \alpha_0|$ small, θ exists for $t \in [0,1]$. Let

$$u(t) = \frac{\partial\theta}{\partial\alpha}(t,\alpha_0)$$

Then

$$u'' - \frac{\partial f}{\partial y}(t,\varphi(t),\varphi'(t))u' - \frac{\partial f}{\partial x}(t,\varphi(t),\varphi'(t))u = 0$$

where $u(0) = 0$, $u'(0) = 1$. Because $\partial f/\partial x > 0$, u is monotone nondecreasing and thus $u(1) = (\partial\theta/\partial\alpha)(1,\alpha_0) > 0$. Thus the equation $\theta(1,\alpha) - \beta = 0$ can be solved for α as a function of β for (α,β) in a neighborhood of (α_0,b).

6. The following problem shows that analyticity with respect to initial conditions may prevail for the solution of a differential equation with a right member which is discontinuous. (This situation arises in practice when a curve may be replaced by a polygonal line to obtain linearization in each of several regions.)

Let F be a real-valued analytic function defined on the $n + 1$ real dimensional space

$$R: \quad |t| \leqq a \qquad |x| \leqq b$$

Here t has one real dimension, and x is n real dimensional, and the analyticity of F means that at each point of R it can be represented by a power series convergent in some domain containing the point. Let the surface S defined by $F(t,x) = 0$, $(t,x) \in R$, divide R into R_1 and R_2 such that

$$\begin{aligned} F(t,x) &< 0 \qquad (t,x) \in R_1 \\ F(t,x) &> 0 \qquad (t,x) \in R_2 \\ F(t,x) &= 0 \qquad (t,x) \in S \end{aligned}$$

Suppose f is a real-valued vector which is analytic for $(t,x) \in R_1 \cup S$, and g is a real-valued vector analytic for $(t,x) \in R_2 \cup S$. Consider the differential equation

$$x' = f(t,x) \qquad (t,x) \in R_1 \tag{1}$$

$$x' = g(t,x) \qquad (t,x) \in R_2 \tag{2}$$

A continuous function φ defined over some t interval I contained in $|t| \leqq a$ is a solution of this differential equation if $(t,\varphi(t)) \in R$ for $t \in I$, and φ satisfies (1) for $(t,\varphi(t)) \in R_1$ and (2) for $(t,\varphi(t)) \in R_2$, and if φ has only a finite number of points in S for $t \in I$. [This definition can be generalized considerably by allowing $(t,\varphi(t)) \in S$ on one or several t intervals contained in I, but this necessitates some additional hypotheses involving J_1 and J_2 defined below.]

Let $(t_1,x_1) \in R_1$ and $(t_2,x_2) \in R_2$, and suppose φ is a solution on $[t_1,t_2]$ such that

$$\varphi(t_1) = x_1 \qquad \varphi(t_2) = x_2$$

Suppose $(t,\varphi(t)) \in S$ for $t = \tau_1, \ldots, \tau_m$, where $t_1 < \tau_1 < \tau_2 < \cdots < \tau_m < t_2$ and for no other values of t on $[t_1,t_2]$. Let J_1 and J_2 be the functions defined on S by

$$J_1 = \frac{\partial F}{\partial t} + \sum_{i=1}^{n} \frac{\partial F}{\partial x_i} f_i$$

$$J_2 = \frac{\partial F}{\partial t} + \sum_{i=1}^{n} \frac{\partial F}{\partial x_i} g_i$$

and suppose $(-1)^j J_k(\tau_j,\varphi(\tau_j)) < 0$ for $k = 1, 2$, and $j = 1, \ldots, m$. Prove that for $(\sigma,\eta) \in R_1$ near (t_1,x_1) there exists a solution $\psi = \psi(t,\sigma,\eta)$ on $\sigma \leqq t \leqq t_2$ satisfying $\psi(\sigma,\sigma,\eta) = \eta$, and prove that the function ψ_2 defined by $\psi_2(\sigma,\eta) = \psi(t_2,\sigma,\eta)$ is analytic for (σ,η) near (t_1,x_1).

HINT: It suffices to take the case $m = 1$ since the same argument is repeated at each τ_j. Thus it can be assumed that there is only one point $(\tau_1,\varphi(\tau_1))$ of the given solution path on S. It is first shown that ψ is analytic in (t,σ,η) for $(t,\psi(t,\sigma,\eta)) \in R_1 \cup S$. The value of t for which $(t,\psi(t,\sigma,\eta))$ intersects S is obtained by solving $F(t,\psi(t,\sigma,\eta)) = 0$ for t. Because $J_1(\tau_1,\varphi(\tau_1)) > 0$ there exists a unique analytic solution $t = \gamma(\sigma,\eta)$, where $\gamma(t_1,x_1) = \tau_1$.

The solution of (2) is then considered with initial value $t = \gamma(\sigma,\eta)$, $x = \psi(\gamma(\sigma,\eta),\sigma,\eta)$. This initial value is analytic in (σ,η). Because $J_2(\tau_1,\varphi(\tau_1)) > 0$ this solution will not again intersect S near $(\tau_1,\varphi(\tau_1))$. Since it stays close to $(t,\varphi(t))$, it can be continued to t_2. The solution is analytic in t and in its initial conditions. The initial conditions are analytic in (σ,η). Thus ψ is analytic in (t,σ,η) for (t,σ,η) sufficiently near (t_2,t_1,x_1).

7. Let f be a continuous function defined on a real t interval $a \leqq t \leqq b$ and for all complex w and μ, where w is complex n-dimensional and μ is complex k-dimensional. For each fixed t let f be analytic in (w,μ) for $|w| + |\mu| < \infty$. For all w, $\tilde{w}$, and $t \in [a,b]$ let

$$|f(t,w,\mu) - f(t,\tilde{w},\mu)| \leqq M(|\mu|)|w - \tilde{w}|$$
$$|f(t,0,\mu)| \leqq M(|\mu|) < \infty$$

where M is a monotone increasing function. (This hypothesis implies f is a linear function of w.) Prove that the solution φ of the initial-value problem $w' = f(t,w,\mu)$, $w(a) = \omega$, is continuous in (t,ω,μ) for $t \in [a,b]$ and $|\omega| + |\mu| < \infty$. Thus φ for fixed t is an entire function of (ω,μ).

REMARK: If the hypothesis is valid for $\mu \in D$, where D is a domain in μ space, instead of for $|\mu| < \infty$, then the result holds for $\mu \in D$.

HINT: The successive approximations φ_j, where $\varphi_0(t,\omega,\mu) = \omega$, satisfy

$$|\varphi_{j+1}(t,\omega,\mu) - \varphi_j(t,\omega,\mu)| \leqq \frac{(1 + |\omega|)(M(|\mu|))^{j+1}(t - a)^{j+1}}{(j + 1)!}$$

and each φ_j is an entire function of (ω,μ) for any fixed t. The result follows from the uniform convergence of φ_j. (The result also follows from Theorem 8.4.)

8. Let F be a real continuous function of (t,x,y) in a real domain D containing the point (t_0,x_0,y_0). Let $\partial F/\partial x$ and $\partial F/\partial y$ exist and be continuous in D, and suppose $F(t_0,x_0,y_0) = 0$, $(\partial F/\partial y)(t_0,x_0,y_0) \neq 0$. Prove that there exists a unique function φ [a solution of $F(t,x,x') = 0$] on some interval containing t_0 satisfying $F(t,\varphi(t),\varphi'(t)) = 0$, $\varphi(t_0) = x_0$, $\varphi'(t_0) = y_0$.

REMARK: The above theorem may fail where $F(t,x,y) = 0$ and $(\partial F/\partial y)(t,x,y) = 0$. A solution of $F(t,x,x') = 0$ may satisfy both these equations but uniqueness may fail. An example is $(x')^2 - 2x' + 4x = 4t - 1$ at $(0,0,1)$ with solutions t and $t - t^2$.

9. In Theorem 7.2 it was shown that $\partial\varphi/\partial\xi_j$, $j = 1, \ldots, n$ were solutions of the linear equation $y' = f_x(t,\varphi(t,\tau,\xi))y$ with initial values e_j at τ. Prove that every solution of this linear system is a linear combination with complex coefficients of these n solutions. Since $\partial\varphi/\partial\tau$ is the solution with initial value $-f(\tau,\xi)$ at τ, prove that

$$\frac{\partial\varphi}{\partial\tau}(t,\tau,\xi) + \sum_{j=1}^{n} f_j(\tau,\xi)\frac{\partial\varphi}{\partial\xi_j}(t,\tau,\xi) = 0$$

HINT: If θ is any solution and $\theta(\tau) = \alpha$, then $\alpha = \Sigma a_j e_j$ for some complex constants a_j. Prove that $\theta = \Sigma a_j(\partial\varphi/\partial\xi_j)$.

10. The following problem illustrates the abstract idea behind the Picard theorem. Consider a Banach space B (a complete normed linear space) with the norm of an element ψ denoted by $\|\psi\|$. Let T be a transformation defined on the set of all ψ

satisfying $\|\psi\| \leqq b$ $(b > 0)$ which is such that $\|T\psi\| \leqq b$, and T satisfies a Lipschitz condition

$$\|T\psi - T\bar{\psi}\| \leqq k\|\psi - \bar{\psi}\|$$

with constant $k < 1$. Prove that there exists an element φ such that $T\varphi = \varphi$, that is, T has a fixed point. Moreover, prove that φ is unique.

HINT: Define the successive approximations $\varphi_0, \varphi_1, \ldots$ by $\varphi_0 = 0$, $\varphi_{j+1} = T\varphi_j$ and using the Lipschitz condition show that $\|\varphi_{j+1} - \varphi_j\| \leqq k^j b$, and hence

$$\|\varphi_{j+m} - \varphi_j\| \leqq (k^j + \cdots + k^{j+m-1})b$$

Since $k^j + \cdots + k^{j+m-1}$ is the Cauchy difference for the convergent geometric series Σk^j, it follows that the sequence $\{\varphi_j\}$ is convergent in B, and hence has a limit $\varphi \varepsilon B$. Since $\|\varphi_j\| \leqq b$, clearly $\|\varphi\| \leqq b$, and thus $T\varphi$ is defined. Also $\|T\varphi - \varphi\| \leqq \|T\varphi - T\varphi_j\| + \|T\varphi_j - \varphi\| \leqq k\|\varphi - \varphi_j\| + \|\varphi_{j+1} - \varphi\| \to 0$ as $j \to \infty$. Uniqueness follows from the Lipschitz condition.

11. Let $f \varepsilon C$ on the $n + 1$ dimensional real rectangle R given by $|t| \leqq a$, $|x| \leqq b$, and assume f satisfies the Lipschitz condition

$$|f(t,x) - f(t,\bar{x})| \leqq \frac{k}{a}|x - \bar{x}|$$

for a constant $k < 1$ in R. Further, suppose $|f(t,x)| \leqq b/a$ for $(t,x) \varepsilon R$. The initial-value problem $x' = f(t,x)$, $x(0) = 0$ is equivalent to the integral equation

$$x(t) = \int_0^t f(s,x(s))\,ds$$

Let B denote the space of all continuous vector functions ψ on $|t| \leqq a$ with a norm given by $\|\psi\| = \max |\psi(t)|$ for $|t| \leqq a$. Show that B is a Banach space. Let T be the transformation defined for $\psi \varepsilon B$ satisfying $\|\psi\| \leqq b$ by

$$T\psi(t) = \int_0^t f(s,\psi(s))\,ds$$

Prove that $\|T\psi\| \leqq b$ if $\|\psi\| \leqq b$, and that T satisfies the condition

$$\|T\psi - T\bar{\psi}\| \leqq k\|\psi - \bar{\psi}\|$$

Apply Prob. 10 to obtain the existence and uniqueness of a solution of the initial-value problem $x' = f(t,x)$, $x(0) = 0$, on $|t| \leqq a$.

12. Let $x' = |x|^{-\frac{3}{4}}x + t \sin(\pi/t)$, $x(0) = 0$. Show that if polygonal approximate solutions are set up as in Theorem 1.1 they need not converge as $\epsilon \to 0$.

HINT: Consider $t \geqq 0$ and let $t_k = k\delta$, $k = 0, 1, 2, \ldots$, where $\delta = (n + \frac{1}{2})^{-1}$ for some large n. If n is even, show that the polygonal solution $\varphi_n(t)$ satisfies $\varphi_n(\delta) = 0$, $\varphi_n(2\delta) = \delta^2$, $\varphi_n(3\delta) > \frac{1}{2}\delta^{\frac{3}{2}}$. Once $\varphi_n(t) \geqq t^{\frac{3}{2}}/6$, it stays there as long as $t < 1/2{,}000$. Indeed, for $t \geqq 4\delta$ and as long as $\varphi_n(t) \geqq t^{\frac{3}{2}}/6$, $\varphi_n'(t) > \varphi_n^{\frac{1}{4}}(t - \delta) - t > \frac{1}{2}(t - \delta)^{\frac{3}{8}} - t > t^{\frac{3}{8}}/10$. Since $\frac{1}{10}t^{\frac{3}{8}} > (d/dt)(t^{\frac{3}{2}}/6)$, the result follows. If n is odd, $\varphi_n(t) < -t^{\frac{3}{2}}/6$ for $3\delta < t < 1/2{,}000$.

CHAPTER 2

EXISTENCE AND UNIQUENESS OF SOLUTIONS (CONTINUED)

1. Extension of the Idea of a Solution, Maximum and Minimum Solutions

It has been seen that if f is a continuous function in some (t,x) domain D, then the differential equation

$$x' = f(t,x) \tag{E}$$

together with an initial condition

$$x(\tau) = \xi \tag{1.1}$$

is equivalent to the integral equation

$$x(t) = \xi + \int_\tau^t f(s,x(s))\,ds \tag{1.2}$$

That is to say, if φ is a solution of (E) on some interval I for which $\varphi(\tau) = \xi$, then $x = \varphi(t)$ will satisfy (1.2) on I, and conversely.

Clearly the integral in (1.2) makes sense for many functions f which are not continuous. Recall that the continuity of f guaranteed that a solution of (E) was of class C^1. Thus, if a continuously differentiable solution of (E) is not demanded, the continuity restriction on f can be relaxed.

Suppose f is a real-valued (not necessarily continuous) function defined in some set S of the (t,x) space. Then one can extend the notion of the differential equation (E) by defining (E) to be the following problem:

Problem. *To find an absolutely continuous function φ defined on a real t interval I such that*

(i) $\quad (t,\varphi(t)) \in S \quad (t \in I)$

(ii) $\quad \varphi'(t) = f(t,\varphi(t))$ *for all $t \in I$, except on a set of Lebesgue-measure zero.*

If such an interval I and function φ exist, then φ is said to be a *solution of* (E) *in the extended sense on* I. Notice that the absolute continuity of a solution guarantees the existence of φ' almost everywhere on I (that is, except on a set of Lebesgue-measure zero), so that (ii) makes sense.

If $f \in C$ on S, and φ is a solution of (E) in the above sense, then from (ii) $\varphi' \in C$ on I, and therefore the more general notion of the equation (E), and of solution φ, reduces to the ordinary definition of (E) when $f \in C$ on S.

It will usually be clear from the context as to the meaning attached to (**E**) and the solution φ, and hence it will rarely be necessary to add the phrase "in the extended sense."

As regards the existence of a solution of (**E**), Caratheodory has proved the following quite general theorem under the assumption that f be bounded by a Lebesgue-integrable function of t. The proof will be carried out for the case $n = 1$ only; it will be clear what modifications are required in the case of a system (**E**). R will denote the rectangle

$$R: \qquad |t - \tau| \leqq a \qquad |x - \xi| \leqq b$$

where (τ,ξ) is a fixed point in the (t,x) plane, and a and b are positive real numbers.

Theorem 1.1 (**Carathéodory**). *Let f be defined on R, and suppose it is measurable in t for each fixed x, continuous in x for each fixed t. If there exists a Lebesgue-integrable function m on the interval $|t - \tau| \leqq a$ such that*

$$|f(t,x)| \leqq m(t) \qquad ((t,x) \in R) \tag{1.3}$$

then there exists a solution φ of (**E**) *in the extended sense on some interval $|t - \tau| \leqq \beta$, $(\beta > 0)$, satisfying $\varphi(\tau) = \xi$.*

Proof. The case $t \geqq \tau$ will be considered; the situation is similar when $t \leqq \tau$. If M is defined by

$$\begin{aligned} M(t) &= 0 \qquad (t < \tau) \\ M(t) &= \int_\tau^t m(s)\,ds \qquad (\tau \leqq t \leqq \tau + a) \end{aligned} \tag{1.4}$$

then it is clear that M is continuous nondecreasing [$m \geqq 0$ by (1.3)], and $M(\tau) = 0$. Therefore $(t, \xi \pm M(t)) \in R$ for some interval $\tau \leqq t \leqq \tau + \beta \leqq \tau + a$, where β is some positive constant. Choose any $\beta > 0$ for which this is true, and define the approximations φ_j $(j = 1, 2, \ldots)$ by

$$\begin{aligned} \varphi_j(t) &= \xi \qquad \left(\tau \leqq t \leqq \tau + \frac{\beta}{j}\right) \\ \varphi_j(t) &= \xi + \int_\tau^{t-\beta/j} f(s,\varphi_j(s))\,ds \qquad \left(\tau + \frac{\beta}{j} < t \leqq \tau + \beta\right) \end{aligned} \tag{1.5}$$

Clearly φ_1 is defined on $\tau \leqq t \leqq \tau + \beta$, for it is the constant ξ. For any fixed $j \geqq 1$, the first formula in (1.5) defines φ_j on $\tau \leqq t \leqq \tau + \beta/j$, and since $(t,\xi) \in R$ for $\tau \leqq t \leqq \tau + \beta/j$, the second formula in (1.5) defines φ_j as a continuous function on the interval $\tau + \beta/j < t \leqq \tau + 2\beta/j$. Further, on this latter interval

$$|\varphi_j(t) - \xi| \leqq M\left(t - \frac{\beta}{j}\right) \tag{1.6}$$

by virtue of (1.3) and (1.4). Assume that φ_j is defined on $\tau \leqq t \leqq \tau + k\beta/j$ for $1 < k < j$. Then the second formula of (1.5) defines φ_j for $\tau + k\beta/j < t \leqq \tau + (k+1)\beta/j$, since knowledge of the measurable integrand is only required on $\tau \leqq t \leqq \tau + k\beta/j$. Also, on $\tau + k\beta/j < t \leqq \tau + (k+1)\beta/j$, the function φ_j satisfies (1.6), because of (1.3) and (1.4). Therefore, by induction, (1.5) defines all φ_j as continuous functions on $\tau \leqq t \leqq \tau + \beta$, which satisfy

$$\begin{aligned} \varphi_j(t) &= \xi & &\left(\tau \leqq t \leqq \tau + \frac{\beta}{j}\right) \\ |\varphi_j(t) - \xi| &\leqq M\left(t - \frac{\beta}{j}\right) & &\left(\tau + \frac{\beta}{j} < t \leqq \tau + \beta\right) \end{aligned} \tag{1.7}$$

If t_1 and t_2 are any two points in the interval $[\tau, \tau + \beta]$, then on account of (1.3), (1.4), and (1.5),

$$|\varphi_j(t_1) - \varphi_j(t_2)| \leqq \left| M\left(t_1 - \frac{\beta}{j}\right) - M\left(t_2 - \frac{\beta}{j}\right) \right| \tag{1.8}$$

Since M is continuous on $[\tau, \tau + \beta]$, it is uniformly continuous there. This implies, by (1.8), that the set $\{\varphi_j\}$ is an equicontinuous set on $[\tau, \tau + \beta]$. Also, by (1.7), the set $\{\varphi_j\}$ is uniformly bounded on $[\tau, \tau + \beta]$. Consequently it follows by the Ascoli lemma that there exists a subsequence $\{\varphi_{j_k}\}$ which converges uniformly on $[\tau, \tau + \beta]$ to a continuous limit function φ, as $k \to \infty$.

From (1.3),

$$|f(t, \varphi_{j_k}(t))| \leqq m(t) \qquad (\tau \leqq t \leqq \tau + \beta)$$

and since f is continuous in x for fixed t,

$$f(t, \varphi_{j_k}(t)) \to f(t, \varphi(t)) \qquad (k \to \infty)$$

for every fixed t in $[\tau, \tau + \beta]$. Therefore the dominated convergence theorem due to Lebesgue may be applied to give

$$\lim_{k \to \infty} \int_\tau^t f(s, \varphi_{j_k}(s))\, ds = \int_\tau^t f(s, \varphi(s))\, ds \tag{1.9}$$

for any t in $[\tau, \tau + \beta]$. But

$$\varphi_{j_k}(t) = \xi + \int_\tau^t f(s, \varphi_{j_k}(s))\, ds - \int_{t-\beta/j_k}^t f(s, \varphi_{j_k}(s))\, ds$$

where it is clear that the latter integral tends to zero as $k \to \infty$. Therefore, letting $k \to \infty$, and using (1.9), it follows that

$$\varphi(t) = \xi + \int_\tau^t f(s, \varphi(s))\, ds$$

from which the theorem follows at once.

It is interesting to remark that the original approximations (1.5) must converge to a solution in the case where a unique solution is known. This situation does not obtain for the ordinary successive approximations; see the example in Sec. 3.

For the case $n = 1$ it can be shown that all solutions of (E) issuing from an initial point (τ,ξ) can be bracketed between two special solutions, the *maximum* and *minimum solutions*. Let f be defined on the rectangle R, as in Theorem 1.1. If φ_M is a solution of (E) passing through (τ,ξ), existing on some interval I containing τ, with the property that every other solution φ of (E) passing through (τ,ξ) and existing on I is such that

$$\varphi(t) \leqq \varphi_M(t) \qquad (t \in I)$$

then φ_M is called a *maximum solution* of (E) on I passing through (τ,ξ). Similarly, if φ_m is a solution of (E) on an interval I for which $\varphi_m(\tau) = \xi$, and such that

$$\varphi(t) \geqq \varphi_m(t) \qquad (t \in I)$$

holds for every other solution of (E) on I for which $\varphi(\tau) = \xi$, then φ_m is called a *minimum solution* of (E) on I passing through (τ,ξ). Clearly, the functions φ_M and φ_m, if they exist, must be unique.

The existence of φ_M and φ_m will now be demonstrated under the Carathéodory assumptions.

Theorem 1.2. *Let the hypothesis of Theorem* 1.1 *be satisfied. Then there exists a maximum solution* φ_M *and a minimum solution* φ_m *of* (E) *on* $|t - \tau| \leqq \beta$ *passing through* (τ,ξ).

Proof. The existence of φ_M on $[\tau, \tau + \beta]$ will be proved. Now any solution φ of (E) passing through (τ,ξ) must satisfy

$$\varphi(t) = \xi + \int_\tau^t f(s,\varphi(s))\, ds \tag{1.10}$$

as far as it exists, and from (1.10) it follows that

$$|\varphi(t_1) - \varphi(t_2)| \leqq |M(t_1) - M(t_2)| \tag{1.11}$$

for any two points t_1, t_2 where φ exists. Recall that M is defined by (1.4). Since M is continuous, (1.11) implies, by the Cauchy criterion for convergence, that the solution φ can be continued, if necessary, to the entire interval $[\tau, \tau + \beta]$, making use of the Carathéodory existence theorem. The details of this argument are entirely similar to those given in Theorem 4.1, Chap. 1. Therefore, *all* solutions of (E) passing through (τ,ξ) exist on $[\tau, \tau + \beta]$, and all must satisfy (1.11) there. From the uniform continuity of M on $[\tau, \tau + \beta]$, it follows from (1.11) that the set of all solutions $\{\varphi\}$ of (E) on $[\tau, \tau + \beta]$ is an equicontinuous set, that is, given any $\epsilon > 0$, there exists a $\delta_\epsilon > 0$, *independent of t and the solution* φ, such that

$$|\varphi(\hat{t}) - \varphi(\tilde{t})| < \epsilon \qquad \text{whenever } |\hat{t} - \tilde{t}| < \delta_\epsilon \tag{1.12}$$

and $\hat{t}$, $\tilde{t}$ are in $[\tau, \tau + \beta]$. Further, from (1.11), putting $t_2 = \tau$, the set $\{\varphi\}$ is uniformly bounded on $[\tau, \tau + \beta]$.

Let Φ be the function defined by

$$\Phi(t) = \sup \{\varphi(t)\} \qquad (t \in [\tau, \tau + \beta])$$

taken over all solutions φ of (E) on $[\tau, \tau + \beta]$ passing through (τ,ξ). Clearly Φ exists on $[\tau, \tau + \beta]$ and is continuous (and hence uniformly continuous) there. Therefore, for any given $\epsilon > 0$, a $\delta_\epsilon > 0$ exists such that, not only is (1.12) true for this δ_ϵ, but also for $\hat{t}$, $\tilde{t}$ in $[\tau, \tau + \beta]$,

$$|\Phi(\hat{t}) - \Phi(\tilde{t})| < \epsilon \qquad \text{whenever } |\hat{t} - \tilde{t}| < \delta_\epsilon \tag{1.13}$$

It will be shown that Φ is a solution of (E) satisfying $\Phi(\tau) = \xi$, and if φ_M is defined to be Φ, it is clear that this φ_M will satisfy the requirements of the theorem on $[\tau, \tau + \beta]$. For a given $\epsilon > 0$, choose δ_ϵ so that (1.12) and (1.13) hold. Subdivide the interval $[\tau, \tau + \beta]$ into n intervals by the points $\tau = t_0 < t_1 < t_2 < \cdots < t_n = \tau + \beta$ in such a way that

$$\max (t_{i+1} - t_i) < \delta_\epsilon$$

For every t_i, $(i = 0, 1, \ldots, n - 1)$, choose a solution φ_i of (E) passing through (τ,ξ) so that

$$0 \leqq \Phi(t_i) - \varphi_i(t_i) < \epsilon$$

and for $i \geqq 1$

$$\varphi_i(t_i) - \varphi_{i-1}(t_i) \geqq 0$$

This is possible from the definition of Φ.

Now, for the given ϵ, define the function φ_ϵ as follows: Let

$$\varphi_\epsilon(t) = \varphi_{n-1}(t) \qquad (t_{n-1} \leqq t \leqq t_n = \tau + \beta)$$

If $\varphi_{n-1}(t_{n-1}) > \varphi_{n-2}(t_{n-1})$, define φ_ϵ to the left of t_{n-1} as φ_{n-1} up to the point τ_{n-2} (if it exists) in (t_{n-2},t_{n-1}) nearest t_{n-1} such that

$$\varphi_\epsilon(\tau_{n-2}) = \varphi_{n-1}(\tau_{n-2}) = \varphi_{n-2}(\tau_{n-2})$$

If τ_{n-2} exists, define $\varphi_\epsilon(t) = \varphi_{n-2}(t)$ for $t_{n-2} \leqq t < \tau_{n-2}$. If τ_{n-2} does not exist, define φ_ϵ on $[t_{n-2},t_{n-1})$ as φ_{n-1}. If $\varphi_{n-1}(t_{n-1}) = \varphi_{n-2}(t_{n-1})$, define $\varphi_\epsilon(t) = \varphi_{n-2}(t)$ on $[t_{n-2},t_{n-1})$. Continuing in this way, one can define a solution φ_ϵ of (E) on $[\tau, \tau + \beta]$ passing through (τ,ξ), obtained by patching together solutions of (E), and having the property

$$0 \leqq \Phi(t_i) - \varphi_\epsilon(t_i) < \epsilon \qquad (i = 0, 1, \ldots, n) \tag{1.14}$$

Since the variation of Φ and φ_ϵ in each interval $[t_i,t_{i+1}]$ is less than ϵ, by (1.12) and (1.13), there results from (1.14)

$$0 \leqq \Phi(t) - \varphi_\epsilon(t) < 3\epsilon \qquad (\tau \leqq t \leqq \tau + \beta) \tag{1.15}$$

Letting $\epsilon = 1/m$, $(m = 1, 2, \ldots)$, one obtains a sequence $\varphi_{1/m}$ of solutions which, by (1.15), converges uniformly to Φ on $[\tau, \tau + \beta]$. From this fact and an application of the Lebesgue dominated convergence theorem to (1.10) with φ replaced by $\varphi_{1/m}$, it follows that

$$\Phi(t) = \xi + \int_\tau^t f(s,\Phi(s))\, ds \qquad (\tau \leqq t \leqq \tau + \beta)$$

that is, Φ is a solution of (E) satisfying $\Phi(\tau) = \xi$, and from its definition it is the maximum solution φ_M on $[\tau, \tau + \beta]$.

Theorem 1.3. *In a domain D of the (t,x) plane let the function f be defined, measurable in t for fixed x and continuous in x for fixed t. Let there exist an integrable function m such that $|f(t,x)| \leqq m(t)$ for $(t,x) \in D$. Then, given a solution φ of* (E) *for $t \in (a,b)$, it is the case that $\varphi(b - 0)$ exists and if $(b,\varphi(b - 0)) \in D$ then φ can be continued over $(a, b + \delta]$ for some $\delta > 0$. A similar result holds at a. Thus the solution φ can be continued up to the boundary of D. Moreover, the same continuation is valid for a maximum solution φ_M or a minimum solution φ_m.*

The proof is very similar to that of Theorem 4.1, Chap. 1.

Corollary to Theorem 1.3. *Let the hypothesis of Theorem 1.1 be satisfied and let φ_M and φ_m, the maximum and minimum solutions through (τ,ξ), exist over $[\tau, \tau + \beta]$, where $\beta \leqq a$. Then for any c satisfying $\varphi_m(\tau + \beta) < c < \varphi_M(\tau + \beta)$ there is at least one solution φ through (τ,ξ) for $\tau \leqq t \leqq \tau + \beta$ and with $\varphi(\tau + \beta) = c$.*

Proof. Start with the solution through $(\tau + \beta, c)$ and continue it to the left. It need not leave the region $\varphi_m(t) \leqq x \leqq \varphi_M(t), \tau \leqq t \leqq \tau + \beta$, since it can always be continued back along one of these extreme solutions if it meets one of them. Thus it can be continued back to (τ,ξ).

Theorem 1.4. *Let the hypothesis of Theorem 1.3 be valid, and suppose the maximum solution $\varphi_{M\xi}$ of* (E) *through (τ,ξ) exists over an interval $[\tau, \tau + \alpha]$. Then there exists a $\delta > 0$ such that* (E) *has a maximum solution $\varphi_{M\eta}$ for each η, $\xi \leqq \eta < \xi + \delta$ on $[\tau, \tau + \alpha]$ with $\varphi_{M\eta}(\tau) = \eta$. Moreover, $\varphi_{M\eta} \to \varphi_{M\xi}$ as $\eta \to \xi + 0$, uniformly over $[\tau, \tau + \alpha]$.*

Proof. By Theorem 1.2, $\varphi_{M\eta}$ certainly exists over some interval with the left end point τ, if $\eta - \xi$ is small enough. From the definition of the maximum solution, it follows readily that, for $\bar{\eta} > \eta > \xi$,

$$\varphi_{M\bar{\eta}}(t) \geqq \varphi_{M\eta}(t) \geqq \varphi_{M\xi}(t)$$

Thus $\varphi_{M\eta}$ is monotone nondecreasing in η and is bounded from below. Therefore, for each t on some interval $[\tau, \tau + \beta]$, there exists

$$\Phi(t) = \varphi_{M\xi+0}(t) \geqq \varphi_{M\xi}(t) \tag{1.16}$$

Since $\varphi_{M\eta}$ satisfies (1.11),

$$|\Phi(t_1) - \Phi(t_2)| \leqq |M(t_1) - M(t_2)|$$

so that Φ is continuous. From

$$\varphi_{M\eta}(t) = \eta + \int_\tau^t f(s,\varphi_{M\eta}(s))\, ds$$

it follows on letting $\eta \to \xi + 0$ that

$$\Phi(t) = \xi + \int_\tau^t f(s,\Phi(s))\, ds$$

But this implies Φ is a solution of (E) through (τ,ξ). Thus, by (1.16), $\Phi(t) = \varphi_{M\xi}(t)$ over $[\tau, \tau + \beta]$. The uniformity of the convergence of $\varphi_{M\eta}$ to $\varphi_{M\xi}$ follows from the equicontinuity of $\varphi_{M\eta}$ in t, as proved by (1.11).

The above argument is clearly valid over the range of existence of Φ on $[\tau, \tau + \alpha]$. Suppose that for some $t_0 \leqq \tau + \alpha$ and for every small $h > 0$, Φ exists over $[\tau, t_0 - h]$ but not over $[\tau, t_0 + h]$. Then for any given $\epsilon > 0$ there exists a $\delta_\epsilon > 0$ such that

$$|\varphi_{M\eta}(t_0 - \epsilon) - \varphi_{M\xi}(t_0 - \epsilon)| \leqq \epsilon \tag{1.17}$$

if $0 \leqq \eta - \xi < \delta_\epsilon$.

Let the region H be the set of points (t,x) which satisfy the inequalities

$$|t - t_0| \leqq \gamma \qquad |x - \varphi_{M\xi}(t_0 - \gamma)| \leqq \gamma + M(t) - M(t_0 - \gamma)$$

By choosing γ small enough, $H \subset D$. Any solution φ of (E) which starts on the left vertical side $t = t_0 - \gamma$ of H [that is, $|\varphi(t_0 - \gamma) - \varphi_{M\xi}(t_0 - \gamma)| \leqq \gamma$] will, by (1.11), remain in H as t increases. Thus any such solution can be continued to $t_0 + \gamma$.

By choosing ϵ in (1.17) so that $\epsilon = \gamma$, it follows that for $0 < \eta - \xi < \delta_\epsilon$ the solutions $\varphi_{M\eta}$ can be continued to $t_0 + \gamma$. This implies the existence of Φ over $[\tau, t_0 + \gamma]$, which contradicts the assumption about t_0. Thus $t_0 > \tau + \alpha$ and therefore Φ exists over $[\tau, \tau + \alpha]$.

2. Further Uniqueness Results

Considerable research has been done on the problem of uniqueness of solutions of the system (E). The following theorem gives a criterion for uniqueness which is sufficient for many practical cases, and includes as special cases many known criteria.

Theorem 2.1. *Let $\psi = \psi(t,r)$ be a continuous nonnegative function defined on*

$$0 < t < a \qquad r \geqq 0 \qquad (a > 0) \tag{2.1}$$

and nondecreasing in r for fixed t there. Suppose that for each α, $0 < \alpha < a$, the function ρ defined by $\rho(t) = 0$, $0 \leqq t < \alpha$, is the only differentiable *function on* $0 \leqq t < \alpha$ *for which* $\rho'_+(0) = \lim (\rho(t) - \rho(0))/t$ *as* $t \to 0+$ *exists,*

$$\rho'(t) = \psi(t,\rho(t)) \qquad (0 < t < \alpha) \tag{2.2}$$

and

$$\rho(0) = \rho'_+(0) = 0 \tag{2.3}$$

Let $f \in C$ *on the* $(n + 1)$*-dimensional region*

$$R: \qquad |t - \tau| \leqq a \qquad |x - \xi| \leqq b \qquad (a,b > 0)$$

and satisfy there, for $t \neq \tau$,

$$|f(t,\hat{x}) - f(t,\tilde{x})| \leqq \psi(|t - \tau|,|\hat{x} - \tilde{x}|) \tag{2.4}$$

Then there exists at most one solution $\varphi \in C^1$ *of* (**E**) *in* R *on* $|t - \tau| \leqq a$ *for which* $\varphi(\tau) = \xi$.

NOTE: If ψ is the function defined on (2.1) by

$$\psi(t,r) = kr$$

where k is a positive constant, then ψ satisfies the conditions of the ψ in Theorem 2.1. It is an easy exercise to see that for each α, $0 < \alpha < a$, the identically zero function is the only differentiable function on $0 \leqq t < \alpha$ satisfying (2.2) and (2.3) for this choice of ψ. In this case, (2.4) just becomes the Lipschitz condition, and thus Theorem 2.1 includes as a special case Theorem 2.2, Chap. 1.

Actually the following generalization of Theorem 2.1 holds and is just as easy to prove. Only its proof will be given.

Theorem 2.2. *Let* ψ *be a nonnegative function defined on* (2.1) *which is Lebesgue measurable in* t *for fixed* r*, and continuous nondecreasing in* r *for fixed* t. *Further, for every bounded subset* B *of* (2.1), *let there exist a function* χ_B *defined on* $0 < t < a$ *such that*

$$\psi(t,r) \leqq \chi_B(t) \qquad ((t,r) \in B) \tag{2.5}$$

and for which χ_B *is Lebesgue integrable on* $\gamma < t < a$ *for every* $\gamma > 0$. *Suppose that for each* α, $0 < \alpha < a$, *the identically zero function is the only absolutely continuous function on* $0 \leqq t < \alpha$ *which satisfies*

$$\rho'(t) = \psi(t,\rho(t)) \tag{2.6}$$

almost everywhere on $0 < t < \alpha$, *and such that* $\rho'_+(0)$ *exists, and*

$$\rho(0) = \rho'_+(0) = 0 \tag{2.7}$$

Then if for this ψ *the function* f *satisfies the same conditions as in Theorem 2.1, the result of Theorem 2.1 is valid.*

Proof. The proof will be given for the interval $\tau \leqq t \leqq \tau + a$; the case $\tau - a \leqq t \leqq \tau$ is similar. Also, it will be assumed that $(\tau,\xi) = (0,0)$ for simplicity.

Suppose that there are two solutions φ_1 and φ_2 of class C^1 of (E) on the interval $0 \leqq t \leqq a$ satisfying

$$\varphi_1(0) = \varphi_2(0) = 0 \tag{2.8}$$

Let p be the function defined by

$$p(t) = |\varphi_1(t) - \varphi_2(t)| \qquad (0 \leqq t \leqq a)$$

Then there exists a σ, $0 < \sigma \leqq a$ such that $p(\sigma) > 0$. From Theorem 1.1 it follows that through the point $(\sigma, p(\sigma))$ there exists an absolutely continuous function ρ satisfying the equation

$$\rho'(t) = \psi(t, \rho(t))$$

on some interval to the left of σ.

As far to the left of σ as ρ exists, it satisfies the inequality

$$\rho(t) \leqq p(t) \tag{2.9}$$

for if this were not the case there would exist a point to the left of σ, say ζ, where $\rho(\zeta) = p(\zeta)$, and $\rho(t) > p(t)$ for $t < \zeta$, and sufficiently near ζ ($\zeta = \sigma$ is not excluded). Now, since φ_1 and φ_2 are both solutions of (E) satisfying (2.8),

$$p(\zeta) = \left| \int_0^\zeta [f(t,\varphi_1(t)) - f(t,\varphi_2(t))]\, dt \right|$$

and for small enough $h > 0$,

$$p(\zeta - h) = \left| \int_0^{\zeta - h} [f(t,\varphi_1(t)) - f(t,\varphi_2(t))]\, dt \right|$$

Since $|u| - |v| \leqq |u - v|$, it follows by subtraction that

$$\begin{aligned} p(\zeta) - p(\zeta - h) &\leqq \left| \int_{\zeta - h}^{\zeta} [f(t,\varphi_1(t)) - f(t,\varphi_2(t))]\, dt \right| \\ &\leqq \int_{\zeta - h}^{\zeta} |f(t,\varphi_1(t)) - f(t,\varphi_2(t))|\, dt \\ &\leqq \int_{\zeta - h}^{\zeta} \psi(t,p(t))\, dt \end{aligned} \tag{2.10}$$

using (2.4). From the definition of ρ one has, since $p(\zeta) = \rho(\zeta)$,

$$p(\zeta) - \rho(\zeta - h) = \int_{\zeta - h}^{\zeta} \psi(t,\rho(t))\, dt \tag{2.11}$$

where h is now assumed so small that ρ exists on $[\zeta - h, \zeta]$. Since ψ is nondecreasing in r,

$$\psi(t,p(t)) \leqq \psi(t,\rho(t)) \qquad (\zeta - h \leqq t \leqq \zeta)$$

and this, together with (2.10) and (2.11), implies that $\rho(\zeta - h) \leqq$

$p(\zeta - h)$, which contradicts the definition of ζ. This establishes the inequality (2.9).

Now $\rho(t) > 0$ on $0 < t \leqq \sigma$, as far as it exists. Otherwise $\rho(\hat{\sigma}) = 0$ for some $\hat{\sigma}$, $0 < \hat{\sigma} < \sigma$, and the function $\hat{\rho}$ defined by

$$\begin{aligned} \hat{\rho}(t) &= 0 \qquad (0 \leqq t \leqq \hat{\sigma}) \\ \hat{\rho}(t) &= \rho(t) \qquad (\hat{\sigma} < t \leqq \sigma) \end{aligned}$$

would be a function on $0 \leqq t \leqq \sigma$ not identically zero, which satisfies (2.6) and (2.7). This contradicts the hypothesis of the theorem. Therefore,

$$0 < \rho(t) \leqq p(t) \tag{2.12}$$

as far to the left of σ as ρ exists. But by (2.12), and an application of Theorem 1.3, it follows that ρ can be continued as a solution, call it ρ again, on the whole interval $0 < t \leqq \sigma$. Hence $\lim \rho(t)$, $t \to +0$, exists, and by (2.12)

$$\lim_{t \to +0} \rho(t) = 0$$

Define $\rho(0)$ to be 0.

From (2.12) it follows further that

$$0 < \frac{\rho(t)}{t} \leqq \frac{p(t)}{t} \qquad (0 < t \leqq \sigma) \tag{2.13}$$

and since $\varphi_1'(0) = \varphi_2'(0) = f(0,0)$, the ratio

$$\frac{p(t)}{t} = \frac{|\varphi_1(t) - \varphi_1(0) - (\varphi_2(t) - \varphi_2(0))|}{t}$$

tends to 0 as $t \to +0$. From (2.13), therefore, $\rho_+'(0) = 0$. This contradicts the hypothesis of the theorem, for ρ is an absolutely continuous solution of (2.6) on $0 < t < \sigma$, and satisfies (2.7), but is not identically zero on $0 < t < \sigma$. Therefore $p(\sigma) \ngtr 0$ for any σ, $0 < \sigma < a$, and this proves the theorem.

It is easily seen from the proof of Theorem 2.2 that this theorem is valid when f is required to satisfy a Carathéodory hypothesis (that is, the hypothesis on f in Theorem 1.1), if instead of (2.7) it is only required that $\rho(0) = 0$. The continuity of f was used only at the origin in order to be able to assert that $\varphi_1'(0) = \varphi_2'(0)$.

The proof of Theorem 2.2 is valid for Theorem 2.1, except that the continuity of ψ now serves to guarantee the existence of the function (of class C^1 here) satisfying $\rho'(t) = \psi(t,\rho(t))$ to the left of $(\sigma,p(\sigma))$.

Theorem 2.1 can be improved if the idea of a minimum solution is used.

Theorem 2.3. *Let f and ψ be functions satisfying the conditions of Theorem* 2.1, *except that ψ need not be monotone in r. Then the conclusion of Theorem* 2.1 *still remains valid.*

Proof. The only place in the proof of Theorem 2.1 where the nondecreasing nature of ψ in r was used was in establishing the inequality

$$\rho(t) \leqq p(t) \tag{2.9}$$

where ρ is a function of class C^1 satisfying

$$\rho'(t) = \psi(t,\rho(t)) \qquad \rho(\sigma) = p(\sigma) > 0 \tag{2.14}$$

A procedure by which (2.9) can be proved for the *minimum* solution ρ_m of (2.14), and which avoids the monotonicity restriction on ψ, is as follows:

Consider the problem of finding a solution to

$$r' = \psi(t,r) + \epsilon \qquad (0 < \epsilon < 1)$$

passing through the point $(\sigma,p(\sigma))$. For every such ϵ there exists at least one solution ρ_ϵ of this problem on some interval $\sigma - \alpha \leqq t \leqq \sigma$, for some positive α which is independent of ϵ. Also

$$\rho_\epsilon(t) \leqq p(t) \qquad (\sigma - \alpha \leqq t \leqq \sigma) \tag{2.15}$$

for if this were not the case there would exist a point to the left of σ, say ζ, where $\rho_\epsilon(\zeta) = p(\zeta)$ and $\rho_\epsilon(t) > p(t)$ for $t < \zeta$ and sufficiently near ζ. At such a ζ, the left-hand derivative $p'_-(\zeta)$ exists, and

$$p'_-(\zeta) \leqq |\varphi_1'(\zeta) - \varphi_2'(\zeta)| \leqq \psi(\zeta,p(\zeta)) < \psi(\zeta,\rho_\epsilon(\zeta)) + \epsilon = \rho_\epsilon'(\zeta)$$

Therefore, for $h > 0$ sufficiently small,

$$\rho_\epsilon(\zeta - h) < p(\zeta - h)$$

which contradicts the definition of ζ. This proves (2.15).

In the same way, it follows that

$$\limsup_{\epsilon \to 0} \rho_\epsilon(t) \leqq \rho_m(t) \tag{2.16}$$

Now

$$p(\sigma) - \rho_\epsilon(t) = \int_t^\sigma \psi(t,\rho_\epsilon(t))\,dt + \epsilon(\sigma - t)$$

and hence on $\sigma - \alpha \leqq t \leqq \sigma$, $(\alpha > 0)$, the set $\{\rho_\epsilon\}$ is an equicontinuous uniformly bounded set of functions. Therefore there exists a subsequence $\{\rho_{\epsilon_k}\}$ such that ρ_{ϵ_k} tends uniformly (as $\epsilon_k \to 0$) to a function ρ on $\sigma - \alpha \leqq t \leqq \sigma$, where ρ satisfies (2.14). But by (2.16), this ρ must be ρ_m, and from (2.15)

$$\rho_m(t) \leqq p(t) \qquad (\sigma - \alpha \leqq t \leqq \sigma)$$

The remainder of the proof is the same as the corresponding part of the proof of Theorem 2.1, only replacing ρ by ρ_m throughout.

3. Uniqueness and Successive Approximations

Let R denote the $(n+1)$-dimensional region

$$R: \qquad |t-\tau| \leqq a \qquad |x-\xi| \leqq b \qquad (a,b>0)$$

Let $f \varepsilon C$ on R, and suppose $M = \max |f|$ on R. In Sec. 3, Chap. 1, it was shown that the successive approximations $\varphi_0, \varphi_1, \varphi_2, \ldots$ defined by

$$\begin{aligned} \varphi_0(t) &= \xi \\ \varphi_{m+1}(t) &= \xi + \int_\tau^t f(s,\varphi_m(s))\,ds \qquad (m = 0, 1, 2, \ldots) \end{aligned} \tag{3.1}$$

converge to a solution φ of (**E**) on the interval

$$|t-\tau| \leqq \alpha \qquad \left(\alpha = \min\left(a, \frac{b}{M}\right)\right)$$

and $\varphi(\tau) = \xi$, *provided* that $f \varepsilon$ Lip on R.

The following example $(n = 1)$ illustrates the fact that the continuity of f alone is not sufficient for the convergence of the successive approximations. Let f be defined by

$$f(t,x) = \begin{cases} 0 & (t = 0, -\infty < x < +\infty) \\ 2t & (0 < t \leqq 1, -\infty < x < 0) \\ 2t - \dfrac{4x}{t} & (0 < t \leqq 1, 0 \leqq x \leqq t^2) \\ -2t & (0 < t \leqq 1, t^2 < x < +\infty) \end{cases} \tag{3.2}$$

On the region $0 \leqq t \leqq 1$, $-\infty < x < +\infty$, this f is continuous and bounded by the constant 2. For the initial point $(\tau,\xi) = (0,0)$, the successive approximations (3.1) become, for $0 \leqq t \leqq 1$,

$$\varphi_0(t) = 0 \qquad \varphi_{2m-1}(t) = t^2 \qquad \varphi_{2m}(t) = -t^2 \qquad (m = 1, 2, \ldots)$$

Therefore the sequence $\{\varphi_m(t)\}$ has two cluster values for each $t \neq 0$, and hence the successive approximations do not converge. Note also that *neither* of the two convergent subsequences $\{\varphi_{2m-1}\}$, $\{\varphi_{2m}\}$ *converge to a solution*, for

$$\varphi'_{2m-1}(t) = 2t \neq f(t,t^2)$$

and

$$\varphi'_{2m}(t) = -2t \neq f(t,-t^2)$$

Since the Lipschitz condition guarantees a unique solution of (**E**), one may ask whether the continuity of f plus uniqueness is sufficient to guarantee the convergence of the successive approximations. The answer to this is in the negative, however, for the above example (3.2) is one for which a unique solution of (E) exists passing through (0,0) and proceeding to the right of the origin. This follows from the fact that this f is monotone nonincreasing in x for fixed t. It is left as an easy exercise to show that this latter condition implies uniqueness to the right of the origin.

It is equally true that if the successive approximations converge, the solution obtained may not be unique. The familiar example

$$x' = x^{\frac{1}{3}}$$

shows this. For the initial point (0,0) the successive approximations are all the zero function, and hence they converge to the identically zero solution. On the other hand, the function φ defined by

$$\varphi(t) = \left(\frac{2t}{3}\right)^{\frac{3}{2}}$$

is another solution which exists to the right of the origin.

Although uniqueness does not imply the convergence of the successive approximations, it is true that the hypotheses of the general uniqueness Theorems 2.1 and 2.2 are sufficient for this convergence.

Theorem 3.1. *Let $f \in C$ on R, and suppose that the hypothesis of Theorem 2.1 or Theorem 2.2 is assumed. Then the successive approximations $\{\varphi_m\}$ defined by (3.1) converge (uniformly) on $|t - \tau| \leqq \alpha$ to the solution φ of* (**E**) *on this interval satisfying*

$$\varphi(\tau) = \xi$$

Proof. The proof will be carried out using the hypothesis of Theorem 2.2, and assuming $(\tau,\xi) = (0,0)$. The latter is clearly no restriction.

It follows from the definition (3.1) of the successive approximations that they satisfy the inequality

$$|\varphi_m(t_1) - \varphi_m(t_2)| \leqq M|t_1 - t_2| \tag{3.3}$$

for any t_1,t_2 in the interval $|t| \leqq \alpha$, where $M = \max |f|$ on R. This implies that the set $\{\varphi_m\}$ is an equicontinuous set there. Letting $t_1 = t$, $t_2 = 0$, in (3.3) one obtains

$$|\varphi_m(t)| \leqq M|t| \leqq M\alpha \leqq b$$

and hence the set $\{\varphi_m\}$ is uniformly bounded on $|t| \leqq \alpha$. By the Ascoli lemma there exists a subsequence $\{\varphi_{m_k}\}$ which is uniformly convergent to a function φ on $|t| \leqq \alpha$, as $k \to \infty$. The subsequence $\{\varphi_{m_k+1}\}$, which satisfies

$$\varphi_{m_k+1}(t) = \int_0^t f(s,\varphi_{m_k}(s))\, ds \qquad (|t| \leqq \alpha)$$

is uniformly convergent on $|t| \leqq \alpha$ to the function φ^* defined by

$$\varphi^*(t) = \int_0^t f(s,\varphi(s))\, ds \qquad (|t| \leqq \alpha)$$

for f is uniformly continuous on $|t| \leqq \alpha$, $|x| \leqq b$. It will be shown below that on $|t| \leqq \alpha$,

$$\varphi_{m+1}(t) - \varphi_m(t) \to 0 \qquad (m \to \infty) \tag{3.4}$$

Assuming this, it follows that

$$\varphi_{m_k+1}(t) - \varphi_{m_k}(t) \to 0 \qquad (k \to \infty)$$

and this implies $\varphi^* = \varphi$, that is, φ is a solution of (**E**). Because of uniqueness, every subsequence of $\{\varphi_m\}$ which is convergent will tend to the same solution, and this proves that the original sequence $\{\varphi_m\}$ is convergent on $|t| \leqq \alpha$ to the solution φ. This convergence is uniform on $|t| \leqq \alpha$, for the set $\{\varphi_m\}$ is equicontinuous and convergent there.

In order to prove (3.4), let w_m be the difference

$$w_m(t) = \varphi_{m+1}(t) - \varphi_m(t) \qquad (|t| \leqq \alpha)$$

and v be defined by

$$v(t) = \limsup_{m \to \infty} |w_m(t)| \qquad (|t| \leqq \alpha)$$

Now $v(0) = 0$, and v is continuous on $|t| \leqq \alpha$, for it is the upper limit of an equicontinuous uniformly bounded sequence of functions. To prove $w_m(t) \to 0$, $(m \to \infty)$, on $|t| \leqq \alpha$, is equivalent to showing $v(t) = 0$ on $|t| \leqq \alpha$. This will be done for $0 \leqq t \leqq \alpha$; the proof for $-\alpha \leqq t \leqq 0$ is similar.

As a matter of notation, for any $\Delta t > 0$, and function g defined at t and $t + \Delta t$, let Δg be defined by

$$\Delta g\,(t) = g(t + \Delta t) - g(t)$$

Then from (3.1) it follows that for any t and $t + \Delta t$ in the interval $[0,\alpha]$

$$|\Delta w_{m+1}\,(t)| \leqq \int_t^{t+\Delta t} |f(s,\varphi_{m+1}(s)) - f(s,\varphi_m(s))|\, ds$$

and because of (2.4)

$$|\Delta w_{m+1}\,(t)| \leqq \int_t^{t+\Delta t} \psi(s,|w_m(s)|)\, ds \tag{3.5}$$

Given any $\delta > 0$ there exists an integer N_δ, independent of s and m, such that

$$|w_m(s)| \leqq v(s) + \delta \qquad (m > N_\delta) \tag{3.6}$$

for all s in the interval $t \leqq s \leqq t + \Delta t$. To see this, note that v is uniformly continuous, and the set $\{w_m\}$ is equicontinuous, on $t \leqq s \leqq t + \Delta t$. Therefore corresponding to any $\delta > 0$, there exists an $\eta_\delta > 0$ such that

$$|v(s) - v(\hat{s})| < \frac{\delta}{3} \qquad |w_m(s) - w_m(\hat{s})| < \frac{\delta}{3} \tag{3.7}$$

whenever $|s - \hat{s}| < \eta_\delta$, and s, $\hat{s}$ are in the interval $[t, t + \Delta t]$. Divide the interval $[t, t + \Delta t]$ into a finite number of subintervals

$$t = s_0 < s_1 < \cdot \cdot \cdot < s_k = t + \Delta t$$

such that $\max (s_{i+1} - s_i) < \eta_\delta$. From the definition of v as a lim sup, it follows that for each s_i, $(i = 0, 1, \ldots, k)$, there exists an integer $N_{\delta i}$ such that

$$|w_m(s_i)| \leqq v(s_i) + \frac{\delta}{3} \qquad (m > N_{\delta i}) \tag{3.8}$$

Define N_δ to be $\max N_{\delta i}$, $(i = 0, 1, \ldots, k)$. Then the inequality (3.8) holds for $m > N_\delta$. For a fixed s in $[t, t + \Delta t]$ there exists an s_i such that $|s - s_i| < \eta_\delta$. Applying (3.7) to this s and $\hat{s} = s_i$, and combining with (3.8), the inequality (3.6) results.

Since ψ is nondecreasing in r, it follows from (3.6) that

$$\psi(s,|w_m(s)|) \leqq \psi(s,v(s) + \delta) \qquad (m > N_\delta) \tag{3.9}$$

and consequently, using (3.5),

$$|\Delta w_{m+1} (t)| \leqq \int_t^{t+\Delta t} \psi(s,v(s) + \delta)\, ds \qquad (m > N_\delta) \tag{3.10}$$

From the definition of v, it is easy to see that

$$|\Delta v (t)| \leqq \limsup_{m\to\infty} |\Delta w_m (t)|$$

and this with (3.10) shows that

$$|\Delta v (t)| \leqq \int_t^{t+\Delta t} \psi(s,v(s) + \delta)\, ds \tag{3.11}$$

On account of the continuity of ψ in r,

$$\psi(s,v(s) + \delta) \to \psi(s,v(s)) \quad (\text{as } \delta \to 0)$$

and from (2.5), and the dominated convergence theorem of Lebesgue, it follows that

$$\lim_{\delta\to 0} \int_t^{t+\Delta t} \psi(s,v(s) + \delta)\, ds = \int_t^{t+\Delta t} \psi(s,v(s))\, ds$$

The last relation, along with (3.11), yields

$$|\Delta v (t)| \leqq \int_t^{t+\Delta t} \psi(s,v(s))\, ds \tag{3.12}$$

The inequality (3.12) implies that v is absolutely continuous over any interval in $[0,\alpha]$, and consequently v' exists almost everywhere on $[0,\alpha]$. From (3.12) this derivative satisfies

$$|v'(t)| \leqq \psi(t,v(t)) \tag{3.13}$$

almost everywhere on $[0,\alpha]$.

Suppose for some σ, $0 < \sigma \leqq \alpha$, that $v(\sigma) \neq 0$. Because of the hypothesis on ψ, there exists a function ρ on some interval $\sigma - \gamma \leqq t \leqq \sigma$, $(\gamma > 0)$, satisfying

$$\rho'(t) = \psi(t,\rho(t)) \qquad \rho(\sigma) = v(\sigma) > 0$$

as shown in Theorem 1.1. Now on this interval

$$\rho(t) \leqq v(t) \tag{3.14}$$

and from this it follows that ρ can be continued to the entire interval $0 < t \leqq \sigma$, and

$$\lim_{t \to 0+} \rho(t) = 0$$

The proof of these facts is entirely similar to the steps between (2.9) and (2.12) in the proof of Theorem 2.2, and so will be omitted.

On account of (3.14) one has

$$0 < \frac{\rho(t)}{t} \leqq \frac{v(t)}{t} \qquad (0 < t \leqq \sigma) \tag{3.15}$$

It will now be shown that $v(t)/t \to 0$ as $t \to 0+$. For this purpose, consider, for $0 < t \leqq \alpha$,

$$w_m(t) = \varphi_{m+1}(t) - \varphi_m(t) = \int_0^t [f(s,\varphi_m(s)) - f(s,\varphi_{m-1}(s))]\, ds \tag{3.16}$$

Since $f \varepsilon C$ on R, and using the fact that $|\varphi_m(t)| \leqq Mt$, from (3.16) it follows that, given any $\epsilon > 0$, there exists an $\eta_\epsilon > 0$ such that

$$|w_m(t)| < \epsilon t \qquad (0 \leqq t \leqq \eta_\epsilon)$$

Hence

$$v(t) = \limsup_{m \to \infty} |w_m(t)| \leqq \epsilon t$$

provided $0 \leqq t \leqq \eta_\epsilon$, or $v(t)/t \to 0$ as $t \to 0+$. From (3.15) it now follows that $\rho(t)/t \to 0$ as $t \to 0+$, or since $\rho(0) = 0$, $\rho'_+(0) = 0$. This contradicts the hypothesis of the theorem, for ρ is an absolutely continuous function satisfying

$$\rho'(t) = \psi(t,\rho(t)) \qquad (0 < t \leqq \sigma)$$

and

$$\rho(0) = \rho'_+(0) = 0$$

although ρ is not identically zero on $0 \leqq t \leqq \sigma$. Therefore $v(\sigma) = 0$ for any σ, $0 < \sigma \leqq \alpha$, and this proves the theorem.

4. Variation of Solutions with Respect to Initial Conditions and Parameters

In Theorem 7.4, Chap. 1, it was shown that if f was a continuous function of (t,x,μ) and satisfied a Lipschitz condition in x, uniformly in μ, then the solution of the initial-value problem

$$x' = f(t,x,\mu) \qquad x(\tau) = \xi$$

was continuous in (t,μ). Actually, the requirement of a Lipschitz condition is too strong; its consequence, the uniqueness of the solution, is sufficient for this important result. The x space is n-dimensional and the μ space is k-dimensional, as in Theorem 7.4.

Theorem 4.1. *Let D be a domain of (t,x) space, I_μ the domain $|\mu - \mu_0| < c$, $c > 0$, and D_μ the set of all (t,x,μ) satisfying $(t,x) \in D$, $\mu \in I_\mu$. Suppose f is a continuous function on D_μ bounded by a constant M there. For $\mu = \mu_0$ let*

$$x' = f(t,x,\mu) \qquad x(\tau) = \xi \tag{4.1}$$

have a unique solution φ_0 on the interval $[a,b]$, where $\tau \in [a,b]$. Then there exists a $\delta > 0$ such that, for any fixed μ satisfying $|\mu - \mu_0| < \delta$, every solution φ_μ of (4.1) exists over $[a,b]$ and as $\mu \to \mu_0$

$$\varphi_\mu \to \varphi_0$$

uniformly over $[a,b]$.

NOTE: Though (4.1) need not have a unique solution for $\mu \neq \mu_0$, nevertheless its solutions are continuous in μ at μ_0.

Proof of Theorem 4.1. The proof will be carried out for the case $\tau \in (a,b)$. The result will first be proved over $|t - \tau| \leqq \alpha$ for some $\alpha > 0$. Choose α small enough so that the region R: $|t - \tau| \leqq \alpha$, $|x - \xi| \leqq M\alpha$ is in D. All solutions of (4.1) with $\mu \in I_\mu$ exist over $[\tau - \alpha, \tau + \alpha]$ and remain in R. Let φ_μ denote a solution. Then the set of functions $\{\varphi_\mu\}$, $\mu \in I_\mu$, is a uniformly bounded and equicontinuous set in $|t - \tau| \leqq \alpha$. This follows from the integral equation

$$\varphi_\mu(t) = \xi + \int_\tau^t f(s,\varphi_\mu(s),\mu)\, ds \qquad (|t - \tau| \leqq \alpha) \tag{4.2}$$

and the inequality $|f| \leqq M$.

Suppose $\varphi_\mu(\bar{t})$ does not tend to $\varphi_0(\bar{t})$ for some $\bar{t} \in [\tau - \alpha, \tau + \alpha]$. Then there exists a sequence $\{\mu_k\}$, $k = 1, 2, \ldots$, for which $\mu_k \to \mu_0$, and corresponding solutions φ_{μ_k} such that φ_{μ_k} converges uniformly over $[\tau - \alpha, \tau + \alpha]$ as $k \to \infty$ to a limit function ψ but $\psi(\bar{t}) \neq \varphi_0(\bar{t})$. From the fact that $f \in C$ on D_μ, that $\psi \in C$ on $[\tau - \alpha, \tau + \alpha]$, and that φ_{μ_k} converges uniformly to ψ, (4.2) for the solutions φ_{μ_k} yields

$$\psi(t) = \xi + \int_\tau^t f(s,\psi(s),\mu_0)\, ds \qquad (|t - \tau| \leqq \alpha)$$

Thus ψ is a solution of (4.1) with $\mu = \mu_0$. By the uniqueness hypothesis, it follows that $\psi(t) = \varphi_0(t)$ on $|t - \tau| \leqq \alpha$. Thus $\psi(\bar{t}) = \varphi_0(\bar{t})$. Thus all solutions φ_μ on $|t - \tau| \leqq \alpha$ tend to φ_0 as $\mu \to \mu_0$. Because of the equicontinuity, the convergence is uniform.

To prove the result over $[a,b]$, a region H similar to that used in Theorem 1.4 will be introduced. The interval $[\tau,b]$ will be treated. Suppose that $t_0 \in [\tau,b)$ and that the result is valid for every small $h > 0$ over $[\tau, t_0 - h]$ but not over $[\tau, t_0 + h]$. It is clear that $t_0 \geqq \tau + \alpha$. By the above assumption, for any small $\epsilon > 0$, there exists a $\delta_\epsilon > 0$ such that

$$|\varphi_\mu(t_0 - \epsilon) - \varphi_0(t_0 - \epsilon)| < \epsilon \tag{4.3}$$

for $|\mu - \mu_0| < \delta_\epsilon$.

Let H denote the region of (t,x) space

$$|t - t_0| \leqq \gamma \qquad |x - \varphi_0(t_0 - \gamma)| \leqq \gamma + M|t - t_0 + \gamma| \tag{4.4}$$

where γ is small enough so that $H \subset D$. Any solution of $x' = f(t,x,\mu)$ starting on $t = t_0 - \gamma$ with initial value x_0, $|x_0 - \varphi_0(t_0 - \gamma)| \leqq \gamma$, will remain in H as t increases. Thus all solutions can be continued to $t_0 + \gamma$.

By choosing $\epsilon = \gamma$ in (4.3), it follows that for $|\mu - \mu_0| < \delta_\epsilon$ the solutions φ_μ can all be continued to $t_0 + \epsilon$. Thus over $[\tau, t_0 + \epsilon]$ these solutions are in D so that the argument that $\varphi_\mu \to \varphi_0$ which has been given for $|t - \tau| \leqq \alpha$, and is based on (4.2), also applies over $[\tau, t_0 + \epsilon]$. Thus the assumption about the existence of $t_0 < b$ is false. The case $t_0 = b$ is treated in similar fashion on $t_0 - \gamma \leqq t \leqq t_0$. A similar argument applies to the left of τ and therefore the theorem is valid over $[a,b]$.

The same result is true if f satisfies a Carathéodory type of hypothesis.

Theorem 4.2. *The conclusion of Theorem 4.1 remains valid if the hypothesis $f \in C$ in D_μ is replaced by the assumptions that on D_μ f is measurable in t for each fixed μ and x; f is continuous in x for each fixed t and μ; for fixed t, f is continuous in (x,μ) at $\mu = \mu_0$; and*

$$|f(t,x,\mu)| \leqq m(t)$$

where m is Lebesgue integrable over $[a,b]$.

The proof is similar to that of Theorem 4.1 with the usual changes necessitated by the Carathéodory type of hypothesis and is left as an exercise for the reader.

Theorem 4.3. *Let the hypothesis of Theorem 4.1 be satisfied. Then there exists a $\delta > 0$ such that for any fixed (σ,η,μ) satisfying*

$$|\sigma - \tau| + |\eta - \xi| + |\mu - \mu_0| < \delta$$

all solutions $\varphi = \varphi(t,\sigma,\eta,\mu)$ of

$$x' = f(t,x,\mu) \qquad x(\sigma) = \eta$$

exist over $[a,b]$. Moreover, as $(\sigma,\eta,\mu) \to (\tau,\xi,\mu_0)$,

$$\varphi(t,\sigma,\eta,\mu) \to \varphi_0(t) = \varphi(t,\tau,\xi,\mu_0)$$

uniformly over $[a,b]$.

Proof. A proof can be constructed with minor changes from that of Theorem 4.1.

An important result in connection with the continuity of solutions with respect to initial conditions is contained in the following theorem.

Theorem 4.4. *Let τ_0, $\tau_1(\tau_0 < \tau_1)$ be fixed real numbers, and (τ_0,ξ_0) a fixed point in the $(n + 1)$-dimensional (t,x) space. Denote by U_0 the set of all points P_0: (τ_0,ξ) such that*

$$|\xi - \xi_0| < b_0 \qquad (b_0 > 0)$$

Suppose that through each point (t,x) in the region

$$V: \qquad \tau_0 \leqq t \leqq \tau_1 \qquad |x - \xi_0| < b \qquad (0 < b_0 \leqq b)$$

there exists a unique solution of

$$\text{(E)} \qquad x' = f(t,x)$$

f being continuous on V. Let $\varphi = \varphi(t,\tau_0,\xi)$ be the solution of (**E**) *passing through $P_0 \in U_0$. Let b be sufficiently large so that $(t,\varphi(t,\tau_0,\xi)) \in V$ for $|\xi - \xi_0| < b_0$, $\tau_0 \leqq t \leqq \tau_1$. Let U_1 denote the set of all points P_1: $(\tau_1,\varphi(\tau_1,\tau_0,\xi))$, where $(\tau_0,\xi) \in U_0$. Then the mapping T which assigns to each point $P_0 \in U_0$ the point $P_1 \in U_1$ is a topological mapping of U_0 onto U_1.*

Proof. Because of the uniqueness assumption, it is clear that T is one-to-one. By the definition of U_1 for every $P_1 \in U_1$, there exists a point $P_0 \in U_0$ such that $TP_0 = P_1$, and thus T is onto. The function ψ defined on $|\xi - \xi_0| < b_0$ by

$$\psi(\xi) = \varphi(\tau_1,\tau_0,\xi)$$

(with τ_1,τ_0 fixed) is a continuous function of ξ on $|\xi - \xi_0| < b_0$ by the uniqueness of φ. Thus T is a continuous mapping. Applying the same continuity argument at τ_1, one gets the continuity of the inverse mapping T^{-1} which assigns to each $P_1 \in U_1$ the point $P_0 \in U_0$. This completes the proof.

PROBLEMS

1. Let Φ be a nonnegative measurable function of t on $0 < t < a$ and let F be a nonnegative measurable function of r on $0 < r < a$. Let $\int_0^a \Phi(t)\,dt < \infty$ and for any $\delta > 0$, $\int_0^\delta dr/F(r) = \infty$. If for small t and $|x|$, $|f(t,x) - f(t,\tilde{x})| \leqq \Phi(t)F(|x - \tilde{x}|)$, show that if f satisfies the hypothesis of Theorem 1.1 then the solution φ of (E) satisfying $\varphi(0) = 0$ is unique.

HINT: If $\psi(t,r) = \Phi(t)F(r)$, then $\rho'(t) = \psi(t,\rho(t))$, $\rho(0) = 0$, implies that $\rho(t) \equiv 0$.

2. Show that Theorem 2.1 is valid with $\psi(t,r) = r/t$.

3. Let $f \in C$ $(n = 1)$ on the rectangle $0 \leqq t \leqq a$, $|x| \leqq b$, where $a,b > 0$, and assume $f(t,x_1) \leqq f(t,x_2)$ if $x_1 \leqq x_2$, and $f(t,0) \geqq 0$ for $0 \leqq t \leqq a$. Prove that the successive approximations (3.1) converge to a solution of $x' = f(t,x)$, $x(0) = 0$, on

$$0 \leqq t \leqq \alpha = \min (a, b/M)$$

where $M = \max |f|$ on the rectangle.

4. Let $f \in C$ on the $(n + 1)$-dimensional (t,x) region $0 \leqq t \leqq a$ $(a > 0)$, $|x| < \infty$, and suppose φ is a solution of the system $x' = f(t,x)$ starting at $(0,\xi)$ and existing for $0 \leqq t < \bar{t} < a$. Prove that either $\varphi(\bar{t} - 0)$ exists as a finite limit, in which case φ can be extended as a solution beyond $\bar{t}$, or else $|\varphi(t)| \to \infty$ as $t \to \bar{t} - 0$.

5. Let f be as in the previous problem, and suppose further that there exists a continuous function ψ on $0 \leqq r < \infty$ such that $|f(t,x)| < \psi(|x|)$, and for some δ, $0 \leqq \delta < \infty$, $\int_\delta^\infty dr/\psi(r) = \infty$. Prove that every solution φ of the system **(E)** such that $\varphi(0) = \xi$ exists on the whole interval $0 \leqq t \leqq a$.

HINT: Use the result of the previous problem.

CHAPTER 3

LINEAR DIFFERENTIAL EQUATIONS

1. Preliminary Definitions and Notations

If A is a matrix of complex numbers (a_{ij}) with n rows and n columns, define the *norm*, $|A|$, of A by

$$|A| = \sum_{i,j=1}^{n} |a_{ij}| \tag{1.1}$$

In case x is an n-dimensional vector, represented as a matrix of n rows and one column, then the vector magnitude as defined in Sec. 5, Chap. 1, coincides with the norm of x as defined by (1.1). It is easily seen that the norm satisfies the following properties:

(i) $$|A + B| \leqq |A| + |B|$$
(ii) $$|AB| \leqq |A|\,|B|$$
(iii) $$|Ax| \leqq |A|\,|x|$$

where A and B are matrices, and x is an n-dimensional vector.

The *distance* between two matrices A and B is defined by $|A - B|$, and this distance satisfies the usual properties of a metric.

The *zero* matrix will be denoted by 0, and the *unit* matrix by E. If there is danger of confusion concerning the dimension, these n-by-n matrices will be denoted by 0_n and E_n, respectively. Note that $|0_n| = 0$, and $|E_n| = n$, and not 1.

The *complex conjugate* matrix of $A = (a_{ij})$, denoted by $\bar{A}$, is defined by $\bar{A} = (\bar{a}_{ij})$, where $\bar{a}_{ij}$ is the complex conjugate of a_{ij}. The *transposed matrix*† of A, denoted by $A^{\backprime}$, is defined by $A^{\backprime} = (a_{ji})$. The *conjugate transposed* matrix of A is $A^* = \bar{A}^{\backprime}$. Note that $|A^*| = |A^{\backprime}| = |\bar{A}| = |A|$. Also $(AB)^* = B^*A^*$. The determinant of A is denoted by det A.

If det $A = 0$, then A is said to be *singular*. A nonsingular matrix A possesses an *inverse* (or *reciprocal*), A^{-1}, which satisfies

$$AA^{-1} = A^{-1}A = E$$

The polynomial in λ of degree n, det $(\lambda E - A)$, is called the *characteristic polynomial* of A, and its roots are the *characteristic roots* of A. If these

† The notation A' will be reserved for differentiation, when A is a matrix function.

roots are denoted by λ_i, $i = 1, \ldots, n$, then clearly

$$\det(\lambda E - A) = \prod_{i=1}^{n} (\lambda - \lambda_i)$$

Two n-by-n complex matrices A and B are said to be *similar* if there exists a nonsingular n-by-n complex matrix P such that

$$B = PAP^{-1}$$

If A and B are similar, then they have the same characteristic polynomial, for

$$\begin{aligned} \det(\lambda E - B) &= \det(P(\lambda E - A)P^{-1}) \\ &= \det P \cdot \det(\lambda E - A) \cdot \det P^{-1} \\ &= \det(\lambda E - A) \end{aligned}$$

In particular, the coefficients of the powers of λ in $\det(\lambda E - A)$ are invariant under similarity transformations. Two of the most important invariants are $\det A$ and $\operatorname{tr} A$, the determinant and trace of A, respectively.

The following fundamental result concerning the canonical form of a matrix is assumed.

Theorem 1.1. *Every complex n-by-n matrix A is similar to a matrix of the form*

$$J = \begin{pmatrix} J_0 & 0 & 0 & \cdots & 0 \\ 0 & J_1 & 0 & \cdots & 0 \\ \cdot & \cdot & \cdot & \cdots & \cdot \\ 0 & 0 & 0 & \cdots & J_s \end{pmatrix}$$

where J_0 *is a diagonal matrix with diagonal* $\lambda_1, \lambda_2, \ldots, \lambda_q$, *and*

$$J_i = \begin{pmatrix} \lambda_{q+i} & 1 & 0 & 0 & \cdots & 0 & 0 \\ 0 & \lambda_{q+i} & 1 & 0 & \cdots & 0 & 0 \\ \cdot & \cdot & \cdot & \cdot & \cdots & \cdot & \cdot \\ 0 & 0 & 0 & 0 & \cdots & \lambda_{q+i} & 1 \\ 0 & 0 & 0 & 0 & \cdots & 0 & \lambda_{q+i} \end{pmatrix} \qquad (i = 1, \ldots, s)$$

The λ_j, $j = 1, \ldots, q + s$, *are the characteristic roots of* A, *which need not all be distinct. If* λ_j *is a simple root, then it occurs in* J_0, *and therefore, if all the roots are distinct,* A *is similar to the diagonal matrix*

$$J = \begin{pmatrix} \lambda_1 & 0 & 0 & \cdots & 0 \\ 0 & \lambda_2 & 0 & \cdots & 0 \\ \cdot & \cdot & \cdot & \cdots & \cdot \\ 0 & 0 & 0 & \cdots & \lambda_n \end{pmatrix}$$

From Theorem 1.1 it follows immediately that

$$\det A = \prod \lambda_i \qquad \operatorname{tr} A = \sum \lambda_i$$

where the product and sum are taken over all roots, each root counted a number of times equal to its multiplicity. The J_i are of the form

$$J_i = \lambda_{q+i} E_{r_i} + Z_i$$

where J_i has r_i rows and columns, and

$$Z_i = \begin{pmatrix} 0 & 1 & 0 & 0 & \cdots & 0 & 0 \\ 0 & 0 & 1 & 0 & \cdots & 0 & 0 \\ \cdot & \cdot & \cdot & \cdot & \cdots & \cdot & \cdot \\ 0 & 0 & 0 & 0 & \cdots & 0 & 1 \\ 0 & 0 & 0 & 0 & \cdots & 0 & 0 \end{pmatrix}$$

An equally valid form of J_i is $\lambda_{q+i} E_{r_i} + \gamma Z_i$, where γ is any constant not zero. Incidentally, the matrix Z_i^2 has its diagonal of 1s moved one element to the right from that of Z_i and all other elements zero. From this it follows that $Z_i^{r_i-1}$ is a matrix which contains all zeros except for a single 1 in the first row and last column. Hence $Z_i^{r_i}$ is the zero matrix, and Z_i is nilpotent.

If $\{A_m\}$ is a sequence of matrices, this sequence is said to be *convergent* if, given any $\epsilon > 0$, there exists a positive integer N_ϵ such that

$$|A_q - A_p| < \epsilon \qquad \text{whenever } p,q > N_\epsilon$$

The sequence $\{A_m\}$ is said to have a *limit* matrix A if, given any $\epsilon > 0$, there exists a positive integer N_ϵ such that

$$|A_m - A| < \epsilon \qquad \text{whenever } m > N_\epsilon$$

Clearly $\{A_m\}$ is convergent if and only if each of the component sequences is, and this implies that $\{A_m\}$ is convergent if and only if there exists a limit matrix to which it tends.

The infinite series

$$\sum_{m=1}^{\infty} A_m$$

is said to be *convergent* if the sequence of partial sums is convergent, and the *sum* of the series is defined to be the limit matrix of the partial sums. A particular series which is of great importance for the study of linear equations is the one defining the *exponential* of a matrix A, namely,

$$e^A = E + \sum_{m=1}^{\infty} \frac{A^m}{m!} \tag{1.2}$$

where A^m represents the mth power of A. The series defining e^A is convergent for all A, since for any positive integers p,q,

$$\left| \sum_{m=p+1}^{p+q} \frac{A^m}{m!} \right| \leq \sum_{m=p+1}^{p+q} \frac{|A|^m}{m!}$$

and the latter represents the Cauchy difference for the series $e^{|A|}$ which is convergent for all finite $|A|$. Also

$$|e^A| \leq (n-1) + e^{|A|} \tag{1.3}$$

For matrices, it is not in general true that $e^{A+B} = e^A e^B$, but this relation is valid if A and B commute. It will be seen in Theorem 4.1 that

$$\det e^A = e^{\operatorname{tr} A} \tag{1.4}$$

and hence e^A is nonsingular for all A. Since $-A$ commutes with A, $e^{-A} = (e^A)^{-1}$.

Every matrix A satisfies its *characteristic equation* $\det(\lambda E - A) = 0$, and this remark is sometimes useful for the actual calculation of e^A. As a simple example, if

$$A = \begin{pmatrix} 0 & 1 \\ 0 & 0 \end{pmatrix}$$

then $\det(\lambda E - A) = \lambda^2 = 0$, and therefore $A^2 = 0$, which implies $A^m = 0$, $m > 1$. Hence,

$$e^A = E + A = \begin{pmatrix} 1 & 1 \\ 0 & 1 \end{pmatrix}$$

If B is a *nonsingular* matrix, then it will be shown that there exists a matrix A (called a *logarithm* of B) such that $e^A = B$. Indeed, if B is in the canonical form J of Theorem 1.1, it is evident that A can be taken as

$$A = \begin{pmatrix} A_0 & 0 & 0 & \cdots & 0 \\ 0 & A_1 & 0 & \cdots & 0 \\ \cdot & \cdot & \cdot & \cdots & \cdot \\ 0 & 0 & 0 & \cdots & A_s \end{pmatrix}$$

provided that $e^{A_j} = J_j$, $j = 0, 1, \ldots, s$. It is also easily verified that a suitable A_0 is given by

$$A_0 = \begin{pmatrix} \log \lambda_1 & 0 & \cdots & 0 \\ 0 & \log \lambda_2 & \cdots & 0 \\ \cdot & \cdot & \cdots & \cdot \\ 0 & 0 & \cdots & \log \lambda_q \end{pmatrix}$$

Clearly

$$J_j = \lambda_{q+j}\left(E_{r_j} + \frac{1}{\lambda_{q+j}} Z_j\right)$$

where Z_j is the nilpotent matrix defined after Theorem 1.1. Since large powers of Z_j all vanish, the series

$$\sum_{k=1}^{\infty} (-1)^{k+1}k^{-1}(\lambda_{q+j})^{-k}Z_j^k$$

has only a finite number of terms, and is thus convergent. Define

$$\log\left(E_{r_j} + \frac{1}{\lambda_{q+j}} Z_j\right)$$

to be this series, which is, of course, a polynomial in $\lambda_{q+j}^{-1}Z_j$. Thus

$$F(\lambda_{q+j}^{-1}Z_j) = \exp\,[\log\,(E_{r_j} + \lambda_{q+j}^{-1}Z_j)]$$

is a polynomial in $\lambda_{q+j}^{-1}Z_j$. On the other hand, from

$$\begin{aligned} 1 + x &= e^{\log\,(1+x)} \\ &= 1 + \left(x - \frac{1}{2}x^2 + \cdots\right) + \frac{1}{2!}\left(x - \frac{1}{2}x^2 + \cdots\right)^2 + \cdots \qquad |x| < 1, \end{aligned}$$

it follows that, when the right member is rearranged, the coefficients of x^k, $k \geqq 2$, are all zero, while the coefficient of x is 1. This implies the same result for F, and proves that

$$\exp\,[\log\,(E_{r_j} + \lambda_{q+j}^{-1}Z_j)] = E_{r_j} + \lambda_{q+j}^{-1}Z_j$$

From this follows readily that a suitable A_j, $j = 1, \ldots, s$, is given by

$$A_j = (\log \lambda_{q+j})E_{r_j} + \log\left(E_{r_j} + \frac{1}{\lambda_{q+j}} Z_j\right)$$

Using the fact that for any matrix M,

$$(PMP^{-1})^k = PM^kP^{-1} \qquad (k = 1, 2, \ldots)$$

one readily sees that

$$Pe^MP^{-1} = e^{PMP^{-1}}$$

From this it follows that the result just sketched for a canonical matrix B is valid for any nonsingular matrix B. Indeed, if $J = e^A$ and $B = PJP^{-1}$, then $B = e^{\tilde{A}}$, where $\tilde{A} = PAP^{-1}$. Naturally A is not unique. For example,

$$e^A = e^Ae^{2\pi ikE} = e^{A+2\pi ikE} \qquad (k = 0, \pm 1, \pm 2, \ldots)$$

If Φ is an n-by-n matrix of functions defined on a real t interval I (the functions may be real or complex), Φ is said to be continuous, differentiable, or analytic on I if every element of Φ is continuous, differentiable, or analytic on I. If Φ is differentiable on I, then Φ' denotes the matrix of derivatives. Note that if Φ, Ψ are differentiable

$$(\Phi\Psi)' = \Phi'\Psi + \Phi\Psi' \tag{1.5}$$

and that $\Phi'\Psi \neq \Psi\Phi'$ in general.

If $\Phi'(t)$ exists and Φ is nonsingular at t, then Φ^{-1} is differentiable at t. This follows from the fact that

$$\Phi^{-1} = \frac{\tilde{\Phi}}{\det \Phi}$$

where $\tilde{\Phi} = (\tilde{\varphi}_{ij})$, and $\tilde{\varphi}_{ij}$ is the cofactor of φ_{ji} in Φ. From (1.5) and the fact that $\Phi\Phi^{-1} = E$, it follows that

$$(\Phi^{-1})' = -\Phi^{-1}\Phi'\Phi^{-1} \qquad \det \Phi \neq 0 \tag{1.6}$$

Recall that in Sec. 7, Chap. 1, it was shown that, if A is a continuous matrix on a t interval I, and Φ satisfies $\Phi'(t) = A(t)\Phi(t)$ on I, then

$$(\det \Phi)' = (\operatorname{tr} A)(\det \Phi) \tag{1.7}$$

or, in integral form,

$$\det \Phi(t) = \det \Phi(\tau) \exp \int_\tau^t \operatorname{tr} A(s)\, ds \qquad (t,\tau \in I) \tag{1.8}$$

2. Linear Homogeneous Systems

Let A be a continuous n-by-n matrix of complex functions on a real t interval I. The linear system

(LH) $$x' = A(t)x \qquad (t \in I)$$

is called a *linear homogeneous system* of the nth order. It was shown in Sec. 5, Chap. 1, that given any ξ, and $\tau \in I$, there exists a unique solution φ of (LH) on I such that $\varphi(\tau) = \xi$.† In the following it will be assumed at least (*) holds for A.

The zero vector function on I is always a solution of (LH). This will be called the *trivial solution* of (LH). If a solution of (LH) is zero for any $\tau \in I$, then, by uniqueness, it must be zero throughout I.

† More generally, if each element of A is measurable on I and

(*) $$|A(t)| \leq m(t) \qquad (t \in I)$$

where m is Lebesgue integrable on I, then an application of Prob. 4, Chap. 1, yields the existence and uniqueness of a solution φ of (LH) satisfying $\varphi(\tau) = \xi$. See also Prob. 1 at the end of this chapter.

Theorem 2.1. *The set of all solutions of* (LH) *on* I *form an* n*-dimensional vector space over the complex field.*‡

Proof. If φ_1, φ_2 are solutions of (LH) and c_1, c_2 are two complex numbers, then $c_1\varphi_1 + c_2\varphi_2$ is again a solution of (LH). This shows that the solutions form a vector space.

To show that the space is n-dimensional, a set of n linearly independent solutions $\varphi_1, \ldots, \varphi_n$ must be exhibited such that every other solution of (LH) is a linear combination (with complex coefficients) of these φ_i. Let ξ_i, $i = 1, \ldots, n$, be linearly independent points in the n-dimensional x space. For example, each ξ_i may be taken as a vector with all components zero except the ith, which is 1. Then, by the existence theorem, if $\tau \varepsilon I$, there exist n solutions φ_i, $i = 1, \ldots, n$, of (LH) such that $\varphi_i(\tau) = \xi_i$. It will be shown that these solutions satisfy the required conditions.

If the φ_i are linearly dependent, there must exist n complex numbers c_i, not all zero, such that

$$\sum_{i=1}^{n} c_i\varphi_i(t) = 0 \qquad (t \varepsilon I)$$

This implies that

$$\sum_{i=1}^{n} c_i\varphi_i(\tau) = \sum_{i=1}^{n} c_i\xi_i = 0$$

and this contradicts the assumption that the ξ_i are linearly independent. This shows that the φ_i are linearly independent.

If φ is any solution of (LH) on I, such that $\varphi(\tau) = \xi$, then for some (unique) constants c_i

$$\xi = \sum_{i=1}^{n} c_i\xi_i$$

for the ξ_i form a basis for the n-dimensional x space. Hence the function

$$\sum_{i=1}^{n} c_i\varphi_i$$

is a solution of (LH) on I which assumes the value ξ at τ, and, by uniqueness, this must be φ, that is,

‡ A reader unfamiliar with the terminology of the statement of the above theorem will find that by reading the proof he can readily restate the result in more familiar terms. See P. R. Halmos, *Finite dimensional vector spaces*, Princeton, for a discussion of vector spaces.

$$\varphi = \sum_{i=1}^{n} c_i \varphi_i$$

Therefore every solution φ is a (unique) linear combination of the φ_i, and this proves Theorem 2.1.

If $\varphi_1, \ldots, \varphi_n$ are a set of n linearly independent solutions of (LH), they are said to form a *basis* or a *fundamental set* of solutions of (LH).

If Φ is a matrix whose n columns are n linearly independent solutions of (LH) on I, then Φ is called a *fundamental matrix* for (LH). Evidently Φ satisfies the matrix equation

$$\Phi'(t) = A(t)\Phi(t) \qquad (t \in I) \tag{2.1}$$

By the *matrix differential equation associated with* (LH) on I is meant the problem of finding an n-by-n matrix Φ whose columns are solutions of (LH) on I. This problem is denoted by

$$X' = A(t)X \qquad (t \in I) \tag{2.2}$$

The matrix Φ is called a *solution* of (2.2) on I, and Φ satisfies (2.1). From Theorem 2.1 it is now evident that a complete knowledge of the set of solutions of (LH) can be obtained if one knows a *fundamental matrix* for (LH), which is, of course, a particular solution of (2.2).

Theorem 2.2. *A necessary and sufficient condition that a solution matrix* Φ *of* (2.2) *be a fundamental matrix is that* $\det \Phi(t) \neq 0$, *for* $t \in I$.

REMARK: If $\det \Phi(t) \neq 0$, for *some* $t \in I$, then by (1.8) $\det \Phi(t) \neq 0$ for *all* $t \in I$.

Proof of Theorem 2.2. Let Φ be a fundamental matrix with column vectors φ_j, and suppose φ is any nontrivial solution of (LH). By Theorem 2.1, there exist unique constants $c_1, \ldots, c_n$, not all zero, such that

$$\varphi = \sum_{j=1}^{n} c_j \varphi_j$$

or, in terms of Φ,

$$\varphi = \Phi c$$

where c is the column vector with components $c_1, \ldots, c_n$. This relation is a system of n linear equations in the n unknowns $c_1, \ldots, c_n$ at any $\tau \in I$, and has a unique solution for any choice of $\varphi(\tau)$. Hence $\det \Phi(\tau) \neq 0$, and by the remark above, $\det \Phi(t) \neq 0$ for any $t \in I$. Notice that this proves that the column vectors of a fundamental matrix are linearly independent at every $t \in I$.

Conversely, let Φ be a solution matrix of (2.2) and suppose $\det \Phi(t) \neq 0$ for $t \in I$. Thus the column vectors of Φ are linearly independent at every $t \in I$.

A matrix of column vectors may have a determinant identically zero on an interval I, although the vectors may be linearly independent. For example, let Φ be defined as

$$\Phi(t) = \begin{pmatrix} t & t^2 \\ 0 & 0 \end{pmatrix}$$

for any real interval I. The content of Theorem 2.2 is that this cannot occur for vectors which are solutions of (LH).

Theorem 2.3. *If Φ is a fundamental matrix of* (LH) *and C a (complex) constant nonsingular matrix, then ΦC is again a fundamental matrix of* (LH). *Every fundamental matrix of* (LH) *is of this type for some nonsingular C.*

Proof. From (2.1), if Φ is a fundamental matrix,

$$\Phi'(t)C = A(t)\Phi(t)C \qquad (t \in I)$$

or

$$(\Phi C)' = A(\Phi C)$$

and hence ΦC is a solution matrix of (2.2). Since

$$\det (\Phi C) = (\det \Phi)(\det C) \neq 0$$

ΦC is a fundamental matrix.

Conversely, if Φ_1 and Φ_2 are fundamental matrices, then $\Phi_2 = \Phi_1 C$ for some constant nonsingular matrix C. To show this, let $\Phi_1^{-1}\Phi_2 = \Psi$. Then $\Phi_2 = \Phi_1\Psi$. Differentiating this equation gives $\Phi_2' = \Phi_1\Psi' + \Phi_1'\Psi$. Using (2.1), this gives $A\Phi_2 = \Phi_1\Psi' + A\Phi_1\Psi$ or $\Phi_1\Psi' = 0$. Thus $\Psi' = 0$ and therefore $\Psi = C$ is a constant. It is nonsingular since Φ_1 and Φ_2 are.

REMARKS: If it is only required that Φ_2 be a solution, then C may be singular.

Observe that, if Φ is a fundamental matrix of (LH) and C is a constant nonsingular matrix, then $C\Phi$ is not in general a fundamental matrix.

Two different homogeneous systems cannot have the same fundamental matrix, for in (LH), $A(t) = \Phi'(t)\Phi^{-1}(t)$. Hence Φ determines A uniquely, although the converse is not true.

Adjoint Systems. If Φ is a fundamental matrix for (LH), then

$$(\Phi^{-1})' = -\Phi^{-1}\Phi'\Phi^{-1} = -\Phi^{-1}A$$

or, taking the conjugate transpose,

$$(\Phi^{*-1})' = -A^*\Phi^{*-1}$$

Therefore Φ^{*-1} is a fundamental matrix for the system

$$x' = -A^*(t)x \qquad (t \in I) \tag{2.3}$$

The system (2.3) is called the *adjoint* to (LH), and the matrix equation

$$X' = -A^*(t)X \qquad (t \in I) \tag{2.4}$$

is called the adjoint to (2.2). The relationship is symmetric, for (LH) and (2.2) are the adjoints to (2.3) and (2.4), respectively.

Theorem 2.4. *If Φ is a fundamental matrix for* (LH), *then Ψ is a fundamental matrix for its adjoint* (2.3) *if and only if*

$$\Psi^*\Phi = C \tag{2.5}$$

where C is a constant nonsingular matrix.

Proof. If Φ is a fundamental matrix for (LH) and Ψ one for (2.3), then since Φ^{*-1} is a particular fundamental matrix for (2.3),

$$\Psi = \Phi^{*-1}D$$

for some constant nonsingular matrix D (Theorem 2.3). Hence

$$\Psi^*\Phi = D^*$$

and let $C = D^*$.

Conversely, if Φ is a fundamental matrix for (LH) and satisfies (2.5), one has $\Psi^* = C\Phi^{-1}$ or $\Psi = \Phi^{*-1}C^*$, and hence, by Theorem 2.3, Ψ is a fundamental matrix of the adjoint system (2.3).

If $A = -A^*$, then since Φ^{*-1} is a fundamental matrix of (2.3) it is also one for (LH). Hence by Theorem 2.3 $\Phi = \Phi^{*-1}\, C$, or

$$\Phi^*\Phi = C \tag{2.6}$$

where C is a constant nonsingular matrix. Equation (2.6) implies, in particular, that the Euclidean length of any solution vector φ of (LH) is constant.

Reduction of the Order of a Homogeneous System. If m $(0 < m < n)$ linearly independent solutions of (LH) are known, it is possible to reduce the order of (LH) by m, and hence a linear system of order $n - m$ only need be solved.

Suppose $\varphi_1, \ldots, \varphi_m$ are m linearly independent vectors which are solutions of (LH) on an interval I. Let φ_j have components φ_{ij} $(i = 1, \ldots, n)$. Then the rank of the n-by-m matrix with elements φ_{ij} $(i = 1, \ldots, n; j = 1, \ldots, m)$ at every $t \in I$ is m, because of the linear independence of its columns. This means that for each $t \in I$ there is an m-by-m determinant in this matrix which does not vanish there. Pick any $t_0 \in I$ and assume for the moment that the determinant of the matrix Φ_m whose elements are φ_{ij} $(i = 1, \ldots, m; j = 1, \ldots, m)$ is not zero at t_0. Then, by the continuity of det Φ_m in its elements φ_{ij}, and the continuity of the functions φ_{ij} near t_0, one has that det $\Phi_m(t) \neq 0$ for t in some interval $\tilde{I}$ containing t_0. Let $\tilde{I}$ be any such interval; the reduction process will be outlined for $\tilde{I}$. (The idea behind the process is a modification of the variation of constants.)

Let the matrix U have the vectors $\varphi_1, \ldots, \varphi_m$ for its first m columns and the vectors $e_{m+1}, \ldots, e_n$ for its last $n - m$ columns, where e_j is the column vector with all elements 0 except for the jth which is 1. Clearly U is nonsingular on $\tilde{I}$. The substitution

$$x = Uy \tag{2.7}$$

is made in (LH). [Note that $x = \varphi_j$ $(j = 1, \ldots, m)$ in (2.7) corresponds to $y = e_j$ $(j = 1, \ldots, m)$. Thus the substitution (2.7) may be expected to yield a system in y which will have e_j, $j = 1, \ldots, m$, as solutions.] The use of (2.7) in (LH) gives

$$U'y + Uy' = AUy$$

Writing this out gives

$$\sum_{j=1}^{m} \varphi'_{ij}y_j + \sum_{j=1}^{m} \varphi_{ij}y'_j = \sum_{j=1}^{m}\sum_{k=1}^{n} a_{ik}\varphi_{kj}y_j + \sum_{k=m+1}^{n} a_{ik}y_k \qquad (i = 1, \ldots, m)$$

$$\sum_{j=1}^{m} \varphi'_{ij}y_j + y'_i + \sum_{j=1}^{m} \varphi_{ij}y'_j = \sum_{j=1}^{m}\sum_{k=1}^{n} a_{ik}\varphi_{kj}y_j + \sum_{k=m+1}^{n} a_{ik}y_k \qquad (i = m + 1, \ldots, n)$$

Expressing the fact that the vectors φ_j with components φ_{ij} are solutions of (LH),

$$\varphi'_{ij} = \sum_{k=1}^{n} a_{ik}\varphi_{kj} \qquad (i = 1, \ldots, n, j = 1, \ldots, m)$$

there results

$$\begin{aligned} \sum_{j=1}^{m} \varphi_{ij}y'_j &= \sum_{k=m+1}^{n} a_{ik}y_k \qquad (i = 1, \ldots, m) \\ y'_i + \sum_{j=1}^{m} \varphi_{ij}y'_j &= \sum_{k=m+1}^{n} a_{ik}y_k \qquad (i = m + 1, \ldots, n) \end{aligned} \tag{2.8}$$

Since $\det \Phi_m \neq 0$ on $\tilde{I}$, the first set of equations in (2.8) may be solved for y'_j $(j = 1, \ldots, m)$ in terms of φ_{ij}, a_{ik}, and y_k $(k = m + 1, \ldots, n)$, and these values of y'_j $(j = 1, \ldots, n)$ may then be put into the second set of formulas of (2.8). This gives a set of first-order equations satisfied by the y_i $(i = m + 1, \ldots, n)$ of the type

$$y'_i = \sum_{k=m+1}^{n} b_{ik}y_k \qquad (i = m + 1, \ldots, n) \tag{2.9}$$

that is, a linear system of order $n - m$.

Working backwards, suppose $\tilde{\psi}_{m+1}, \ldots, \tilde{\psi}_n$ [$\tilde{\psi}_j$ having components ψ_{ij} ($i,j = m + 1, \ldots, n$)] is a fundamental set on $\tilde{I}$ for the system (2.9). Let $\tilde{\Psi}_{n-m}$ denote the matrix with elements ψ_{ij} ($i,j = m + 1, \ldots, n$). Clearly $\det \tilde{\Psi}_{n-m}(t) \neq 0$ on $\tilde{I}$. For each $j = m + 1, \ldots, n$, let ψ_{ij} ($i = 1, \ldots, m$) be solved for by quadratures (that is, by integration) from the relations

$$\sum_{j=1}^{m} \varphi_{ij}\psi'_{jp} = \sum_{k=m+1}^{n} a_{ik}\psi_{kp} \tag{2.10}$$
$$i = 1, \ldots, m \qquad p = m + 1, \ldots, n$$

Let ψ_p ($p = m + 1, \ldots, n$) denote the vectors having components ψ_{ip} ($i = 1, \ldots, n$), and let

$$\psi_p = e_p \qquad (p = 1, \ldots, m)$$

Since ψ_p, $p = 1, \ldots, n$, satisfy (2.9) and the first set of equations of (2.8), they must also satisfy the second set of equations of (2.8), and therefore ψ_p, $p = 1, \ldots, n$, are solutions of (2.8). Thus, if now Ψ is the matrix with columns ψ_p, $p = 1, \ldots, n$, and if

$$\Phi = U\Psi$$

then Φ is a matrix solution of (LH) on $\tilde{I}$. U is nonsingular. Since $\det \Psi = \det \tilde{\Psi}_{n-m}$ on $\tilde{I}$, it follows that Φ is nonsingular on $\tilde{I}$ and hence a fundamental solution of (LH) on $\tilde{I}$.

The above procedure is summarized in the following theorem.

Theorem 2.5. *Let $\varphi_1, \ldots, \varphi_m$ ($m < n$) be m known linearly independent solutions of* (LH) *with φ_j ($j = 1, \ldots, m$) having components φ_{ij} ($i = 1, \ldots, n$). Assume the determinant of the matrix with elements φ_{ij} ($i,j = 1, \ldots, m$) is not zero on some subinterval $\tilde{I}$ of I. Then the construction of a set of n linearly independent solutions of* (LH) *on $\tilde{I}$ can be reduced to the solution of a linear system* (2.9) *of order $n - m$, plus quadratures* (2.10), *using the substitution* (2.7).

The restriction that the matrix Φ_m should be nonsingular on an interval will now be removed. It is clear that the n-by-m matrix with elements φ_{ij} ($i = 1, \ldots, n$; $j = 1, \ldots, m$), has rank m because of the independence of the solutions φ_j, $j = 1, \ldots, m$. Thus, at any $t = t_0$, there is a nonsingular m-by-m matrix obtained by taking m rows, $i_1, \ldots, i_m$, of the n-by-m matrix. By continuity, this matrix is nonsingular over some interval $\tilde{I}$.

It is well known, and readily proved, that there exists a constant nonsingular matrix T which has the property that, applied to any vector x with n components, Tx has the $i_1, \ldots, i_m$ components of x as its first m components. Setting $\tilde{x} = Tx$, the equation (LH) is replaced by a similar equation where now the original restriction holds. Since $x = T^{-1}\tilde{x}$, the result for x follows from that for $\tilde{x}$.

3. Nonhomogeneous Linear Systems

Suppose A is an n-by-n matrix of continuous functions on a real t interval I, and b is a continuous vector on I which is not identically zero there. The system

$$\text{(NH)} \qquad x' = A(t)x + b(t) \qquad (t \in I)$$

is called a *nonhomogeneous linear system* of the nth order. If the elements of A and b are continuous, or are just measurable and majorized by integrable functions on I, there exists a unique solution φ of (NH) for which

$$\varphi(\tau) = \xi$$

where $\tau \in I$ and $|\xi| < \infty$. The continuous case follows from Sec. 5, Chap. 1, and the more general case from Prob. 1. That the solution is unique also follows from the fact that if there were two such solutions φ_1, φ_2, then their difference $\varphi = \varphi_1 - \varphi_2$ would be a solution of (LH) on I and would satisfy $\varphi(\tau) = 0$. But, by the uniqueness theorem for (LH), φ must be the zero function on I, and thus $\varphi_1 = \varphi_2$.

If a fundamental matrix Φ for (LH) is known, then there is a simple method for calculating a solution of (NH).

Theorem 3.1. *If Φ is a fundamental matrix for* (LH), *then the function φ defined by*

$$\varphi(t) = \Phi(t) \int_\tau^t \Phi^{-1}(s)b(s)\,ds \qquad (t \in I) \tag{3.1}$$

is that solution of (NH) *satisfying*

$$\varphi(\tau) = 0 \qquad (\tau \in I)$$

Proof. The proof follows at once by direct verification.

An intuitive idea of how one obtains the expression (3.1) is given in the following: For any constant vector c, the function Φc is a solution of (LH). The method here consists of considering c as a function, or parameter, on I and determining what c must be (if it exists) in order that the function $\varphi = \Phi c$ be a solution of the nonhomogeneous system (NH).

Suppose $\varphi = \Phi c$ is a solution of (NH). Then

$$\varphi' = \Phi' c + \Phi c' = A\Phi c + \Phi c' = A\varphi + \Phi c' = A\varphi + b$$

the latter following from (NH). Therefore $\Phi c' = b$, or

$$c' = \Phi^{-1}b$$

This equation is always solvable and one gets

$$c(t) = \int_\tau^t \Phi^{-1}(s)b(s)\, ds \qquad (t \in I)$$

as that function for which $c(\tau) = 0$. Thus φ is given by (3.1).

It is a simple matter to see, under the assumptions of Theorem 3.1, that the solution φ of (NH) satisfying

$$\varphi(\tau) = \xi \qquad (\tau \in I,\ |\xi| < \infty)$$

is given by

$$\varphi(t) = \varphi_h(t) + \Phi(t) \int_\tau^t \Phi^{-1}(s)b(s)\, ds \qquad (t \in I) \tag{3.2}$$

where φ_h is that solution of (LH) on I satisfying

$$\varphi_h(\tau) = \xi$$

The formula (3.1) [or (3.2)] is called the *variation-of-constants formula* for (NH).

Note that (3.1) may be written as

$$\varphi(t) = \Psi^{*-1}(t) \int_\tau^t \Psi^*(s)b(s)\, ds \qquad (t \in I)$$

where Ψ is a fundamental matrix of the adjoint system

$$x' = -A^*(t)x$$

to (LH). Another form of (3.1) is

$$\varphi(t) = \Phi(t) \int_\tau^t \Psi^*(s)b(s)\, ds$$

but here the restriction $\Psi^*(t)\Phi(t) = E$ is needed.

4. Linear Systems with Constant Coefficients

Let A be an n-by-n *constant* matrix, and consider the corresponding homogeneous system

$$x' = Ax \tag{4.1}$$

If $n = 1$, then it is trivial that (4.1) has as a solution e^{tA}, and the solution assuming the value ξ at τ ($|\tau| < \infty$, $|\xi| < \infty$) is given by $e^{(t-\tau)A}\xi$. It turns out that the form of the solution is the same when x, ξ are vectors of arbitrary finite dimension n, and A is an n-by-n matrix.

Theorem 4.1. *A fundamental matrix Φ for* (4.1) *is given by*

$$\Phi(t) = e^{tA} \qquad (|t| < \infty) \tag{4.2}$$

and the solution φ *of* (4.1) *satisfying*

$$\varphi(\tau) = \xi \qquad (|\tau| < \infty, |\xi| < \infty)$$

is given by

$$\varphi(t) = e^{(t-\tau)A}\xi \qquad (|t| < \infty) \tag{4.3}$$

Proof. Since $e^{(t+\Delta t)A} = e^{tA}e^{\Delta tA}$, it follows easily from the definition of derivative that

$$(e^{tA})' = Ae^{tA}$$

Thus the Φ defined by $\Phi(t) = e^{tA}$ is a solution. Since $\Phi(0) = E$, it follows from (1.8) that $\det \Phi(t) = e^{t \operatorname{tr} A}$. Thus Φ is a fundamental matrix. The formula (4.3) is obvious.

REMARK: Notice that $\exp\left(\int_\tau^t A(s)\,ds\right)$ need not be a solution of $x' = A(t)x$ unless $A(t)$ and $\int_\tau^t A(s)\,ds$ commute. They do commute if A is constant or if $A(t)$ is diagonal.

It is of interest to investigate the form of the fundamental matrix (4.2). Let J be the canonical form of A, as given by Theorem 1.1, and suppose P is a nonsingular constant matrix such that

$$AP = PJ$$

Then

$$e^{tA} = e^{tPJP^{-1}} = Pe^{tJ}P^{-1} \tag{4.4}$$

and J has the form

$$J = \begin{pmatrix} J_0 & 0 & 0 & \cdots & 0 \\ 0 & J_1 & 0 & \cdots & 0 \\ \cdot & \cdot & \cdot & \cdots & \cdot \\ 0 & 0 & 0 & \cdots & J_s \end{pmatrix} \tag{4.5}$$

where J_0 is a diagonal matrix with diagonal $\lambda_1, \ldots, \lambda_q$, and

$$J_i = \begin{pmatrix} \lambda_{q+i} & 1 & 0 & \cdots & 0 & 0 \\ 0 & \lambda_{q+i} & 1 & \cdots & 0 & 0 \\ \cdot & \cdot & \cdot & \cdots & \cdot & \cdot \\ 0 & 0 & 0 & \cdots & \lambda_{q+i} & 1 \\ 0 & 0 & 0 & \cdots & 0 & \lambda_{q+i} \end{pmatrix} \qquad (i = 1, \ldots, s) \tag{4.6}$$

It follows that

$$e^{tJ} = \begin{pmatrix} e^{tJ_0} & 0 & \cdots & 0 \\ 0 & e^{tJ_1} & \cdots & 0 \\ \cdot & \cdot & \cdots & \cdot \\ 0 & 0 & \cdots & e^{tJ_s} \end{pmatrix} \tag{4.7}$$

and it is an easy calculation to see that

$$e^{tJ_0} = \begin{pmatrix} e^{t\lambda_1} & 0 & \cdots & 0 \\ 0 & e^{t\lambda_2} & \cdots & 0 \\ \cdot & \cdot & \cdots & \cdot \\ 0 & 0 & \cdots & e^{t\lambda_q} \end{pmatrix} \tag{4.8}$$

Since $J_i = \lambda_{q+i}E_{r_i} + Z_i$, $e^{tJ_i} = e^{t\lambda_{q+i}}e^{tZ}$. Thus

$$e^{tJ_i} = e^{t\lambda_{q+i}} \begin{pmatrix} 1 & t & \dfrac{t^2}{2!} & \cdots & \dfrac{t^{r_i-1}}{(r_i-1)!} \\ 0 & 1 & t & \cdots & \dfrac{t^{r_i-2}}{(r_i-2)!} \\ \cdot & \cdot & \cdot & \cdots & \cdot \\ 0 & 0 & 0 & \cdots & 1 \end{pmatrix} \tag{4.9}$$

where J_i is an r_i-by-r_i matrix ($n = q + r_1 + \cdots + r_s$). Therefore, if the canonical form (4.5), (4.6), of A is known, a fundamental matrix e^{tA} of (4.1) is given explicitly by (4.4), where e^{tJ} can be calculated from (4.7), (4.8), and (4.9).

Another fundamental matrix of (4.1) is given by Ψ, where

$$\Psi(t) = e^{tA}P = Pe^{tJ} \tag{4.10}$$

Suppose P has as columns the vectors $p_1, \ldots, p_n$. The columns of Ψ, which are $\psi_1, \ldots, \psi_n$, make up a set of n linearly independent solutions of (4.1), and from (4.10), and the form of J, one obtains

$$\begin{aligned}
\psi_1(t) &= e^{t\lambda_1}p_1,\ \psi_2(t) = e^{t\lambda_2}p_2,\ \ldots,\ \psi_q(t) = e^{t\lambda_q}p_q \\
\psi_{q+1}(t) &= e^{t\lambda_{q+1}}p_{q+1} \\
\psi_{q+2}(t) &= e^{t\lambda_{q+1}}(tp_{q+1} + p_{q+2}) \\
&\cdots\cdots\cdots\cdots\cdots\cdots \\
\psi_{q+r_1}(t) &= e^{t\lambda_{q+1}}\left(\frac{t^{r_1-1}}{(r_1-1)!}p_{q+1} + \cdots + tp_{q+r_1-1} + p_{q+r_1}\right) \\
&\cdots\cdots\cdots\cdots\cdots\cdots \\
\psi_{n-r_s+1}(t) &= e^{t\lambda_{q+s}}p_{n-r_s+1} \\
&\cdots\cdots\cdots\cdots\cdots\cdots \\
\psi_n(t) &= e^{t\lambda_{q+s}}\left(\frac{t^{r_s-1}}{(r_s-1)!}p_{n-r_s+1} + \cdots + tp_{n-1} + p_n\right)
\end{aligned}$$

Since $AP = PJ$, $p_1, \ldots, p_n$ satisfy the relations

$$\begin{aligned}
Ap_1 &= \lambda_1 p_1,\ \ldots,\ Ap_q = \lambda_q p_q \\
Ap_{q+1} &= \lambda_{q+1}p_{q+1} \\
Ap_{q+2} &= p_{q+1} + \lambda_{q+1}p_{q+2} \\
&\cdots\cdots\cdots\cdots \\
Ap_{q+r_1} &= p_{q+r_1-1} + \lambda_{q+1}p_{q+r_1}
\end{aligned}$$

$$\begin{aligned}
&\cdots\cdots\cdots\cdots\cdots\cdots\\
Ap_{n-r_s+1} &= \lambda_{q+s}p_{n-r_s+1}\\
Ap_{n-r_s+2} &= p_{n-r_s+1} + \lambda_{q+s}p_{n-r_s+2}\\
&\cdots\cdots\cdots\cdots\cdots\cdots\\
Ap_n &= p_{n-1} + \lambda_{q+s}p_n
\end{aligned}$$

The solutions ψ_j are expressed in terms of the independent vectors p_1, p_2, . . . , p_n in the preceding set of equations. (For another derivation, see Prob. 13.)

The variation-of-constants formula (3.1) applied to the nonhomogeneous system

$$x' = Ax + b(t) \qquad (t \in I) \tag{4.11}$$

where A is a constant matrix, gives for the solution φ of (4.11) satisfying $\varphi(\tau) = 0$, $\tau \in I$, the following formula

$$\varphi(t) = e^{tA}\int_\tau^t e^{-sA}b(s)\,ds = \int_\tau^t e^{(t-s)A}b(s)\,ds \qquad (t \in I)$$

The solution φ of (4.11) satisfying $\varphi(\tau) = \xi$, where $\tau \in I$, $|\xi| < \infty$, is given by

$$\varphi(t) = e^{(t-\tau)A}\xi + \int_\tau^t e^{(t-s)A}b(s)\,ds \qquad (t \in I)$$

5. Linear Systems with Periodic Coefficients

Consider the linear homogeneous system

$$x' = A(t)x \qquad (-\infty < t < +\infty) \tag{5.1}$$

where A is a matrix of complex continuous functions, and

$$A(t + \omega) = A(t) \tag{5.2}$$

for some constant $\omega \neq 0$. In this case, (5.1) is called a periodic system, and ω a period of A. The fundamental result for such systems concerns the representation of a fundamental matrix as the product of a periodic matrix with the same period ω and a solution matrix for a system with constant coefficients.

Theorem 5.1. *If Φ is a fundamental matrix for (5.1), then so is Ψ, where*

$$\Psi(t) = \Phi(t + \omega) \qquad (-\infty < t < \infty)$$

Corresponding to every such Φ, there exists a periodic nonsingular matrix P with period ω, and a constant matrix R such that

$$\Phi(t) = P(t)e^{tR} \tag{5.3}$$

Proof. Since

$$\Phi'(t) = A(t)\Phi(t) \qquad (-\infty < t < +\infty)$$

one has

$$\Psi'(t) = \Phi'(t+\omega) = A(t+\omega)\Phi(t+\omega) = A(t)\Psi(t) \qquad (-\infty < t < +\infty)$$

using (5.2). Thus Ψ is a solution matrix of (5.1), and it is a fundamental matrix since $\det \Psi(t) = \det \Phi(t+\omega) \neq 0$ for $-\infty < t < +\infty$.

Therefore there exists a constant nonsingular matrix C such that

$$\Phi(t+\omega) = \Phi(t)C \tag{5.4}$$

and, moreover, there exists a constant matrix R such that

$$C = e^{\omega R} \tag{5.5}$$

as shown in Sec. 1. From (5.4) and (5.5) one obtains

$$\Phi(t+\omega) = \Phi(t)e^{\omega R} \tag{5.6}$$

Let P be defined by

$$P(t) = \Phi(t)e^{-tR} \tag{5.7}$$

Then, using (5.6),

$$P(t+\omega) = \Phi(t+\omega)e^{-(t+\omega)R} = \Phi(t)e^{\omega R}e^{-(t+\omega)R} = \Phi(t)e^{-tR} = P(t)$$

Since $\Phi(t)$ and e^{-tR} are nonsingular for $-\infty < t < \infty$, so is $P(t)$, and this completes the proof.

The significance of Theorem 5.1 is that the determination of a fundamental matrix Φ over an interval of length ω, for example, $0 \leqq t \leqq \omega$, leads at once to the determination of Φ over $(-\infty, \infty)$. Thus C in (5.5) is given by $\Phi^{-1}(0)\Phi(\omega)$ and from this, R is given by $(\log C)/\omega$. $P(t)$ is then determined by (5.7) over $(0,\omega)$. However, since $P(t)$ has ω as a period, it is determined at once over $(-\infty, \infty)$. Then Φ is determined over $(-\infty, \infty)$ by (5.3).

If Φ_1 is any other fundamental matrix for (5.1), where (5.2) holds, then

$$\Phi = \Phi_1 T$$

for some constant nonsingular matrix T. From (5.6) one has

$$\Phi_1(t+\omega)T = \Phi_1(t)Te^{\omega R}$$

or

$$\Phi_1(t+\omega) = \Phi_1(t)(Te^{\omega R}T^{-1}) \tag{5.8}$$

Thus, by (5.8), every fundamental matrix Φ_1 determines a matrix $Te^{\omega R}T^{-1}$, which is similar to $e^{\omega R}$. Conversely, if T is any constant nonsingular matrix, there exists a fundamental matrix Φ_1 of (5.1) such that (5.8) holds. Consequently, although Φ does not determine R uniquely, the set of all fundamental matrices of (5.1), and hence A, determines uniquely all quantities associated with R which are invariant under a similarity transformation. In particular, the set of all fundamental matrices of (5.1) determine a *unique set of characteristic roots*, namely, those of $C = e^{\omega R}$. Denote these roots by $\lambda_1, \ldots, \lambda_n$, and call them the *multipliers associated with A*. None of the multipliers vanishes for $\Pi\lambda_i = \det e^{\omega R} \neq 0$. The characteristic roots of R are called *characteristic exponents*.

It is of interest to see the explicit form that a set of n linearly independent solution vectors of (5.1) assumes. Let T be a constant nonsingular matrix such that $T^{-1}RT = J$ has the canonical form given in Theorem 1.1, and put $\Phi_1 = \Phi T$, $P_1 = PT$. Then from (5.3) one has

$$\Phi_1(t) = P_1(t)e^{tJ} \qquad P_1(t + \omega) = P_1(t) \tag{5.9}$$

Therefore, if the characteristic roots of R are ρ_j, then e^{tJ} will have the form

$$e^{tJ} = \begin{pmatrix} e^{tJ_0} & 0 & \cdots & 0 \\ 0 & e^{tJ_1} & \cdots & 0 \\ \cdot & \cdot & \cdots & \cdot \\ 0 & 0 & \cdots & e^{tJ_s} \end{pmatrix}$$

where

$$e^{tJ_0} = \begin{pmatrix} e^{t\rho_1} & 0 & \cdots & 0 \\ 0 & e^{t\rho_2} & \cdots & 0 \\ \cdot & \cdot & \cdots & \cdot \\ 0 & 0 & \cdots & e^{t\rho_q} \end{pmatrix}$$

and

$$e^{tJ_i} = e^{t\rho_{q+i}} \begin{pmatrix} 1 & t & \cdots & \dfrac{t^{r_i-1}}{(r_i-1)!} \\ 0 & 1 & \cdots & \dfrac{t^{r_i-2}}{(r_i-2)!} \\ \cdot & \cdot & \cdots & \cdot \\ 0 & 0 & \cdots & 1 \end{pmatrix} \qquad (i = 1, \ldots, s; q + \Sigma r_i = n)$$

Clearly $\lambda_i = e^{\omega\rho_i}$, and therefore, while the ρ_i are not uniquely determined, their real parts are. From (5.9) it follows that the columns $\varphi_1, \ldots, \varphi_n$ of Φ_1, which form a set of n linearly independent solutions of (5.1), are of the form

$$\begin{aligned} \varphi_1(t) &= e^{t\rho_1}p_1(t) \\ \varphi_2(t) &= e^{t\rho_2}p_2(t) \\ &\cdots\cdots\cdots \\ \varphi_q(t) &= e^{t\rho_q}p_q(t) \\ \varphi_{q+1}(t) &= e^{t\rho_{q+1}}p_{q+1}(t) \\ \varphi_{q+2}(t) &= e^{t\rho_{q+1}}(tp_{q+1}(t) + p_{q+2}(t)) \\ &\cdots\cdots\cdots\cdots\cdots\cdots\cdots\cdots \end{aligned} \tag{5.10}$$

$$\varphi_{q+r_1}(t) = e^{t\rho_{q+1}}\left(\frac{t^{r_1-1}}{(r_1-1)!}p_{q+1}(t) + \cdots + tp_{q+r_1-1}(t) + p_{q+r_1}(t)\right)$$

. . . .

$$\varphi_{n-r_s+1}(t) = e^{t\rho_{q+s}}p_{n-r_s+1}(t)$$

. . . .

$$\varphi_n(t) = e^{t\rho_{q+s}}\left(\frac{t^{r_s-1}}{(r_s-1)!}p_{n-r_s+1}(t) + \cdots + tp_{n-1}(t) + p_n(t)\right)$$

In the above, $p_1, \ldots, p_n$ are the periodic column vectors of P_1.

From (5.10) it is clear that if $\Re\rho_i < 0$, or equivalently $|\lambda_i| < 1$, then

$$\varphi_i(t) \to 0 \qquad (t \to +\infty)$$

exponentially fast.

From (5.6), $\Phi(\omega) = \Phi(0)e^{\omega R}$, and hence the λ_i may be thought of as the characteristic roots of the matrix $\Phi^{-1}(0)\Phi(\omega)$. In particular, if $\Phi(0) = E$, then $e^{\omega R} = \Phi(\omega)$, and the λ_i are the characteristic roots of $\Phi(\omega)$. Since

$$\det \Phi(\omega) = \lambda_1\lambda_2 \cdots \lambda_n = \exp \int_0^\omega \operatorname{tr} A(s)\, ds \tag{5.11}$$

it follows that, if $n - 1$ of the λ_i are known, the remaining one is determined from (5.11).

A real nonsingular matrix C need not have a real logarithm; that is, there need exist no real B such that $e^B = C$. Indeed, the matrix of one row and column $C = -1$ is an example. However, it is the case that if C is real then there is always a real matrix B such that $C^2 = e^B$; see Prob. 41.

The above used in the proof of Theorem 5.1 yields readily that *if $A(t)$ is real in (5.1) and of period ω, then corresponding to any real fundamental matrix Φ there exists a real matrix P of period 2ω and a real constant matrix R such that*

$$\Phi(t) = P(t)e^{tR}$$

6. Linear Differential Equations of Order n

Suppose $a_0, a_1, \ldots, a_n$ are $n + 1$ continuous (complex) functions defined on a real t interval I, and let L_n denote the formal differential operator

$$L_n = a_0\frac{d^n}{dt^n} + a_1\frac{d^{n-1}}{dt^{n-1}} + \cdots + a_n$$

that is, if g is any function possessing n derivatives on I,

$$L_n g = a_0 g^{(n)} + a_1 g^{(n-1)} + \cdots + a_n g$$

Further suppose $a_0(t) \neq 0$ for any $t \in I$. Then the equation

$$L_n x = 0 \qquad (t \in I)$$

[written out as $a_0(t)x^{(n)} + a_1(t)x^{(n-1)} + \cdots + a_n(t)x = 0, \ (t \in I)$] is defined to be the differential equation

$$x^{(n)} + \frac{a_1(t)}{a_0(t)} x^{(n-1)} + \cdots + \frac{a_n(t)}{a_0(t)} x = 0 \qquad (t \in I)$$

and is called a *linear homogeneous differential equation of order n*. The system associated with this equation (see Sec. 6, Chap. 1) is then the vector equation

$$\hat{x}' = A(t)\hat{x} \tag{6.1}$$

where

$$A = \begin{pmatrix} 0 & 1 & 0 & 0 & \cdots & 0 \\ 0 & 0 & 1 & 0 & \cdots & 0 \\ \cdot & \cdot & \cdot & \cdot & \cdots & \cdot \\ 0 & 0 & 0 & 0 & \cdots & 1 \\ \dfrac{-a_n}{a_0} & \dfrac{-a_{n-1}}{a_0} & \dfrac{-a_{n-2}}{a_0} & \dfrac{-a_{n-3}}{a_0} & \cdots & \dfrac{-a_1}{a_0} \end{pmatrix} \tag{6.2}$$

Since (6.1) is a linear system with a continuous coefficient matrix A on I, there exists a unique vector solution $\hat{\varphi}$ of (6.1) on I satisfying

$$\hat{\varphi}(\tau) = \hat{\xi}$$

where $\tau \in I$, $|\hat{\xi}| < \infty$. Thus φ_1, the first component of $\hat{\varphi}$, satisfies

$$\varphi_1(\tau) = \xi_1, \ \varphi_1'(\tau) = \xi_2, \ \ldots, \ \varphi_1^{(n-1)}(\tau) = \xi_n \tag{6.3}$$

Since φ_1 is a solution of $L_n x = 0$, it is the solution satisfying (6.3).

The remainder of the results so far obtained for linear systems will be interpreted for $L_n x = 0$.

If $\varphi_1, \ldots, \varphi_n$ are n solutions of $L_n x = 0$, then the matrix

$$\Phi = \begin{pmatrix} \varphi_1 & \varphi_2 & \cdots & \varphi_n \\ \varphi_1' & \varphi_2' & \cdots & \varphi_n' \\ \cdot & \cdot & \cdots & \cdot \\ \varphi_1^{(n-1)} & \varphi_2^{(n-1)} & \cdots & \varphi_n^{(n-1)} \end{pmatrix} \tag{6.4}$$

is a solution matrix for (6.1). The determinant of this matrix is called the *Wronskian* of $L_n x = 0$ with respect to $\varphi_1, \ldots, \varphi_n$, and is denoted by $W(\varphi_1, \ldots, \varphi_n)$. It is a function of t on I for fixed $\varphi_1, \ldots, \varphi_n$; its value at t is denoted by $W(\varphi_1, \ldots, \varphi_n)(t)$. From the fact that for a linear system, such as (6.1),

$$\det \Phi(t) = \det \Phi(\tau) \exp \int_\tau^t \operatorname{tr} A(s)\, ds \qquad (t \in I)$$

one obtains, noting from (6.2) that $\operatorname{tr} A = -a_1/a_0$, ✓

$$W(\varphi_1, \ldots, \varphi_n)(t) = W(\varphi_1, \ldots, \varphi_n)(\tau) \exp \int_\tau^t -\frac{a_1(s)}{a_0(s)}\, ds \qquad (t \in I) \tag{6.5}$$

Theorem 6.1. *A necessary and sufficient condition that n solutions* $\varphi_1, \ldots, \varphi_n$ *of* $L_n x = 0$ *on an interval I be linearly independent there is that*

$$W(\varphi_1, \ldots, \varphi_n)(t) \neq 0 \qquad (t \in I)$$

Every solution of $L_n x = 0$ *is a linear combination with complex coefficients of any n linearly independent solutions.*

Proof. If $\varphi_1, \ldots, \varphi_n$ are linearly dependent on I, there exist constants $c_1, \ldots, c_n$ not all zero such that

$$\sum_{i=1}^{n} c_i \varphi_i = 0$$

This implies that

$$\sum_{i=1}^{n} c_i \varphi_i^{(k)}(t) = 0 \qquad (k = 0, 1, \ldots, n-1)$$

and hence the vectors $\hat{\varphi}_i$ with components $\varphi_i, \varphi_i', \ldots, \varphi_i^{(n-1)}$, $(i = 1, \ldots, n)$ are linearly dependent on I. Conversely, if the vectors $\hat{\varphi}_i$ are linearly dependent, so are the solutions $\varphi_1, \ldots, \varphi_n$ of $L_n x = 0$. From Theorem 2.2 a necessary and sufficient condition that the vectors $\hat{\varphi}_1, \ldots, \hat{\varphi}_n$ be linearly independent is that $\det \Phi(t) \neq 0$ on I, where Φ is the matrix (6.4). But this is just the condition $W(\varphi_1, \ldots, \varphi_n)(t) \neq 0$ on I. By (6.5), if $W(\varphi_1, \ldots, \varphi_n)(\tau) \neq 0$ for *some* $\tau \in I$, $W(\varphi_1, \ldots, \varphi_n)(t) \neq 0$ for *any* $t \in I$.

Since every solution vector of (6.1) and (6.2) is a linear combination of n linearly independent vector solutions, every solution of $L_n x = 0$ is a linear combination of n linearly independent solutions of $L_n x = 0$. This proves the theorem.

Because of the properties exhibited in Theorem 6.1, a set of n linearly independent solutions of $L_n x = 0$ is called a *basis*, or a *fundamental set*, for $L_n x = 0$.

Theorem 6.2. *Suppose* $\varphi_1, \ldots, \varphi_n$ *are n functions which possess continuous nth-order derivatives on a real t interval I, and* $W(\varphi_1, \ldots, \varphi_n)(t) \neq 0$ *on I. Then there exists a unique homogeneous differential equation of order n (with coefficient of* $x^{(n)}$ *one) for which these functions form a fundamental set, namely*

$$(-1)^n \frac{W(x, \varphi_1, \ldots, \varphi_n)}{W(\varphi_1, \ldots, \varphi_n)} = 0 \tag{6.6}$$

NOTE: The Wronskian $W(x, \varphi_1, \ldots, \varphi_n)$ is the determinant of the matrix with the first row consisting of the elements $x, \varphi_1, \ldots, \varphi_n$ and the other rows being the derivatives of the first row up to the order n for the last row.

Proof of Theorem 6.2. Clearly $W(\varphi_i, \varphi_1, \ldots, \varphi_n) = 0$, $(i = 1, \ldots, n)$, for two columns of this determinant are equal. An expansion of the numerator, $W(x, \varphi_1, \ldots, \varphi_n)$, of (6.6) by the first column shows that (6.6) is a differential equation of order n, and the coefficient of $x^{(n)}$ in $W(x, \varphi_1, \ldots, \varphi_n)$ is just $(-1)^n W(\varphi_1, \ldots, \varphi_n)$, which proves the coefficient of $x^{(n)}$ in (6.6) is one. Since $W(\varphi_1, \ldots, \varphi_n) \neq 0$, it follows from Theorem 6.1 that $\varphi_1, \ldots, \varphi_n$ form a fundamental set for (6.6).

The uniqueness of (6.6) follows from the fact that the corresponding vectors $\hat{\varphi}_i$ with components $\varphi_i, \varphi_i', \ldots, \varphi_i^{(n-1)}$ determine the coefficient matrix (6.2) of the associated system (6.1) uniquely. Since there is a one-to-one correspondence between linear equations of order n and linear systems of the type (6.1), (6.2), the proof is complete.

If one or more solutions of $L_n x = 0$ are known, then using the associated system (6.1) it follows that a reduction of order can be effected. A more direct procedure is suggested by the following process, which is the variation of constants adapted to $L_n x = 0$. Let $L_n \varphi_1 = 0$, and set $x = y\varphi_1$. Then $L_n x = 0$ yields a linear differential equation of the nth order in y which has $y = 1$ as a solution, since φ_1 is a solution of $L_n x = 0$. Thus the coefficient of y in the new equation must vanish. Considered as an equation in $u = y'$, it is of order $n - 1$. If φ_2 is independent of φ_1 and $L_n \varphi_2 = 0$, then $(\varphi_2/\varphi_1)'$ is a solution of the $(n - 1)$st-order equation in u, which can, by a repetition of the above, be reduced to an equation of order $n - 2$, etc.

Adjoint Equations. Intimately connected with the formal operator L_n is another linear operator L_n^+ of order n, called the *adjoint* of L_n, given by

$$L_n^+ = (-1)^n \left(\frac{d^n}{dt^n}\right)(\bar{a}_0 \cdot) + (-1)^{n-1}\left(\frac{d^{n-1}}{dt^{n-1}}\right)(\bar{a}_1 \cdot) + \cdots + \bar{a}_n \cdot$$

that is, if g is any function on I which is such that $\bar{a}_k g$ $(k = 0, 1, \ldots, n)$ has $n - k$ derivatives on I, then

$$L_n^+ g = (-1)^n (\bar{a}_0 g)^{(n)} + (-1)^{n-1}(\bar{a}_1 g)^{(n-1)} + \cdots + \bar{a}_n g$$

The equation

$$L_n^+ x = 0 \qquad (t \in I)$$

[written out as

$$(-1)^n(\bar{a}_0(t)x)^{(n)} + (-1)^{n-1}(\bar{a}_1(t)x)^{(n-1)} + \cdots + \bar{a}_n(t)x = 0]$$

called the *adjoint equation* to $L_n x = 0$ on I, is defined to be the problem of finding a function φ (a solution) on I such that $\bar{a}_k\varphi$ ($k = 0, 1, \ldots, n$) has $n - k$ derivatives on I and satisfying

$$(-1)^n(\bar{a}_0\varphi)^{(n)} + (-1)^{n-1}(\bar{a}_1\varphi)^{(n-1)} + \cdots + \bar{a}_n\varphi = 0$$

on I.

If $a_k \in C^{n-k}$ on I and φ is a solution of $L_n^+x = 0$ with n derivatives on I, then by using the product rule of differentiation

$$L_n^+\varphi = (-1)^n\bar{a}_0\varphi^{(n)} + \cdots = 0$$

and by dividing by $(-1)^n\bar{a}_0$ one sees φ is a solution of a differential equation of order n of the type considered previously.

Consider the special case of an L_n where $a_0 = 1$. For the system (6.1), (6.2) associated with the equation

$$L_nx = x^{(n)} + a_1x^{(n-1)} + \cdots + a_nx = 0 \tag{6.7}$$

the adjoint system is

$$\hat{x}' = -A^*(t)\hat{x} \qquad (t \in I) \tag{6.8}$$

where, from (6.2),

$$-A^* = \begin{pmatrix} 0 & 0 & \cdots & 0 & \bar{a}_n \\ -1 & 0 & \cdots & 0 & \bar{a}_{n-1} \\ 0 & -1 & \cdots & 0 & \bar{a}_{n-2} \\ \cdot & \cdot & \cdots & \cdot & \cdot \\ \cdot & \cdot & \cdots & 0 & \bar{a}_2 \\ 0 & 0 & \cdots & -1 & \bar{a}_1 \end{pmatrix} \tag{6.9}$$

In terms of components, (6.8) and (6.9) give

$$x_1' = \bar{a}_nx_n \qquad x_k' = -x_{k-1} + \bar{a}_{n-k+1}x_n \qquad (k = 2, \ldots, n) \tag{6.10}$$

Thus if $\varphi_1, \ldots, \varphi_n$ is a solution of (6.10) for which $\varphi_k^{(k)}$ and

$$(\bar{a}_{n-k+1}\varphi_n)^{(k-1)}$$

exist, one obtains, by differentiating the kth relation in (6.10) $(k - 1)$ times and solving for $\varphi_n^{(n)}$,

$$\varphi_n^{(n)} - (\bar{a}_1\varphi_n)^{(n-1)} + \cdots + (-1)^n(\bar{a}_n\varphi_n) = 0$$

Therefore φ_n satisfies the equation

$$L_n^+x = (-1)^nx^{(n)} + (-1)^{n-1}(\bar{a}_1x)^{(n-1)} + \cdots + \bar{a}_nx = 0$$

which is just the adjoint equation to (6.7).

The importance of L_n^+ is due to an interesting relation connecting L_n and L_n^+, which is indispensable for the study of boundary-value problems (see Chaps. 7–12).

Theorem 6.3 (Lagrange Identity). *In L_n suppose $a_k \varepsilon C^{n-k}$ on I $(k = 0, 1, \ldots, n)$. If u,v are any two (complex) functions on I possessing n derivatives there, then*

$$\bar{v}L_n u - u\overline{L_n^+ v} = [uv]' \qquad \left(' = \frac{d}{dt}\right) \tag{6.11}$$

where $[uv]$ is a form in $(u, u', \ldots, u^{(n-1)})$ and $(v, v', \ldots, v^{(n-1)})$ given by

$$[uv] = \sum_{m=1}^{n} \sum_{\substack{j+k=m-1 \\ j \geq 0, k \geq 0}} (-1)^j u^{(k)} (a_{n-m}\bar{v})^{(j)} \tag{6.12}$$

Proof. Using the product rule for differentiation,

$$\bar{v}u^{(m)} = (-1)^m \bar{v}^{(m)} u + (u^{(m-1)}\bar{v} - u^{(m-2)}\bar{v}' + \cdots + (-1)^{m-1} u\bar{v}^{(m-1)})'$$

for $m = 0, 1, \ldots, n$. Thus one obtains

$$\begin{aligned}\bar{v}L_n u &= \bar{v} \sum_{m=1}^{n} a_{n-m} u^{(m)} + \bar{v}a_n u \\ &= \sum_{m=1}^{n} (a_{n-m}\bar{v}) u^{(m)} + \bar{v}a_n u \\ &= \sum_{m=1}^{n} (-1)^m (a_{n-m}\bar{v})^{(m)} u + \bar{v}a_n u \\ &\qquad + \Big[\sum_{m=1}^{n} \sum_{\substack{j+k=m-1 \\ j \geq 0, k \geq 0}} (-1)^j u^{(k)} (a_{n-m}\bar{v})^{(j)} \Big]'\end{aligned}$$

proving the result.

Corollary (Green's Formula). *If the a_k in L_n, and u,v are the same as in Theorem 6.3, then for any $t_1, t_2 \varepsilon I$,*

$$\int_{t_1}^{t_2} (\bar{v}L_n u - u\overline{L_n^+ v})\, dt = [uv](t_2) - [uv](t_1) \tag{6.13}$$

where $[uv](t)$ is the value at t of $[uv]$.

Proof. Integrate the Lagrange identity (6.11) from t_1 to t_2.

If ψ is a known solution of $L_n^+ x = 0$ on I, the solution of the equation $L_n x = 0$ is reduced by (6.11) to finding a function φ on I satisfying an equation of order $n - 1$, namely,

$$\sum_{m=1}^{n} \sum_{\substack{j+k=m-1 \\ j \geq 0, k \geq 0}} (-1)^j x^{(k)} (a_{n-m}\bar{\psi})^{(j)} = \text{constant}$$

The Nonhomogeneous Linear Equation of Order n. On a real t interval I, suppose $a_0 \neq 0, a_1, \ldots, a_n$ and b are continuous functions, and consider the equation

$$L_n x = a_0(t)x^{(n)} + a_1(t)x^{(n-1)} + \cdots + a_n(t)x = b(t) \qquad (t \in I)$$

which is defined to be the same as

$$x^{(n)} + \frac{a_1(t)}{a_0(t)} x^{(n-1)} + \cdots + \frac{a_n(t)}{a_0(t)} x = \frac{b(t)}{a_0(t)}$$

This is called (in case $b \not\equiv 0$) a *nonhomogeneous linear equation of order n.* The system associated with this equation (see Chap. 1, Sec. 6) is given by

$$\hat{x}' = A(t)\hat{x} + \hat{b}(t) \qquad (t \in I) \tag{6.14}$$

where A is the matrix (6.2), and $\hat{b}$ is the column vector with all elements zero except the last which is b/a_0. Thus the system (6.14) associated with $L_n x = b(t)$ is a linear nonhomogeneous system, and the existence and uniqueness of solutions of (6.14) can be interpreted, as usual, as existence and uniqueness results for $L_n x = b(t)$.

It is of interest to determine the explicit form that the variation-of-constants formula (3.2) takes for the special system (6.14). Only the first component $\psi = \psi_1$ of any vector solution $\hat{\psi}$ of (6.14) is of interest, since this component is a solution of the equation $L_n x = b(t)$.

Theorem 6.4. *If $\varphi_1, \ldots, \varphi_n$ is a fundamental set for the homogeneous equation*

$$L_n x = x^{(n)} + a_1 x^{(n-1)} + \cdots + a_n x = 0 \qquad (a_k \in C \text{ on } I)$$

then the solution ψ of the nonhomogeneous equation

$$L_n x = b(t) \qquad (b \in C \text{ on } I)$$

satisfying

$$\hat{\psi}(\tau) = \hat{\xi} \qquad (\tau \in I, |\hat{\xi}| < \infty)$$

is given by

$$\psi(t) = \psi_h(t) + \sum_{k=1}^{n} \varphi_k(t) \int_\tau^t \left\{ \frac{W_k(\varphi_1, \ldots, \varphi_n)(s)}{W(\varphi_1, \ldots, \varphi_n)(s)} \right\} b(s)\, ds \tag{6.15}$$

where ψ_h is the solution of $L_n x = 0$ for which $\hat{\psi}_h(\tau) = \hat{\xi}$ and $W_k(\varphi_1, \ldots, \varphi_n)$ is the determinant obtained from $W(\varphi_1, \ldots, \varphi_n)$ by replacing the kth column by $(0, \ldots, 0, 1)$.

Proof. By (3.1) the first component $\psi = \psi_1$ of the vector solution $\hat{\psi}$ of (6.14) for which $\hat{\psi}(\tau) = 0$ is given by

$$\psi(t) = \int_\tau^t \gamma_{1n}(t,s) b(s)\, ds$$

where $\gamma_{1n}(t,s)$ is the element in the first row and nth column of the matrix $\Phi(t)\Phi^{-1}(s)$. Recall that the element in the ith row and jth column of $\Phi(t)$ is $\varphi_j^{(i-1)}$, and $\det \Phi(t) = W(\varphi_1, \ldots, \varphi_n)(t)$. Now the element in the ith row and nth column of Φ^{-1} is given by

$$\frac{\tilde{\varphi}_{in}}{W(\varphi_1, \ldots, \varphi_n)}$$

where $\tilde{\varphi}_{in}$ is the cofactor of $\varphi_i^{(n-1)}$ in Φ. Therefore,

$$W(\varphi_1, \ldots, \varphi_n)(s)\gamma_{1n}(t,s) = \sum_{k=1}^{n} \varphi_k(t) W_k(\varphi_1, \ldots, \varphi_n)(s)$$

where $W_k(\varphi_1, \ldots, \varphi_n)(s)$ is defined as in the statement of the theorem. Thus the solution ψ of $L_n x = b(t)$ satisfying $\hat{\psi}(\tau) = 0$ is given by

$$\psi(t) = \sum_{k=1}^{n} \varphi_k(t) \int_\tau^t \left\{ \frac{W_k(\varphi_1, \ldots, \varphi_n)(s)}{W(\varphi_1, \ldots, \varphi_n)(s)} \right\} b(s)\, ds$$

and obviously (6.15) gives the solution satisfying $\hat{\psi}(\tau) = \xi$, if $\hat{\psi}_h(\tau) = \xi$.

The Linear Equation of Order n with Constant Coefficients. Consider the case where in L_n the functions $a_0 = 1, a_1, \ldots, a_n$ are all constants. Then I may be assumed to be the entire real t axis. In this case,

$$L_n x = x^{(n)} + a_1 x^{(n-1)} + \cdots + a_n x = 0 \tag{6.16}$$

has as its associated system

$$\hat{x}' = A\hat{x} \tag{6.17}$$

where A is the constant matrix

$$A = \begin{pmatrix} 0 & 1 & 0 & \cdots & 0 \\ 0 & 0 & 1 & \cdots & 0 \\ \cdot & \cdot & \cdot & \cdots & \cdot \\ 0 & 0 & 0 & \cdots & 1 \\ -a_n & -a_{n-1} & -a_{n-2} & \cdots & -a_1 \end{pmatrix} \tag{6.18}$$

As is to be expected, a fundamental set of solutions of (6.16) can be exhibited, and the precise form of these functions depends on the characteristic polynomial $f(\lambda) = \det(\lambda E - A)$ of the constant matrix A in (6.18).

Lemma. *The characteristic polynomial for A in* (6.18) *is given by*

$$f(\lambda) = \lambda^n + a_1\lambda^{n-1} + \cdots + a_n \tag{6.19}$$

Note that $f(\lambda)$ can be obtained from $L_n x$ by formally changing $x^{(k)}$ to λ^k.

Proof. The proof proceeds by induction. For $n = 1$, $A = -a_1$, and hence $\det(\lambda E_1 - A) = \lambda + a_1$, and therefore (6.19) is true for $n = 1$. Assume the result for $n - 1$. Then expand

$$\det(\lambda E_n - A) = \begin{vmatrix} \lambda & -1 & 0 & \cdots & 0 & 0 \\ 0 & \lambda & -1 & \cdots & 0 & 0 \\ \cdot & \cdot & \cdot & \cdots & \cdot & \cdot \\ 0 & 0 & 0 & \cdots & \lambda & -1 \\ a_n & a_{n-1} & a_{n-2} & \cdots & a_2 & \lambda + a_1 \end{vmatrix}$$

by the first column, and notice that the coefficient of λ is a determinant of order $n - 1$ equal to $\det(\lambda E_{n-1} - A_1)$, where

$$A_1 = \begin{pmatrix} 0 & 1 & 0 & \cdots & 0 \\ 0 & 0 & 1 & \cdots & 0 \\ \cdot & \cdot & \cdot & \cdots & \cdot \\ 0 & 0 & 0 & \cdots & 1 \\ -a_{n-1} & -a_{n-2} & -a_{n-3} & \cdots & -a_1 \end{pmatrix}$$

Hence $\lambda \det(\lambda E_{n-1} - A_1) = \lambda^n + a_1\lambda^{n-1} + \cdots + a_{n-1}\lambda$. The only other nonzero element in the first column is a_n, and the contribution to $\det(\lambda E - A)$ due to a_n is a_n itself, since the cofactor of a_n is 1. Hence $\det(\lambda E - A) = \lambda^n + a_1\lambda^{n-1} + \cdots + a_{n-1}\lambda + a_n$, which was to be proved.

Theorem 6.5. *Let $\lambda_1, \ldots, \lambda_s$ be the distinct roots of the characteristic equation*

$$f(\lambda) = \lambda^n + a_1\lambda^{n-1} + \cdots + a_n = 0$$

and suppose λ_i has multiplicity m_i $(i = 1, \ldots, s)$. Then a fundamental set for (6.16) *is given by the n functions*

$$t^k e^{t\lambda_i} \qquad (k = 0, 1, \ldots, m_i - 1;\ i = 1, \ldots, s) \tag{6.20}$$

Proof. The proof can be based on the corresponding result for linear systems with constant coefficients. However, a direct proof will be given here, which depends upon the fact that, if λ_i is a root of $f(\lambda) = 0$ with multiplicity m_i, then it is also a root of the equations $f'(\lambda) = 0, \ldots, f^{(m_i-1)}(\lambda) = 0$. Now clearly

$$L_n(e^{t\lambda}) = f(\lambda)e^{t\lambda}$$

and in general

$$L_n(t^k e^{t\lambda}) = L_n\left(\frac{\partial^k}{\partial\lambda^k} e^{t\lambda}\right) = \frac{\partial^k}{\partial\lambda^k} L_n(e^{t\lambda}) = \frac{\partial^k}{\partial\lambda^k} (f(\lambda)e^{t\lambda})$$

$$= \left[f^{(k)}(\lambda) + kf^{(k-1)}(\lambda)t + \frac{k(k-1)}{2!} f^{(k-2)}(\lambda)t^2 + \cdots + f(\lambda)t^k \right] e^{t\lambda}$$

From this it is now obvious that, for any fixed i,

$$L_n(t^k e^{t\lambda_i}) = 0 \qquad (k = 0, 1, \ldots, m_i - 1)$$

thus proving that the functions (6.20) are solutions of $L_n x = 0$.

Suppose the functions (6.20) are not linearly independent. Then there exist constants c_{ik} not all zero such that

$$\sum_{i=1}^{s} \sum_{k=0}^{m_i-1} c_{ik} t^k e^{t\lambda_i} = 0$$

or

$$\sum_{i=1}^{\sigma} P_i(t) e^{t\lambda_i} = 0$$

where the $P_i(t)$ are polynomials and $\sigma \leqq s$ is chosen so that $P_\sigma \not\equiv 0$ while $P_{\sigma+i}(t) \equiv 0$, $i \geqq 1$. Divide the above expression by $e^{t\lambda_1}$ and differentiate enough times so that the polynomial $P_1(t)$ becomes zero. Note that the degrees and the nonidentically vanishing nature of the polynomials multiplying $e^{(\lambda_i - \lambda_1)t}$, $i > 1$, do not change under this operation. Thus there results

$$\sum_{i=2}^{\sigma} Q_i(t) e^{t\lambda_i} = 0$$

where $Q_i(t)$ has the same degree as $P_i(t)$ for $i \geqq 2$. Repeating the procedure results finally in a polynomial $F(t)$ of a degree equal to that of $P_\sigma(t)$ such that $F(t) = 0$ for all t. This is impossible, since a polynomial can vanish only at isolated points. Thus the solutions are linearly independent.

7. Linear Equations with Analytic Coefficients

Suppose A is an n-by-n matrix and b an n-dimensional vector defined and analytic on a simply connected domain D of the z plane, and let $z_0 \in D$. Using the method of successive approximations, it is readily shown that the linear system

$$\hat{w}' = A(z)\hat{w} + \hat{b}(z) \tag{7.1}$$

has a unique analytic solution $\hat{\varphi}$ on D such that

$$\hat{\varphi}(z_0) = \hat{\omega}$$

where $|\hat{\omega}| < \infty$.

Indeed, let $z_1 \in D$ and let C be an arc from z_0 to z_1 which lies in D, has a continuously turning tangent, and is of length L. Let the arc length along C starting from z_0 be denoted by s. Let the constant K be large enough so that $|A(z)| < K$ and $|\hat{b}(z)| < K$ for z on C. Let $\hat{\varphi}_0(z) = \hat{\omega}$ and

$$\hat{\varphi}_n(z) = \hat{\omega} + \int_{z_0}^{z} A(\zeta)\hat{\varphi}_{n-1}(\zeta)\, d\zeta + \int_{z_0}^{z} \hat{b}(\zeta)\, d\zeta$$

where the integration is carried out along C so that $\hat{\varphi}_n$ is defined on C. It follows readily that

$$|\hat{\varphi}_1 - \hat{\varphi}_0| \leqq K(|\hat{\omega}| + 1)s \leqq KL(|\hat{\omega}| + 1), \ldots,$$

$$|\hat{\varphi}_n - \hat{\varphi}_{n-1}| \leqq K^n(|\hat{\omega}| + 1)\frac{s^n}{n!} \leqq \frac{K^nL^n}{n!}(|\hat{\omega}| + 1)$$

Clearly these appraisals are valid at all points z in D which can be reached from z_0 on an arc of length L on which $|A(z)|$ and $|\hat{b}(z)|$ are bounded by K. This implies that they are valid in any fixed closed region R contained in D. Since each $\hat{\varphi}_n$ is analytic in R, it follows from the uniform convergence of $\hat{\varphi}_n$ that the limiting function $\hat{\varphi}$ is also analytic in R. It also follows that

$$\hat{\varphi}(z) = \hat{\omega} + \int_{z_0}^{z} A(\zeta)\hat{\varphi}(\zeta)\, d\zeta + \int_{z_0}^{z} \hat{b}(\zeta)\, d\zeta$$

This proves the result in R and therefore in D.

Moreover, all the theorems proved in Secs. 2 and 3, being essentially algebraic in nature, are valid for the system (7.1).

Correspondingly, if $a_1, \ldots, a_n, b$ are n analytic functions on D, then the linear equation of order n,

$$w^{(n)} + a_1(z)w^{(n-1)} + \cdots + a_n(z)w = b(z) \tag{7.2}$$

has a unique analytic solution φ on D satisfying

$$w(z_0) = \omega_1,\ w'(z_0) = \omega_2, \ldots, w^{(n-1)}(z_0) = \omega_n$$

where $\omega_1, \omega_2, \ldots, \omega_n$ are any given n complex numbers. In addition, all the results of Sec. 6 carry over to the case (7.2) in an obvious way.

8. Asymptotic Behavior of the Solutions of Certain Linear Systems

If the coefficients of a linear system of differential equations tend to constants as $t \to \infty$, it is sometimes possible to characterize the behavior of the solutions. In the analytic case, this problem is treated in Secs. 4 and 5, Chap. 5.

A real variable problem will be considered here. Simpler cases are treated in Probs. 29 and 35 at the end of this chapter. First consider the example

$$x'' + [1 + v(t) + r(t)]x = 0$$

where v is a real-valued differentiable function with $\lim_{t\to\infty} v(t) = 0$, r is integrable, and

$$\int_{t_0}^{\infty} |v'(t)|\, dt < \infty \qquad \int_{t_0}^{\infty} |r(t)|\, dt < \infty$$

for some t_0. [Actually it is sufficient for v to be of bounded variation on (t_0, ∞).] With no real restriction, t_0 will be taken as zero in what follows. A consequence of the theorem stated below is that the equation has two solutions φ and ψ such that

$$\varphi(t) - \exp\left[i \int_0^t \sqrt{1 + v(\tau)}\, d\tau\right] \to 0$$
$$\varphi'(t) - i \exp\left[i \int_0^t \sqrt{1 + v(\tau)}\, d\tau\right] \to 0$$

as $t \to \infty$, and ψ has similar behavior with i above replaced by $-i$.

This result indicates that r does not affect the gross asymptotic behavior at all. However, the case

$$v(t) = t^{-\alpha} \qquad (0 < \alpha < 1)$$

shows that v enters in an essential way. The result shows that, if the equation were treated as though $r(t)$ were zero and $1 + v(t)$ constant, the result would be accurate to within a term which is $o(1)$ as $t \to \infty$.

In what follows, a linear system

$$x' = (A + V(t) + R(t))x \tag{8.1}$$

will be considered, which includes the above example as a special case.

Theorem 8.1. *Let A be a constant matrix with characteristic roots μ_j, $j = 1, \ldots, n$, all of which are distinct. Let the matrix V be differentiable and satisfy*

$$\int_0^{\infty} |V'(t)|\, dt < \infty \tag{8.2}$$

and let $V(t) \to 0$ as $t \to \infty$. Let the matrix R be integrable and let

$$\int_0^{\infty} |R(t)|\, dt < \infty \tag{8.3}$$

Let the roots of $\det\,(A + V(t) - \lambda E) = 0$ *be denoted by $\lambda_j(t)$, $j = 1, \ldots, n$. Clearly, by reordering the μ_j if necessary, $\lim_{t \to \infty} \lambda_j(t) = \mu_j$. For a given k, let*

$$D_{kj}(t) = \Re(\lambda_k(t) - \lambda_j(t))$$

Suppose all j, $1 \leqq j \leqq n$, fall into one of two classes I_1 and I_2, where

$$j \in I_1 \quad \text{if } \int_0^t D_{kj}(\tau)\, d\tau \to \infty \quad \text{as } t \to \infty \quad \text{and} \\ \int_{t_1}^{t_2} D_{kj}(\tau)\, d\tau > -K \qquad (t_2 \geqq t_1 \geqq 0) \tag{8.4}$$

$$j \in I_2 \qquad \textit{if} \qquad \int_{t_1}^{t_2} D_{kj}(\tau)\, d\tau < K \qquad (t_2 \geqq t_1 \geqq 0) \tag{8.5}$$

where k is fixed and where K is a constant. Let p_k be a characteristic vector of A associated with μ_k, so that

$$Ap_k = \mu_k p_k \tag{8.6}$$

Then there is a solution φ_k of (8.1) *and a t_0, $0 \leqq t_0 < \infty$, such that*

$$\lim_{t \to \infty} \varphi_k(t) \exp\left[-\int_{t_0}^{t} \lambda_k(\tau)\, d\tau\right] = p_k \tag{8.7}$$

If the hypothesis is satisfied for all k, $1 \leqq k \leqq n$, and if Φ is the matrix with columns $\varphi_1, \varphi_2, \ldots, \varphi_n$, then Φ is a fundamental matrix because $\det \Phi(t) \neq 0$ for large t since the p_j are independent.

Suppose first that $A + V(t)$ is in diagonal form $\Lambda(t)$ for $t \geqq t_0$, where t_0 is chosen so that

$$e^K \int_{t_0}^{\infty} |R(\tau)|\, d\tau < \tfrac{1}{2} \tag{8.8}$$

Let $\Psi(t)$ be the diagonal matrix

$$\Psi(t) = \exp\left[\int_{t_0}^{t} \Lambda(s)\, ds\right]$$

so that

$$\Psi' = \Lambda\Psi \tag{8.9}$$

Let e_k denote the column vector with all components zero except the kth, which is 1, and let ψ_k be the vector defined by

$$\psi_k(t) = \Psi(t)e_k = \exp\left[\int_{t_0}^{t} \lambda_k(s)\, ds\right] e_k$$

With k fixed and I_1 and I_2 defined as in (8.4) and (8.5), let

$$\Psi = \Psi_1 + \Psi_2$$

where the diagonal matrices Ψ_1 and Ψ_2 contain those elements of Ψ associated with columns of index j belonging to I_1 and I_2, respectively. Then

$$\Psi_j' = \Lambda\Psi_j \qquad (j = 1, 2) \tag{8.10}$$

Consider next the equation

$$\varphi(t) = \psi_k(t) + \int_{t_0}^{t} \Psi_1(t)\Psi^{-1}(\tau)R(\tau)\varphi(\tau)\, d\tau - \int_{t}^{\infty} \Psi_2(t)\Psi^{-1}(\tau)R(\tau)\varphi(\tau)\, d\tau \tag{8.11}$$

If the equation (8.11) has a solution φ, it may be verified directly that

$$\varphi' = (\Lambda + R)\varphi \tag{8.12}$$

which is the form of Eq. (8.1) under consideration here.

Let $\varphi^0(t) = 0$ and

$$\varphi^{j+1}(t) = \psi_k(t) + \int_{t_0}^{t} \Psi_1(t)\Psi^{-1}(\tau)R(\tau)\varphi^j(\tau)\,d\tau - \int_{t}^{\infty} \Psi_2(t)\Psi^{-1}(\tau)R(\tau)\varphi^j(\tau)\,d\tau \quad (8.13)$$

Then $\varphi^1(t) = \psi_k(t)$ and for $t \geqq t_0$

$$|\varphi^1(t) - \varphi^0(t)| = \left| \exp\left[\int_{t_0}^{t} \lambda_k(s)\,ds\right]\right| \quad (8.14)$$

Each element of the diagonal matrix $\Psi_1(t)\Psi^{-1}(\tau)$ is of the form

$$h_l(t) = \exp\left[\int_{\tau}^{t} \lambda_l(s)\,ds\right] \qquad (l \varepsilon I_1)$$

or else is zero. But for $t_0 \leqq \tau \leqq t$

$$\begin{aligned}|h_l(t)| &= \exp\left[-\int_{\tau}^{t} D_{kl}(s)\,ds\right]\exp\left[\int_{\tau}^{t} \Re\lambda_k(s)\,ds\right]\\ &\leqq e^K \exp\left[\int_{\tau}^{t} \Re\lambda_k(s)\,ds\right]\end{aligned}$$

Thus for $t_0 \leqq \tau \leqq t$

$$|\Psi_1(t)\Psi^{-1}(\tau)R(\tau)| \leqq e^K|R(\tau)| \exp\left[\int_{\tau}^{t} \Re\lambda_k(s)\,ds\right]$$

In the same way for $\tau \geqq t$,

$$|\Psi_2(t)\Psi^{-1}(\tau)R(\tau)| \leqq e^K\,|R(\tau)| \exp\left[-\int_{t}^{\tau} \Re\lambda_k(s)\,ds\right]$$

Using these inequalities in (8.13)

$$\begin{aligned}&|\varphi^{j+1}(t) - \varphi^j(t)| \exp\left[-\int_{t_0}^{t} \Re\lambda_k(s)\,ds\right]\\ &\qquad \leqq e^K\left(\int_{t_0}^{t} + \int_{t}^{\infty}\right)|R(\tau)|\,|\varphi^j(\tau) - \varphi^{j-1}(\tau)| \exp\left[-\int_{t_0}^{\tau} \Re\lambda_k(s)\,ds\right] d\tau\end{aligned}$$

Using (8.8) and (8.14), it now follows by induction that

$$|\varphi^{j+1}(t) - \varphi^j(t)| \exp\left[-\int_{t_0}^{t} \Re\lambda_k(s)\,ds\right] \leqq (\tfrac{1}{2})^j$$

From this follows the uniform convergence of $\{\varphi^j\}$ on every finite subinterval of $[t_0, \infty)$. Since each φ^j is continuous, the limit function φ is also continuous and clearly

$$|\varphi(t)| \leqq 2 \exp\left[\int_{t_0}^{t} \Re\lambda_k(s)\,ds\right] \quad (8.15)$$

Clearly φ is a solution of (8.11). It will be shown that

$$\lim_{t\to\infty}\left\{\varphi(t)\exp\left[-\int_{t_0}^{t}\lambda_k(s)\,ds\right]-e_k\right\}=0 \tag{8.16}$$

This will follow by showing that as $t\to\infty$

$$\exp\left[-\int_{t_0}^{t}\Re\lambda_k(s)\,ds\right]\int_{t_0}^{t}\Psi_1(t)\Psi^{-1}(\tau)R(\tau)\varphi(\tau)\,d\tau\to 0 \tag{8.17}$$

and

$$\exp\left[-\int_{t_0}^{t}\Re\lambda_k(s)\,ds\right]\int_{t}^{\infty}\Psi_2(t)\Psi^{-1}(\tau)R(\tau)\varphi(\tau)\,d\tau\to 0 \tag{8.18}$$

The proof of (8.18) follows at once from (8.15) and (8.5). The proof of (8.17) requires

$$\lim_{t\to\infty}|\Psi_1(t)|\exp\left[-\int_{t_0}^{t}\Re\lambda_k(s)\,ds\right]=0 \tag{8.19}$$

which is a consequence of (8.4). Given any $\epsilon>0$ it is possible to choose t_1 so that

$$2e^{K}\int_{t_1}^{\infty}|R(\tau)|\,d\tau<\epsilon$$

Thus, denoting the left side of (8.17) by $J(t)$,

$$|J(t)|\leqq\epsilon+\exp\left[-\int_{t_0}^{t}\Re\lambda_k(s)\,ds\right]|\Psi_1(t)|\int_{t_0}^{t_1}|\Psi^{-1}(\tau)R(\tau)\varphi(\tau)|\,d\tau$$

As $t\to\infty$, it follows from (8.19) that

$$\limsup_{t\to\infty}|J(t)|\leqq\epsilon$$

Since ϵ is arbitrary, (8.17) is proved. Thus if φ_k is taken as φ, the theorem is proved for the case $A+V(t)=\Lambda(t)$.

The proof of Theorem 8.1 is a consequence of the following lemma.

Lemma. *Suppose A and V satisfy the requirements of Theorem* 8.1. *Then there exists a matrix $S(t)$, which as $t\to\infty$ tends to a constant nonsingular matrix T, such that*

$$S(A+V)=\Lambda S \tag{8.20}$$

where $\Lambda(t)$ is a diagonal matrix with diagonal elements $\lambda_j(t)$, $j=1,2,\ldots,n$. As $t\to\infty$, $\lambda_j(t)\to\mu_j$, where the μ_j are the characteristic roots of A. Moreover, for some t_0,

$$\int_{t_0}^{\infty}|S'(t)|\,dt<\infty \tag{8.21}$$

The proof of the lemma will follow that of Theorem 8.1.

Proof of Theorem 8.1. Since $S(t)\to T$ as $t\to\infty$, and T is nonsingular, $S(t)$ is nonsingular for all sufficiently large t. Choose t_0 so large that not only (8.21) is valid but $S^{-1}(t)$ exists for $t \geqq t_0$. Then, letting $y = S(t)x$ in (8.1),

$$y' = \Lambda y + (SRS^{-1} + S'S^{-1})y \qquad (t \geqq t_0) \tag{8.22}$$

Let $\tilde{R} = SRS^{-1} + S'S^{-1}$. Then, by (8.3) and (8.21), it follows that $|\tilde{R}|$ is integrable. Thus the proof for the special case of Theorem 8.1 given above is valid for (8.22) so that (8.22) has as a solution θ_k, where

$$\lim_{t\to\infty} \theta_k(t) \exp\left[-\int_{t_0}^t \lambda_k(s)\,ds\right] = e_k$$

Thus (8.1) has as a solution $S^{-1}\theta_k = \varphi_k$. Since, as $t \to \infty$, $S^{-1}(t) \to T^{-1}$, it follows that

$$\varphi_k(t) \exp\left[-\int_{t_0}^t \lambda_k(s)\,ds\right] \to p_k \qquad (t \to \infty)$$

where p_k is the kth column of T^{-1}. Since $AT^{-1} = T^{-1}\Lambda(\infty)$, it follows that $Ap_k = \mu_k p_k$. This completes the proof of Theorem 8.1.

Proof of Lemma. There exists a constant matrix T such that

$$TAT^{-1} = B$$

where B is a diagonal matrix with diagonal elements μ_j.

Let $S = \tilde{S}T$. Then it is required that

$$\begin{aligned} S(A+V)S^{-1} &= \tilde{S}T(A+V)T^{-1}\tilde{S}^{-1} \\ &= \tilde{S}(B + TVT^{-1})\tilde{S}^{-1} \\ &= \tilde{S}(B+\tilde{V})\tilde{S}^{-1} = \Lambda \end{aligned}$$

where $\tilde{V} = TVT^{-1}$. Because $\tilde{V}$ is linear in the elements of V it satisfies

$$\int_{t_0}^\infty |\tilde{V}'(t)|\,dt < \infty$$

and $\tilde{V}(\infty) = 0$.

Consider the matrix

$$M(\lambda,t) = B + \tilde{V}(t) - \lambda E$$

Thus $\det M(\lambda,t) = 0$ has roots $\lambda_j(t)$, where $\lambda_j(\infty) = \mu_j$. Denote the cofactor of the element $m_{ik}(\lambda,t)$ of $M(\lambda,t)$ by $C_{ik}(\lambda,t)$. Let

$$\tilde{s}_{ik}(t) = \frac{C_{ki}(\lambda_i(t),t)}{\prod_{j=1}^{n}{}' (\mu_j - \mu_i)} \tag{8.23}$$

where the prime on the product denotes that $j = i$ is omitted. Because $C_{ki}(\lambda_i(t),t)$ tends to the cofactor of the element in the kth row and ith column of $(B - \mu_i E)$ as $t \to \infty$, it follows that

$$\tilde{s}_{ik}(\infty) = \delta_{ik}$$

which is 1 if $i = k$, and 0 otherwise. Let the matrix with elements $\tilde{s}_{ik}(t)$ be $\tilde{S}(t)$. Then clearly

$$\tilde{S}(\infty) = E$$

Also

$$\sum_{j=1}^{n} C_{ji}(\lambda_i(t),t)[b_{jk} + \tilde{v}_{jk}(t) - \lambda_i(t)\delta_{jk}] = 0 \tag{8.24}$$

for $k = 1, \ldots, n$. For $i \neq k$ this is true because a determinant with two columns identical is zero. For $i = k$ it is true because $\lambda_i(t)$ is a characteristic root of $B + \tilde{V}$ (and thus also of $A + V$).

Thus (8.23) and (8.24) imply that

$$\tilde{S}(B + \tilde{V}) = \Lambda\tilde{S}$$

Because $\tilde{S}(\infty) = E$, clearly $\tilde{S}^{-1}$ exists for large t and

$$\tilde{S}(B + \tilde{V})\tilde{S}^{-1} = \Lambda$$

Finally

$$\int_{t_0}^{\infty} |\tilde{S}'(t)|\, dt < \infty \tag{8.25}$$

This follows from the fact that s'_{ik} is linear homogeneous in the elements $\tilde{v}'_{ik}$ and λ'_i. The former are absolutely integrable. Thus it remains only to show that the λ'_i are absolutely integrable. Let $F(\lambda,t) = \det M(\lambda,t)$. Then since $F(\lambda_i(t),t) = 0$,

$$\frac{\partial F}{\partial \lambda}(\lambda_i(t),t)\lambda'_i(t) + \frac{\partial F}{\partial t}(\lambda_i(t),t) = 0$$

Because the characterisic roots of B are distinct, $(\partial F/\partial\lambda)(\lambda_i(t),t)$ tends to a nonvanishing limit as $t \to \infty$. The term $(\partial F/\partial t)(\lambda_i(t),t)$ is linear homogeneous in $\tilde{v}'_{ij}$ and so is absolutely integrable. Thus λ'_i is absolutely integrable and the proof of (8.25) is completed.

Clearly $S = \tilde{S}T$ satisfies the lemma.

PROBLEMS

1. Let the matrix A and the vector b be integrable functions of t over $[a,b]$. Let

$$|A(t)| \leq k(t) \qquad |b(t)| \leq k(t)$$

where

$$\int_a^b k(t)\, dt < \infty$$

Let $\tau \in [a,b]$ and consider the initial-value problem

$$x' = A(t)x + b(t) \qquad x(\tau) = \xi$$

Prove that there is a unique solution φ over $[a,b]$ in the sense that $\varphi \in C$ and

$$\varphi(t) = \xi + \int_\tau^t A(s)\varphi(s)\,ds + \int_\tau^t b(s)\,ds$$

on $[a,b]$.

HINT: Use successive approximations. Let $\varphi_0(t) = \xi$ and

$$\varphi_{j+1}(t) = \xi + \int_\tau^t A(s)\varphi_j(s)\,ds + \int_\tau^t b(s)\,ds \qquad (j \geqq 0)$$

Prove that, if $\int_\tau^t k(s)\,ds = K(t)$, then

$$|\varphi_j(t) - \varphi_{j-1}(t)| \leqq (1 + |\xi|)\frac{|K(t)|^j}{j!}$$

so that $\{\varphi_j\}$ converges uniformly over $[a,b]$. If the above holds for all $b < \tilde{b}$, then the solution exists over $[a,\tilde{b})$. The case $\tilde{b} = \infty$ is allowed. A similar situation prevails at the left end point. For uniqueness, use Prob. 1, Chap. 1.

2. In this problem let the norm of a matrix A be defined by

$$|A| = \max_j \sum_{i=1}^n |a_{ij}|$$

Then $|A + B| \leqq |A| + |B|$ and $|AB| \leqq |A|\,|B|$. Let A be of class C on $[a,b]$. Product integration of $x' = A(t)x$ is defined as follows:

Divide $[a,b]$ into m parts $a = t_0 < t_1 < \cdots < t_m = b$. For given t, choose k so that $t_k < t \leqq t_{k+1}$. Let E be the unit matrix and

$$\Phi_m(t) = [E + (t - t_k)A(t_k)][E + (t_k - t_{k-1})A(t_{k-1})] \cdots [E + (t_1 - t_0)A(t_0)]$$

Clearly Φ_m is continuous and Φ_m' is piecewise continuous on $[a,b]$. If $|A(t)| \leqq K$, then

$$|E + (t_j - t_{j-1})A(t_{j-1})| \leqq 1 + (t_j - t_{j-1})K < e^{K(t_j - t_{j-1})}$$

Thus

$$|\Phi_m(t)| \leqq e^{K(t-a)} \leqq e^{K(b-a)}$$

From the definition of $\Phi_m(t)$,

$$\Phi_m'(t) = A(t_k)[E + (t - t_k)A(t_k)]^{-1}\Phi_m(t)$$

Show that

$$\Phi_m' = A(t)\Phi_m + J_m(t)$$

where, given any $\epsilon > 0$, m can be chosen large enough so that $|J_m(t)| \leqq \epsilon$. Thus $\Phi_m(a) = E$ and Φ_m is an ϵ-approximate solution of $x' = A(t)x$. Use this to prove the existence of the fundamental solution Φ, $\Phi(a) = E$.

3. Let the matrix A be continuous over $[0, \infty]$. A fundamental solution Φ of $x' = A(t)x$ is uniformly bounded over $[0, \infty]$ and

$$\liminf_{t\to\infty} \Re \int_0^t \operatorname{tr} A(s)\,ds > -\infty$$

Prove that Φ^{-1} is uniformly bounded over $[0, \infty]$. Moreover, prove that no solution φ not identically zero can satisfy $\varphi(t) \to 0$ as $t \to \infty$.

HINT: Use (1.8).

4. Consider the differential equation of Prob. 3 and also the differential equation $x' = B(t)x$ ($B \in C$ on $(0, \infty)$), a solution of which will be designated by ψ. Suppose

$$\int_0^\infty |A(t) - B(t)|\, dt < \infty$$

Prove that ψ is bounded over $(0, \infty)$. (Here φ is a solution of $x' = Ax$.)

HINT: Use $\psi(t) = \varphi(t) + \int_{t_0}^t \Phi(t)\Phi^{-1}(s)(B(s) - A(s))\psi(s)\, ds$ and Prob. 1, Chap. 1.

5. Show in Prob. 4 that corresponding to any given φ there exists a unique ψ such that $\varphi(t) - \psi(t) \to 0$ as $t \to \infty$.

HINT: Use $\psi(t) = \varphi(t) - \int_t^\infty \Phi(t)\Phi^{-1}(s)(B(s) - A(s))\psi(s)\, ds$.

6. If $\int_0^\infty |B(t)|\, dt < \infty$, then any solution of $x' = B(t)x$ not identically zero tends to a limit different from zero as $t \to \infty$. Moreover, given any constant vector c, there is a unique solution ψ which tends to c as $t \to \infty$.

HINT: Use $A \equiv 0$ and $\Phi(t) = E$.

7. Let a_1, a_2 be continuous and periodic of period ω. Let φ_1 and φ_2 be solutions of $x'' + a_1(t)x' + a_2(t)x = 0$, where $\varphi_1(0) = 1$, $\varphi_1'(0) = 0$, $\varphi_2(0) = 0$, $\varphi_2'(0) = 1$. Use the system formulation and show that the multipliers (or characteristic roots) are solutions of $\lambda^2 - A\lambda + B = 0$, where $A = \varphi_1(\omega) + \varphi_2'(\omega)$ and $B = \exp\left[-\int_0^\omega a_1(t)\, dt\right]$.

8. Let a and b be real constants and p a real continuous function of t of period ω. Consider $x'' + [a + bp(t)]x = 0$. Let φ_1 and φ_2 be defined as in Prob. 7. Let $F(a,b) = \varphi_1(\omega) + \varphi_2'(\omega)$. Show that F is an entire function of (a,b). Show that if $-2 < F(a,b) < 2$ then the multipliers are complex conjugate and of magnitude 1 and that all solutions are uniformly bounded, together with their first derivatives, on $(-\infty, \infty)$.

Show that if $F(a,b) > 2$ or $F(a,b) < -2$ then no solution is uniformly bounded on $(-\infty, \infty)$.

9. If in the previous problem $F(a,b) = 2$, show that there is at least one solution of period ω, and that if $F(a,b) = -2$ there is at least one solution of period 2ω.

10. If in Prob. 8, $a \neq n^2$ for any integer n, $a \geqq 0$, $b = 0$, and $\omega = \pi$, then show that $-2 < F(a,0) < 2$. From the continuity of $F(a,b)$ show that if $a \neq n^2$ and b is sufficiently small, all solutions are uniformly bounded on $(-\infty, \infty)$.

11. In Prob. 8 let $p(t) = \cos 2t$ and consider the case where a is near $4n^2$ and b is small. This may be formulated as

$$x'' + [4n^2 + \gamma\mu + \mu \cos 2t]x = 0$$

where γ is real and μ is a small real parameter. Determine the behavior of the curves on which $F(a,b) = F(4n^2 + \gamma\mu, \mu) = 2$ in the neighborhood of $(\mu = 0, \gamma = 0)$.

HINT: In vector form, $\hat{x}' = (A + \mu P(t))\hat{x}$, where

$$A = \begin{pmatrix} 0 & 1 \\ -4n^2 & 0 \end{pmatrix} \qquad P(t) = \begin{pmatrix} 0 & 0 \\ -\gamma - \cos 2t & 0 \end{pmatrix}$$

The fundamental solution Φ which is E at $t = 0$ is an entire function of μ and therefore

$$\Phi(t,\mu) = e^{At} + \mu\Phi_1(t) + \mu^2\Phi_2(t) + \cdots$$

where $F(4n^2 + \gamma\mu, \mu) = \operatorname{tr} \Phi(\pi,\mu)$. Show that

$$\Phi_j(t) = \int_0^t e^{A(t-s)}P(s)\Phi_{j-1}(s)\, ds$$

where

$$\Phi_1(t) = e^{At} \int_0^t e^{-As}P(s)e^{As}\, ds$$

At $t = \pi$

$$\Phi_1(\pi) = \int_0^\pi e^{-As}P(s)e^{As}\, ds$$

and therefore

$$\operatorname{tr} \Phi_1(\pi) = \int_0^\pi \operatorname{tr} [e^{-As}P(s)e^{As}]\, ds$$

$$= \int_0^\pi \operatorname{tr} P(s)\, ds = 0$$

Thus

$$F(4n^2 + \gamma\mu, \mu) = 2 + \mu^2 \operatorname{tr} \Phi_2(\pi) + \mu^3 \operatorname{tr} \Phi_3(\pi) + \cdots$$

and from the behavior of $\operatorname{tr} \Phi_2(\pi)$ obtain the result for small μ.

12. Give a direct proof of (6.5) by showing that $W' = (-a_1/a_0)W$ with the use of (6.4) and $L_n x = 0$.

13. Let A be a constant square matrix. Give the analogue of the proof of Theorem 6.5 for the system $x' = Ax$.

HINT: Let p be a constant vector. Then

$$\left(E\frac{d}{dt} - A\right)(e^{\lambda t}p) = e^{\lambda t}(E\lambda - A)p$$

Taking the partial derivative with respect to λ,

$$\left(E\frac{d}{dt} - A\right) te^{\lambda t}p = te^{\lambda t}(E\lambda - A)p + e^{\lambda t}p$$

Using the above relations, show that if

$$Ap_1 = \lambda_1 p_1 \qquad \text{and} \qquad Ap_2 = \lambda_1 p_2 + p_1$$

then $e^{\lambda_1 t}p_1$ and $e^{\lambda_1 t}p_2 + te^{\lambda_1 t}p_1$ are solutions. Generalize the procedure to include the general result of Sec. 4.

14. Let $L_n x = x^{(n)} + a_1 x^{(n-1)} + \cdots + a_n x$, where the a_j are periodic functions of period ω on $(-\infty, \infty)$. Find the form of the solutions over $(-\infty, \infty)$.

15. If φ_1 and φ_2 are solutions of $x'' + a_1(t)x' + a_2(t)x = 0$, show that

$$\varphi_1(t)\varphi_2'(t) - \varphi_2(t)\varphi_1'(t) = c \exp\left[-\int^t a_1(s)\, ds\right]$$

where c is a constant. If φ_1 is a solution, show that

$$\varphi_1(t) \int^t \exp\left[-\int^s a_1(u)\, du\right] \frac{ds}{\varphi_1^2(s)}$$

is an independent solution on an interval where $\varphi_1(t) \neq 0$.

16. In $x'' + a_1(t)x' + a_2(t)x = 0$, make the change of variable $s = F(t)$, where $F'(t) = \exp\left[-\int^t a_1(s)\, ds\right]$ and let $t = G(s)$. Show that this leads to

$$\frac{d^2x}{ds^2} + g(s)x = 0 \qquad \text{where } g(s) \text{ is } \frac{a_2(t)}{(F'(t))^2}$$

evaluated at $t = G(s)$.

17. Let $x = y \exp\left(-\frac{1}{2}\int^t a_1(\tau)\, d\tau\right)$ in the equation in line 1 of Prob. 16. Show that it becomes

$$\frac{d^2y}{dt^2} + \left(a_2 - \frac{a_1^2}{4} - \frac{a_1'}{2}\right) y = 0$$

18. Let $a(t) > 0$ and let $a \in C^2$. Consider $x'' + a^2(t)x = 0$. Let $s = F(t)$, where $F'(t) = a(t)$, and let $t = G(s)$. Then the equation becomes

$$\frac{d^2x}{ds^2} + a_1(s)\frac{dx}{ds} + x = 0 \qquad a_1(s) = \frac{a'(G(s))}{a^2(G(s))}$$

Let $x = y \exp\left(-\frac{1}{2}\int^s a_1(s)\, ds\right)$ and show the above becomes

$$\frac{d^2y}{ds^2} + [1 + b(s)]y = 0$$

where $b(s)$ is

$$\frac{3}{4}\frac{a'^2(t)}{a^4(t)} - \frac{1}{2}\frac{a''(t)}{a^3(t)} \qquad \text{at } t = G(s)$$

If b is of class C^2, the above may be repeated. If $a^2(t)$ is a polynomial in t, note that $b(s) \to 0$ as $s \to \infty$ and, indeed, that b is of bounded variation so that the result of Theorem 8.1 applies.

19. Show that the conjugates of $W_k(\varphi_1, \ldots, \varphi_n)/W(\varphi_1, \ldots, \varphi_n)$, as defined in Theorem 6.4, are solutions of the adjoint equation $L_n^+x = 0$.

HINT: Use the system formulation.

20. Obtain the result (6.15) by using the "variation-of-constants" procedure directly. That is, assume

$$\psi = c_1\varphi_1 + \cdots + c_n\varphi_n$$

where the c_j are to be regarded as functions of t subject to the requirement

$$\sum_{j=1}^{n} c_j'\varphi_j^{(i)} = 0 \qquad (i = 0, 1, \ldots, n-2)$$

where $\varphi_j^{(i)}$ denotes the ith derivative of φ_j.

21. Let $f = f(t,s)$ denote the solution of $L_nx = 0$ which satisfies the initial conditions $x^{(j)}(s) = 0$, $j = 0, 1, \ldots, n-2$, and $x^{(n-1)}(s) = 1/a_0(s)$. Show that

$$\int_\tau^t f(t,s)b(s)\, ds$$

is a solution of $L_nx = b(t)$ which vanishes with its first $n-1$ derivatives at $t = \tau$ Comparing this solution with (6.15), show that

$$f(t,s) = \sum_{k=1}^{n} \varphi_k(t) \frac{W_k(\varphi_1, \ldots, \varphi_n)(s)}{W(\varphi_1, \ldots, \varphi_n)(s)a_0(s)}$$

22. Let $a_j \in C^{n-j}[a,b]$ so that L_n^+ is defined. Let $K(t,s) = f(t,s)$, as defined in Prob. 21 for $s < t$ and $K = 0$ for $s > t$. Show that K is of class $C^{n-2}[a,b]$ as a function of (s,t) and that $\partial^{n-1}K/\partial s^{n-1}$ has a simple discontinuity, $(-1)^n/a_0(t)$, at $s = t$ but is continuous for $a \leqq s \leqq t \leqq b$ and $a \leqq t \leqq s \leqq b$. Also show that, as a function of s, $\bar{K}$ is a solution of $L_n^+ x = 0$ for $s < t$.

HINT: Let $H = H(t,s)$ be the solution of $L_n^+ x = 0$ for $t \leqq s$ which satisfies

$$\frac{\partial^k H}{\partial t^k} = 0$$

at $t = s$ for $k = 0, \ldots, n-2$ and let $\partial^{n-1}H/\partial t^{n-1} = (-1)^{n-1}/\bar{a}_0(s)$ at $t = s$. Let $H = 0$ for $t > s$. Then for any $p, q \in C[a,b]$ the functions u and v defined by

$$u(t) = \int_a^t K(t,s)p(s)\,ds \qquad v(t) = \int_t^b H(t,s)q(s)\,ds$$

satisfy $L_n u = p$ and $L_n^+ v = q$, respectively, and $u^{(j)}(a) = v^{(j)}(b) = 0$, $j = 0, \ldots, n-1$. Thus by Green's formula,

$$\int_a^b (p\bar{v} - u\bar{q})\,dt = [uv](b) - [uv](a)$$

Since $u^{(j)}(a) = v^{(j)}(b) = 0$, both terms on the right above vanish and therefore

$$\int_a^b \bar{q}(t)\,dt \int_a^t (K(t,s) - \bar{H}(s,t))p(s)\,ds = 0$$

Since this holds for all p and q, it follows that $K(t,s) = \bar{H}(s,t)$ and the differentiability of K with respect to s follows from that of H with respect to t.

23. If the form (6.12) is written as

$$[uv] = \sum_{j,k=0}^{n-1} B_{jk} u^{(k)} \bar{v}^{(j)}$$

determine the form of the matrix $B = (B_{jk})$ and prove that it is nonsingular for all $t \in I$. In fact, compute its determinant.

24. If u is a solution of $L_n x = 0$ and v is a solution of $L_n^+ x = 0$, show that $[uv](t)$ is a constant $[uv]$, independent of $t \in I$. Let $\varphi_1, \ldots, \varphi_n$ be a fundamental set for $L_n x = 0$ on I, and let $\psi_1, \ldots, \psi_n$ be a similar set for $L_n^+ x = 0$ on I. Show that the matrix $S = (s_{jk}) = ([\varphi_j \psi_k])$ is nonsingular on I. Let $S^{-1} = (s_{jk}^{-1})$ be the inverse matrix to S. Define K by

$$K(t,s) = \sum_{j,k=1}^{n} s_{jk}^{-1} \varphi_k(t) \bar{\psi}_j(s) \qquad (s \leqq t)$$

Prove that the function u given by

$$u(t) = \int_\tau^t K(t,s)b(s)\,ds \qquad (\tau, t \in I)$$

is a solution of $L_n x = b$ which vanishes with its first $n-1$ derivatives at τ. Compare this result with Probs. 21 and 22.

25. If $L_n = L_n^+$, prove that $[uv](t)$ is skew Hermitian, that is,

$$[uv](t) = -\overline{[vu](t)}$$

What does this imply concerning the matrix B in Prob. 23, and the matrix S in Prob. 24?

26. Let P_j be polynomials and λ_j be constants and

$$f(t) = \sum_{j=1}^{m} P_j(t)e^{\lambda_j t}$$

Let $m \geqq 1$ and $\lambda_j \neq \lambda_k$, $j \neq k$, and let none of the P_j vanish identically. If

$$\sigma = \max (\Re \lambda_j)$$

then show that

$$\limsup_{t \to \infty} (e^{-\sigma t}|f(t)|) > 0$$

REMARK: This proves the linear independence of the terms $P_j(t)e^{\lambda_j t}$.

HINT: *Case* 1. Let the P_j all be constants and the $\lambda_j = i\mu_j$, where the μ_j are real so that

$$f(t) = \sum_{j=1}^{m} c_j e^{i\mu_j t}$$

From

$$\lim_{T \to \infty} \frac{1}{T} \int_0^T f(t)e^{-i\mu_1 t}\, dt = c_1$$

prove that $\limsup_{t \to \infty} |f(t)| > 0$.

Case 2. Let $\lambda_j = i\mu_j$ as above and the highest power of t in any polynomial P_j be M. Then

$$f(t) = t^M f_1(t) + t^{M-1}f_2(t) + \cdots + f_M(t)$$

where the f_j are as in Case 1 above and f_1 does not vanish identically Thus, for large t,

$$t^{-M}f(t) = f_1(t) + 0\left(\frac{1}{t}\right)$$

and by Case 1

$$\limsup_{t \to \infty} (t^{-M}|f(t)|) > 0$$

General Case. Here

$$f(t) = e^{\sigma_1 t}f_1(t) + e^{\sigma_2 t}f_2(t) + \cdots + e^{\sigma_p t}f_p(t)$$

where $\sigma_1 > \sigma_2 > \cdots > \sigma_p$ and the f_j are as in Case 2 and f_1 is not identically zero. Clearly,

$$e^{-\sigma_1 t}f(t) = f_1(t) + 0(t^Q e^{-(\sigma_1 - \sigma_2)t})$$

for some constant Q. Thus, by Case 2,

$$\limsup_{t \to \infty} |e^{-\sigma_1 t}f(t)| > 0$$

27. Consider the system of linear equations

$$w^{(n)} + A_1 w^{(n-1)} + \cdots + A_n w = 0$$

where w is an m-dimensional vector and the A_i are m-by-m constant matrices. Define a fundamental matrix for this equation and compute one such.

28. Let f be integrable and let

$$\int_1^\infty t|f(t)|\, dt < \infty$$

Prove that $x'' + f(t)x = 0$ has a solution φ such that

$$\lim_{t\to\infty} \varphi(t) = 1 \qquad \lim_{t\to\infty} \varphi'(t) = 0$$

Prove that there is a solution ψ such that

$$\lim_{t\to\infty} \frac{\psi(t)}{t} = 1 \qquad \lim_{t\to\infty} \psi'(t) = 1$$

HINT: Use successive approximations on

$$\varphi(t) = 1 + \int_t^\infty (t - s)f(s)\varphi(s)\,ds$$

and

$$\psi(t) = t + \int_a^t sf(s)\psi(s)\,ds + t\int_t^\infty f(s)\psi(s)\,ds$$

where a is chosen so that $\int_a^\infty t|f(t)|\,dt < \frac{1}{2}$. (The relation $\varphi\psi' - \varphi'\psi = 1$ can be used as an alternative way to get ψ, once φ has been shown to exist.)

This problem is a special case of Prob. 35.

29. Let A be a constant matrix and R an integrable matrix such that

$$\int_1^\infty |R(t)|\,dt < \infty$$

It is assumed that the canonical matrix J similar to A is diagonal, that is, $J = J_0$. (In particular, this is always the case if the characteristic roots of A are distinct.) If λ_j is a characteristic root of A and p_j is the characteristic vector, so that $Ap_j = \lambda_j p_j$, then prove that

$$x' = Ax + R(t)x$$

has a solution φ_j such that

$$\lim_{t\to\infty} \varphi_j(t)e^{-\lambda_j t} = p_j \qquad (j = 1, 2, \ldots, n)$$

[In other words, for large t the solution acts like the corresponding one for the case $R(t) \equiv 0$.]

HINT: Let j be fixed, let $\Re\lambda_j = \sigma$, and let $e^{At} = Y_1(t) + Y_2(t)$, where the elements of $Y_1(t)$ are sums of terms of the form $e^{\lambda_k t}$, $\Re\lambda_k < \sigma$, and $Y_2(t)$ contains only terms of the form $e^{\lambda_k t}$, $\Re\lambda_k \geqq \sigma$. Then there exists $\delta > 0$ and constants K_1 and K_2 such that

$$|Y_1(t)| \leqq K_1 e^{(\sigma-\delta)t} \qquad (t \geqq 0)$$
$$|Y_2(t)| \leqq K_2 e^{\sigma t} \qquad (t \leqq 0)$$

Let $\psi_0(t) = e^{\lambda_j t}p_j$ and

$$\psi_{l+1}(t) = e^{\lambda_j t}p_j + \int_a^t Y_1(t - s)R(s)\psi_l(s)\,ds - \int_t^\infty Y_2(t - s)R(s)\psi_l(s)\,ds$$

where a is chosen so large that

$$(K_1 + K_2)\int_a^\infty |R(t)|\,dt < \tfrac{1}{2}$$

Let $|\psi_0(t)| \leqq K_0 e^{\sigma t}$, $t \geqq 0$. Then show

$$|\psi_{l+1}(t) - \psi_l(t)| \leqq \frac{K_0 e^{\sigma t}}{2^{l+1}}$$

thus proving the existence of a limit function which is denoted by φ_j and satisfies

$$|\varphi_j(t)| \leqq 2K_0e^{\sigma t}$$

$$\varphi_j(t) = e^{\lambda_j t}p_j + \int_a^t Y_1(t-s)R(s)\varphi_j(s)\,ds - \int_t^\infty Y_2(t-s)R(s)\varphi_j(s)\,ds$$

From this,

$$e^{-\sigma t}|\varphi_j(t) - e^{\lambda_j t}p_j| \leqq 2K_0K_1\int_a^t e^{-\delta(t-s)}|R(s)|\,ds + 2K_0K_2\int_t^\infty |R(s)|\,ds$$

$$\leqq 2K_0K_1e^{-\frac{1}{2}\delta t}\int_a^{\frac{1}{2}t} |R(s)|\,ds + 2K_0(K_1+K_2)\int_{\frac{1}{2}t}^\infty |R(s)|\,ds$$

which gives the result as $t \to \infty$.

30. Let $x'' + (1 + r(t))x = 0$, where $\int_1^\infty |r(t)|\,dt < \infty$. Show that the equation has solutions φ_1 and φ_2 such that

$$\lim_{t\to\infty} (\varphi_1(t) - e^{it}) = 0 \qquad \lim_{t\to\infty} (\varphi_1'(t) - ie^{it}) = 0$$

and similarly for φ_2 with i replaced by $-i$.

31. Formulate and prove a result similar to the above for $x'' - (1 + r(t))x = 0$.

32. Let $L_nx = x^{(n)} + [a_1 + r_1(t)]x^{(n-1)} + \cdots + [a_n + r_n(t)]x = 0$, where the a_k are constants and

$$\int_1^\infty |r_k(t)|\,dt < \infty \qquad (k = 1, \ldots, n)$$

Let the roots of $\lambda^n + a_1\lambda^{n-1} + \cdots + a_n = 0$ be distinct and let λ_j be a root. Then $L_nx = 0$ has a solution φ_j such that

$$\lim_{t\to\infty} (\varphi_j^{(k)}(t) - \lambda_j^k e^{\lambda_j t})e^{-\lambda_j t} = 0 \qquad (k = 0, 1, \ldots, n-1)$$

for $j = 1, 2, \ldots, n$.

HINT: Use Prob. 29.

33. Let A be continuous and periodic of period ω,

$$x' = [A(t) + R(t)]x$$

where R is as in Prob. 29. Suppose the equation $y' = A(t)y$ has n independent solutions of the form $e^{\rho_j t}p_j(t)$, where the p_j have period ω. Then prove that the given equation has n solutions φ_j such that

$$\lim_{t\to\infty} [\varphi_j(t)e^{-\rho_j t} - p_j(t)] = 0 \qquad (j = 1, \ldots, n)$$

HINT: The equation $y' = A(t)y$ has as a fundamental solution $P(t)e^{Bt}$, where B is in diagonal form, $B = J_0$. Clearly, $P' + PB = AP$. Let $x = P(t)z$. Then the equation for z becomes

$$z' = Bz + P^{-1}RPz$$

Now Prob. 29 may be used since B is constant and this yields the required result. Note that $u' = Bu$ has as solutions $e^{\rho_j t}e_j$, where e_j is the constant vector with jth row 1 and all other rows 0.

34. Formulate and prove a result similar to the above for the equation

$$L_nx = x^{(n)} + [a_1(t) + r_1(t)]x^{(n-1)} + \cdots + [a_n(t) + r_n(t)]x = 0$$

where the a_j are periodic of period ω.

35. Consider the case $x' = Ax + R(t)x$, where A is constant but where now the canonical form of A has, in the terminology of Theorem 1.1, submatrices J_k, $k \geqq 1$, and where $r + 1$ is the maximum number of rows in any matrix J_k, $k \geqq 1$. Then no polynomial multiplying an exponential term in any element of e^{At} is of degree higher than r. Here the case $r \geqq 1$ is considered. (For $r = 0$, see Prob. 29.) Assume that $\int_1^\infty t^r|R(t)|\,dt < \infty$. Let λ_j be a characteristic root of A and let $y' = Ay$ have a solution of the form

$$e^{\lambda_j t}t^k c + O(e^{\lambda_j t}t^{k-1})$$

where c is a vector. Clearly, $0 \leqq k \leqq r$. Then show that $x' = Ax + R(t)x$ has a solution φ such that

$$\lim_{t\to\infty}\,[\varphi(t)e^{-\lambda_j t}t^{-k} - c] = 0$$

HINT: Let $\Re\lambda_j = \sigma$. The elements of $e^{A(t-s)}$ are sums of terms of the form $e^{\lambda_p(t-s)}t^l s^m$, $0 \leqq l + m \leqq r$. Let

$$e^{A(t-s)} = Y_1(t,s) + Y_2(t,s)$$

where

$$|Y_1(t,s)| \leqq K_1 e^{\sigma(t-s)}t^{k-1}s^{r-k+1} \qquad (t \geqq s \geqq 1)$$
$$|Y_2(t,s)| \leqq K_2 e^{\sigma(t-s)}t^k s^{r-k} \qquad (s \geqq t \geqq 1)$$

That is, $Y_1(t,s)$ has all terms for which the exponential factor $e^{\lambda_p t}$ satisfies $\Re\lambda_p < \sigma$. If $\Re\lambda_p = \sigma$, then Y_1 has the terms which have as factor a power of t less than k. The proof is analogous to the case $r = 0$. In the final step, $\int_a^t$ involving Y_1 is now written as $\int_a^{t^{\frac{1}{2}}} + \int_{t^{\frac{1}{2}}}^t$.

36. Formulate and prove the analogue of the above result for an equation of the nth order, $L_n x = 0$.

37. Formulate and prove the analogue of Prob. 35 for $r \geqq 1$, for the case where A is replaced by a periodic matrix $A(t)$.

38. Formulate and prove the analogue of the above result for the nth-order equation $L_n x = 0$.

39. Let A be the n-by-n matrix $A = \lambda E + Z$, where $Z = (z_{ij})$ and $z_{ij} = 1$ if $j = i + 1$, and $z_{ij} = 0$ otherwise. Show that A is similar to a matrix $B = \lambda E + \gamma Z$, where $\gamma \neq 0$.

HINT: Let $P = (p_{ij})$, where $p_{ij} = \gamma^{i-1}\delta_{ij}$, and prove $B = P^{-1}AP$.

40. Let A be a real n-by-n matrix. Prove that there exists a *real* nonsingular matrix P such that $\tilde{A} = P^{-1}AP$ has the real canonical form consisting of real square matrices $A_1, \ldots, A_k, B_1, \ldots, B_m$ down the main diagonal. Each A_j has the form

$$A_j = \begin{pmatrix} S_j & 0_2 & \cdots & 0_2 & 0_2 \\ E_2 & S_j & \cdots & 0_2 & 0_2 \\ 0_2 & E_2 & \cdots & 0_2 & 0_2 \\ \cdot & \cdot & \cdots & \cdot & \cdot \\ 0_2 & 0_2 & \cdots & E_2 & S_j \end{pmatrix}$$

where 0_2 is the 2-by-2 zero matrix, E_2 the 2-by-2 unit matrix, and

$$S_j = \begin{pmatrix} \alpha_j & -\beta_j \\ \beta_j & \alpha_j \end{pmatrix}$$

The B_j have the form

$$B_j = \begin{pmatrix} \lambda_j & 0 & \cdots & 0 & 0 \\ 1 & \lambda_j & \cdots & 0 & 0 \\ 0 & 1 & \cdots & 0 & 0 \\ \cdot & \cdot & \cdots & \cdot & \cdot \\ 0 & 0 & \cdots & 1 & \lambda_j \end{pmatrix}$$

41. If C is a real nonsingular n-by-n matrix, prove there exists a real matrix A such that $e^A = C^2$.

HINT: Use Prob. 40, and consider the two cases $\lambda_j > 0$, $\lambda_j < 0$. Note that

$$S_j = \exp \begin{pmatrix} \log (\alpha_j^2 + \beta_j^2)^{\frac{1}{2}} & -\tan^{-1}\left(\dfrac{\beta_j}{\alpha_j}\right) \\ \tan^{-1}\left(\dfrac{\beta_j}{\alpha_j}\right) & \log (\alpha_j^2 + \beta_j^2)^{\frac{1}{2}} \end{pmatrix}$$

CHAPTER 4

LINEAR SYSTEMS WITH ISOLATED SINGULARITIES: SINGULARITIES OF THE FIRST KIND

1. Introduction

In this and the next chapter the linear system

$$w' = A(z)w \qquad (z \text{ complex}) \tag{1.1}$$

will be analyzed, where A is an n-by-n (complex-valued) matrix with at most an isolated singularity at some point z_0, but is otherwise single-valued and analytic near z_0. If A is assumed to have only a pole at z_0, certain very specific results can be derived concerning the nature of a solution matrix Φ of (1.1) near z_0. However, there is one general result which gives a qualitative picture of Φ even when A has an arbitrary isolated singularity at z_0.

Suppose the domain in which A is analytic and single-valued is $0 < |z - z_0| < a$, where a is some positive constant. This domain is not simply connected, and because of this the solutions of (1.1) need not be single-valued. For example, consider the equation $w' = w/(2z)$, where w is one-dimensional. Then $(wz^{-\frac{1}{2}})' = 0$ or $w = cz^{\frac{1}{2}}$, where c is a constant. Thus the solution, except for the case $c = 0$, is not single-valued for $0 < |z| < a$.

The problem can again be considered in a simply connected domain if the domain is allowed to be many-sheeted. Let $z - z_0 = \rho e^{i\theta}$, where $\rho \geqq 0$ and θ is real. Let the domain D be given by

$$D: \qquad 0 < \rho < a, \ -\infty < \theta < \infty$$

This domain is simply connected. The method of successive approximations, as stated in Sec. 7, Chap. 3, leads readily to the existence of an analytic fundamental matrix in D for (1.1).

An alternative procedure is to set $z - z_0 = e^{\zeta}$. Then (1.1) becomes

$$\frac{dw}{d\zeta} = B(\zeta)w \qquad B(\zeta) = e^{\zeta}A(z_0 + e^{\zeta})$$

Clearly B is analytic for $\zeta \varepsilon \tilde{D}$, where $\tilde{D}$ is the half plane, $-\infty < \Re\zeta <$

$\log a$. Since $\tilde{D}$ is simply connected, there exists a fundamental matrix solution Ψ analytic for $\zeta \in \tilde{D}$. Thus $\Phi(z) = \Psi(\log (z - z_0))$ is an analytic fundamental matrix for (1.1) for $z \in D$. Since $\log (z - z_0)$ is not single-valued, Φ need not be a single-valued function of z.

If M is any matrix of complex numbers, let the exponential matrix z^M be defined by

$$z^M = e^{(\log z)M} \tag{1.2}$$

Note that for $z \neq 0$, z^M is nonsingular for all M, and $(z^M)^{-1} = z^{-M}$.

Theorem 1.1. *If A in (1.1) is single-valued and analytic in a punctured vicinity of z_0, $0 < |z - z_0| < a$, then every fundamental matrix Φ of (1.1) has the form*

$$\Phi(z) = S(z)(z - z_0)^P \qquad (0 < |z - z_0| < a) \tag{1.3}$$

where S is single-valued, analytic on $0 < |z - z_0| < a$, and P is a constant matrix.

Proof. The proof is essentially the same as that of Theorem 5.1, Chap. 3, and will be given for the case $z_0 = 0$. Consider a fundamental matrix Φ on the infinitely sheeted domain D described above. On D,

$$\Phi'(z) = A(z)\Phi(z)$$

Since $A(ze^{2\pi i}) = A(z)$, it follows that

$$\Phi'(ze^{2\pi i}) = A(z)\Phi(ze^{2\pi i})$$

Therefore $\tilde{\Phi}$, where $\tilde{\Phi}(z) = \Phi(ze^{2\pi i})$, is a fundamental matrix, and hence

$$\Phi(ze^{2\pi i}) = \Phi(z)C \tag{1.4}$$

where C is a constant nonsingular matrix. Since C is nonsingular, there exists a constant matrix P so that

$$C = e^{2\pi iP} \tag{1.5}$$

(note that P is not unique), and from (1.4) it follows that

$$\Phi(ze^{2\pi i}) = \Phi(z)e^{2\pi iP} \tag{1.6}$$

Now let S be defined by the requirement that

$$\Phi(z) = S(z)z^P \qquad (0 < |z| < a) \tag{1.7}$$

Clearly S is analytic for $0 < |z| < a$; it will be shown that it is also single-valued there.

On the one hand, from (1.7),

$$\Phi(ze^{2\pi i}) = S(ze^{2\pi i})(ze^{2\pi i})^P = S(ze^{2\pi i})z^Pe^{2\pi iP} \tag{1.8}$$

and on the other, from (1.6),

$$\Phi(ze^{2\pi i}) = S(z)z^Pe^{2\pi iP} \tag{1.9}$$

A comparison of (1.8) and (1.9) shows that $S(ze^{2\pi i}) = S(z)$, and therefore S is single-valued on $0 < |z| < a$, thus proving the theorem.

There is a fundamental matrix in which P is replaced by its canonical form J, where J and P are related by $PT = TJ$ for some nonsingular constant matrix T. Clearly $S(z - z_0)^P T$ is also a fundamental matrix and is equal to $STT^{-1}(z - z_0)^P T = ST(z - z_0)^J$. Since T is constant, $U = ST$ is analytic and single-valued for $0 < |z - z_0| < a$. The explicit form of $(z - z_0)^J$ is given by (4.7), (4.8), and (4.9), Chap. 3, if t is there replaced by $\log (z - z_0)$. If the column vectors of U are denoted by u_j, $j = 1, \ldots, n$, then the u_j are analytic and single-valued for $0 < |z - z_0| < a$, and the columns φ_j of the fundamental matrix $U(z - z_0)^J$ are given, much as below (4.10), Chap. 3, by

$$\begin{aligned}
\varphi_j(z) &= (z - z_0)^{\lambda_j} u_j(z) \qquad (j = 1, 2, \ldots, q) \\
\varphi_{q+1}(z) &= (z - z_0)^{\lambda_{q+1}} u_{q+1}(z) \\
\varphi_{q+2}(z) &= (z - z_0)^{\lambda_{q+1}} [u_{q+1}(z) \log (z - z_0) + u_{q+2}(z)] \\
&\;\;\vdots \\
\varphi_{q+r_1}(z) &= (z - z_0)^{\lambda_{q+1}} \left[\frac{u_{q+1}(z)}{(r_1 - 1)!} \log^{r_1 - 1} (z - z_0) + \cdots \right. \\
&\qquad \left. + u_{q+r_1}(z) \right] \qquad (1.10) \\
\varphi_{q+r_1+1}(z) &= (z - z_0)^{\lambda_{q+2}} u_{q+r_1+1}(z) \\
&\;\;\vdots \\
\varphi_n(z) &= (z - z_0)^{\lambda_{q+s}} \left[\frac{u_{n-r_s+1}(z)}{(r_s - 1)!} \log^{r_s - 1} (z - z_0) + \cdots + u_n(z) \right]
\end{aligned}$$

In any case corresponding to any characteristic root λ_i of P, there is always at least one vector solution

$$(z - z_0)^{\lambda_i} u \qquad (1.11)$$

where u is analytic and single-valued on $0 < |z - z_0| < a$.

As in Theorem 7.3, Chap. 1,

$$(\det \Phi)' = (\det \Phi)(\operatorname{tr} A) \qquad (1.12)$$

Since

$$\det \Phi(z) = \det S(z) \det (z - z_0)^P = \det S(z)(z - z_0)^{\operatorname{tr} P}$$

there follows

$$\frac{(\det S)'}{\det S} + \frac{1}{z - z_0} \operatorname{tr} P = \operatorname{tr} A$$

Integrating around a circle Γ with its center at z_0 and radius less than a,

$$m + \operatorname{tr} P = \frac{1}{2\pi i} \int_\Gamma \operatorname{tr} A(z)\, dz \tag{1.13}$$

where m is an integer. If $\det S(z_0) \neq 0$ or ∞, then $m = 0$.

From the integration of (1.12) there results

$$\det \Phi(z) = \det \Phi(z_1) \exp \left(\int_{z_1}^{z} \operatorname{tr} A(\zeta)\, d\zeta \right) \tag{1.14}$$

2. Classification of Singularities

If A has a singularity at z_0, then z_0 is called a *singular point* for the system

$$w' = A(z)w \tag{2.1}$$

If A has at most a pole at z_0 (that is, either A is analytic at z_0, or has a pole there), but is analytic for $0 < |z - z_0| < a$, $a > 0$, then A may be written as

$$A(z) = (z - z_0)^{-\mu-1}\tilde{A}(z) \tag{2.2}$$

where μ is an integer, $\tilde{A}$ analytic for $|z - z_0| < a$, $a > 0$, and $\tilde{A}(z_0) \neq 0$. When $\mu \leqq -1$, it is clear that A is then analytic at z_0 and hence every fundamental matrix of (2.1) is analytic for $|z - z_0| < a$. Because of this, if $\mu \leqq -1$, the point z_0 is called an analytic point for (2.1). If $\mu \geqq 0$, the integer μ is called (after Poincaré) the *rank* of the singularity. It turns out that there is a significant difference between the cases $\mu = 0$ and $\mu \geqq 1$. Therefore, according as $\mu = 0$, or $\mu \geqq 1$, the point z_0 will be called a *singular point of the first kind*, or a *singular point of the second kind*, for (2.1). The case where $z_0 = \infty$ will be treated in Sec. 6.

The above classification of the systems (2.1), (2.2) is without regard to the nature of the solution matrices of (2.1) at z_0. From Sec. 1 it follows that any fundamental matrix Φ of (2.1), where A has an isolated singularity at z_0, is of the form $\Phi(z) = S(z)(z - z_0)^P$, where S is single-valued, analytic for $0 < |z - z_0| < a$, and P is a constant matrix. If S has at most a pole at z_0, then z_0 is called a *regular singular point* for (2.1); otherwise z_0 is called an *irregular singular point* for (2.1). These names are not very suggestive, but they have been in common usage, and therefore will be retained here. If z_0 is a regular singular point for (2.1), then S may be written as $S(z) = (z - z_0)^{-k}\tilde{S}(z)$, where k is an integer, $\tilde{S}$ analytic at z_0, $\tilde{S}(z_0) \neq 0$. Consequently, Φ may be written as

$$\Phi(z) = \tilde{S}(z)(z - z_0)^{P-kE} \tag{2.3}$$

Theorem 2.1. *If z_0 is a singular point of the first kind for* (2.1), *then it is a regular singular point for* (2.1).

Proof. The proof will be given for the case $z_0 = 0$. By hypothesis, the system (2.1) may be written

$$w' = z^{-1}\tilde{A}(z)w \tag{2.4}$$

where $\tilde{A}$ is analytic for $0 \leqq |z| < a$, $a > 0$, and $\tilde{A}(0) \neq 0$. If Φ is any fundamental matrix for (2.4), it must be shown that in the representation $\Phi = Sz^P$ (see Theorem 1.1) S is either analytic or has a pole at $z = 0$. This will be done by showing that there exists a positive integer m such that $z^m S$ is bounded in a neighborhood of $z = 0$, and, by a theorem due to Riemann, this implies the result.

Let φ be any nonzero vector solution of (2.4), and let $\tilde{\varphi}(\rho,\theta) = \varphi(\rho e^{i\theta})$, $r = \|\tilde{\varphi}\|$. Then

$$\frac{\partial \tilde{\varphi}}{\partial \rho}(\rho,\theta) = \frac{d\varphi}{dz}(\rho e^{i\theta})e^{i\theta}$$

and thus

$$\left\|\frac{\partial \tilde{\varphi}}{\partial \rho}(\rho,\theta)\right\| = \left\|\frac{d\varphi}{dz}(\rho e^{i\theta})\right\| \leqq \|\tilde{A}(\rho e^{i\theta})\| \frac{r(\rho,\theta)}{\rho}$$

But as was seen in Sec. 5, Chap. 1, with t in place of ρ,

$$\left|\frac{\partial r}{\partial \rho}\right| \leqq \left\|\frac{\partial \tilde{\varphi}}{\partial \rho}\right\|$$

Therefore, if $\|\tilde{A}(z)\| \leqq c$ for $|z| \leqq \rho_1 < a$,

$$\left|\frac{\partial r}{\partial \rho}\right| \leqq \frac{cr}{\rho} \qquad (0 < \rho \leqq \rho_1)$$

From this follows

$$\frac{\partial r}{\partial \rho} + \frac{cr}{\rho} \geqq 0 \qquad (0 < \rho \leqq \rho_1)$$

and hence for $0 < \rho \leqq \rho_1$

$$\rho_1^c r(\rho_1,\theta) - \rho^c r(\rho,\theta) \geqq 0$$

If M denotes the maximum of $r(\rho_1,\theta)$ for $0 \leqq \theta \leqq 2\pi$, then

$$\|\varphi(\rho e^{i\theta})\| = r(\rho,\theta) \leqq \frac{\rho_1^c r(\rho_1,\theta)}{\rho^c} \leqq \frac{M\rho_1^c}{\rho^c}$$

Thus, if Φ is a fundamental matrix for (2.4), there exists a constant $d > 0$ such that if $z = \rho e^{i\theta}$

$$|\Phi(z)| \leqq \frac{d}{\rho^c} \qquad (0 \leqq \theta \leqq 2\pi,\ 0 < |z| \leqq \rho_1) \tag{2.5}$$

It remains to appraise the term z^{-P} in the representation $S = \Phi z^{-P}$. One has $z^{-P} = e^{-(\log z)P} = e^{-(\log \rho)P}e^{-i\theta P}$, and hence

$$|z^{-P}| \leqq |e^{-(\log \rho)P}|\,|e^{-i\theta P}| \tag{2.6}$$

Now

$$|e^{-(\log \rho)P}| \leqq (n-1) + e^{|\log \rho|\,|P|}$$

and if $0 < \rho < 1$,

$$|e^{-(\log \rho)P}| \leqq (n-1) + e^{-(\log \rho)|P|} \leqq n\rho^{-|P|} \tag{2.7}$$

Also, if $0 \leqq \theta \leqq 2\pi$,

$$|e^{-i\theta P}| \leqq (n-1) + e^{2\pi|P|} \tag{2.8}$$

Therefore, from (2.6) through (2.8), there results

$$|z^{-P}| \leqq n\rho^{-|P|}((n-1) + e^{2\pi|P|})$$

provided $0 < \rho < 1$, $0 \leqq \theta \leqq 2\pi$. Combining this with (2.5), one obtains finally

$$\rho^{c+|P|}|S(z)| \leqq \tilde{d}, \qquad 0 < \rho < \min(1,\rho_1), \qquad 0 \leqq \theta \leqq 2\pi$$

where $\tilde{d}$ is a constant independent of z in the range $0 < |z| < \min(1,\rho_1)$. Therefore a positive integer m can be chosen so large that $z^m S$ is bounded in a neighborhood of $z = 0$, thus completing the proof of the theorem.

For systems ($n > 1$) the converse of Theorem 2.1 is not in general true. For example, let $n = 2$, and consider the system

$$w' = (z^{-2}C_1 + C_2)w$$

where

$$C_1 = \begin{pmatrix} 0 & 0 \\ \frac{-3}{16} & 0 \end{pmatrix} \qquad C_2 = \begin{pmatrix} 0 & 1 \\ 0 & 0 \end{pmatrix}$$

This system has at $z = 0$ a singularity of the second kind with rank $\mu = 1$. A fundamental matrix Φ for this system is readily seen to be given by

$$\Phi(z) = \begin{pmatrix} z^{\frac{1}{4}} & z^{\frac{3}{4}} \\ \frac{1}{4}z^{-\frac{3}{4}} & \frac{3}{4}z^{-\frac{1}{4}} \end{pmatrix}$$

If S and R are defined by

$$S = \begin{pmatrix} z & z \\ \frac{1}{4} & \frac{3}{4} \end{pmatrix} \qquad R = \begin{pmatrix} -\frac{3}{4} & 0 \\ 0 & -\frac{1}{4} \end{pmatrix}$$

it is seen that $\Phi = Sz^R$, and from this representation of Φ it follows that $z = 0$ is a regular singular point.

For an equation of the nth order, however, it is possible to give a necessary and sufficient condition on the coefficients of the equation for a point z_0 to be a regular singular point; see Sec. 5 of this chapter, especially Theorems 5.1 and 5.2.

It may happen that, even though the coefficient matrix A in (2.1) has a singularity at z_0, every fundamental matrix is analytic at z_0. In this case, z_0 is referred to as an *apparent singularity* for (2.1). For example, consider the system

$$w' = z^{-1}Ew$$

Clearly a fundamental matrix for this is given by $\Phi = z^E = zE$, which is analytic at $z = 0$. Notice that $\det \Phi(0) = 0$. This is the general situation under these circumstances.

Theorem 2.2. *In* (2.1) *let A be single-valued and analytic in a vicinity of z_0 but have a singularity at z_0. If Φ is any fundamental matrix, then either Φ has a singularity at z_0, or* $\det \Phi(z_0) = 0$.

Proof. Suppose Φ is analytic at z_0, and, if possible, $\det \Phi(z_0) \neq 0$. Then Φ^{-1} exists at z_0, and is an analytic function of z in a neighborhood of z_0. Hence $\Phi'\Phi^{-1}$ is analytic at z_0. But $\Phi'\Phi^{-1} = A$, and this gives a contradiction.

3. Formal Solutions

Although Theorem 2.1 gives a qualitative idea of the solutions for a system with a singularity of the first kind at a point z_0, it does not give explicit information concerning the matrix $P - kE$ in (2.3), or, for that matter, a constructive procedure for calculating the solutions. This will be done in the present section. The case $z_0 = 0$ will be treated; the modifications necessary for any z_0 will be obvious.

As an example, consider the case

$$\begin{pmatrix} w_1' \\ w_2' \end{pmatrix} = \begin{pmatrix} 0 & 1 \\ \frac{1}{z} & 0 \end{pmatrix} \begin{pmatrix} w_1 \\ w_2 \end{pmatrix}$$

This leads easily to the second-order equation $w_1'' - w_1/z = 0$. Using the fact that, by (1.11), there is at least one solution of the form $z^p(s_0 + s_1 z + \cdots)$, where $p, s_0, \ldots$ are constants, it follows that

$$p(p-1)s_0 z^{p-2} + (p+1)ps_1 z^{p-1} + \cdots - s_0 z^{p-1} - s_1 z^p - \cdots = 0$$

Thus

$$p(p-1)s_0 z^{p-2} + [(p+1)ps_1 - s_0]z^{p-1} + \cdots + [(p+k)(p+k-1)s_k - s_{k-1}]z^{p+k-2} + \cdots = 0$$

From this follows as one possible solution

$$p = 1,\ s_0 = 1,\ s_1 = \frac{1}{2},\ s_2 = \frac{1}{(2)^2 3},\ s_3 = \frac{1}{2^2 3^2 4},\ \dots,$$

$$s_k = \frac{1}{(k!)^2(k+1)},\ \dots$$

Thus the series

$$\sum_{k=0}^{\infty} \frac{z^{k+1}}{(k!)^2(k+1)}$$

satisfies $w_2'' - w_2/z = 0$. The question arises as to whether the series represents an actual solution or, what is equivalent, whether the series is convergent. In this case, it is obvious that the series does converge. Indeed, it is always the case that a series which satisfies (2.4) formally is an actual solution and this will be proved.

[It is not always the case that a formal series satisfying a more general class of equations than (2.4) converges. Indeed, the divergent series

$$\sum_{k=0}^{\infty} k! z^k$$

is a formal solution of the second-order equation

$$z^2 w'' + (3z - 1)w' + w = 0$$

where w is a scalar.]

It is necessary to define the notion of a formal series in sufficient generality to include all actual solutions of (2.4).

By a *formal* (*Laurent*) *series* f will be meant an expression of the form

$$f = \sum_{m=-\infty}^{\infty} c_m z^m$$

where the c_m are complex numbers, and *all but a finite number of the* c_m *with negative indices are zero.* If

$$g = \sum_{m=-\infty}^{\infty} d_m z^m$$

is another formal series, then f is defined to be *equal* to g if and only if $c_m = d_m$ for all m. The *sum*, $f + g$, and *product*, fg, of two such formal series are defined by the relations

$$f + g = \sum_{m=-\infty}^{\infty} (c_m + d_m) z^m$$

$$fg = \sum_{m=-\infty}^{\infty} h_m z^m \qquad h_m = \sum_{k+l=n} c_k d_l$$

Note that the sum involved in the definition of h_m is a finite one, and hence fg is defined for all formal series f and g. (If the c_{-m} do not all vanish for sufficiently large m, then the sum which defines h_m would not be finite and hence need not converge. Thus products fg would not be defined.) If a formal series f is such that $c_{-m} = 0$ for $m = 1, 2, \ldots$, then f is called a *formal power series*. The derivative f' of a formal series f is defined to be the formal series

$$f' = \sum_{m=-\infty}^{\infty} (m+1)c_{m+1}z^m$$

If the f_{jk} are formal Laurent series and μ_j are complex numbers, the finite sum

$$p = \sum_{j,k=0}^{\infty} f_{jk}z^{\mu_j} (\log z)^k \qquad f_{jk} = 0 \text{ for } j + k \text{ large}$$

is said to be a *formal logarithmic sum*. Let

$$q = \sum_{j,k=0}^{\infty} g_{jk}z^{\mu_j} (\log z)^k$$

also be a formal logarithmic sum. The sum $p + q$ and the product pq are defined by proceeding as though the coefficients f_{jk} and g_{jk} were scalars. The resulting coefficients may then be combined, and yield formal Laurent series. Thus the addition or multiplication of formal logarithmic sums results in formal logarithmic sums.† The derivative of a formal logarithmic sum p is defined by

$$p' = \sum_{j,k=0}^{\infty} [f'_{jk} + \mu_j f_{jk}z^{-1} + (k+1)f_{j\,k+1}z^{-1}]z^{\mu_j}(\log z)^k \tag{3.1}$$

which is again a formal logarithmic sum.

A formal logarithmic sum is said to be reduced if none of the differences $\mu_i - \mu_j$, $i \neq j$, is an integer. Clearly a formal logarithmic sum can always be reduced. A reduced sum p is said to be zero if and only if all the coefficients, f_{jk}, are zero. A formal logarithmic sum is said to be zero if and only if its reduced sum is zero. Two formal logarithmic sums are said to be equal if their difference is zero.

† Algebraically speaking, a formal logarithmic sum is an element of the algebra over the complex numbers generated by formal Laurent series, powers of z, and integer powers of $\log z$.

A *formal logarithmic matrix* L is defined to be a matrix with elements l_{ij} $(i, j = 1, \ldots, n)$ which are formal logarithmic sums. The sum, product, and equality of two such matrices are defined to be the usual formal matrix sum, product, and equality. The derivative L' of such a matrix is defined to be the matrix with elements l'_{ij}.

Now, to return to differential equations, consider a system having a singularity of the first kind at $z = 0$,

$$w' = A(z)w \tag{3.2}$$

where $A(z) = \sum_{m=-1}^{\infty} z^m A_m$ is a convergent Laurent series about $z = 0$. Clearly A can be regarded as a formal logarithmic matrix. By a *formal solution* of (3.2) is meant a formal logarithmic matrix Φ which satisfies $\Phi' = A\Phi$ (3.2) considered as an equality for formal logarithmic matrices.

Theorem 3.1. *If Φ is a formal solution of* (3.2), *then Φ is an actual solution, that is, all formal series occurring in Φ are convergent in a region $0 < |z| < a$, for some $a > 0$.*

Proof. There exists an actual fundamental matrix $\tilde{\Phi}$ of (3.2) which, by Theorem 2.1, has the form

$$\tilde{\Phi} = Sz^P$$

where P is in canonical form and where S is single-valued, analytic for $0 < |z| < a$, and has at most a pole at $z = 0$. Hence S can be expanded in a convergent Laurent series for $0 < |z| < a$ with a finite number of negative terms. From the structure of S and z^P it is clear that $\tilde{\Phi}$ can be regarded as a formal logarithmic matrix also. Since $\tilde{\Phi}^{-1} = z^{-P}S^{-1}$ exists for $0 < |z| < a$, it is true that $\tilde{\Phi}^{-1}$ may also be regarded as a formal logarithmic matrix.

If Φ is any formal solution of (3.2), then in a formal sense

$$(\tilde{\Phi}^{-1}\Phi)' = -\tilde{\Phi}^{-1}\tilde{\Phi}'\tilde{\Phi}^{-1}\Phi + \tilde{\Phi}^{-1}\Phi' = -\tilde{\Phi}^{-1}A\Phi + \tilde{\Phi}^{-1}A\Phi = 0$$

since both Φ and $\tilde{\Phi}$ are formal solutions of (3.2). It will now be shown that this implies that the formal logarithmic matrix $\tilde{\Phi}^{-1}\Phi$ is a constant matrix. It is sufficient to prove that if p is any formal (scalar) logarithmic sum and $p' = 0$, then p is a constant.

Let p be reduced. Since $p' = 0$, it follows from (3.1) that

$$f'_{jk} + \mu_j f_{jk} z^{-1} + (k + 1)f_{j\,k+1} z^{-1} = 0 \tag{3.3}$$

for all j and k. Let the highest power of $\log z$ in p with nonzero coefficient be N, and suppose

$$f_{0N} z^{\mu_0} + f_{1N} z^{\mu_1} + \cdots + f_{rN} z^{\mu_r}$$

is this coefficient. For $k = N$, (3.3) yields

$$f'_{jN} + \mu_j f_{jN} z^{-1} = 0 \qquad (j = 0, 1, \ldots, r)$$

since $f_{jN+1} = 0$. Using the fact that f_{jN} is a formal Laurent series $\Sigma c_{jm}^{(N)} z^m$, there follows

$$(m + \mu_j) c_{jm}^{(N)} = 0$$

for all m and $j = 0, 1, \ldots, r$. This implies μ_j has to be an integer for some j, for otherwise $c_{jm}^{(N)} = 0$ for all m and $j = 0, 1, \ldots, r$, which would imply $f_{jN} = 0$ for all $j = 0, 1, \ldots, r$, contradicting the choice of N. There is at most one μ_j which is an integer, for p is reduced. Let this be μ_0, and assume $\mu_0 = 0$ with no restriction. Then it follows that

$$\begin{aligned} f_{jN} &= 0 \qquad (j \geqq 1) \\ f_{0N} &= c_{00}^{(N)} \end{aligned}$$

Now assume $N \geqq 1$. Then from (3.3), with $k = N - 1$, it follows as above that

$$\begin{aligned} f_{jN-1} &= 0 \qquad (j \geqq 1) \\ f'_{0N-1} + N c_{00}^{(N)} z^{-1} &= 0 \end{aligned}$$

But the last relation cannot hold unless $c_{00}^{(N)} = 0$, again contradicting the choice of N. Thus $N = 0$ and $p = c_{00}^{(0)}$, a constant.

Now $\tilde{\Phi}^{-1}\Phi = C$, C a constant matrix, implies that $\Phi = \tilde{\Phi}C$ formally. But since $\tilde{\Phi}$ is an actual fundamental matrix for (3.2), $\tilde{\Phi}C$ is an actual solution matrix for (3.2). Therefore Φ itself must be an actual solution matrix for (3.2), and all formal series in Φ must be convergent for $0 < |z| < a$. This proves Theorem 3.1. In particular, any formal vector solution is an actual solution since Φ can have all its columns identical.

4. Structure of Fundamental Matrices

The form of the solutions of (2.4) is known from (1.10) and Theorem 2.1. Here will be given an explicit way of finding the solutions by recursion formulas for the coefficients of the series. A system $w' = A(z)w$ with a singularity of the first kind at $z = 0$ may be written as

$$w' = \left(z^{-1}R + \sum_{m=0}^{\infty} z^m A_m\right) w \tag{4.1}$$

where $R \neq 0$, A_m are constant matrices, and the power series in (4.1) converges for $|z| < a$, $a > 0$. If all the $A_m = 0$, then the equation reduces to the system $w' = z^{-1}Rw$ which has a fundamental matrix $\Phi = z^R$, as can readily be checked. The essential effect of the power-series perturbation in (4.1) is to introduce a power-series term in the solution, that is, a fundamental matrix for (4.1) is given by $\Phi = Pz^{\hat{R}}$, where

P is a power series and $\hat{R}$ is a constant matrix (see Theorem 2.1). In a special case, $\hat{R}$ turns out to be the R appearing in (4.1).

Theorem 4.1. *In the system* (4.1), *if R has characteristic roots which do not differ by positive integers, then* (4.1) *has a fundamental matrix Φ of the form*

$$\Phi = Pz^R \qquad (0 < |z| < c,\ c > 0) \tag{4.2}$$

where P is a power series

$$P(z) = \sum_{m=0}^{\infty} z^m P_m \qquad P_0 = E \tag{4.3}$$

REMARK: From (4.2) and (4.3) follows at once the fact that a fundamental matrix is also given by Sz^{R_0}, where R_0 is the canonical form of R, and S is a power series with $S(0)$ nonsingular. This puts the solutions in the form (1.10) with the u_j analytic in a vicinity of $z_0 = 0$.

Proof of Theorem 4.1. It will be proved that (4.1) has a formal solution of the type (4.2), (4.3), and, by Theorem 3.1, this implies that (4.2) is an actual solution. Since $P_0 = E$, it follows that $P(z)$ is nonsingular on $|z| < c$, for some $c > 0$, and this implies Φ is nonsingular for $0 < |z| < c$, and hence is a fundamental matrix in this region.

Let J be the canonical form of R. Then there is a nonsingular constant matrix T such that $RT = TJ$. J has the form given in Theorem 1.1, Chap. 3. Let the Q_m be constant matrices and let

$$\Phi(z) = Q(z)z^J = (Q_0 + zQ_1 + \cdots)z^J \tag{4.4}$$

be a formal logarithmic matrix. Substituting it in (4.1), there results

$$\sum_{m=0}^{\infty} (m+1)z^m Q_{m+1} = z^{-1}(RQ_0 - Q_0J) + \sum_{m=0}^{\infty} z^m(RQ_{m+1} - Q_{m+1}J) + \sum_{m=0}^{\infty} z^m C_m \tag{4.5}$$

where

$$C_m = \sum_{k=0}^{m} A_k Q_{m-k}$$

For (4.5) to hold, it is necessary and sufficient for

$$\begin{aligned} RQ_0 &= Q_0J \\ Q_{m+1}[J + (m+1)E] &= RQ_{m+1} + C_m \qquad (m = 0, 1, 2, \ldots) \end{aligned} \tag{4.6}$$

The first equation of (4.6) is satisfied by taking $Q_0 = T$. To satisfy the other equations, it is convenient to treat the matrix equation column by column. Let the columns of Q_m be denoted by $q_m^{(j)}$, $j = 1, \ldots, n$.

The jth column of J contains two elements which may be different from zero, λ_j (the jth characteristic root of R) in the jth row of this jth column and, for $j \geqq 2$, δ_j^* in the $(j-1)$st row of this jth column, where δ_j^* is either 0 or 1. In what follows δ_1^* is always zero. Taking the jth column of (4.6) yields

$$[\lambda_j + (m+1)]q_{m+1}^{(j)} + \delta_j^* q_{m+1}^{(j-1)} = Rq_{m+1}^{(j)} + c_m^{(j)} \qquad (j = 1, \ldots, n;\ m = 0, 1, 2, \ldots) \tag{4.7}$$

where $c_m^{(j)}$ is the jth column of C_m. The equation (4.7) can be written as

$$[(\lambda_j + m + 1)E - R]q_{m+1}^{(j)} = c_m^{(j)} - \delta_j^* q_{m+1}^{(j-1)} \qquad (j = 1, \ldots, n;\ m = 0, 1, 2, \ldots) \tag{4.8}$$

The $c_m^{(j)}$ depend only on $q_k^{(l)}$, $l \leqq j$, $k \leqq m$. Taking $m = 0$, the equations (4.8) are a recursive set for $q_1^{(j)}$, $j = 1, 2, \ldots, n$, because $\lambda_j + 1$ is not a characteristic root of R. Taking $m = 1$, (4.8) is again a recursive set for $q_2^{(j)}$, $j = 1, \ldots, n$, and by an induction it follows that the formal solution (4.4) is determined recursively by (4.8). (It may actually be found column by column, which means that n vector solutions may be found which then comprise the matrix.)

Clearly ΦT^{-1} is also a solution of (4.1). This may be written as

$$QT^{-1}(Tz^J T^{-1}) = Pz^R$$

where $P(z) = Q(z)T^{-1} = (T + zQ_1 + \cdots)T^{-1} = E + zP_1 + \cdots$. This completes the proof.

The general case, where R may have characteristic roots which differ by positive integers, may be reduced to Theorem 4.1 by means of the following lemma.

Lemma. *Let the distinct characteristic roots of R (disregarding their multiplicity) in* (4.1) *be $\rho_1, \ldots, \rho_k$ $(k \leqq n)$. There exists a matrix function V of z, nonsingular for $z \neq 0$, and linear in z, such that the transformation $w = V\tilde{w}$ transforms* (4.1) *into a system for $\tilde{w}$ with the same properties as* (4.1),

$$\tilde{w}' = \left(z^{-1}\tilde{R} + \sum_{m=0}^{\infty} z^m \tilde{A}_m\right)\tilde{w} \tag{4.9}$$

and where $\tilde{R}$ has the characteristic roots $\rho_1 - 1, \rho_2, \ldots, \rho_k$.

Proof. It will be assumed to begin with that R is in canonical form and

$$R = \begin{pmatrix} R_1 & 0 \\ 0 & R_2 \end{pmatrix}$$

where R_1 is a p_1-by-p_1 matrix which contains all the terms involving the root ρ_1 in R,

$$R_1 = \begin{pmatrix} \rho_1 & \delta_2^* & 0 & \cdots & 0 \\ 0 & \rho_1 & \delta_3^* & \cdots & 0 \\ \cdot & \cdot & \cdot & \cdots & \cdot \\ 0 & 0 & 0 & \cdots & \rho_1 \end{pmatrix}$$

δ_i^* being either 0 or 1. The matrix U is defined by

$$U = \begin{pmatrix} zE_{p_1} & 0 \\ 0 & E_{n-p_1} \end{pmatrix} \tag{4.10}$$

Clearly U is nonsingular for $z \neq 0$, and

$$U^{-1} = \begin{pmatrix} z^{-1}E_{p_1} & 0 \\ 0 & E_{n-p_1} \end{pmatrix}$$

Then $w = U\tilde{w}$ implies, by (4.1),

$$\tilde{w}' = [z^{-1}U^{-1}RU - U^{-1}U' + \sum_{m=0}^{\infty} z^m(U^{-1}A_mU)]\tilde{w} \tag{4.11}$$

But $U^{-1}RU = R$, and after some calculation one obtains

$$z^{-1}U^{-1}RU - U^{-1}U' = z^{-1}\begin{pmatrix} R_1 - E_{p_1} & 0 \\ 0 & R_2 \end{pmatrix}$$

If

$$A_0 = \begin{pmatrix} A_{11} & A_{12} \\ A_{21} & A_{22} \end{pmatrix}$$

where A_{11} is the block in A_0 of length and width p_1, then (4.11) may be written as (4.9), where

$$\tilde{R} = \begin{pmatrix} R_1 - E_{p_1} & A_{12} \\ 0 & R_2 \end{pmatrix}$$

This $\tilde{R}$ has the required properties. In case R is not in the assumed form, the transformation U can be replaced by TU, where T is chosen so that $T^{-1}RT$ is in the desired form. Setting $V = TU$, the lemma is proved.

Theorem 4.2. *The system* (4.1) *has a fundamental matrix* Φ *of the form*

$$\Phi = Pz^{\hat{R}} \qquad (0 < |z| < c,\ c > 0) \tag{4.12}$$

where P is a power series

$$P(z) = \sum_{m=0}^{\infty} z^m P_m \tag{4.13}$$

and $\hat{R}$ is a constant matrix with characteristic roots which do not differ by positive integers.

Proof. The proof follows directly by applying successively the above lemma and finally Theorem 4.1. Indeed, by using sufficiently many transformations V_i, $i = 1, \ldots, l$, of the type determined in the above lemma, there results finally $\Phi = V_1 \cdots V_l \tilde{P} z^{\hat{R}}$, where $\tilde{P}(0) = E$, and $\hat{R}$ is derived from R in an explicit fashion. The matrix $P = V_1 \cdots V_l \tilde{P}$ and is thus a power series.

5. The Equation of the nth Order

Consider an equation of the nth order

$$\sum_{m=0}^{n} a_{n-m}(z) w^{(m)} = 0 \qquad (a_0(z) \equiv 1) \tag{5.1}$$

where the a_k are single-valued and analytic in a punctured vicinity of a point z_0. If any of the a_k have a singularity at z_0, then z_0 is called a *singular point* for (5.1); otherwise z_0 is called an analytic point for (5.1). Analogous to the definition of a singular point of the first kind for a system of the first order, one says z_0 is a *singular point of the first kind* for (5.1) if z_0 is a singular point for (5.1) and the coefficients in (5.1) have the form

$$a_k(z) = (z - z_0)^{-k} b_k(z) \qquad (k = 1, \ldots, n) \tag{5.2}$$

where the b_k are analytic at z_0. The equation (5.1) is said to have *at most a singularity of the first kind* at z_0 if z_0 is either an analytic point or a singular point of the first kind for (5.1).

The simplest equation of the nth order having the origin as a singularity of the first kind is

$$w^{(n)} + b_1 z^{-1} w^{(n-1)} + b_2 z^{-2} w^{(n-2)} + \cdots + b_n z^{-n} w = 0$$

where the b_i are *constants.* This equation is equivalent, in an obvious way, to the equation

$$z^n w^{(n)} + b_1 z^{n-1} w^{(n-1)} + \cdots + b_n w = 0$$

which is called *Euler's equation.* It can be transformed into an equation with constant coefficients by the substitution $z = e^s$, for if $\tilde{w}(s) = w(e^s)$, then

$$(zw')_{z=e^s} = \frac{d\tilde{w}}{ds}(s) \qquad (z^2 w'')_{z=e^s} = \frac{d^2\tilde{w}}{ds^2}(s) - \frac{d\tilde{w}}{ds}(s), \text{ etc.}$$

The transformed equation

$$\tilde{w}^{(n)} + c_1 \tilde{w}^{(n-1)} + \cdots + c_n \tilde{w} = 0$$

with constants c_i, has a fundamental set of solutions consisting of functions of the form

$$s^k e^{\mu s}$$

where μ is a root of the characteristic equation

$$\lambda^n + c_1\lambda^{n-1} + \cdots + c_n = 0$$

and k is a nonnegative integer less than the multiplicity of μ. The original Euler equation then has a fundamental set of solutions of the form

$$z^\mu (\log z)^k$$

A short calculation shows that the characteristic equation which μ satisfies is given in terms of the b_i by

$$\lambda(\lambda - 1) \cdots (\lambda - n + 1) + \lambda(\lambda - 1) \cdots (\lambda - n + 2)b_1 + \cdots + b_n = 0$$

It is called the *indicial equation* for the Euler equation.

Another way of obtaining the solutions of the Euler equation is to observe that if one puts

$$L(w) = z^n w^{(n)} + b_1 z^{n-1} w^{(n-1)} + \cdots + b_n w$$

then

$$L(z^\lambda) = f(\lambda) z^\lambda$$

where f is the indicial polynomial

$$f(\lambda) = \lambda(\lambda - 1) \cdots (\lambda - n + 1) + \lambda(\lambda - 1) \cdots (\lambda - n + 2)b_1 + \cdots + b_n$$

Therefore z^μ is a solution if $f(\mu) = 0$. If all the roots $\lambda_1, \ldots, \lambda_n$ of $f(\lambda) = 0$ are distinct, then $z^{\lambda_1}, \ldots, z^{\lambda_n}$ is a fundamental set for the Euler equation. If μ is a root of double multiplicity, then

$$f(\mu) = f'(\mu) = 0$$

But

$$L\left(\frac{\partial}{\partial\lambda} z^\lambda\right) = L(z^\lambda \log z) = \frac{\partial}{\partial\lambda} L(z^\lambda) = [f'(\lambda) + (\log z) f(\lambda)] z^\lambda$$

and hence $z^\mu \log z$ is another solution in this case. Continuing in this way, one can obtain a fundamental set for the Euler equation. This idea can be generalized so as to yield a fundamental set for an arbitrary nth order equation having the origin as a regular singular point; see Sec. 8.

It is to be observed that if z_0 is a singular point of the first kind for (5.1), then z_0 may not be a singular point of the first kind for the first-order system associated with (5.1). (See Chap. 1, Sec. 6.) Indeed, only in the case where the coefficients a_k have at most simple poles at z_0 will this be true. However, there does exist a first-order system connected with (5.1) with the property that, if z_0 is a singular point of the first kind, then z_0 is a singular point of the first kind for the system.

Suppose (5.1) has, at most, a singularity of the first kind at z_0, and let φ be any solution of (5.1). Define the vector $\hat{\varphi}$ with components $\varphi_1, \ldots, \varphi_n$ by

$$\varphi_k = (z - z_0)^{k-1}\varphi^{(k-1)} \qquad (k = 1, \ldots, n) \tag{5.3}$$

Then clearly

$$\begin{aligned} (z - z_0)\varphi_k' &= (k-1)\varphi_k + \varphi_{k+1} \qquad (k = 1, \ldots, n-1) \\ (z - z_0)\varphi_n' &= (n-1)\varphi_n - \sum_{m=1}^{n} b_{n-m+1}(z)\varphi_m \end{aligned} \tag{5.4}$$

Therefore the vector $\hat{\varphi}$ is a solution of the linear system

$$w' = A(z)w \tag{5.5}$$

where A has the structure

$$A(z) = (z - z_0)^{-1}\begin{pmatrix} 0 & 1 & 0 & 0 & \cdots & 0 \\ 0 & 1 & 1 & 0 & \cdots & 0 \\ 0 & 0 & 2 & 1 & \cdots & 0 \\ 0 & 0 & 0 & 3 & \cdots & 0 \\ \cdot & \cdot & \cdot & \cdot & \cdots & \cdot \\ 0 & 0 & 0 & 0 & \cdots & 1 \\ -b_n & -b_{n-1} & \cdot & \cdot & \cdots & (n-1) - b_1 \end{pmatrix} \tag{5.6}$$

Obviously $(z - z_0)A$ is analytic at z_0 and does not vanish there, and hence the system (5.5), (5.6) has at z_0 a singularity of the first kind. From Theorem 2.1 the point z_0 is a regular singular point for (5.5). Since the elements of the first row of any fundamental matrix for (5.5) constitute n linearly independent solutions of (5.1) [see (5.3), (5.4)], it follows that every solution of (5.1) near z_0 is a finite linear combination of terms of the form

$$(z - z_0)^r(\log (z - z_0))^k p(z) \tag{5.7}$$

where r is a constant (in general, complex), k is a nonnegative integer which cannot exceed $n - 1$, and p is analytic at z_0, $p(z_0) \neq 0$.

If every solution of (5.1) can be expressed in a vicinity of z_0 as a finite linear combination of terms of the form (5.7), where r and p are as above, then z_0 is said to be a *regular singular point* for (5.1). Thus the above argument proves the following analogue of Theorem 2.1.

Theorem 5.1. *If* (5.1) *has at most a singularity of the first kind at* z_0*, then* z_0 *is a regular singular point for* (5.1).

From the result of Sec. 1, it follows that in any case the solution of (5.1) will be a finite linear combination of terms of the form (5.7) but with p having a possible essential singularity at z_0, so that it is represented by a Laurent series and not necessarily a power series. In case the p cannot all be chosen as analytic at z_0, the equation (5.1) is said to have an *irregular singularity* at z_0.

The converse of Theorem 5.1 also holds.

Theorem 5.2. *If z_0 is a regular singular point for* (5.1), *then* (5.1) *has at most a singularity of the first kind at z_0.*

Proof. Suppose the b_k are related to the a_k in (5.1) via (5.2). Here it is not assumed the b_k are analytic *at* z_0, but it is true that the b_k are analytic and single-valued in a punctured vicinity of z_0. It is clear then that the system (5.5), (5.6) meets the requirements of Theorem 1.1. Since the element in the first row of any solution vector of (5.5) is a solution of (5.1), it follows from this theorem, and (1.10), (1.11), that there exists a solution φ_1 of (5.1) near z_0 of the form

$$\varphi_1(z) = (z - z_0)^r p(z)$$

where p is single-valued and analytic in a punctured vicinity of z_0. But since z_0 is a regular singular point, this solution must be of the form

$$\varphi_1(z) = (z - z_0)^s q(z) \tag{5.8}$$

where s is a constant and q is analytic at z_0, $q(z_0) \neq 0$.

If φ is any solution of (5.1) near z_0, and

$$\varphi = \varphi_1 \psi$$

(variation of parameters), then ψ must be a solution of an equation

$$\sum_{m=0}^{n} c_{n-m}(z) \varpi^{(m)} = 0 \tag{5.9}$$

where

$$c_{n-m} = a_{n-m}\varphi_1 + (m+1)a_{n-m-1}\varphi_1' + \cdots + \binom{n-1}{n-m-1} a_1 \varphi_1^{(n-m-1)} + \binom{n}{m} \varphi_1^{(n-m)} \qquad (m = 0, 1, \ldots, n) \tag{5.10}$$

However, from (5.10),

$$c_n = a_n\varphi_1 + a_{n-1}\varphi_1' + \cdots + a_1\varphi_1^{(n-1)} + \varphi_1^{(n)}$$

which is zero, for φ_1 satisfies (5.1). Hence (5.9) actually is a linear equation of order $n - 1$ for ϖ'. Letting $u = \varpi'$, and dividing (5.9) through by φ_1, there results an equation

$$\sum_{m=0}^{n-1} d_{n-m-1}(z) u^{(m)} = 0 \tag{5.11}$$

where

$$d_0 = 1$$

$$d_k = \frac{c_k}{\varphi_1} = a_k + (n - k + 1)a_{k-1}\frac{\varphi_1'}{\varphi_1} + \cdots + \binom{n}{n-k}\frac{\varphi_1^{(k)}}{\varphi_1} \qquad (k = 1, \ldots, n-1) \quad (5.12)$$

The proof will now proceed by induction. Consider the case $n = 1$,

$$w' + a_1(z)w = 0 \qquad (5.13)$$

where a_1 is analytic and single-valued in a punctured vicinity of z_0. If the solution φ_1 of the form (5.8) is substituted back into (5.13), one obtains

$$(z - z_0)a_1(z) = -s - (z - z_0)\frac{q'(z)}{q(z)}$$

Therefore $(z - z_0)a_1$ is analytic at z_0, which proves the theorem for $n = 1$.

Assume the theorem for equations of order $n - 1$. Since z_0 is a regular singular point for (5.1), it is also one for (5.11), for (5.11) has as solutions the functions $(\varphi_i/\varphi_1)'$, $(i = 2, \ldots, n)$, where $\varphi_1, \ldots, \varphi_n$ are n linearly independent solutions of (5.1), φ_1 being the function in (5.8). If the functions $(\varphi_i/\varphi_1)'$ are dependent, then there are constants c_i such that $\sum_{i=2}^{n} c_i(\varphi_i/\varphi_1)' = 0$. Integrating, there follows the linear dependence of φ_i, $i = 1, 2, \ldots, n$, which is impossible. Thus $(\varphi_i/\varphi_1)'$ are a fundamental set for (5.11). These derivatives $(\varphi_i/\varphi_1)'$ are, by hypothesis, sums of expressions of the type

$$(z - z_0)^a(\log(z - z_0))^b\frac{\tilde{p}(z)}{p(z)}$$

where a is a constant, b an integer, $p(z_0) \neq 0$, $\tilde{p}$ analytic at z_0. By the induction assumption, therefore, the coefficients d_k in (5.11) have at z_0 at most a pole of order k. Putting $k = 1$ in (5.12), it follows that a_1 has at most a pole of order 1. From (5.12) it follows by an induction, and noting that $\varphi_1^{(k)}/\varphi_1$ has at most a pole of order k at z_0, that a_k must have at z_0 at most a pole of order k, thus proving the theorem. [The formula (5.12) is valid for $k = n$ if d_n is defined as zero.]

If z_0 is a regular singular point for (5.1), the actual calculation of a fundamental set may be carried out by considering the corresponding system (5.5), (5.6) and then applying Theorems 4.1 and 4.2. If (5.5) is written in the form

$$w' = \left[(z - z_0)^{-1}R + \sum_{m=0}^{\infty}(z - z_0)^m A_m\right]w$$

where R and the A_m are constant matrices, then R is the residue of A at z_0. If the b_k in (5.2) are of the form

$$b_k(z) = \sum_{m=0}^{\infty} b_{km} z^m \qquad (k = 1, \ldots, n)$$

then the characteristic equation of R, $\det(\lambda E - R) = 0$, is calculated to be

$$\lambda(\lambda - 1) \cdots (\lambda - n + 1) + b_{10}\lambda(\lambda - 1) \cdots (\lambda - n + 2) + \cdots + b_{n-1\,0}\lambda + b_{n0} = 0 \quad (5.14)$$

This equation is called the *indicial equation* for (5.1) relative to the regular singular point z_0. As shown in Sec. 4, the nature of the roots of the indicial equation determines the complexity of the solutions of (5.1). If the roots $\lambda_1, \ldots, \lambda_n$ of (5.14) are distinct and do not differ by positive integers, then a set of n linearly independent solutions of (5.1) is given by

$$\varphi_i = (z - z_0)^{\lambda_i} p_i \qquad (i = 1, \ldots, n)$$

where the p_i can be expanded in power series convergent in a vicinity of z_0, and $p_i(z_0) \neq 0$. In more complicated situations, where logarithms are involved in the solutions, the actual labor can be lightened by using methods such as that due to Frobenius, which is sketched in Sec. 8.

6. Singularities at Infinity

A function f is said to be analytic at ∞ if it can be represented by a power series

$$f(z) = \sum_{j=0}^{\infty} \frac{c_j}{z^j}$$

which converges for $|z|$ sufficiently large. The function f has a zero of order m at ∞ if $c_m \neq 0$ and $c_j = 0$, $j < m$, and has a pole of order m at ∞ if $z^{-k}f$ is analytic at ∞ for $k = m$ but not for $k < m$. Thus f is analytic at ∞ if the function g given by $g(z) = f(1/z)$ is analytic at 0, and has a zero or pole at ∞ of a certain order if g has a zero or pole at 0 of the same order.

In order to study the behavior of a system

$$w' = A(z)w \tag{6.1}$$

or an nth-order equation

$$\sum_{m=0}^{n} a_{n-m}(z) w^{(m)} = 0 \tag{6.2}$$

in the vicinity of an isolated singularity at $z = \infty$, one makes the substitution $z = 1/\zeta$, and obtains a new system or equation, with solutions functions of ζ, called the system or equation *induced* by the substitution

$z = 1/\zeta$. The point $z = \infty$ is said to be a singularity of a given type for (6.1) or (6.2), if $\zeta = 0$ is the same type of singularity for the corresponding induced system or equation. For example, in the case of the system (6.1) if $z = 1/\zeta$, $\tilde{w}(\zeta) = w(1/\zeta)$, $\tilde{A}(\zeta) = A(1/\zeta)$, then the induced system corresponding to (6.1) is

$$\frac{d\tilde{w}}{d\zeta} = \frac{-\tilde{A}(\zeta)}{\zeta^2}\tilde{w} \tag{6.3}$$

Theorem 6.1. *In order that the system* (6.1) *have at most a singular point of the first kind at $z = \infty$, it is necessary and sufficient that A be analytic at $z = \infty$ and $A(\infty) = 0$.*

Proof. The induced system (6.3) has at most a singularity of the first kind at $\zeta = 0$ if and only if $\tilde{A}$ is analytic at $\zeta = 0$ and $\tilde{A}(0) = 0$. Since $A(1/\zeta) = \tilde{A}(\zeta)$, this proves the theorem.

Theorem 6.2. *Necessary and sufficient that $z = \infty$ be a regular singular point for the equation*

$$\sum_{m=0}^{n} a_{n-m}(z)w^{(m)} = 0 \qquad (a_0(z) \equiv 1)$$

is that each a_k be analytic at $z = \infty$ and have a zero there of order at least k.

Proof. If $b_k(z) = z^k a_k(z)$, then the above condition on a_k is equivalent to the condition that the b_k all be analytic at $z = \infty$. The differential equation can be written as

$$\sum_{m=0}^{n} z^m b_{n-m}(z)w^{(m)} = 0 \qquad (b_0(z) \equiv 1) \tag{6.4}$$

Let $z = 1/\zeta$, $\tilde{w}(\zeta) = w(1/\zeta)$, $\tilde{b}_{n-m}(\zeta) = b_{n-m}(1/\zeta)$. Then it is easily seen by induction that

$$(z^m w^{(m)})_{z=1/\zeta} = (-1)^m \zeta^m \tilde{w}^{(m)}(\zeta) + \sum_{j=1}^{m-1} \alpha_{jm}\zeta^j \tilde{w}^{(j)}(\zeta)$$

where the α_{jm} are constants. Therefore (6.4) is transformed by the substitution $z = 1/\zeta$ into the equation

$$\sum_{m=0}^{n} \zeta^m c_{n-m}(\zeta)\tilde{w}^{(m)} = 0 \qquad (c_0(\zeta) \equiv 1) \tag{6.5}$$

where

$$(-1)^n c_{n-m} = (-1)^m \tilde{b}_{n-m} + \sum_{j=m+1}^{n} \alpha_{mj}\tilde{b}_{n-j} \qquad (m = 0, 1, \ldots, n-1) \tag{6.6}$$

Now $\zeta = 0$ is a regular singular point for (6.5) if and only if every c_k is analytic at $\zeta = 0$, and by (6.6) this is true if and only if every $\tilde{b}_k$ is analytic at $\zeta = 0$. But the latter holds when and only when every b_k is analytic at $z = \infty$, which proves the theorem.

It is of interest to know the structure of A in (6.1) when the $k + 1$ distinct points $z_1, z_2, \ldots, z_k, \infty$ are isolated singularities of the first kind for (6.1), and (6.1) has no further singular points. Such a system is said to be of the *Fuchsian type.*

Theorem 6.3. *The system* (6.1) *has isolated singularities of the first kind at the distinct points* $z_1, \ldots, z_k, \infty$, *and no other singular points, if and only if* A *is of the form*

$$A(z) = \sum_{m=1}^{k} (z - z_m)^{-1} R_m \tag{6.7}$$

where the R_m *are constant matrices,* $R_m \neq 0$ *for any* m.

Proof. First, it is clear that if A has the form (6.7) then (6.1) has isolated singularities of the first kind at $z_1, \ldots, z_k$, and since A is analytic at $z = \infty$ and $A(\infty) = 0$, it follows from Theorem 6.1 that (6.1) has a singularity of the first kind at $z = \infty$. Obviously, these are the only singular points for (6.1) in this case.

Conversely, suppose (6.1) has isolated singularities of the first kind at $z_1, \ldots, z_k, \infty$, and no other singular points. Thus A has a simple pole at each of the points $z_1, \ldots, z_k$; let R_m denote the residue of A at z_m. Then the matrix function F defined by

$$F(z) = A(z) - \sum_{m=1}^{k} (z - z_m)^{-1} R_m \tag{6.8}$$

must be an entire function. Since $z = \infty$ is also a singularity of the first kind, by Theorem 6.1 A is analytic at $z = \infty$ and $A(\infty) = 0$. From (6.8) it follows that F must be analytic at $z = \infty$. By Liouville's theorem F must be a constant, and since $F(\infty) = 0$, one has $F(z) \equiv 0$. This proves the theorem.

For the case $k = 1$, the system (6.1), where A is given by (6.7), becomes

$$w' = (z - z_1)^{-1} R_1 w$$

which has a fundamental matrix $\Phi = (z - z_1)^{R_1}$. For the cases $k \neq 1$, the nonlocal problem is much more difficult, and will not be treated here.

The corresponding result for an equation of the nth order is given in the following theorem.

Theorem 6.4. *In order that the equation*

$$\sum_{m=0}^{n} a_{n-m}(z) w^{(m)} = 0 \qquad (a_0(z) \equiv 1) \tag{6.9}$$

have regular singular points at the distinct points $z_1, \ldots, z_k, \infty$, and no other singularities, it is necessary and sufficient that the coefficients a_h be of the form

$$a_h(z) = \prod_{m=1}^{k} (z - z_m)^{-h} b_h(z) \qquad (h = 1, \ldots, n) \tag{6.10}$$

where b_h is a polynomial of degree at most $h(k-1)$.

Proof. From Theorems 5.1, 5.2, it follows that a necessary and sufficient condition for $z_1, \ldots, z_k$ to be regular singular points for (6.9) is that the a_h be such that the $b_h = \prod_{m=1}^{k} (z - z_m)^h a_h$ are analytic for all finite z. From Theorem 6.2, a necessary and sufficient condition that $z = \infty$ be a regular singular point is that $\tilde{a}_h = z^h a_h$ be analytic at $z = \infty$. Therefore, $z = \infty$ is a regular singular point for (6.9) if and only if

$$b_h = z^{-h} \prod_{m=1}^{k} (z - z_m)^h \tilde{a}_h \tag{6.11}$$

where $\tilde{a}_h$ is analytic at $z = \infty$, and b_h analytic in the finite part of the z plane. But (6.11) is equivalent to

$$b_h = z^{h(k-1)} \prod_{m=1}^{k} \left(1 - \frac{z_m}{z}\right)^h \tilde{a}_h$$

and since $\prod_{m=1}^{k} (1 - z_m/z)^h \tilde{a}_h$ is analytic at $z = \infty$, this can hold if and only if b_h is a polynomial in z of degree less than or equal to $h(k-1)$. This proves the theorem.

7. An Example: the Second-order Equation

The previous material will be illustrated by the case of the second-order linear equation

$$w'' + f(z)w' + g(z)w = 0 \tag{7.1}$$

In order that the distinct points $z_1, \ldots, z_k, \infty$, be regular singular points for (7.1), by Theorem 6.4, it is necessary and sufficient that $f = \prod_{m=1}^{k} (z - z_m)^{-1}\tilde{f}$, where $\tilde{f}$ is a polynomial of degree at most $k - 1$ and g is as below. Hence, f may be expanded by partial fractions

$$f(z) = \sum_{m=1}^{k} \frac{a_m}{(z - z_m)} \qquad (a_m \text{ constants}) \tag{7.2}$$

Similarly,

$$g(z) = \sum_{m=1}^{k} \frac{b_m}{(z - z_m)^2} + \sum_{m=1}^{k} \frac{c_m}{(z - z_m)} \tag{7.3}$$

where b_m, c_m are constants. It is easily seen that in order for $z^2 g$ to be analytic at $z = \infty$, it is necessary and sufficient that $\sum_{m=1}^{k} c_m = 0$.

In the case that (7.1) has two regular singular points, say at $z = 0$, $z = \infty$, then $k = 1$, and (7.1) becomes

$$z^2 w'' + a_1 z w' + b_1 w = 0$$

where a_1, b_1 are *constants*. This is an Euler equation considered in the introduction to Sec. 5.

Suppose (7.1) has exactly three regular singular points at $z_1 = 0$, $z_2 = 1$, and at $z = \infty$. Then (7.1) has the form [see (7.2), (7.3)]

$$z^2(z-1)^2 w'' + (az + b)z(z-1)w' + (cz^2 + dz + e)w = 0 \tag{7.4}$$

where $a, \ldots, e$ are constants. It is usual to consider (7.4) in a normalized form. The indicial equation for (7.4) relative to $z = 0$ is given by [see (5.14)]

$$\lambda(\lambda - 1) - \lambda b + e = 0 \tag{7.5}$$

and the indicial equation relative to $z = 1$ is

$$\lambda(\lambda - 1) + \lambda(a + b) + (c + d + e) = 0 \tag{7.6}$$

Let r be a root of (7.5) such that (7.4) has a solution of the form

$$z^r + c_1 z^{r+1} + \cdots \tag{7.7}$$

That there is always such a solution follows from Theorem 4.1 (or more directly by the considerations at the beginning of Sec. 8). Let s be a similar root of (7.6). Let $\tilde{w} = w z^{-r}(z-1)^{-s}$. Then the differential equation for $\tilde{w}$ obtained from (7.4) must have the same form as (7.4) itself, since the substitution takes all analytic solutions into analytic solutions, except possibly at $z = 0, 1,$ or ∞, and preserves the regular singular character of solutions at $z = 0, 1,$ and ∞. Moreover, since the $\tilde{w}$ equation has corresponding to (7.7) a solution $1 + c_1 z + \cdots$, it follows that zero is a root of the indicial equation at $z = 0$, which corresponds to (7.5). Thus the constant corresponding to e must be zero in the $\tilde{w}$ equation. A similar result holds for the constant corresponding to $c + d + e$ in (7.6).

With the above substitution carried out, then the equation (7.4) will have the form

$$z(z - 1)w'' + (az + b)w' + cw = 0$$

and in terms of new constants α, β, γ, this has the form

$$z(1 - z)w'' + [\gamma - (\alpha + \beta + 1)z]w' - \alpha\beta w = 0 \tag{7.8}$$

This is the *hypergeometric equation* whose theory has been investigated in detail.†

In (7.8) let $\zeta = \beta z, \bar{w}(\zeta) = w(\zeta/\beta)$. Then (7.8) is transformed into the following equation for $\bar{w}$:

$$\zeta\left(1 - \frac{\zeta}{\beta}\right)\bar{w}'' + \left[\gamma - \zeta - \frac{(\alpha + 1)}{\beta}\zeta\right]\bar{w}' - \alpha\bar{w} = 0, \qquad \left(' = \frac{d}{d\zeta}\right) \tag{7.9}$$

Now (7.9) has regular singular points at $\zeta = 0, \beta, \infty$, and if $\beta \to \infty$ formally in (7.9), what results is

$$\zeta\bar{w}'' + (\gamma - \zeta)\bar{w}' - \alpha\bar{w} = 0 \tag{7.10}$$

This has $\zeta = 0$ as a regular singular point, but now $\zeta = \infty$ is an irregular singular point. There are no other singular points for (7.10). The equation (7.10) is one of the forms of an equation which for obvious reasons is called the *confluent hypergeometric equation.*

8. The Frobenius Method

The generalization to arbitrary nth-order equations of the second method of obtaining the solutions of the Euler equation (Sec. 5) is called the Frobenius method. If the origin is taken as a regular singular point, the nth-order equation assumes the form

$$z^n w^{(n)} + z^{n-1} b_1 w^{(n-1)} + \cdots + b_n w = 0 \tag{8.1}$$

where the b_j are analytic in a neighborhood of the origin. Let

$$L(w) = z^n w^{(n)} + z^{n-1} b_1 w^{(n-1)} + \cdots + b_n w$$

and

$$b_j(z) = \sum_{k=0}^{\infty} b_{jk} z^k \qquad (j = 1, \ldots, n)$$

The indicial equation associated with (8.1) is

$$\lambda(\lambda - 1) \cdots (\lambda - n + 1) + b_{10}\lambda(\lambda - 1) \cdots (\lambda - n + 2) + \cdots + b_{n-1\,0}\lambda + b_{n0} = 0$$

Let $f(\lambda)$ denote the polynomial on the left of this equation. If for $j = 1, \ldots, n$ it is true that

† E. T. Copson, *An introduction to the theory of functions of a complex variable*, New York, 1935, chap. 10.

$$b_{jk} = 0 \qquad (k = 1, 2, \ldots) \tag{8.2}$$

then (8.1) becomes the Euler equation. It was seen in this case that

$$L(z^\lambda) = f(\lambda)z^\lambda$$

and z^λ was a solution of $L(w) = 0$ if $f(\lambda) = 0$. In the more general case (8.1), one tries to find a formal series

$$\varphi(z) = z^\lambda \sum_{j=0}^{\infty} c_j z^j \qquad (c_0 = 1)$$

such that

$$L(\varphi) = f(\lambda)z^\lambda$$

This is the basic idea behind the Frobenius method.

The formal series for φ substituted into L yields

$$L(\varphi) = f(\lambda)z^\lambda + [f(\lambda + 1)c_1 - g_1]z^{\lambda+1} + \cdots + [f(\lambda + j)c_j - g_j]z^{\lambda+j} + \cdots \tag{8.3}$$

where the g_j are linear in $c_1, \ldots, c_{j-1}$ with coefficients that are polynomials in λ. The equations

$$f(\lambda + j)c_j = g_j \qquad (j = 1, 2, \ldots) \tag{8.4}$$

form a recursive system which can be solved for $c_1, c_2, \ldots,$ as functions of λ, except possibly at the zeros of $f(\lambda + j)$. Clearly the c_j thus determined are rational functions of λ, and (8.3) becomes

$$L(\varphi) = f(\lambda)z^\lambda \tag{8.5}$$

If λ_1 is a root of the indicial equation $f(\lambda) = 0$ and $f(\lambda_1 + j) \neq 0$, $j \geqq 1$, then from (8.5) it follows that φ is a formal, and therefore an actual solution of $L(w) = 0$ which will be denoted by φ_1.

Consider the relation (8.5) near λ_1 and differentiate both sides with respect to λ. This results in

$$\frac{\partial}{\partial\lambda} L(\varphi) = (f'(\lambda) + (\log z)f(\lambda))z^\lambda$$

and if one takes into account the formal commutativity

$$\frac{\partial L(\varphi)}{\partial\lambda} = L\left(\frac{\partial\varphi}{\partial\lambda}\right)$$

formally one obtains

$$L\left(\frac{\partial\varphi}{\partial\lambda}\right) = (f'(\lambda) + (\log z)f(\lambda))z^\lambda \tag{8.6}$$

If λ_1 is a double root of $f(\lambda) = 0$, then $f(\lambda_1) = f'(\lambda_1) = 0$, and (8.6) shows that $\partial\varphi/\partial\lambda$ evaluated at $\lambda = \lambda_1$ is a formal, and hence an actual, solution of $L(w) = 0$. This solution is

$$(\log z)\varphi_1 + \varphi_2$$

where

$$\varphi_2(z) = z^{\lambda_1} \sum_{j=1}^{\infty} \left(\frac{\partial c_j}{\partial \lambda}\right)_{\lambda=\lambda_1} z^j$$

If λ_1 is a root of multiplicity m, it is readily seen that $m - 1$ differentiations with respect to λ may be carried out to yield m solutions.

In case λ_2 is also a root of $f(\lambda) = 0$ and $\lambda_1 - \lambda_2 = k$ is a positive integer, then the above argument cannot be used for the root λ_2 since $f(\lambda_2 + j)$ vanishes when $j = k$. Let $f(\lambda_2 + j) \neq 0$ for $1 \leqq j < k$ and for $j > k$. Let m be the multiplicity of λ_1 as a root of $f(\lambda) = 0$. Consider now the formal series

$$\varphi(z) = (\lambda - \lambda_2)^m z^\lambda + c_1 z^{\lambda+1} + c_2 z^{\lambda+2} + \cdots$$

Then the same procedure which gave (8.5) now gives

$$L(\varphi) = f(\lambda)(\lambda - \lambda_2)^m z^\lambda \tag{8.7}$$

Moreover, the equations (8.4) now yield $c_1, c_2, \ldots, c_{k-1}$ with $(\lambda - \lambda_2)^m$ as a factor. However, for c_k the equation is

$$f(\lambda + k)c_k = g_k$$

and not only is $(\lambda - \lambda_2)^m$ a factor of g_k but also of $f(\lambda + k)$. Thus c_k is determined as a rational function of λ, and it does not have λ_2 as a pole. The terms c_j, $j > k$, are now readily obtained and also will not have λ_2 as a pole.

The series for φ now has $(\lambda - \lambda_2)^m$ as a factor of its first k terms but not necessarily of the later terms. If λ is taken as λ_2, then (8.7) shows that φ is a solution. However, the first k terms of φ vanish so that φ can have only z^{λ_1} as its leading term in z. Indeed, the solution found in this way is merely a multiple of φ_1 found above.

To find a solution really associated with the indicial root λ_2, the mth derivative of (8.7) with respect to λ is considered. This is

$$L\left(\frac{\partial^m \varphi}{\partial \lambda^m}\right) = m!f(\lambda)z^\lambda + I \tag{8.8}$$

where I has $\lambda - \lambda_2$ as a factor. Letting $\lambda = \lambda_2$, there follows

$$L(\varphi_{m+1}) = 0$$

where φ_{m+1} is $\partial^m\varphi/\partial\lambda^m$ at $\lambda = \lambda_2$. The leading term of φ_{m+1} is $m!z^{\lambda_2}$ and thus a solution different from any associated with the root λ_1 has been found. Note that in φ_{m+1} the powers z^{λ_2+j}, $j \geqq k$, may occur multiplied by powers of log z of order up to m.

If $f(\lambda)$ has λ_2 as a multiple root, then higher derivatives of φ with respect to λ will clearly yield further solutions.

The procedure for the case of three roots λ_1, λ_2, λ_3 differing by integers is left as an exercise, as is the general formulation of the method.

PROBLEMS

1. Consider the system

$$z^n w^{(n)} + z^{n-1}B_1 w^{(n-1)} + \cdots + B_n w = 0$$

where the B_i are m-by-m constant matrices and w is an m-dimensional vector. Calculate a fundamental set for this system.

2. Treat in detail the system

$$z^n w^{(n)} + z^{n-1}B_1 w^{(n-1)} + \cdots + B_n w = 0$$

where the B_i are analytic (near the origin) m-by-m matrices and w is an m-dimensional vector.

3. Suppose (5.1) has at most a singularity of the first kind at z_0. Let $z - z_0 = e^s$, and then find the system associated with the transformed equation. Show that it has the form

$$(z - z_0)w' = A(z)w$$

where

$$A(z) = A_0 + (z - z_0)A_1 + (z - z_0)^2 A_2 + \cdots$$

Compute the characteristic equation of A_0 and show that it is the same as the indicial equation (5.14).

4. Consider the second-order equation

$$(*) \qquad w'' + f(z)w' + g(z)w = 0$$

What conditions on f and g must hold if ∞ is to be an analytic point for (*)? Show that if f and g are not both identically zero (and are analytic throughout the whole plane minus the origin) and ∞ is an analytic point, then the origin must be a singular point for (*). Discuss the possible nature of the singularity at the origin.

5. Show that

$$F(\alpha,\beta;\gamma;z) = 1 + \frac{\alpha\cdot\beta}{1\cdot\gamma}z + \frac{\alpha(\alpha+1)\beta(\beta+1)}{1\cdot 2\cdot\gamma(\gamma+1)}z^2 + \cdots$$

is a solution of the hypergeometric equation. Show from (7.8) that as a function of z, $F(\alpha,\beta;\gamma;z)$ can have singularities only at $z = 1$ and ∞.

6. For appropriately restricted ranges of β, γ, and z show that

$$\frac{\Gamma(\beta)\Gamma(\gamma-\beta)}{\Gamma(\gamma)}F(\alpha,\beta;\gamma;z) = \int_0^1 t^{\beta-1}(1-t)^{\gamma-\beta-1}(1-zt)^{-\alpha}\,dt$$

7. In (7.10) let $w = \zeta^{\frac{1}{2}\gamma} e^{-\frac{1}{2}\zeta} \tilde{w}$. Show that the equation for w is of the form

$$\frac{d^2 w}{d\zeta^2} + \left[-\frac{1}{4} + \frac{k}{\zeta} + \frac{\frac{1}{4} - m^2}{\zeta^2} \right] w = 0$$

where $m = \frac{1}{2}(\gamma - 1)$ and $k = \frac{1}{2}\gamma - \alpha$. (See Prob. 17, Chap. 3.)

8. The Bessel equation is

$$w'' + \frac{1}{z} w' + \left(1 - \frac{n^2}{z^2}\right) w = 0$$

If $w = z^{-\frac{1}{2}} u$, show that

$$u'' + \left(1 + \frac{\frac{1}{4} - n^2}{z^2}\right) u = 0$$

9. Show that, if $w = z^n v$, then the Bessel equation becomes

$$z v'' + (2n + 1) v' + z v = 0$$

10. Find two series solutions for the Bessel equation valid for small $|z|$ in the case where n is not an integer.

11. Classify the singular points of the Legendre equation

$$(1 - z^2) w'' - 2 z w' + n(n + 1) w = 0$$

and the associated Legendre equation

$$(1 - z^2) w'' - 2 z w' + \left[n(n + 1) - \frac{m^2}{1 - z^2} \right] w = 0$$

12. The regular singular point is related to the equation with "nearly" constant coefficients considered in Probs. 29 and 35, Chap. 3. Show this for the regular point at $z = \infty$ by transforming

$$\frac{dw}{dz} = \left(\frac{A_0}{z} + \frac{A_1}{z^2} + \cdots \right) w$$

to

$$\frac{dw}{ds} = (A_0 + A_1 e^{-s} + \cdots) w$$

13. Let $A(z) = R/z + A_0 + A_1 z + \cdots$, where R, A_0, . . . are constant square matrices. Let ψ denote the formal series $s_0 z^\lambda + s_1 z^{\lambda+1} + \cdots$, where the s_j are vectors. Show that $s_1, s_2, \ldots$ can be chosen as rational functions of λ so that

$$\left[E \frac{d}{dz} - A(z) \right] \psi = (\lambda E - R) s_0 z^{\lambda - 1}$$

As in the Frobenius treatment of the nth-order equation, show that, if λ_1 is a characteristic root of R and $\lambda_1 + j$, $j \geq 1$, is not a characteristic root, then choosing $\lambda = \lambda_1$ and s_0 as p_1, where p_1 is the characteristic vector

$$R p_1 = \lambda_1 p_1$$

ψ becomes an actual solution, ψ_1, of $w' - A(z) w = 0$. If, in the above problem, λ_1 is a multiple root and if

$$R p_j = \lambda_1 p_j + p_{j-1}, \qquad j = 2, 3, \ldots, l$$

a further solution can be obtained by considering

$$\left[E\frac{d}{dz} - A(z)\right]\frac{\partial\psi}{\partial\lambda} = (\lambda E - R)s_0 z^{\lambda-1}\log z + s_0 z^{\lambda-1}$$

If, in the above, s_0 is taken as p_1 and λ as λ_1 and $\partial\psi/\partial\lambda$ is then denoted by φ_1, it follows that

$$\left[E\frac{d}{dz} - A(z)\right]\varphi_1 = p_1 z^{\lambda_1-1}$$

(Note that φ_1 contains a series with log z as a factor.) From the equation

$$\left[E\frac{d}{dz} - A(z)\right]\psi = (\lambda E - R)s_0 z^{\lambda-1}$$

it now follows on putting $s_0 = p_2$ and $\lambda = \lambda_1$ and calling ψ, $\bar{\psi}_2$ that

$$\left[E\frac{d}{dz} - A(z)\right]\bar{\psi}_2 = (\lambda_1 E - R)p_2 z^{\lambda_1-1} = -p_1 z^{\lambda_1-1}$$

Thus $\varphi_1 + \bar{\psi}_2$ is a solution of $w' - A(z)w = 0$. Extend the above procedure to the case where $l > 2$.

Let λ_2 be a characteristic root such that $\lambda_1 - \lambda_2 = k$ is a positive integer and $\lambda_2 + j$, $1 \leqq j < k$ and $j > k$, is not a characteristic root of R. Show that, if λ_1 is a root of multiplicity m of det $(\lambda E - R) = 0$, then replacing s_0 by $s_0(\lambda - \lambda_2)^m$ in $\psi(z)$ leads to the determination of a solution with leading term z^{λ_2}.

CHAPTER 5

LINEAR SYSTEMS WITH ISOLATED SINGULARITIES: SINGULARITIES OF THE SECOND KIND

1. Introduction

According to the classification of singular points for linear systems given in Chap. 4, the point $z = 0$ is a singularity of rank ρ if the system is of the form

$$w' = z^{-\rho-1}B(z)w \tag{1.1}$$

where B is analytic at $z = 0$, and $B(0) \neq 0$. This chapter will be concerned with the study of the behavior of solutions of linear systems in the neighborhood of a *singularity of the second kind*, that is, where ρ is a positive integer. It will be convenient to consider this singular point at $z = \infty$ instead of at the origin. In this case, the system to be considered is the one induced by the substitution $z = 1/\zeta$ (see Chap. 4, Sec. 6), which has the form, after relabeling,

$$w' = z^r A(z)w \tag{1.2}$$

where r is a nonnegative integer, and A is analytic at $z = \infty$, $A(\infty) \neq 0$. It turns out that the study of (1.2) with $r \geqq 0$ is much more complicated than the study of (1.2) with $r = -1$, the case of a singularity of the *first* kind at $z = \infty$. Although it is not easy to prove in general (only a special case will be considered here), there do exist "formal" solutions of (1.2). The real difficulty now enters because there is no analogue of Theorem 3.1, Chap. 4. This was demonstrated by a simple example in Chap. 4 which showed that a formal solution of (1.2) may actually be a divergent series. It was apparently Poincaré who first realized that even these "formal" divergent expressions have a meaning. He showed, for the case of an nth-order equation, that corresponding to the formal solutions actual solutions of (1.2) exist which have the formal solutions as "asymptotic expansions." These facts will be made more precise in what follows.

The following example will give some indication of the method to be used in this chapter. The equation†

† The first-order system associated with (1.3) is given by $x' = A(t)x$, where

$$x'' + \left(1 - \frac{a}{t^2}\right)x = 0 \tag{1.3}$$

where a is a constant and t is real, behaves for large t almost like the constant coefficient case with $a = 0$. This fact, combined with the results in the case of the regular singular point, suggests trying as solutions for large t

$$\varphi(t) = e^{it}(t^\sigma + c_1t^{\sigma-1} + c_2t^{\sigma-2} + \cdots) \tag{1.4}$$

and a similar expression with i replaced by $-i$. The use of (1.4) in (1.3) formally leads to $\sigma = 0$ and to

$$c_{k+1} = \frac{i}{2}\left(\frac{a - k(k+1)}{k+1}\right)c_k \qquad (k \geqq 0,\ c_0 = 1) \tag{1.5}$$

Unless $a = m(m+1)$ for some integer m, the c_k form a nonterminating sequence with $|c_{k+1}/c_k| \to \infty$, $k \to \infty$. Thus the series in (1.4) is divergent for all $t \neq 0$. However, since (1.4), with the c_k given by (1.5), formally satisfies (1.3), it will be called a formal solution of (1.3).

If two distinct formal solutions of (1.3) were truncated, that is, the infinite series replaced by finite sums containing the early terms, it might be expected that the second-order differential equation satisfied by these truncated functions deviates from (1.3) only in terms involving large powers of $1/t$. In this way, an equation "close" to (1.3) could be found with the help of the formal solutions. For this example, however, this refined procedure will be omitted. The equation $x'' + x = 0$ is close enough to (1.3) for the purpose of getting a representation of the actual solutions of (1.3).

The equation (1.3) can be written as

$$x'' + x = \frac{a}{t^2}x \tag{1.6}$$

If (1.6) has a solution φ which acts like e^{it} as $t \to \infty$, the variation of constants suggests that

$$\varphi(t) = e^{it} - a\int_t^\infty \sin(t - \tau)\varphi(\tau)\tau^{-2}\,d\tau \tag{1.7}$$

Indeed, if φ is a continuous function which is uniformly bounded as $t \to \infty$, and satisfies (1.7), then a direct calculation shows φ must satisfy (1.6), and (1.7) shows that

$$\varphi(t) - e^{it} \to 0 \qquad (t \to \infty)$$

$$A(t) = \begin{pmatrix} 0 & 1 \\ at^{-2} - 1 & 0 \end{pmatrix}$$

and if this is considered for *complex* t, this system is of the type (1.2) for $r = 0$.

To show that (1.7) has a solution, the successive approximation procedure

$$\begin{aligned} \varphi_0(t) &= 0 \\ \varphi_{n+1}(t) &= e^{it} - a \int_t^\infty \sin (t - \tau)\varphi_n(\tau)\tau^{-2}\, d\tau \quad (n \geqq 0) \end{aligned} \tag{1.8}$$

can be used. Clearly

$$|\varphi_1(t) - \varphi_0(t)| \leqq 1$$

and an induction shows that each of the integrals on the right of (1.8) exists for $1 \leqq t < \infty$, and

$$|\varphi_{n+1}(t) - \varphi_n(t)| \leqq \frac{|a|^n}{n!t^n} \qquad (n \geqq 0,\ 1 \leqq t < \infty)$$

Thus the sequence $\{\varphi_n\}$ converges uniformly on $1 \leqq t < \infty$ to a continuous limit function φ. Since

$$|\varphi_n(t)| = \left| \sum_{k=0}^{n-1} (\varphi_{k+1}(t) - \varphi_k(t)) \right| \leqq \sum_{k=0}^{n-1} \frac{|a|^k}{k!t^k} < e^{|a|/t} \leqq e^{|a|}$$

for $1 \leqq t < \infty$, it follows that φ is uniformly bounded, and

$$|\varphi(t)| \leqq e^{|a|} \qquad (1 \leqq t < \infty) \tag{1.9}$$

Now, letting $n \to \infty$ in (1.8), one obtains (1.7).

This solution φ of (1.7), which has already been shown to be a solution of (1.6), satisfies, by virtue of (1.9) and (1.7),

$$|\varphi(t) - e^{it}| \leqq \frac{|a|e^{|a|}}{t}$$

Used in the right side of (1.7) again, this yields

$$\left| \varphi(t) - e^{it} + a \int_t^\infty \sin (t - \tau)e^{i\tau}\tau^{-2}\, d\tau \right| \leqq \frac{|a|^2e^{|a|}}{2!t^2}$$

or, writing $\sin (t - \tau)$ in terms of exponentials and integrating by parts, there results

$$\varphi(t) = e^{it}\left(1 + \frac{ia}{2t}\right) + 0(t^{-2}) \qquad (t \to \infty) \tag{1.10}$$

where $0(t^{-2})$ represents a function g such that $t^2g(t)$ is bounded as $t \to \infty$. Formula (1.10) shows that the sum of the first two terms of the formal series (divergent) given by (1.4) and (1.5) is a better approximation to φ for large t than is the first term. Use of (1.10) in the right side of (1.7) shows that three terms of (1.4) give an even better approximation to φ

for large t. Indeed, although (1.4) diverges it yields information about the solution φ in the sense that for any integer $n \geqq 0$,

$$\varphi(t) = e^{it} \sum_{k=0}^{n} c_k t^{-k} + O(t^{-n-1}) \qquad (t \to \infty)$$

where the c_k are given by (1.5).

In what follows it will be seen that formal-series solutions of the type just considered are typical for a singularity of the second kind and that by the variation-of-constants formula the formal series can be shown to be related to actual solutions as in the case above.

Incidentally, (1.3) has as solutions $t^{\frac{1}{2}} J_\alpha(t)$ and $t^{\frac{1}{2}} Y_\alpha(t)$, where

$$\alpha = (a + \tfrac{1}{4})^{\frac{1}{2}}$$

and J_α and Y_α are solutions of the Bessel equation

$$(tx')' + \left(t - \frac{\alpha^2}{t}\right) x = 0$$

In case $\alpha = m + \frac{1}{2}$ for any integer $m \geqq 0$, it follows from (1.5) that the series (1.4) terminates, and in this case (1.4) yields an actual solution of (1.3) in terms of elementary functions.

2. Formal Solutions

The formal solutions for (1.2) involve exponentials of polynomials as well as the formal logarithmic sums introduced in Chap. 4, Sec. 3. A formal *log-exponential* sum, u, is defined to be a finite expression of the form

$$u = \sum_{j=1}^{k} p_j e^{\mu_j} \tag{2.1}$$

where the p_j are formal logarithmic sums in powers of $1/z$ and the μ_j are distinct polynomials in z and the $\mu_j(z)$ vanish at $z = 0$. It is assumed that u is identified with the sum obtained by any rearrangement of the terms in the sum (2.1). If

$$v = \sum_{j=1}^{m} q_j e^{\nu_j}$$

is another formal log-exponential sum, then u is defined to be *equal* to v if and only if $k = m$ and for some permutation $i_1, \ldots, i_k$ of $1, \ldots, k$ one has $\mu_j = \nu_{i_j}$ and $p_j = q_{i_j}$ for $j = 1, \ldots, k$.

If $\omega_1, \ldots, \omega_n$ are the distinct polynomials occurring in the set $\mu_1, \ldots, \mu_k, \nu_1, \ldots, \nu_m$, then clearly u and v may be written as

$$u = \sum_{j=1}^{n} p_j e^{\omega_j} \qquad v = \sum_{j=1}^{n} q_j e^{\omega_j}$$

where some of the coefficients p_j and q_j may be zero. The *sum* $u + v$ is defined to be

$$u + v = \sum_{j=1}^{n} (p_j + q_j) e^{\omega_j}$$

If $\sigma_1, \ldots, \sigma_r$ denotes the set of distinct polynomials obtained from all sums $\mu_i + \nu_j$ $(i = 1, \ldots, k;\ j = 1, \ldots, m)$, then the *product* uv is defined by

$$uv = \sum_{k=1}^{r} \Big(\sum_{\mu_i + \nu_j = \sigma_k} p_i q_j \Big) e^{\sigma_k}$$

The *derivative*, u', of the formal log-exponential sum (2.1) is defined to be the formal log-exponential sum

$$u' = \sum_{j=1}^{k} (p_j' + p_j \mu_j') e^{\mu_j}$$

It is not difficult to verify that these definitions imply that the usual algebraic and differentiation rules hold for sums of the type (2.1).

A *formal log-exponential matrix* is defined to be a matrix, U, with elements u_{ij} which are formal log-exponential sums. The sum and products of two such matrices are defined to be the usual formal matrix sum and product. The derivative, U', of U is defined to be the matrix with elements u_{ij}'. Clearly the set of formal log-exponential matrices, by definition, are closed under addition, multiplication, and differentiation. If $V = (v_{ij})$ is another formal log-exponential matrix, then V is defined to be equal to U if and only if $u_{ij} = v_{ij}$, $(i, j = 1, \ldots, n)$.

A *formal-solution matrix* of the system (1.2) is defined to be a formal log-exponential matrix whose columns satisfy (1.2) in the sense of equality for such matrices. It is, of course, clear that in (1.2) $z^r A(z)$ can be considered as a formal log-exponential matrix; in fact, it can be represented as a Laurent series in $1/z$ near $z = \infty$.

Theorem 2.1. *For nonnegative integral r consider the linear system*

$$w' = z^r A(z) w \tag{2.2}$$

where A is a convergent power series in z^{-1} in some neighborhood of ∞,

$$A(z) = \sum_{k=0}^{\infty} z^{-k} A_k \tag{2.3}$$

Assume $A_0 \neq 0$ has distinct characteristic roots $\lambda_1, \ldots, \lambda_n$. Then there exists a formal solution matrix for (2.2) of the form

$$\hat{\Phi} = P z^R e^Q \tag{2.4}$$

where P is a formal power series in z^{-1},

$$P = \sum_{k=0}^{\infty} z^{-k} P_k \qquad \det P_0 \neq 0$$

R is a diagonal matrix of complex constants, and Q is a matrix polynomial

$$Q = \frac{z^{r+1}}{r+1} Q_0 + \frac{z^r}{r} Q_1 + \cdots + z Q_r \tag{2.5}$$

with complex diagonal matrices

$$Q_i = \begin{pmatrix} \lambda_1^{(i)} & 0 & \cdots & 0 \\ 0 & \lambda_2^{(i)} & \cdots & 0 \\ \cdot & \cdot & \cdots & \cdot \\ 0 & \cdot & \cdots & \lambda_n^{(i)} \end{pmatrix} \qquad (i = 0, 1, \ldots, r) \tag{2.6}$$

($\lambda_j^{(0)} = \lambda_j$) *as coefficients.*

REMARK: The simplest case of a system with a singularity of the second kind at ∞ is the system

$$w' = Aw$$

where A is a *constant* matrix. A solution matrix Φ is given by

$$\Phi = e^{zA}$$

Perhaps the next simplest case is the system

$$w' = z^r A w$$

where r is a positive integer and A is a constant matrix. It is readily verified that a solution matrix of this equation is given by

$$\Phi = e^{(z^{r+1}/r+1)A}$$

which indicates that the lower-order terms in Q in (2.5), R, and the formal power series P in (2.4) are completely due to the effect of the terms $z^{-1}A_1 + z^{-2}A_2 + \cdots$ in the A of (2.3).

Proof of Theorem 2.1. First, it is clear that, if P, R, Q are matrices as described, the product $Pz^R e^Q$ is a formal log-exponential matrix, for each of the factors P, z^R, e^Q is one. Actually, there are no logarithmic terms in this case since R is diagonal. Also, it may be assumed at the outset that A_0 is a diagonal matrix with elements $\lambda_1, \ldots, \lambda_n$, for a simple substitution $\tilde{w} = Tw$ in (2.2) would effect this, if T is that constant nonsingular matrix such that TA_0T^{-1} has the diagonal form with $\lambda_1, \ldots, \lambda_n$ as diagonal elements. Note that when A_0 is assumed to be diagonal, the assertion (2.6) says, in particular, that $Q_0 = A_0$.

Suppose $\hat{\Phi}$ in (2.4) is a formal-solution matrix of (2.2), where P, Q, R have the properties stated in the theorem. Then differentiation yields

$$\hat{\Phi}' = P'z^R e^Q + z^{-1}PRz^R e^Q + Pz^R(z^r Q_0 + z^{r-1}Q_1 + \cdots + Q_r)e^Q$$

and using the fact that the Q_i and z^R are diagonal one obtains

$$\hat{\Phi}' = [P' + z^{-1}PR + P(z^r Q_0 + z^{r-1}Q_1 + \cdots + Q_r)]z^R e^Q$$

But from (2.2) there results

$$\hat{\Phi}' = z^r A P z^R e^Q$$

and hence

$$P' + z^{-1}PR + P(z^r Q_0 + z^{r-1}Q_1 + \cdots + Q_r) = z^r A P$$

Using the power-series nature of P and A, this gives

$$\sum_{k=0}^{\infty} z^{-k-1}P_k(R - kE) + \left(\sum_{k=0}^{\infty} z^{-k}P_k\right)(z^r Q_0 + z^{r-1}Q_1 + \cdots + Q_r)$$
$$= z^r\left(\sum_{k=0}^{\infty} z^{-k}A_k\right)\left(\sum_{k=0}^{\infty} z^{-k}P_k\right)$$

Comparing coefficients of the various powers of z^{-1} yields

$$\begin{aligned}
P_0Q_0 - A_0P_0 &= 0 \\
P_kQ_0 - A_0P_k &= \sum_{l=1}^{k} (A_lP_{k-l} - P_{k-l}Q_l) \qquad (1 \leqq k \leqq r) \\
P_{k+r+1}Q_0 - A_0P_{k+r+1} &= \sum_{l=1}^{r} (A_lP_{k+r+1-l} - P_{k+r+1-l}Q_l) \\
&\quad + \sum_{l=r+1}^{k+r+1} A_lP_{k+r+1-l} + P_k(kE - R) \qquad (k \geqq 0)
\end{aligned} \tag{2.7}$$

Thus a necessary condition that $\hat{\Phi}$ in (2.4) be a formal solution matrix of (1.2) is that the matrices P_k, Q_k, R satisfy the relations (2.7). Conversely, if a set of matrices P_k, Q_k, R exist which satisfy (2.7), then $\hat{\Phi}$ given by (2.4), (2.5), and (2.6) will be a formal-solution matrix of (1.2).

This follows by a retracing of steps. Thus all that remains to be proved is to show that the relations (2.7) can be solved for matrices P_k, Q_k, R.

Since A_0 is assumed to be diagonal, a solution of the first equation in (2.7) is given by

$$Q_0 = A_0 \qquad P_0 = E \tag{2.8}$$

where E is the identity matrix.

The second equation in (2.7) for $k = 1$ is

$$P_1Q_0 - A_0P_1 = A_1P_0 - P_0Q_1$$

or, using (2.8),

$$P_1A_0 - A_0P_1 = A_1 - Q_1 \tag{2.9}$$

Since A_0 is diagonal, the diagonal terms of the left side of (2.9) are zero, and thus the diagonal elements of Q_1 must be identical with those of A_1. This determines the diagonal matrix Q_1 uniquely. The nondiagonal terms of P_1 are determined from (2.9) by

$$(\lambda_j - \lambda_i)p_{ij}^{(1)} = a_{ij}^{(1)} \qquad (i \neq j) \tag{2.10}$$

where $p_{ij}^{(1)}$, $a_{ij}^{(1)}$ are the elements in the ith row and jth column of the matrices P_1, A_1, respectively. Since $\lambda_i \neq \lambda_j$ $(i \neq j)$, Eq. (2.10) determines the nondiagonal elements of P_1 uniquely. Let $\tilde{P}_1$ denote the matrix with diagonal elements zero and $p_{ij}^{(1)}$ in the ith row and jth column $(i \neq j)$. Then a solution of (2.9) is

$$P_1 = \tilde{P}_1 + D_1 = \tilde{P}_1 + P_0D_1$$

where D_1 is any diagonal matrix. Here use is made of the fact that $D_1A_0 - A_0D_1 = 0$ since A_0 is diagonal. Note that $\tilde{P}_1$ satisfies

$$\tilde{P}_1A_0 - A_0\tilde{P}_1 = A_1P_0 - P_0Q_1 = A_1 - Q_1$$

Let $1 < k \leq r$, and assume the existence of diagonal matrices Q_0, Q_1, . . . , Q_{k-1} and matrices P_1, . . . , P_{k-1} of the form

$$P_i = \tilde{P}_i + \tilde{P}_{i-1}D_1 + \cdots + P_0D_i \tag{2.11}$$

where D_1, . . . , D_{k-1} are arbitrary diagonal matrices, the diagonal elements of the $\tilde{P}_i$ are zero, and the $\tilde{P}_i$ satisfy

$$\tilde{P}_iA_0 - A_0\tilde{P}_i = S_i \qquad (i = 1, \ldots, k-1) \tag{2.12}$$

where

$$S_i = \sum_{l=1}^{i} (A_l\tilde{P}_{i-l} - \tilde{P}_{i-l}Q_l) \qquad (i = 1, \ldots, k-1; \tilde{P}_0 = E)$$

Since A_0 is diagonal, it follows from (2.12) that the diagonal elements of each S_i are zero. Placing (2.11) into the second relation of (2.7) for k, one obtains, on collecting terms,

$$P_kA_0 - A_0P_k = \sum_{l=1}^{k-1} (A_l\tilde{P}_{k-l} - \tilde{P}_{k-l}Q_l) + (A_k - Q_k) + S_{k-1}D_1 + S_{k-2}D_2 + \cdots + S_1D_{k-1} \quad (2.13)$$

Since the diagonal terms of $P_kA_0 - A_0P_k$ are zero, as is the case for $S_1, \ldots, S_{k-1}$, one sees that (2.13) determines the diagonal elements of Q_k uniquely, and thus the diagonal matrix Q_k uniquely. As in the passage (2.9) and (2.10), a solution $\tilde{P}_k$ of

$$\tilde{P}_kA_0 - A_0\tilde{P}_k = \sum_{l=1}^{k-1} (A_l\tilde{P}_{k-l} - \tilde{P}_{k-l}Q_l) + (A_k - Q_k) \quad (2.14)$$

is determined uniquely as regards the elements off the main diagonal. The elements of $\tilde{P}_k$ on the main diagonal are taken to be zero. Then the matrix

$$P_k = \tilde{P}_k + \tilde{P}_{k-1}D_1 + \cdots + P_0D_k \quad (2.15)$$

where D_k is any diagonal matrix, will be a solution of (2.13), for

$$\begin{aligned} P_kA_0 - A_0P_k &= \tilde{P}_kA_0 - A_0\tilde{P}_k + (\tilde{P}_{k-1}D_1A_0 - A_0\tilde{P}_{k-1}D_1) + \cdots + (P_0D_kA_0 - A_0P_0D_k) \\ &= \tilde{P}_kA_0 - A_0\tilde{P}_k + (\tilde{P}_{k-1}A_0 - A_0\tilde{P}_{k-1})D_1 + \cdots + (P_0A_0 - A_0P_0)D_k \end{aligned}$$

since A_0 and the D_i are diagonal. Using (2.12) and (2.14), one readily sees P_k given by (2.15) satisfies (2.13). By induction, corresponding to the choice $Q_0 = A_0$, $P_0 = E$, this proves the existence of diagonal matrices $Q_1, \ldots, Q_r$, and matrices $\tilde{P}_1, \ldots, \tilde{P}_r$ with diagonal elements all zero, satisfying (2.12) for $i = 0, 1, \ldots, r$, and such that the matrices P_i in (2.11) satisfy the second relation in (2.7) for $k = i$.

For $k = 0$ in the third relation of (2.7) one obtains

$$P_{r+1}A_0 - A_0P_{r+1} = \sum_{l=1}^{r} (A_lP_{r+1-l} - P_{r+1-l}Q_l) + A_{r+1}P_0 - P_0R \quad (2.16)$$

and here is where R enters. If the P_i as given by (2.11) are put in the right side of (2.16), one gets

$$P_{r+1}A_0 - A_0P_{r+1} = \sum_{l=1}^{r} (A_l\tilde{P}_{r+1-l} - \tilde{P}_{r+1-l}Q_l) + (A_{r+1} - R) + S_rD_1 + S_{r-1}D_2 + \cdots + S_1D_r \quad (2.17)$$

The diagonal elements of the left side of (2.17) are zero, as well as those in $S_rD_1 + \cdots + S_1D_r$. Therefore (2.17) determines uniquely the diagonal matrix R, and as in (2.13) one obtains a solution of (2.17) of the form

$$P_{r+1} = \tilde{P}_{r+1} + \tilde{P}_r D_1 + \cdots + P_0 D_{r+1} \tag{2.18}$$

where $\tilde{P}_{r+1}$ satisfies

$$\tilde{P}_{r+1}A_0 - A_0\tilde{P}_{r+1} = \sum_{l=1}^{r} (A_l\tilde{P}_{r+1-l} - \tilde{P}_{r+1-l}Q_l) + A_{r+1} - R \tag{2.19}$$

and has diagonal elements zero, and D_{r+1} is an arbitrary diagonal matrix.

The last relation in (2.7) for $k = 1$ brings a change in that no new terms involving Q_k or R enter. This equation is

$$P_{r+2}A_0 - A_0P_{r+2} = \sum_{l=1}^{r} (A_lP_{r+2-l} - P_{r+2-l}Q_l) + A_{r+1}P_1 + A_{r+2}P_0 + P_1(E - R) \tag{2.20}$$

and, using the expressions (2.11), (2.18) for the P_i, this yields

$$\begin{aligned} P_{r+2}A_0 - A_0P_{r+2} = {} & \sum_{l=1}^{r} (A_l\tilde{P}_{r+2-l} - \tilde{P}_{r+2-l}Q_l) + A_{r+1}\tilde{P}_1 + A_{r+2} \\ & + \tilde{P}_1(E - R) + \left[\sum_{l=1}^{r} (A_l\tilde{P}_{r+1-l} - \tilde{P}_{r+1-l}Q_l) + A_{r+1} - R\right] D_1 \\ & + S_rD_2 + \cdots + S_1D_{r+1} + D_1 \end{aligned} \tag{2.21}$$

But by (2.19) the expression in the brackets [ ] has a diagonal consisting of zeros, and since the diagonal terms of the left side of (2.21) are all zero, it follows that the diagonal matrix D_1 is uniquely determined by (2.21). Just as before, a solution P_{r+2} of (2.20) can be found of the form

$$P_{r+2} = \tilde{P}_{r+2} + \tilde{P}_{r+1}D_1 + \cdots + P_0D_{r+2}$$

where $\tilde{P}_{r+2}$ is a solution of (2.20) with the P_k replaced by $\tilde{P}_k$ everywhere and D_1 added to the right side, and the diagonal elements of $\tilde{P}_{r+2}$ all zero, and D_{r+2} is an arbitrary diagonal matrix.

In the next step, D_2 is determined uniquely and a new diagonal matrix D_{r+3} is introduced. Thus $r + 1$ matrices D_k are always being carried in the procedure. Using another induction, it follows that all $\tilde{P}_k$ and D_k are determined uniquely from the equations (2.7), and hence all the P_k are determined uniquely, once the initial choice $Q_0 = A_0$, $P_0 = E$ is made. This completes the proof of the theorem.

3. Asymptotic Series

Recall that in Sec. 1 the series

$$e^{it}(1 + c_1t^{-1} + c_2t^{-2} + \cdots) \tag{1.4}$$

was a formal solution† of the equation

$$x'' + \left(1 - \frac{a}{t^2}\right)x = 0 \tag{1.3}$$

provided the c_k were determined recursively by

$$c_{k+1} = \frac{i}{2}\left(\frac{a - k(k+1)}{k+1}\right)c_k \qquad (k \geqq 0,\ c_0 = 1) \tag{1.5}$$

If a is not of the form $n(n + 1)$, the series in (1.4) diverges for all $t \neq 0$. However, it was seen that corresponding to this formal solution there existed an actual solution φ of (1.3) such that

$$\varphi(t) = e^{it} \sum_{k=0}^{n} c_k t^{-k} + 0(t^{-n-1}) \qquad (t \to \infty)$$

and in particular

$$t^n \left[\varphi(t) - e^{it} \sum_{k=0}^{n} c_k t^{-k}\right] \to 0 \qquad (t \to \infty) \tag{3.1}$$

The relation (3.1), which represents the usual situation as regards singular points of the second kind, expresses the fact that the formal series (1.4) is an *asymptotic series* for the solution φ of (1.3).

To be more precise, let S denote a connected set in the complex z plane containing ∞. A formal power series in z^{-1},

$$p = \sum_{k=0}^{\infty} p_k z^{-k} \tag{3.2}$$

with partial sums

$$s_k = \sum_{j=0}^{k} p_j z^{-j} \qquad (k = 0, 1, \ldots)$$

is said to be an *asymptotic series* (or *expansion*) in S for a function f (as $|z| \to \infty$) which is defined in S, if for every $k = 0, 1, 2, \ldots,$

$$z^k(f - s_k) \to 0 \qquad (|z| \to \infty)$$

uniformly for $z \varepsilon S$.

† Strictly speaking, the notion of a formal solution of a second-order equation has not been defined. It can be defined directly in an obvious manner, or it can be taken as the first component of any vector formal solution of the first-order system associated with (1.3).

If p is an asymptotic series for f in S, then this relation is written

$$f \sim p \text{ in } S$$

Often S is a part of a sector of the z plane

$$S: \quad \varphi_1 \leqq \arg z \leqq \varphi_2 \qquad |z| \geqq r$$

For example, if the formal series p converges in this S, it represents in S an analytic function f, and it is clear that $f \sim p$ in S.

If f has an asymptotic expansion p in a set S, the expansion is unique, for the coefficients p_k in (3.2) are uniquely determined by the conditions

$$f \to p_0, \; z(f - p_0) \to p_1, \; z^2(f - p_0 - p_1 z^{-1}) \to p_2, \text{ etc.}$$

However, different functions may have the same asymptotic series. For example, the function $g = e^{-z}$, defined for the set S: $|z| > 0$, $-\alpha \leqq \arg z \leqq \alpha$, where $\alpha < \pi/2$, has the identically zero formal power series as an asymptotic series in S, that is, $e^{-z} \sim 0$ in S. Hence, if f is any function with an asymptotic series in S, $f + e^{-z}$ has the same asymptotic expansion in S as f.

If f, g, h are three functions defined for $z \varepsilon S$, $h \neq 0$, and if

$$(f - g)h^{-1} \sim \sum_{k=0}^{\infty} p_k z^{-k} \text{ in } S$$

then this is sometimes written as

$$f \sim g + h \sum_{k=0}^{\infty} p_k z^{-k} \text{ in } S$$

For example, it was shown for real $t > 0$ (that is, S is the region $|z| > 0$, $\arg z = 0$), that there exists a solution φ of (1.3) such that

$$\varphi \sim e^{it}(1 + c_1 t^{-1} + c_2 t^{-2} + \cdots) \text{ in } S$$

where the c_k are defined by (1.5).

Theorem 3.1. *Suppose f and g are functions defined in a connected set S including ∞, and*

$$f \sim p = \sum_{k=0}^{\infty} p_k z^{-k}, \qquad g \sim q = \sum_{k=0}^{\infty} q_k z^{-k} \textit{ in } S$$

If α, β are any two complex numbers, then in S,

(a) $$\alpha f + \beta g \sim \alpha p + \beta q = \sum_{k=0}^{\infty} (\alpha p_k + \beta q_k) z^{-k}$$

(b) $$fg \sim pq = \sum_{k=0}^{\infty} c_k z^{-k} \qquad c_k = \sum_{j=0}^{k} p_j q_{k-j}$$

(c) $f^{-1} \sim \frac{1}{p_0} - \left(\frac{p_1}{p_0^2}\right) z^{-1} + \left(\frac{p_1^2 - p_0 p_2}{p_0^3}\right) z^{-2} + \cdots$ **(if $p_0 \neq 0$)**

Proof. The proof follows easily from the definition of an asymptotic series, and is left to the reader.

Corollary. *If f_i $(i = 1, \ldots, m)$, are m functions, $f_i \sim p_i$, $z \varepsilon S$, and $g(z_1, \ldots, z_m)$ is a polynomial, then $F(z) = g(f_1(z), \ldots, f_m(z))$, has an asymptotic expansion in S, and this is calculated as if all the expansions were convergent series.*

Proof. The proof follows by repeated application of (a), (b) in Theorem 3.1.

Application. If A is a matrix of functions whose components have asymptotic expansions, $A \sim \sum_{k=0}^{\infty} z^{-k} A_k$ in S, then det A has an asymptotic expansion there, and the first term in this expansion is det A_0. Thus if det $A_0 \neq 0$, $(\det A)^{-1}$ has an asymptotic expansion in S with the first term $(\det A_0)^{-1}$. Since the elements of A^{-1} are composed of $(n - 1)$-rowed minors of A (which have asymptotic expansions) divided by det A, it follows that, if det $A_0 \neq 0$, A^{-1} has an asymptotic expansion in S.

Theorem 3.2. *(a) If $f \sim \sum_{k=0}^{\infty} p_k t^{-k}$, and f is continuous for $t \geqq t_0$ (t real), then*

$$F(t) = \int_t^\infty (f(\tau) - p_0 - p_1 \tau^{-1})\, d\tau \sim \sum_{k=1}^{\infty} \frac{p_{k+1}}{k} t^{-k}$$

(b) If, further, f' exists and is continuous, and f' has an asymptotic expansion, then $f' \sim -\sum_{k=2}^{\infty} (k - 1) p_{k-1} t^{-k}$.

Proof. (a) $t^2(f - p_0 - p_1 t^{-1}) \to p_2$, $t \to +\infty$, and therefore $F(t)$ exists for $t > t_0$. Also, for fixed $m \geqq 1$,

$$f - \left(\sum_{k=0}^{m+1} p_k t^{-k}\right) = \epsilon(t) t^{-(m+1)}$$

where $\epsilon(t) \to 0$, $t \to +\infty$. Hence

$$\left| F - \left(\sum_{k=1}^{m} \frac{p_{k+1}}{k} t^{-k}\right) \right| = \left| \int_t^\infty \epsilon(\tau) \tau^{-(m+1)}\, d\tau \right| \leqq \epsilon_M(t) \int_t^\infty \tau^{-(m+1)}\, d\tau$$

where $\epsilon_M(t) = \sup_{t \leq \tau < \infty} |\epsilon(\tau)|$. But $\epsilon_M(t) \to 0$, $t \to +\infty$, and since

$$\int_t^\infty \tau^{-(m+1)}\, d\tau = \frac{1}{m} t^{-m}$$

one has $t^m \left(F - \sum_{k=1}^{m} \frac{p_{k+1}}{k} t^{-k} \right) \to 0$, $t \to +\infty$. This proves (*a*).

(*b*) Let $f' \sim \sum_{k=0}^{\infty} q_k t^{-k}$. Then

$$f = \int_{t_0}^t f'(\tau)\, d\tau + f(t_0) = \int_{t_0}^t (q_0 + q_1\tau^{-1})\, d\tau$$
$$+ \int_{t_0}^t (f'(\tau) - q_0 - q_1\tau^{-1})\, d\tau + f(t_0)$$

or

$$f = q_0 t + q_1 \log t - \int_t^\infty (f'(\tau) - q_0 - q_1\tau^{-1})\, d\tau + c$$

where c is a constant. Since f has a unique asymptotic expansion, it follows from (*a*) that $q_0 = q_1 = 0$ and $q_k = -(k-1)p_{k-1}$, $k \geqq 2$. This proves (*b*).

If f has an asymptotic expansion, f' need not have one. For example, if $f = e^{-t} \sin e^t$, then $f \sim 0$, but $f' = -e^{-t} \sin e^t + \cos e^t$ does not have an expansion, for $\lim \cos e^t$, $t \to +\infty$, does not exist.

4. Existence of Solutions Which Have the Formal Solutions as Asymptotic Expansions—the Real Case

It will now be shown that corresponding to every formal-solution vector of (2.2) there exists an actual solution with the formal solution as an asymptotic expansion which is valid in some sector in the complex z plane, for z sufficiently large. In order to do this, certain appraisals will have to be made.

It is important to distinguish, in the following, between formal solutions and actual solutions. An actual-solution matrix (or vector) of the system (2.2) will be denoted by Φ (or φ), whereas formal solutions will always be denoted by $\hat{\Phi}$ (or $\hat{\varphi}$). In this section, unless otherwise stated, it will always be assumed that the system under consideration is the one considered in Theorem 2.1, namely

$$w' = z^r A w \qquad (r \geqq 0) \tag{4.1}$$

where A_0 ***has distinct characteristic roots.***

If P is a formal power series in z^{-1},

$$P = \sum_{k=0}^{\infty} z^{-k} P_k$$

denote by $P_{(m)}$ the polynomial in z^{-1},

$$P_{(m)} = \sum_{k=0}^{m} z^{-k} P_k \qquad (m = 0, 1, 2, \ldots)$$

If $\hat{\Phi} = P z^R e^Q$ [(2.4)] is a formal-solution matrix of (4.1), denote by $\hat{\Phi}_{(m)}$ the "truncated formal solution,"

$$\hat{\Phi}_{(m)} = P_{(m)} z^R e^Q \tag{4.2}$$

Clearly $\hat{\Phi}_{(m)}$ can be regarded as a function of z also.

A sketch of the method to be used here will now be given. Since $\hat{\Phi}$ is a formal solution of (4.1), it is clear that formally

$$\hat{\Phi}' \hat{\Phi}^{-1} = z^r A$$

For the truncated formal solutions it might be expected that, if $B_{(m)}$ is defined by the equation

$$\hat{\Phi}'_{(m)} \hat{\Phi}^{-1}_{(m)} = z^r B_{(m)}$$

then the early terms in $B_{(m)}$ are identical with those of A. This will be shown to be the case in Lemma 4.1 below.

Since $\hat{\Phi}_{(m)}$, $\hat{\Phi}'_{(m)}$, and $\hat{\Phi}^{-1}_{(m)}$ all exist as well-defined functions of z for all z sufficiently large, $\hat{\Phi}_{(m)}$ is an actual- (and not only a formal-) solution matrix of the system

$$w' = z^r B_{(m)} w \tag{4.3}$$

If (4.1) is written as

$$w' = z^r B_{(m)} w + z^r (A - B_{(m)}) w \tag{4.4}$$

then, since $A - B_{(m)}$ is small for large z, the equation (4.4) can be recast, by treating the last term as though it were a given function of z and using the variation-of-constants formula much as in (1.7), to get an integral-equation formulation. Since a solution $\hat{\varphi}_{(m)}$ of the homogeneous equation (4.3) corresponding to (4.4) is known, it will be shown that the integral equation can be dealt with by using the method of successive approximations to obtain a solution of (4.4) [and hence of (4.1)] which behaves like $\hat{\varphi}_{(m)}$ for large z.

Lemma 4.1. *The matrices $\hat{\Phi}'_{(m)}$, $\hat{\Phi}^{-1}_{(m)}$ exist for z sufficiently large, and if $B_{(m)}$ is defined by*

$$z^r B_{(m)} = \hat{\Phi}'_{(m)} \hat{\Phi}^{-1}_{(m)}$$

then

$$z^r A = z^r B_{(m)} + E_{(m)} \tag{4.5}$$

where $B_{(m)}$, $z^{-r}E_{(m)}$ *are analytic for all* z *sufficiently large (including* ∞*), and*†

$$E_{(m)}(z) = 0(|z|^{r-m-1}) \qquad (|z| \to +\infty) \tag{4.6}$$

Proof. It is obvious that $\hat{\Phi}'_{(m)}$ exists. Since $P_{(m)}$ is a polynomial in z^{-1}, and $\det P_0 \neq 0$, $(\det P_{(m)})^{-1}$ exists, for z sufficiently large, as a convergent power series in z^{-1}. Therefore $P_{(m)}^{-1}$ exists and is analytic for all z sufficiently large. From (4.2), it follows that

$$\hat{\Phi}_{(m)}^{-1} = e^{-Q}z^{-R}P_{(m)}^{-1} \tag{4.7}$$

exists for z sufficiently large. Also

$$\hat{\Phi}'_{(m)} = (P'_{(m)} + z^{-1}P_{(m)}R + P_{(m)}Q')z^R e^Q \tag{4.8}$$

since R, Q are diagonal. From (4.7) and (4.8) one has

$$\hat{\Phi}'_{(m)}\hat{\Phi}_{(m)}^{-1} = (P'_{(m)} + z^{-1}P_{(m)}R + P_{(m)}Q')P_{(m)}^{-1} \tag{4.9}$$

and since Q' is the polynomial matrix

$$Q' = z^r Q_0 + z^{r-1}Q_1 + \cdots + Q_r$$

it is clear that $B_{(m)} = z^{-r}\hat{\Phi}'_{(m)}\hat{\Phi}_{(m)}^{-1}$ is analytic for z sufficiently large.

Since $\det P_0 \neq 0$, the formal power series P has a formal reciprocal P^{-1}, and hence $\hat{\Phi}$ has a formal reciprocal which is given by

$$\hat{\Phi}^{-1} = e^{-Q}z^{-R}P^{-1}$$

Now $\hat{\Phi}' = (P' + z^{-1}PR + PQ')z^R e^Q$, and therefore

$$\hat{\Phi}'\hat{\Phi}^{-1} = (P' + z^{-1}PR + PQ')P^{-1} \tag{4.10}$$

But from (4.1), $\hat{\Phi}'\hat{\Phi}^{-1} = z^r A$, and since A is analytic for z sufficiently large, $z^{-r}\hat{\Phi}'\hat{\Phi}^{-1}$ must be a convergent power series in z^{-1} for z sufficiently large, and hence analytic for large z.

It remains to compare the expressions (4.9) and (4.10). The formal series P may be written as

$$P = P_{(m)} + z^{-(m+1)}R_m \tag{4.11}$$

where $R_m = R_0 + z^{-1}R_1 + \cdots$ is another formal power series in z^{-1}. It will be useful to let J_k denote any formal-matrix power series in z^{-1} having z^{-k} as a factor, and thus J_k is such that

$$z^k J_k = \tilde{J}_0 + z^{-1}\tilde{J}_1 + \cdots$$

for some constant matrices $\tilde{J}_0, \tilde{J}_1, \ldots$. Using this notation, (4.11) may be written as

$$P = P_{(m)} + J_{m+1} \tag{4.12}$$

† By (4.6) is meant $|E_{(m)}(z)| \cdot |z|^{m+1-r} = 0(1)$, $|z| \to +\infty$, where the bound depends on m.

Now

$$\det P = \det P_{(m)} + J_{m+1}$$

and this implies that

$$(\det P)^{-1} = (\det P_{(m)})^{-1} + J_{m+1}$$

If adj P is the formal matrix such that

$$P \operatorname{adj} P = (\det P)E$$

then, since the matrix adj P has as its elements the cofactors of the elements of P, it follows that

$$\operatorname{adj} P = \operatorname{adj} P_{(m)} + J_{m+1}$$

Therefore

$$P^{-1} = P_{(m)}^{-1} + J_{m+1} \tag{4.13}$$

From (4.12) one obtains

$$P' = P'_{(m)} + J_{m+2} \tag{4.14}$$

and combining (4.12), (4.13), and (4.14) there results from (4.9) and (4.10),

$$\hat{\Phi}'\hat{\Phi}^{-1} = \hat{\Phi}'_{(m)}\hat{\Phi}_{(m)}^{-1} + J_{m+1-r} \tag{4.15}$$

The r in the last term is due to the term $PQ'P^{-1}$ in (4.10). But (4.15) implies that

$$z^r A = z^r B_{(m)} + J_{m+1-r} \tag{4.16}$$

and since A and $B_{(m)}$ are both analytic for z sufficiently large, so is $z^{-r}J_{m+1-r}$. Denoting J_{m+1-r} in (4.16) by $E_{(m)}$, one sees that this $E_{(m)}$ satisfies the conditions of the lemma.

The asymptotic nature of the formal solutions will first be deduced for the case when $z = t$ is real. Theorem 2.1 and Lemma 4.1 apply to this particular case. In order to prove Lemma 4.2 below, some notation will be required.

For fixed integral $m \geqq 0$, let the column vectors of $\hat{\Phi}_{(m)}$ be denoted by $\hat{\varphi}_{(m)i}$ $(i = 1, \ldots, n)$. Then

$$\hat{\varphi}_{(m)i} = p_{(m)i}t^{\rho_i}e^{q_i} \tag{4.17}$$

where

$$q_i(t) = \lambda_i \frac{t^{r+1}}{r+1} + \lambda_i^{(1)}\frac{t^r}{r} + \cdots + \lambda_i^{(r)}t \tag{4.18}$$

$p_{(m)i}$ is the ith column of $P_{(m)}$, and ρ_i is the element in the ith row and column of R (see Theorem 2.1).

Consider, for the following, a *fixed* i. Since $\Re q_j$ is a polynomial, its behavior as $t \to \infty$ is determined by the term of highest power in t. Divide the integers $j = 1, \ldots, n$ into two classes I_1, I_2 according to the following rule:

$$\begin{array}{lll} j \in I_1 & \text{if } \Re(q_i - q_j) \to +\infty & (t \to +\infty) \\ j \in I_2 & \text{if } \Re(q_i - q_j) \text{ is bounded above} & (t \to +\infty) \end{array} \tag{4.19}$$

Of course I_1, I_2 depend on the i chosen. Let, further, $\rho = \max_j \Re\rho_j$.

Lemma 4.2. *Let m be any positive integer such that $m - r - \rho + \Re\rho_j > 0$, for all $j = 1, \ldots, n$. Corresponding to any column vector $\hat{\varphi}_{(m)i}$ of $\hat{\Phi}_{(m)}$, there exists an actual solution vector $\varphi_{(m)i}$ of the system*

$$w' = t^r A(t) w \tag{4.20}$$

such that

$$|\varphi_{(m)i}(t)| = 0(t^\rho e^{\Re q_i(t)}) \qquad (t \to +\infty)$$

Proof. The solution $\varphi_{(m)i}$ will be constructed using the method of successive approximations, combined with a version of the variation-of-constants formula.

Consider the two systems

$$w' = t^r A w = t^r B_{(m)} w + E_{(m)} w \tag{4.21}$$

$$w' = t^r B_{(m)} w \tag{4.22}$$

From the definition of $B_{(m)}$, the matrix $\hat{\Phi}_{(m)}(t)$ is a fundamental matrix for (4.22), if t is sufficiently large ($\det P_0 \neq 0$). Thus if (4.21) is regarded as a nonhomogeneous system with (4.22) as the corresponding homogeneous system, the variation-of-constants formula can be applied to express solutions of (4.21) in terms of a quadrature of solutions of (4.22). In doing this, the limits of integration must be set correctly. Let t_0 be so large that $\hat{\Phi}^{-1}_{(m)}(t)$ exists for $t \geqq t_0$, and split $\hat{\Phi}^{-1}_{(m)}(t)$ into two parts

$$\hat{\Phi}^{-1}_{(m)} = e^{-Q} t^{-R} P^{-1}_{(m)} = \hat{\Psi}^{(1)}_{(m)} + \hat{\Psi}^{(2)}_{(m)}$$

where the jth row of $\hat{\Psi}^{(1)}_{(m)}$ is identical with the jth row of $\hat{\Phi}^{-1}_{(m)}$, or identically zero, according as $j \in I_1$ or $j \in I_2$; similarly for $\hat{\Psi}^{(2)}_{(m)}$. Thus the nonzero rows of $\hat{\Psi}^{(k)}_{(m)}$ consist of those rows of $\hat{\Phi}^{-1}_{(m)}$ which have as factor $e^{-q_j(t)}$ for $j \in I_k$, $k = 1, 2$.

The integral equation to be considered is the following:

$$w(t) = \hat{\varphi}_{(m)i}(t) + \int_{t_0}^{t} K_1(t,\tau) w(\tau)\, d\tau + \int_{\infty}^{t} K_2(t,\tau) w(\tau)\, d\tau \qquad (t_0 \leqq t < \infty) \tag{4.23}$$

where

$$K_1(t,\tau) = \hat{\Phi}_{(m)}(t) \hat{\Psi}^{(1)}_{(m)}(\tau) E_{(m)}(\tau), \qquad K_2(t,\tau) = \hat{\Phi}_{(m)}(t) \hat{\Psi}^{(2)}_{(m)}(\tau) E_{(m)}(\tau) \tag{4.24}$$

By direct verification, it follows that if $w = \varphi(t)$ satisfies (4.23), where the integral $\int_\infty^t$ converges, then φ satisfies (4.21).

In order to solve (4.23), define the successive approximations for $t \geqq t_0$ by

$$\varphi^0(t) \equiv 0$$
$$\varphi^{k+1}(t) = \hat{\varphi}_{(m)i}(t) + \int_{t_0}^{t} K_1(t,\tau)\varphi^k(\tau)\,d\tau + \int_{\infty}^{t} K_2(t,\tau)\varphi^k(\tau)\,d\tau, \qquad (k = 0, 1, \ldots) \tag{4.25}$$

It is a matter of proof that each of the approximations exists. This will be omitted since it is entirely similar to the proof which will be given below concerning the magnitudes of the successive differences. In order to appraise these, it is necessary to appraise the kernels K_1 and K_2. Now

$$|K_1(t,\tau)| \leqq |\hat{\Phi}_{(m)}(t)\hat{\Psi}^{(1)}_{(m)}(\tau)|\,|E_{(m)}(\tau)| \tag{4.26}$$

and from Lemma 4.1,

$$|E_{(m)}(\tau)| = 0(\tau^{r-m-1}) \qquad (\tau \to +\infty) \tag{4.27}$$

The i,jth element of the matrix $\hat{\Phi}_{(m)}(t)\hat{\Psi}^{(1)}_{(m)}(\tau)$ is given by

$$(\hat{\Phi}_{(m)}(t)\hat{\Psi}^{(1)}_{(m)}(\tau))_{ij} = \sum_{l \in I_1} (P_{(m)}(t))_{il}(P^{-1}_{(m)}(\tau))_{lj} \left(\frac{t}{\tau}\right)^{\rho_l} e^{q_l(t)-q_l(\tau)} \tag{4.28}$$

Because of the convergence of $P^{-1}_{(m)}$ for large enough t,

$$|P_{(m)}(t)|,\ |P^{-1}_{(m)}(t)| = 0(1) \qquad (t \to +\infty) \tag{4.29}$$

From (4.26) through (4.29), and a similar consideration for $K_2(t,\tau)$, it then follows that there exists a constant $c = c(m) > 0$ and a t_0 sufficiently large [which can be taken to be the t_0 in (4.25)] such that

$$|K_j(t,\tau)| \leqq c \sum_{l \in I_j} t^{\Re\rho_l}\,\tau^{r-m-1-\Re\rho_l}\,e^{\Re(q_l(t)-q_l(\tau))} \qquad (t,\tau \geqq t_0,\ j = 1, 2) \tag{4.30}$$

Further assume that t_0 is so large that†

$$\sum_{l=1}^{n} \int_{t_0}^{\infty} \tau^{r+\rho-m-1-\Re\rho_l}\,d\tau < \frac{1}{4c} \tag{4.31}$$

and that

$$\begin{aligned} &\Re(q_i(t) - q_l(t)) \text{ is increasing} && (l \in I_1,\ t \geqq t_0) \\ &\Re(q_i(t) - q_l(t)) \text{ is nonincreasing} && (l \in I_2,\ t \geqq t_0) \end{aligned} \tag{4.32}$$

From (4.25)

$$\varphi^1(t) = \hat{\varphi}_{(m)i}(t) = p_{(m)i}(t)t^{\rho_i}e^{q_i(t)}$$

and therefore

$$|\varphi^1(t) - \varphi^0(t)| \leqq ct^{\rho}e^{\Re q_i(t)} \qquad (t \geqq t_0)$$

† All the integrals in (4.31) exist because of the assumption on m.

Assume

$$|\varphi^k(t) - \varphi^{k-1}(t)| \leqq c2^{-(k-1)}t^{\rho}e^{\Re q_i(t)} \qquad (t \geqq t_0) \tag{4.33}$$

Then from (4.25)

$$|\varphi^{k+1}(t) - \varphi^k(t)| \leqq \Gamma_1 + \Gamma_2 \tag{4.34}$$

where

$$\Gamma_1 = \int_{t_0}^{t} |K_1(t,\tau)|\,|\varphi^k(\tau) - \varphi^{k-1}(\tau)|\,d\tau$$
$$\Gamma_2 = \int_{t}^{\infty} |K_2(t,\tau)|\,|\varphi^k(\tau) - \varphi^{k-1}(\tau)|\,d\tau$$

On account of (4.30) and (4.33) one has

$$\Gamma_1 \leqq c^2 2^{-(k-1)} t^{\rho} e^{\Re q_i(t)} \sum_{l \in I_1} \int_{t_0}^{t} \tau^{r+\rho-m-1-\Re\rho_l} e^{\Re(q_i(\tau)-q_l(\tau)+q_l(t)-q_i(t))}\,d\tau$$

and by virtue of (4.32), $\Re(q_i(\tau) - q_l(\tau) + q_l(t) - q_i(t)) \leqq 0$ for $t_0 \leqq \tau \leqq t$.
Therefore,

$$\Gamma_1 \leqq c^2 2^{-(k-1)} t^{\rho} e^{\Re q_i(t)} \sum_{l \in I_1} \int_{t_0}^{\infty} \tau^{r+\rho-m-1-\Re\rho_l}\,d\tau$$

and using (4.31)

$$\Gamma_1 \leqq c2^{-(k+1)} t^{\rho} e^{\Re q_i(t)}$$

A similar argument shows that Γ_2 satisfies the same inequality as Γ_1. Thus (4.33) is true with k replaced by $k + 1$, and since (4.33) is true for $k = 1$, (4.33) is established for all k by induction.

Therefore the series

$$\varphi^0(t) + \sum_{k=1}^{\infty} (\varphi^k(t) - \varphi^{k-1}(t))$$

is absolutely uniformly convergent to a vector function $\varphi = \varphi_{(m)i}(t)$ on every finite interval $t_0 \leqq t < T < \infty$. Also

$$|\varphi_{(m)i}(t)| \leqq ct^{\rho}e^{\Re q_i(t)} \sum_{k=1}^{\infty} 2^{-(k-1)} = 2ct^{\rho}e^{\Re q_i(t)} \qquad (t \geqq t_0) \tag{4.35}$$

Using the standard argument for successive approximations, it now follows that $\varphi_{(m)i}$ satisfies (4.23), and hence the differential system (4.21), thus proving Lemma 4.2.

Lemma 4.3. *For any sufficiently large m, the solution $\varphi_{(m)i}$ of Lemma 4.2 satisfies*

$$|\varphi_{(m)i}(t) - \hat{\varphi}_{(m)i}(t)| = 0(t^{\Re\rho_i+\mu-m}e^{\Re q_i(t)}) \qquad (t \to \infty) \tag{4.36}$$

where μ is a positive integer independent of m and of i.

Proof. Just how large m must be will be seen in the following. From

(4.23), if $t > 2t_0$, and $m > r + \rho - \Re\rho_j$, $(j = 1, \ldots, n)$, there exists a solution $\varphi_{(m)i}$ such that

$$|\varphi_{(m)i}(t) - \hat{\varphi}_{(m)i}(t)| \leqq \Lambda_1 + \Lambda_2 + \Lambda_3 \tag{4.37}$$

where

$$\Lambda_1 = \int_{t_0}^{t/2} |K_1(t,\tau)|\,|\varphi_{(m)i}(\tau)|\,d\tau, \qquad \Lambda_2 = \int_{t/2}^{t} |K_1(t,\tau)|\,|\varphi_{(m)i}(\tau)|\,d\tau$$
$$\Lambda_3 = \int_t^\infty |K_2(t,\tau)|\,|\varphi_{(m)i}(\tau)|\,d\tau \tag{4.38}$$

Consider Λ_2 first. From (4.30), (4.32), and (4.35),

$$\Lambda_2 \leqq 2c^2 t^\rho e^{\Re q_i(t)} \sum_{l=1}^{n} \int_{t/2}^{\infty} \tau^{r-m-1+\rho-\Re\rho_l}\,d\tau$$

Now

$$t^\rho \int_{t/2}^{\infty} \tau^{r-m-1+\rho-\Re\rho_l}\,d\tau = 0(t^{r-m+2\rho-\Re\rho_l}) = 0(t^{\mu-m+\Re\rho_i}) \qquad (t \to +\infty)$$

where μ is any positive integer exceeding $r + 2\rho - 2\min_l \Re\rho_l$. Choose m so large that $\mu - m < 0$.
Then

$$\Lambda_2 = 0(t^{\mu-m+\Re\rho_i} e^{\Re q_i(t)}) \qquad (t \to +\infty) \tag{4.39}$$

A similar argument shows that this also holds with Λ_3 replacing Λ_2.

Turning now to Λ_1, apply (4.30), (4.31), (4.32), (4.35) to Λ_1. What results is

$$\Lambda_1 \leqq \frac{c}{2} t^\rho e^{\Re q_i(t)} \sum_{l \in I_1} e^{\Re(q_l(t) - q_i(t) + q_i(t/2) - q_l(t/2))} \tag{4.40}$$

Let σ_{il} denote the highest power of t appearing in $\Re(q_i(t) - q_l(t))$. Since $\Re(q_i - q_l)$ is increasing for $t \geqq t_0 (l \in I_1)$, the coefficient β_{il} of $t^{\sigma_{il}}$ in $\Re(q_i(t) - q_l(t))$ is such that $\beta_{il} > 0$. The coefficient of $t^{\sigma_{il}}$ in

$$\Re\left(q_l(t) - q_i(t) + q_i\left(\frac{t}{2}\right) - q_l\left(\frac{t}{2}\right)\right)$$

is then given by

$$\frac{\beta_{il}}{2^{\sigma_{il}}} - \beta_{il} = \beta_{il}\left(\frac{1}{2^{\sigma_{il}}} - 1\right) < 0$$

for $\sigma_{il} > 0$. From (4.40), therefore, the term under the summation sign is $0(e^{-\gamma t})$ for some $\gamma > 0$, and hence, in particular,

$$\Lambda_1 = 0(t^{\mu-m+\Re\rho_i} e^{\Re q_i(t)}) \qquad (t \to +\infty) \tag{4.41}$$

Combining (4.37) through (4.41), one obtains (4.36), which proves the lemma.

Lemma 4.4. *If m is sufficiently large, then for any fixed integer $m' > m$,*

$$|\varphi_{(m')i}(t) - \varphi_{(m)i}(t)| = 0(e^{\Re q_i(t) - at}) \qquad (t \to +\infty) \tag{4.42}$$

where a is a positive constant independent of m' and m.

Proof. Choose m so large that Lemma 4.3 holds, and let $m' = m + l$, l a positive integer. From Lemma 4.3,

$$\varphi_{(m)i}(t) = \hat{\varphi}_{(m)i}(t) + 0(t^{\Re \rho_i + \mu - m} e^{\Re q_i(t)}) \qquad (t \to +\infty) \tag{4.43}$$

where m is such that $\mu - m < 0$, and hence, if $\Phi_{(m)}$ is the matrix with the column vectors $\varphi_{(m)1}, \ldots, \varphi_{(m)n}$

$$\begin{aligned}\Phi_{(m)}(t)e^{-Q(t)}t^{-R} &= \hat{\Phi}_{(m)}(t)e^{-Q(t)}t^{-R} + 0(t^{\mu - m}) \\ &= P_{(m)}(t) + 0(t^{\mu-m}) = P_0 + 0(t^{-1}) \qquad (t \to \infty)\end{aligned}$$

Since by Theorem 2.1 det $P_0 \neq 0$, this proves that det $\Phi_{(m)}(t) \neq 0$ for all t sufficiently large, and hence $\Phi_{(m)}$ is a fundamental matrix for $w' = t^r A w$ for all sufficiently large t.

But $\Phi_{(m+l)}$ is also a fundamental matrix, and hence

$$\varphi_{(m)i}(t) = \sum_{j=1}^{n} c_{ij}\varphi_{(m+l)j}(t) \tag{4.44}$$

where the c_{ij} are constants. Recall that

$$\hat{\varphi}_{(m)i}(t) = p_{(m)i}(t)t^{\rho_i}e^{q_i(t)} \tag{4.45}$$

where $p_{(m)i}(t)$ is the ith column of $P_{(m)}(t) = \sum_{k=0}^{m} t^{-k}P_k$, and hence is of the form

$$p_{(m)i}(t) = \sum_{k=0}^{m} t^{-k}p_{ik} \tag{4.46}$$

where the p_{ik} are constant vectors. Recall also that if A_0 is assumed to be in diagonal form, then P_0 can be chosen to be the unit matrix E. This is clearly no loss of generality, and so this will be assumed in the following. Then p_{i0} is a vector with 1 in the ith row, and 0 elsewhere. By virtue of (4.43), (4.44) is equivalent to

$$\hat{\varphi}_{(m)i}(t) + 0(t^{\Re\rho_i + \mu - m}e^{\Re q_i(t)}) = \sum_{j=1}^{n} c_{ij}\hat{\varphi}_{(m+l)j}(t) + 0\left(\sum_{j=1}^{n} |c_{ij}| t^{\Re\rho_j + \mu - m - l}e^{\Re q_j(t)}\right)$$

and by (4.45), (4.46), this gives

$$t^{\rho_i}e^{q_i(t)}(c_{ii} - 1)p_{(m)i}(t) + \sum_{j \neq i} c_{ij}t^{\rho_j}e^{q_j(t)}p_{(m)j}(t) = 0(t^{\Re\rho_i+\mu-m}e^{\Re q_i(t)})$$

$$+ 0\left(\sum_{j=1}^{n} |c_{ij}|t^{\Re\rho_j+\mu-m}e^{\Re q_j(t)}\right) \qquad (4.47)$$

Let, as before, I_1 denote the set of all integers k, $(k = 1, \ldots, n)$, such that $\Re(q_i - q_k) \to +\infty$, $(t \to +\infty)$, and I_2 the complementary set in 1, 2, . . . , n. From the structure of the q_i as polynomials with no constant term, it follows that $k \in I_2$ if and only if either $\Re q_i = \Re q_k$ or $\Re(q_i - q_k) \to -\infty$, $(t \to +\infty)$. It will now be shown that $c_{ii} = 1$ and if $k \neq i$, $k \in I_2$, then $c_{ik} = 0$ in (4.44).

Let I_{21} be the set of all k such that $\Re(q_i - q_k) \to -\infty$, $(t \to +\infty)$. Suppose I_{21}^* is the set of all $k' \in I_{21}$ such that $\Re(q_k - q_{k'})$ is bounded above as $t \to +\infty$, for all $k \in I_{21}$. Let k'' be any integer in I_{21}^* such that $\Re\rho_{k''} \geqq \Re\rho_{k'}$, for all $k' \in I_{21}^*$. Divide (4.47) by $t^{\rho_{k''}}e^{q_{k''}(t)}$ and let $t \to +\infty$. If attention is confined to the k''th row, what results is $c_{ik''} = 0$. Continuing in this fashion, one shows $c_{ik} = 0$ for all $k \in I_{21}$.

Let I_{22} be the set of all $k \neq i$ such that $\Re q_i = \Re q_k$, and let $k' \in I_{22}$ be such that $\Re\rho_{k'} \geqq \Re\rho_k$, for all $k \in I_{22}$. Divide (4.47) by $t^{\rho_{k'}}e^{q_{k'}(t)}$ and let $t \to +\infty$. By observing the k'th row, in passing to the limit, it is found that $c_{ik'} = 0$, if $\Re\rho_{k'} > \Re\rho_i$. It is true that for $\Re\rho_{k'} = \Re\rho_i$ the k'th row shows again that $c_{ik'} = 0$. Then it follows easily that $c_{ii} = 1$, by dividing by $t^{\rho_i}e^{q_i(t)}$ and letting $t \to +\infty$. After that, the argument goes nicely for $\Re\rho_{k'} < \Re\rho_i$, $k' \in I_{22}$. Here m must be assumed so large that $\Re\rho_i - \Re\rho_{k'} + \mu - m < 0$ for all $k' \in I_{22}$ such that $\Re\rho_{k'} < \Re\rho_i$.

Therefore, from (4.44) one obtains

$$\varphi_{(m)i}(t) = \varphi_{(m+l)i}(t) + \sum_{j \in I_1} c_{ij}\varphi_{(m+l)j}(t)$$

From this it follows, using (4.43) and (4.45), that

$$\varphi_{(m)i}(t) = \varphi_{(m+l)i}(t) + 0(e^{\Re q_i(t)}E(t))$$

where $E(t) = 0(e^{-at})$ for some constant $a > 0$, which does not depend on m or m'. This proves the estimate (4.42).

It is now possible to prove the asymptotic nature of the formal solutions in the real case.

Theorem 4.1. *Let $\hat{\varphi}_i = p_i t^{\rho_i} e^{q_i}$ be any column vector of the formal solution matrix $\hat{\Phi} = Pt^R e^Q$ of (4.20), where A satisfies the conditions of Theorem 2.1 for $z = t$. Then there exists for all sufficiently large t an actual-solution vector of this system, φ_i, such that*

$$|\varphi_i(t) - \hat{\varphi}_{(m)i}(t)| = O(t^{\Re\rho_i - m - 1} e^{\Re q_i(t)}) \qquad (t \to +\infty) \tag{4.48}$$

holds for all $m = 0, 1, 2, \ldots$. *In particular,* $\varphi_i \sim \hat{\varphi}_i$.

Proof. From Lemma 4.3, for every m' sufficiently large, say $m' \geqq m_1$, there exists a solution $\varphi_{(m')i}$ such that

$$\varphi_{(m')i}(t) = \hat{\varphi}_{(m')i}(t) + O(t^{\Re\rho_i + \mu - m'} e^{\Re q_i(t)}) \qquad (t \to +\infty) \tag{4.49}$$

where μ is a positive integer independent of m'. The integer m_1 can be chosen so that $m_1 > \mu$. By Lemma 4.4, for m sufficiently large, say $m \geqq m_2 \geqq m_1$, and $m' > m$,

$$\varphi_{(m)i}(t) = \varphi_{(m')i}(t) + O(e^{\Re q_i(t) - at}) \qquad (t \to +\infty) \tag{4.50}$$

where a is a positive constant. Combining (4.49) and (4.50), there results for $m = m_2$, and all $m' > m_2$,

$$\varphi_{(m_2)i}(t) = \hat{\varphi}_{(m')i}(t) + O(t^{\Re\rho_i + \mu - m'} e^{\Re q_i(t)}) \qquad (t \to +\infty) \tag{4.51}$$

But, by the definition of $\hat{\varphi}_{(m')i}(t)$, it follows that

$$\hat{\varphi}_{(m')i}(t) = \hat{\varphi}_{(m' - \mu - 1)i}(t) + O(t^{\Re\rho_i + \mu - m'} e^{\Re q_i(t)}) \qquad (t \to +\infty) \tag{4.52}$$

Letting $m = m' - \mu - 1$, and combining (4.51) and (4.52), one gets for all $m > m_2 - \mu - 1$

$$\varphi_{(m_2)i}(t) = \hat{\varphi}_{(m)i}(t) + O(t^{\Re\rho_i - m - 1} e^{\Re q_i(t)}) \qquad (t \to +\infty) \tag{4.53}$$

It remains to prove (4.53) for $m = 0, 1, \ldots, m_2 - \mu - 1$. Since

$$\hat{\varphi}_{(m)i}(t) = \hat{\varphi}_{(m-1)i}(t) + O(t^{\Re\rho_i - m} e^{\Re q_i(t)}) \qquad (t \to +\infty)$$

it follows that (4.53) holds for $m = m_2 - \mu - 1$. Using an induction, it is easy to see that (4.53) must be valid for $m = 0, 1, 2, \ldots$, thus proving the theorem if φ_i is chosen to be the solution $\varphi_{(m_2)i}$.

5. The Asymptotic Nature of the Formal Solutions in the Complex Case

The result given in Theorem 4.1 holds not only for real $z = t$, but it is also obviously valid on each radial line $z = te^{i\theta}$, for any fixed θ. However, the theorem does not relate the solutions along one radial line with the solutions along another radial line. Here it will be shown that, with the aid of certain theorems in (complex) function theory, the result of Theorem 4.1 can be used to prove that a solution φ_i with the asymptotic expansion $\hat{\varphi}_i$ along an appropriate radial direction actually has $\hat{\varphi}_i$ as an asymptotic expansion in a sector of the z plane.

The results required from function theory are due to Phragmen and Lindelöf, and are extensions of the maximum-modulus theorem.† A

† For proofs of these theorems, see E. C. Titchmarsh, *The theory of functions*, Oxford, 1939, pp. 177–180.

statement of the results needed here follows. With $z = |z|e^{i\theta}$, the notation $\arg z = \theta$ will be used.

Theorem A. *Let f be an analytic function for*

$$k \leqq |z| < \infty, \theta_1 \leqq \arg z \leqq \theta_2$$

where k, θ_1, θ_2 are real constants. Let

$$f(z) = 0(e^{c|z|^m}) \qquad (|z| \to \infty)$$

uniformly in $\theta_1 \leqq \arg z \leqq \theta_2$, for some constants c and m such that

$$m(\theta_2 - \theta_1) < \pi$$

If f is bounded as $|z| \to \infty$ on the lines $\arg z = \theta_1$ and $\arg z = \theta_2$, then f is bounded uniformly as $|z| \to \infty$ in $\theta_1 \leqq \arg z \leqq \theta_2$.

Theorem B. *Let f be analytic and uniformly bounded for the region in Theorem A. Moreover, suppose that there exist constants a and b such that $f(z) \to a$ as $|z| \to \infty$ on $\arg z = \theta_1$, and $f(z) \to b$ as $|z| \to \infty$ on $\arg z = \theta_2$. Then $a = b$ and $f(z) \to a$ uniformly in $\theta_1 \leqq \arg z \leqq \theta_2$ as $|z| \to \infty$.*

Recall that the system under consideration is

$$w' = z^r A(z)w \qquad (r \geqq 0) \tag{5.1}$$

where

$$A(z) = \sum_{k=0}^{\infty} z^{-k} A_k \tag{5.2}$$

and the latter series converges for $|z| > d$, for some $d > 0$. An immediate consequence of (5.1) is the existence of a constant $c_1 > 0$ such that any solution φ of (5.1) satisfies $|\varphi'| \leqq c_1|z|^r|\varphi|$ for large $|z|$. If $c = c_1/(r + 1)$, this implies

$$\varphi(z) = 0(e^{c|z|^{r+1}}) \qquad (|z| \to \infty) \tag{5.3}$$

uniformly in any fixed sector of the z plane bounded by two radial lines.

Let the integer i $(1 \leqq i \leqq n)$ be chosen and kept fixed in the discussion that follows. Since the characteristic roots λ_j $(j = 1, \ldots, n)$ of the matrix A_0 are assumed to be distinct, it follows that the equation

$$\Re[(\lambda_i - \lambda_j)z^{r+1}] = 0 \qquad (j \neq i) \tag{5.4}$$

determines a finite number of directions in the z plane. These are the directions $\arg z = \theta \pmod{2\pi}$ for which

$$\cos [\arg (\lambda_i - \lambda_j) + (r + 1)\theta] = 0$$

Let S_i be a sector

$$S_i: \qquad \alpha \leqq \arg z \leqq \beta$$

such that all the directions determined by (5.4) are exterior to S_i. The following result will be proved.

Theorem 5.1. *If $\hat{\varphi}_i = p_i z^{\rho_i} e^{q_i}$ is any formal-solution vector of* (5.1), *then there exists in the sector S_i, determined above, an actual solution φ_i of* (5.1) *for all sufficiently large $|z|$, such that*

$$\varphi_i \sim \hat{\varphi}_i$$

uniformly in S_i.

An easy corollary to the above theorem is the following result.

Theorem 5.2. *If a sector S in the z plane contains no direction for which*

$$\Re[(\lambda_l - \lambda_j)z^{r+1}] = 0 \qquad (l, j = 1, \ldots, n; l \neq j)$$

then there exists a fundamental set of solutions φ_i $(i = 1, \ldots, n)$ of (5.1) *in S for all z sufficiently large such that*

$$\varphi_i \sim \hat{\varphi}_i \qquad (i = 1, \ldots, n)$$

uniformly in S.

Proof of Theorem 5.1. For any $j = 1, \ldots, n$,

$$q_i(z) - q_j(z) = (\lambda_i - \lambda_j)\frac{z^{r+1}}{r+1} + (\lambda_i^{(1)} - \lambda_j^{(1)})\frac{z^r}{r} + \cdots + (\lambda_i^{(r)} - \lambda_j^{(r)})z$$

and it is clear that the behavior of $\Re(q_i - q_j)$ in S_i, for $j \neq i$, as $|z| \to \infty$, depends on the first term

$$\Re\left[\frac{(\lambda_i - \lambda_j)z^{r+1}}{r+1}\right]$$

since $|\lambda_i - \lambda_j| \neq 0$. It follows from the definition of S_i that the integers $j = 1, 2, \ldots, i-1, i+1, \ldots, n$ fall into two classes I_1 and I_2, where

$$\Re(q_i - q_j) \to \infty \qquad (j \in I_1) \tag{5.5}$$

uniformly in S_i as $|z| \to \infty$, and

$$\Re(q_i - q_j) \to -\infty \qquad (j \in I_2) \tag{5.6}$$

uniformly in S_i as $|z| \to \infty$.

Theorem 4.1 gives the existence of a solution φ_i of (5.1) on the line $\arg z = \alpha$ for all sufficiently large $|z|$ having the property that

$$\varphi_i \sim \hat{\varphi}_i \qquad (\arg z = \alpha) \tag{5.7}$$

By uniqueness, it follows that φ_i can be continued off the line $\arg z = \alpha$, and hence it may be assumed that φ_i exists for all sufficiently large $|z|$ and satisfies (5.7) on $\arg z = \alpha$. Now let γ be chosen so that

$$\alpha < \gamma \leqq \beta \qquad \text{and} \qquad \gamma - \alpha < \frac{\pi}{r+1} \tag{5.8}$$

It can also be assumed that γ is chosen so that on $\arg z = \gamma$ one has, for any l and j, such that $l \neq j$, either

$$\Re(q_l - q_j) \to \infty \qquad (\arg z = \gamma)$$

or

$$\Re(q_l - q_j) \to -\infty \qquad (\arg z = \gamma)$$

From Theorem 4.1 it follows that along the line $\arg z = \gamma$ there exists for sufficiently large $|z|$ a fundamental set of solutions $\psi_1, \ldots, \psi_n$ of (5.1) such that

$$\psi_j \sim \hat{\varphi}_j \qquad (\arg z = \gamma) \tag{5.9}$$

Thus, for some constants $c_1, \ldots, c_n$,

$$\varphi_i = \sum_{j=1}^{n} c_j \psi_j \tag{5.10}$$

for all sufficiently large $|z|$.

Assume that there exists $k \varepsilon I_2$ such that $c_k \neq 0$ and such that for all $l \varepsilon I_2$, with $l \neq k$ and $c_l \neq 0$, it is the case that

$$\Re(q_k - q_l) \to \infty \qquad (\arg z = \gamma) \tag{5.11}$$

as $|z| \to \infty$. If $m > 0$ is any fixed integer, then by (5.9) and (5.10) one has in the notation of Theorem 4.1

$$\varphi_i = c_k \hat{\varphi}_{(m)k} + 0(|z|^{\Re \rho_k - m - 1} e^{\Re q_k(z)}) \qquad (\arg z = \gamma)$$

and in particular, if

$$f(z) = \varphi_i(z) e^{-q_k(z)} z^{-\rho_k}$$

then

$$f(z) = a + 0(|z|^{-1}) \qquad (|z| \to \infty, \arg z = \gamma)$$

where a is a constant vector, not identically zero. On $\arg z = \alpha$ it follows, since $k \varepsilon I_2$, and (5.7) holds, that

$$f(z) = 0(|z|^{-1}) \qquad (|z| \to \infty, \arg z = \alpha)$$

Recalling (5.3) and using Theorem A and then Theorem B for each component of f, it follows that $a = 0$, which is impossible. Hence k does not exist and the nonvanishing terms on the right of (5.10) are from $j \varepsilon I_1$ and $j = i$.

From (5.7) it follows that

$$g(z) = \varphi_i(z) e^{-q_i(z)} z^{-\rho_i}$$

is such that

$$g(z) \to c \qquad (|z| \to \infty, \arg z = \alpha)$$

where c is the constant vector given by

$$c = \lim_{z \to \infty} \hat{\varphi}_{(m)i}(z) e^{-q_i(z)} z^{-\rho_i}$$

From (5.9) and (5.10) one has also that

$$g(z) \to c_i c \qquad (|z| \to \infty,\ \arg z = \gamma)$$

and by Theorem B this implies that $c_i = 1$. Thus (5.10) gives

$$(\varphi_i - \hat{\varphi}_{(m)i})e^{-q_i z - \rho_i z^{m+1}} = 0(1) \qquad (|z| \to \infty,\ \arg z = \gamma)$$

This relation is obviously valid for $\arg z = \alpha$ by (5.7). Thus, by applying Theorem A, it is valid uniformly in the region $\alpha \leqq \arg z \leqq \gamma$. This is equivalent to

$$\varphi_i \sim \hat{\varphi}_i \qquad (\alpha \leqq \arg z \leqq \gamma)$$

If $\gamma < \beta$, then repeating the above argument a finite number of times in sectors with angle less than $\pi/(r + 1)$ yields the theorem for the sector S_i.

The sector S_i can usually be increased in size by observing the following facts. A curve along which $\Re(q_i - q_j) = 0$ for some $j \neq i$ serves quite as well as a radial line in the hypothesis of Theorems A and B. Moreover, the results proved in Sec. 4 for z along a radial line will generally be valid along such a curve. By minor variations of the preceding argument, asymptotic relations can be proved in a sector S_i bounded by two such curves, but containing none such in its interior. Generally such relationships can be extended into an adjoining sector. This will be shown by an example. The method is quite general in scope.

Consider the equation (1.3) with $a = -\frac{1}{4}$ and the variables complex so that it becomes as a system

$$\frac{dw}{dz} = \begin{pmatrix} 0 & 1 \\ -\left(1 + \frac{1}{4z^2}\right) & 0 \end{pmatrix} w$$

with roots $\lambda_1 = i$ and $\lambda_2 = -i$. Taking the first component w_1 of the vector w and denoting it by u,

$$\frac{d^2u}{dz^2} + \left(1 + \frac{1}{4z^2}\right) u = 0 \tag{5.12}$$

and the two formal solutions of (5.12), as already seen in (1.5), are

$$\hat{\varphi}_1 = e^{iz}\left(1 + \frac{1}{8iz} + \cdots\right)$$

$$\hat{\varphi}_2 = e^{-iz}\left(1 - \frac{1}{8iz} + \cdots\right)$$

Consider now the solution of (5.12), φ_1, which by Theorem 5.1 satisfies for $\delta \leqq \arg z \leqq \pi - \delta$, where $\delta > 0$,

$$\varphi_1 \sim \hat{\varphi}_1 \tag{5.13}$$

Let ψ_1 and ψ_2 be the solutions of (5.12) which on $\arg z = 0$ are asymptotic to $\hat{\varphi}_1$ and $\hat{\varphi}_2$, respectively. Then for some constants c_1 and c_2

$$\varphi_1 = c_1\psi_1 + c_2\psi_2$$

Multiplying the above by e^{iz}, it follows that

$$e^{iz}\varphi_1(z) = c_1 e^{2iz} + c_2 + 0\left(\frac{1}{|z|}\right) \tag{5.14}$$

on $\arg z = 0$ as $z \to +\infty$. By (5.13) it follows that

$$e^{iz}\varphi_1(z) = e^{2iz}\left(1 + 0\left(\frac{1}{|z|}\right)\right) \tag{5.15}$$

on $\arg z = \pi/2$ as $|z| \to \infty$. Let

$$F(z) = \frac{1}{z}\int_{x_0}^{z} e^{is}\varphi_1(s)\,ds$$

where x_0 is large and positive and the integral extends in the upper half plane on the arc of the circle $|s| = x_0$ until $\arg s = \arg z$ and then the integral is taken along the radius $\arg s = \arg z$. Clearly (5.14) and (5.15) imply

$$\lim_{z \to \infty} F(z) = c_2 \qquad (\arg z = 0)$$

$$\lim_{z \to \infty} F(z) = 0 \qquad \left(\arg z = \frac{\pi}{2}\right)$$

Thus by Theorem A followed by Theorem B, $c_2 = 0$. Considering next $e^{-iz}\varphi_1(z)$ on $\arg z = 0$ and $\pi/2$, it follows that $c_1 = 1$. Thus

$$[\varphi_1(z) - \hat{\varphi}_{(m)1}(z)]e^{-iz}z^{m+1} = 0(1)$$

on $\arg z = 0$ and $\pi/2$, from which follows $\varphi_1 \sim \hat{\varphi}_1$ uniformly on $0 \leqq \arg z \leqq \pi/2$. The argument can now be repeated for $[\pi/2,\pi]$. Moreover, it can be repeated for $[\pi, 2\pi - \delta]$ and $[-\pi + \delta, 0]$ for any $\delta > 0$, so that finally

$$\varphi_1 \sim \hat{\varphi}_1 \qquad (-\pi + \delta \leqq \arg z \leqq 2\pi - \delta) \tag{5.16}$$

[Indeed, $z^{-\frac{1}{2}}\varphi_1(z)$ is, except for a constant factor, a Hankel function.]

Similar results hold for the solution φ_2,

$$\varphi_2 \sim \hat{\varphi}_2 \qquad (-2\pi + \delta \leqq \arg z \leqq \pi - \delta) \tag{5.17}$$

A solution of (5.12) is $z^{\frac{1}{2}}J_0(z)$, where J_0 is the Bessel function of zero order. Since φ_1 and φ_2 are independent

$$z^{\frac{1}{2}}J_0(z) = c_1\varphi_1(z) + c_2\varphi_2(z)$$

for some constants c_1 and c_2. [That $c_1 = \bar{c}_2 = (2/\pi)^{\frac{1}{2}}e^{-\pi i/4}$ is readily verifiable.] From (5.16) and (5.17) follows

$$z^{\frac{1}{2}}J_0(z) \sim c_1\hat{\varphi}_1(z) + c_2\hat{\varphi}_2(z) \qquad (-\pi + \delta \leqq \arg z \leqq \pi - \delta)$$

Since J_0 is an entire function of z, it is single-valued. The formal series on the right above is also single-valued. Because $z^{\frac{1}{2}}$ occurs on the left, the left side is multiple-valued so that the above asymptotic formula cannot be valid on $-\pi \leqq \arg z \leqq \pi$. Thus the result obtained is in a sense the best possible. By using obvious generalizations of the method of this example, asymptotic results can often be extended into sectors larger than S_i of Theorem 5.1.

6. The Case Where A_0 Has Multiple Characteristic Roots

This case is considerably more complicated than the simple situation treated in Sec. 2. The proof of the existence of formal solutions involves essential new difficulties. This can be illustrated in the real case $z = t$ by the example

$$tw'' + w' + w = 0 \tag{6.1}$$

The system associated with (6.1) is

$$\begin{aligned} w_1' &= w_2 \\ w_2' &= -t^{-1}w_1 - t^{-1}w_2 \end{aligned}$$

If w is the vector with components w_1, w_2, then

$$w' = (A_0 + t^{-1}A_1)w \tag{6.2}$$

where

$$A_0 = \begin{pmatrix} 0 & 1 \\ 0 & 0 \end{pmatrix} \qquad A_1 = \begin{pmatrix} 0 & 0 \\ -1 & -1 \end{pmatrix} \tag{6.3}$$

Therefore A_0 has a double root $\lambda = 0$, with nonsimple elementary divisor. If $t = s^2$, the equation (6.1) is transformed into

$$w'' + s^{-1}w' + 4w = 0 \qquad \left(' = \frac{d}{ds}\right) \tag{6.4}$$

with associated system

$$w' = (B_0 + s^{-1}B_1)w \qquad \left(' = \frac{d}{ds}\right) \tag{6.5}$$

where

$$B_0 = \begin{pmatrix} 0 & 1 \\ -4 & 0 \end{pmatrix} \qquad B_1 = \begin{pmatrix} 0 & 0 \\ 0 & -1 \end{pmatrix} \tag{6.6}$$

Thus the characteristic roots of B_0 are $\lambda = \pm 2i$, and hence Theorem 2.1 can be applied to (6.5). An implication of Theorem 2.1 is that (6.4) has a formal solution of the form

$$e^{2is}s^r\left(1 + \frac{c_1}{s} + \cdots\right) \tag{6.7}$$

By substituting (6.7) into (6.4), r is found to be $-\frac{1}{2}$. Since (6.4) is real, the complex conjugate of (6.7) must also be a formal solution. Setting $s = t^{\frac{1}{2}}$, this shows that (6.1) has formal solutions of the form

$$\hat{w} = c_1p_1(t^{\frac{1}{2}})t^{-\frac{1}{4}}e^{2it^{\frac{1}{2}}} + c_2p_2(t^{\frac{1}{2}})t^{-\frac{1}{4}}e^{-2it^{\frac{1}{2}}} \tag{6.8}$$

where c_1, c_2 are constants, and p_1, p_2 are formal power series in t^{-1}. Thus it is seen that fractional powers of t can enter into the exponential term and the formal series.

The following theorem can be shown to apply in the general case when A_0 has multiple characteristic roots, but because of the complexity of the proof, the latter will be omitted.

Theorem 6.1. *Consider the system*

$$w' = z^rA(z)w \tag{6.9}$$

where r is a nonnegative integer,

$$A(z) = \sum_{k=0}^{\infty} z^{-k}A_k$$

and the latter series converges for $|z| > a$ for some $a > 0$. Then there exists a formal-solution matrix of (6.9) of the form

$$\Phi = Se^Q \tag{6.10}$$

where Q is a diagonal matrix with diagonal elements q_i which are polynomials of the type

$$q_i(z) = q_{i0}(z^{1/h})^{l_i} + q_{i1}(z^{1/h})^{l_i-1} + \cdots + q_{i\,l_i-1}z^{1/h}$$

l_i and h being integers, and S is a matrix whose elements s_{ij} are formal expressions of the type

$$s_{ij} = z^{r_{ij}} \sum_{m=0}^{m_{ij}} \sigma_{ijm} \log^m z$$

Here the r_{ij} are constants and the σ_{ijm} are formal series

$$\sigma_{ijm} = \sum_{l=0}^{\infty} \sigma_{ijml}z^{-l/h}$$

where the σ_{ijml} are constants. Moreover, the formal determinant of S does not vanish for large $|z| < \infty$.

It can further be shown that there exist solutions of (6.9) which have these formal solutions as asymptotic expansions valid in certain sectors of the z plane. The proofs of Secs. 4 and 5 can be readily adapted to this more general case.

7. Irregular Singular Points of an nth-order Equation

Consider the nth-order equation of the form

$$\sum_{m=0}^{n} z^{mr} a_m(z) w^{(n-m)} = 0 \qquad (a_0(z) \equiv 1) \tag{7.1}$$

where $r \geqq 0$ is an integer, and the coefficients a_m are analytic in a neighborhood of $z = \infty$, that is,

$$a_m(z) = \sum_{p=0}^{\infty} a_{mp} z^{-p}$$

and the series are convergent for $|z| > a$ for some $a > 0$. If φ is any solution of (7.1), let the components φ_k of a vector be given by

$$\varphi_k = z^{-(k-1)r}\varphi^{(k-1)} \qquad (k = 1, \ldots, n) \tag{7.2}$$

Then it is easy to check that

$$\begin{aligned} \varphi_k' &= -(k-1)rz^{-1}\varphi_k + z^r\varphi_{k+1} \qquad (k = 1, \ldots, n-1) \\ \varphi_n' &= -(n-1)rz^{-1}\varphi_n - z^r(a_n\varphi_1 + a_{n-1}\varphi_2 + \cdots + a_1\varphi_n) \end{aligned} \tag{7.3}$$

Therefore, if φ is a solution of (7.1), the vector with components φ_k given by (7.2) is a solution of the system

$$w' = z^r A(z) w \tag{7.4}$$

where

$$A(z) = \begin{pmatrix} 0 & 1 & 0 & \cdots & 0 \\ 0 & -rz^{-r-1} & 1 & \cdots & \cdot \\ \cdot & \cdot & -2rz^{-r-1} & \cdots & \cdot \\ \cdot & \cdot & \cdot & \cdots & \cdot \\ \cdot & \cdot & \cdot & \cdots & 0 \\ 0 & \cdot & \cdot & \cdots\ -(n-2)rz^{-r-1} & 1 \\ -a_n(z) & -a_{n-1}(z) & \cdot & \cdots\ -a_2(z) & -(n-1)rz^{-r-1} - a_1(z) \end{pmatrix} \tag{7.5}$$

Conversely, the first component of any solution vector of (7.4), (7.5) will be a solution of (7.1). Therefore Theorems 2.1, 4.1, and 5.1 can be applied to (7.4), (7.5) to obtain the formal solutions and their asymptotic character, and, observing the first components, one gets the corresponding information for the equation (7.1).

If A is written as

$$A(z) = \sum_{k=0}^{\infty} z^{-k} A_k$$

where the A_k are constant matrices, then

$$A_0 = \begin{pmatrix} 0 & 1 & 0 & \cdots & 0 \\ 0 & 0 & 1 & \cdots & \cdot \\ \cdot & \cdot & \cdot & \cdots & \cdot \\ \cdot & \cdot & \cdot & \cdots & \cdot \\ \cdot & \cdot & \cdot & \cdots & 0 \\ 0 & \cdot & \cdot & \cdot\ 0\ 0 & 1 \\ -a_{n0} & \cdot & \cdot & \cdots & -a_{10} \end{pmatrix}$$

The characteristic equation of A_0, as was shown in (6.19), Chap. 3, is

$$\lambda^n + a_{10}\lambda^{n-1} + \cdots + a_{n0} = 0 \tag{7.6}$$

and this can be immediately read off by replacing $w^{(k)}$ in (7.1) by λ^k, and $z^{kr}a_k(z)$ by a_{k0}, the constant term in the expansion of a_k.

8. The Laplace Integral and Asymptotic Series

The Laplace integral can be made the basis for proving the existence of actual solutions of (1.2) which are represented asymptotically by formal log-exponential series.†

Here the special case

$$(a_0z + b_0)w^{(n)} + (a_1z + b_1)w^{(n-1)} + \cdots + (a_nz + b_n)w = 0 \tag{8.1}$$

will be treated. The a_j and b_j are constants. Let

$$P(s) = a_0s^n + a_1s^{n-1} + \cdots + a_n$$
$$Q(s) = b_0s^n + b_1s^{n-1} + \cdots + b_n$$

Let F be an analytic function and let

$$\varphi(z) = \int_C F(s)e^{sz}\,ds$$

where C is a path to be determined in the complex s plane. Suppose φ is a solution of (8.1). Then, since formally

$$\varphi^{(k)}(z) = \int_C F(s)s^k e^{sz}\,ds$$

† See for example E. L. Ince, *Ordinary differential equations*, London, 1927, chap. 19.

the equation (8.1) becomes

$$\int_C F(s)[zP(s) + Q(s)]e^{sz}\,ds = 0$$

Clearly, integration by parts yields

$$\int_C FPze^{sz}\,ds = F(s)P(s)e^{sz}\Big]_C - \int_C \frac{d}{ds}(FP)e^{sz}\,ds$$

where $FPe^{sz}\Big]_C$ represents the variation of the function on C. Thus

$$\int_C [FQ - FP' - F'P]e^{sz}\,ds + FPe^{sz}\Big]_C = 0$$

Choose F so that

$$F'P + F(P' - Q) = 0$$

Thus

$$F = \frac{1}{P}\exp\left[\int^s \frac{Q(\sigma)}{P(\sigma)}\,d\sigma\right] \tag{8.2}$$

The condition that φ satisfy (8.1) now becomes simply

$$V = \exp\left[\int^s \frac{Q(\sigma)}{P(\sigma)}\,d\sigma\right] e^{sz}\Big]_C = 0 \tag{8.3}$$

If the roots of $P(s) = 0$ are $s_1, \ldots, s_m$ and these are simple, then

$$\int^s \frac{Q(\sigma)}{P(\sigma)}\,d\sigma = R(s) + \sum_{j=1}^{m} \alpha_j \log(s - s_j)$$

where $R(s)$ is a polynomial and the α_j are constants. Thus

$$V = e^{sz+R(s)} \prod_{j=1}^{m} (s - s_j)^{\alpha_j}\Big]_C$$

and C is chosen so that $V = 0$. This may require that z be restricted.

In case the degree of $P(s)$ is n, then $R(s) = as$, where a is a constant and

$$V = e^{(z+a)s} \prod_{j=1}^{n} (s - s_j)^{\alpha_j}\Big]_C$$

If $\Re\alpha_k > 0$, let

$$\varphi_k(z) = \int_{s_k}^{\infty} e^{s(z+a)} \prod_{j=1}^{n} (s - s_j)^{\alpha_j - 1}\,ds \tag{8.4}$$

where the integral is along a line from $s = s_k$ to $s = \infty$. If the line of integration makes an angle γ with the positive real direction in the s plane, then the integral for φ_k converges for $\pi/2 < \arg z + \gamma < 3\pi/2$, and φ_k is a solution of (8.1). The range of validity of the representation may be changed by varying γ.

It is rather simple to show that such integrals (which are, in fact, Laplace transforms) are represented by asymptotic series.

As an example, take the equation

$$zw'' + (\gamma - z)w' - \alpha w = 0 \tag{8.5}$$

(considered in Chap. 4). Here

$$P(s) = s^2 - s \qquad Q(s) = \gamma s - \alpha$$

Thus

$$F(s) = s^{\alpha-1}(s - 1)^{\gamma-\alpha-1}$$

and

$$V = s^{\alpha}(s - 1)^{\gamma-\alpha}e^{sz}\Big]_C$$

If $\Re\alpha > 0$ and $\Re(\gamma - \alpha) > 0$, then a solution is given by

$$\int_0^1 s^{\alpha-1}(s - 1)^{\gamma-\alpha-1}e^{sz}\,ds$$

Another solution valid for $\Re z < 0$, $\Re(\gamma - \alpha) > 0$ is

$$\int_1^\infty s^{\alpha-1}(s - 1)^{\gamma-\alpha-1}e^{sz}\,ds \tag{8.6}$$

Solutions may also be represented by closed-loop integrals that make a positive turn about $s = 0$, a negative one about $s = 1$, a negative one about $s = 0$, and a positive one about $s = 1$.

If $s = \sigma + 1$, then the solution (8.6) takes the form

$$e^z \int_0^\infty (1 + \sigma)^{\alpha-1}\sigma^{\gamma-\alpha-1}e^{\sigma z}\,d\sigma \tag{8.7}$$

For all $\sigma \geqq 0$ (indeed, for all σ such that $|\arg \sigma| < \pi - \delta$),

$$(1 + \sigma)^{\alpha-1} = 1 + (\alpha - 1)\sigma + \frac{(\alpha - 1)(\alpha - 2)}{2!}\sigma^2 + \cdots + \frac{(\alpha - 1)\cdots(\alpha - k)}{k!}\sigma^k + F_k(\sigma)\sigma^{k+1}$$

where $F_k(\sigma)$ is uniformly bounded.

Thus the solution (8.7) is given asymptotically by

$$e^z\left[\frac{\Gamma(\gamma - \alpha)}{z^{\gamma-\alpha}} + \frac{(\alpha - 1)\Gamma(\gamma - \alpha + 1)}{z^{\gamma-\alpha+1}} + \frac{(\alpha - 1)(\alpha - 2)\Gamma(\gamma - \alpha + 2)}{2!z^{\gamma-\alpha+2}} + \cdots\right]$$

By changing the direction of the path of integration in the σ plane, the range of validity of the asymptotic expansion may be increased to the range $-\pi/2 < \arg z < 5\pi/2$.

PROBLEMS

1. Suppose that A in (4.1) is analytic in some domain D of the complex z plane and that

$$A \sim A_0 + z^{-1}A_1 + \cdots$$

in D. Prove that Theorems 4.1, 5.1, and 5.2 are valid, with the added restriction that $z \varepsilon D$.

2. Carry out the analogue of the treatment (5.12), at the end of Sec. 5, for the equation

$$\frac{d^2u}{dz^2} + \left[1 + \frac{a}{z} + \frac{b}{z^2}\right] u = 0$$

where a and b are real constants.

3. Show the relationship between the problem above and (8.5) treated at the end of Sec. 8. (See Prob. 7, Chap. 4.)

4. For large $|z|$, let

$$f(z) = e^{g(z)}z^{\mu}\left(b_0 + \frac{b_1}{z} + \cdots\right)$$

where $g(z) = g_0 z^{\pi+1}/(\pi + 1) + \cdots + g_\pi z$ is a polynomial of degree $\pi + 1$, μ is a constant, and the b_j are constant vectors. Let $A(z)$ be as in (2.3). Show that

$$w' = z^r A(z)w + f(z)$$

has a formal solution

$$\psi = e^{g}z^{k}\left[c_0 + \frac{c_1}{z} + \cdots\right]$$

where $k = \mu - \pi$ if $r \leqq \pi$ and $k = \mu - r$ if $r > \pi$, provided (1) no characteristic root of A_0 is equal to g_0 if $r = \pi$, or (2) no characteristic root of A_0 is zero if $r > \pi$.

5. Let q_i be defined as in (4.18). Let S be a sector of the z plane where for each $j = 1, 2, \ldots, n$ either

(a) $$\Re(q_j - g) \to \infty$$

or else

(b) $$\Re(q_j - g) \to -\infty$$

as $|z| \to \infty$ in S. Show that the differential equation of Prob. 4 has a solution φ such that

$$\varphi \sim \psi \qquad (z \varepsilon S)$$

HINT: Let $\hat{\psi}_m$ be the truncated sum consisting of the first $m + 1$ terms of ψ. Let $\tilde{w} = w - \hat{\psi}_m$. Then

$$\tilde{w}' = z^r A(z)\tilde{w} + f_m(z)$$

where

$$e^{-g}f_m = 0(|z|^{l+k-m-1}) \qquad (l = \max(\pi, r))$$

Show that there is a solution $\tilde{w} = \chi(z,m)$, $\chi(z,m)e^{-g} = 0(|z|^{k-m-1})$ along any fixed radius in S by use of integral equations similar to (4.23) but with $\hat{\Psi}^{(1)}_{(m)}$ and $\hat{\Psi}^{(2)}_{(m)}$ determined by (a) and (b) above rather than by (4.19). Then use the devices of Sec. 5.

CHAPTER 6

ASYMPTOTIC BEHAVIOR OF LINEAR SYSTEMS CONTAINING A LARGE PARAMETER

1. Introduction

Here will be considered the system of linear differential equations

$$x' = \rho^r A(t,\rho)x \qquad (a \leqq t \leqq b) \tag{1.1}$$

where $r \geqq 1$ is an integer, A is a matrix continuous in (t,ρ) for $a \leqq t \leqq b$ and $|\rho|$ large, and analytic in ρ for large $|\rho|$ so that

$$A(t,\rho) = \sum_{k=0}^{\infty} \rho^{-k} A_k(t) \tag{1.2}$$

for large $|\rho|$ with A_k continuous. Such systems arise in eigenvalue problems, as will be seen in Chap. 7. (The results of this chapter will not be required for Chap. 7.) These systems also arise in cases where the highest derivative in an nth-order linear differential equation is multiplied by a small parameter—boundary-layer theory, for example. The relationship (1.2) can be asymptotic with the series divergent without changing the results and methods of this chapter.

In some cases, the solutions of (1.1) are studied with t real and ρ complex and large. In other cases, ρ may be real and large while t is complex, or both may be complex. The method of this chapter has much in common with that of Chap. 5. The case of real t and complex ρ will be considered here. The modifications required to handle the other cases are sufficiently close to the procedures of Chap. 5 and the present chapter that they will not be dealt with.

A requirement here will be that the matrix $A_0(t)$ have distinct characteristic roots for $t \in [a,b]$, or at least that the number of distinct characteristic roots of $A_0(t)$ does not change as t goes from a to b. This eliminates from consideration here certain problems of great interest. Such a problem arises in the case of the second-order equation

$$\frac{d^2w}{dt^2} + [\rho^2 t + q_0(t) + q_1(t)\rho^{-1} + \cdots]w = 0 \tag{1.3}$$

in the neighborhood of $t = 0$. If a system formulation were used by setting $w = w_1$ and $w' = \rho w_2$, then (1.3) would have $r = 1$ and

$$A_0(t) = \begin{pmatrix} 0 & 1 \\ -t & 0 \end{pmatrix}$$

Clearly $A_0(t)$ has distinct characteristic roots except at $t = 0$. To treat (1.3), it is replaced by an integral equation based on the use of the variation-of-constants formula and the solutions of

$$w'' + \rho^2 tw = 0$$

which are given explicitly by certain Bessel functions. In method this is very similar to what has already been done in Chap. 5 and to what will be done here. However, there is a more complicated asymptotic formula here because of the appearance of the Bessel functions. The equation (1.3) is said to have a *transition point* at $t = 0$. The treatment of transition points will not be given here.

2. Formal Solutions

For the case that will first be treated it suffices to consider *formal (Laurent) series* in ρ^{-1} with continuous functions as coefficients, that is, a series of the type

$$p = \sum_{k=-\infty}^{\infty} p_k(t)\rho^{-k}$$

where the p_k are continuous functions of t on $a \leqq t \leqq b$ and all but a finite number of the p_k with negative indices are zero on $a \leqq t \leqq b$. The series need not be convergent. If each of the p_k is differentiable, then the derivative p' of p is defined as the formal series

$$p' = \sum_{k=-\infty}^{\infty} p_k'(t)\rho^{-k}$$

Two formal series are said to be equal if the coefficients of like powers of ρ^{-1} are equal. Sums, products, etc., of formal series are defined in the expected manner.

Let q denote the polynomial in ρ

$$q = q_0(t)\rho^r + q_1(t)\rho^{r-1} + \cdots + q_r(t)$$

where the $q_k \in C$ on $a \leqq t \leqq b$. Formal expressions pe^q will be considered. Two such expressions are equal if and only if the polynomials q are equal for $a \leqq t \leqq b$ and the formal series p are equal. If p and q are differentiable in t, then by definition

$$(pe^q)' = (p' + pq')e^q$$

Clearly $(pe^q)'$ is an expression of the same type as pe^q.

Formal matrices Pe^Q will be considered where the elements of P are formal series and Q is a diagonal matrix whose elements are polynomials in ρ of the type q above. Two such matrices are said to be equal if and only if the series matrices P are equal and the diagonal polynomial matrices Q are equal. Since Q is diagonal, the derivative of e^Q is $Q'e^Q$.

From (1.1) and (1.2) it is clear that $\rho^r A(t,\rho)$ can be considered as a formal series. The formal matrix Pe^Q is said to be a *formal solution* of (1.1) if formally

$$(Pe^Q)' = (P' + PQ')e^Q = \rho^r A(t,\rho)Pe^Q$$

that is, if

$$P' + PQ' = \rho^r A(t,\rho)P$$

Theorem 2.1. *Let the A_k in (1.2) be infinitely differentiable on $a \leqq t \leqq b$, and assume the characteristic roots $\lambda_i(t)$ $(i = 1, \ldots, n)$ of $A_0(t)$ are distinct on $a \leqq t \leqq b$, so that*

$$\lambda_i(t) - \lambda_j(t) \neq 0 \qquad (i \neq j,\ a \leqq t \leqq b) \tag{2.1}$$

Then (1.1) has a formal matrix solution Pe^Q, where

$$P = \sum_{k=0}^{\infty} \rho^{-k} P_k(t) \qquad Q = \rho^r Q_0(t) + \cdots + Q_r(t)$$

Moreover, $P_0(t)$ is nonsingular on $a \leqq t \leqq b$, $Q_0'(t) = \Lambda(t)$, where $\Lambda(t)$ is the diagonal matrix with diagonal elements $\lambda_i(t)$, $i = 1, \ldots, n$.

Proof. The proof is similar to that of Theorem 2.1 of Chap. 5. From (2.1) it follows easily that the existence of all derivatives of A_0 implies the same for λ_i, and therefore for Λ. For each t in $a \leqq t \leqq b$ a nonsingular matrix $B_0(t)$ exists such that $B_0^{-1}(t)A_0(t)B_0(t) = \Lambda(t)$. It is important to observe that $A_0B_0 = B_0\Lambda$ implies, with the use of (2.1), that B_0 can be chosen so that it has all derivatives on $[a,b]$. Indeed, each column of B_0 is unique except for a scalar factor. On the other hand, the kth column of B_0 can be taken as a multiple of the cofactors of a row of the matrix $A_0 - \lambda_k E$. Since the roots λ_j are distinct by (2.1), it follows that for any fixed t the cofactors of every row cannot all vanish. If the cofactors of a particular row of $A_0 - \lambda_k E$ are not all zero at t, by continuity this is true for an interval containing t. Using the Heine-Borel theorem, there are a finite number of intervals whose union is $[a,b]$ such that on each of the intervals the cofactors of some row of $A_0 - \lambda_k E$ are not all zero. Choosing one such row for each of the finite number of intervals, it is clear that the cofactors are infinitely differentiable. By patching together the cofactors of these rows by use of infinitely differentiable scalar factors, it is thus possible to find a kth column of B_0 which does not vanish over $[a,b]$ and has all derivatives.

The transformation $x = B_0 y$ yields a system for y of the same type as (1.1) with A_0 replaced by Λ. Thus, with no restriction, it can be assumed A_0 is diagonal. The requirement that Pe^Q be a formal solution of (1.1) is

$$\sum_{k=0}^{\infty} \rho^{-k} P_k' + \left(\sum_{k=0}^{\infty} \rho^{-k} P_k \right) (\rho^r Q_0' + \cdots + Q_r') = \rho^r \left(\sum_{k=0}^{\infty} \rho^{-k} A_k \right) \left(\sum_{k=0}^{\infty} \rho^{-k} P_k \right) \tag{2.2}$$

Thus, setting the coefficients of ρ^r equal, there follows

$$P_0 Q_0' = A_0 P_0$$

Assuming that the equation (1.1) has been transformed as indicated so that A_0 is in diagonal form, it is clear that $Q_0' = \Lambda = A_0$ and $P_0 = E$ is a solution.

The coefficients of ρ^{r-1} in (2.2) yield

$$P_1 Q_0' + P_0 Q_1' = A_0 P_1 + A_1 P_0 \tag{2.3}$$

Since $P_0 = E$ and $A_0 = Q_0'$ and Q_1' are diagonal, the elements $p_{ij}^{(1)}$ of P_1 with $i \neq j$ satisfy

$$p_{ij}^{(1)}(\lambda_j - \lambda_i) = a_{ij}^{(1)}$$

where $a_{ij}^{(1)}$ are the elements of A_1. By (2.1) this determines $p_{ij}^{(1)}$, $i \neq j$, uniquely. Let $\tilde{P}_1$ be the matrix with elements $p_{ij}^{(1)}$, $i \neq j$, determined as above and with diagonal elements zero. For $i = j$, (2.3) yields

$$(q_{ii}^{(1)})' = a_{ii}^{(1)}$$

Thus Q_1' is uniquely determined. The determination of $Q_j(t)$ from $Q_j'(t)$ may be made unique by requiring $Q_j(a) = 0$. Let $P_1 = \tilde{P}_1 + P_0 D_0$, where $D_0(t)$ is an undetermined diagonal matrix. Then clearly (2.3) is satisfied.

This procedure continues until the coefficients of ρ^{-1} are equated and the term P_1' enters. That $\tilde{P}_1'$ exists is clearly a consequence of the differentiability of A_0 and A_1. As the coefficients of later powers of ρ are considered, the existence of higher derivatives of the A_j will enter. From the equation resulting from the ρ^{-1} term, it also follows that D_0' is determined. The strong similarity with the proof of Theorem 2.1, Chap. 5, is evident and further details are omitted.

REMARK: In case the A_k are of class C_m but not of class C_{m+1}, then the above process is valid only up to the point in the argument where the

mth derivatives enter. Thus in this case the existence of the early terms only of the formal series can be established.

3. Asymptotic Behavior of Solutions

The concept of asymptotic series has already been introduced in Chap. 5. There asymptotic expansions in the variable z were obtained. Here the asymptotic expansions will be with respect to the parameter ρ. A region in the ρ plane bounded by two arcs each of which tends toward ∞ and which do not intersect except at their common initial point will be denoted by S. A function $f = f(t,\rho)$ is said to be represented asymptotically in S by the formal series below,

$$f(t,\rho) \sim \sum_{j=0}^{\infty} c_j(t)\rho^{-j}$$

for $a \leqq t \leqq b$ if for every nonnegative integer m there exists a constant K_m such that

$$\left| f(t,\rho) - \sum_{j=0}^{m} c_j(t)\rho^{-j} \right| \leqq \frac{K_m}{|\rho|^{m+1}}$$

for all sufficiently large $|\rho|$, ρ being in S, and for $a \leqq t \leqq b$. Let q be continuous in (t,ρ) for t in $[a,b]$ and ρ in S. A function f will be said to be represented asymptotically in S by the series below,

$$f(t,\rho) \sim e^{q(t,\rho)} \sum_{j=0}^{\infty} c_j(t)\rho^{-j}$$

if for each m there exists a constant K_m such that

$$\left| f(t,\rho) - e^{q(t,\rho)} \sum_{j=0}^{m} c_j(t)\rho^{-j} \right| \leqq \frac{|e^{q(t,\rho)}|K_m}{|\rho|^{m+1}} \tag{3.1}$$

for ρ in S, $|\rho|$ sufficiently large, and t in $[a,b]$. Similarly, f is represented asymptotically by

$$\sum_{k=1}^{N} e^{q_k(t,\rho)} \sum_{j=0}^{\infty} c_{kj}(t)\rho^{-j}$$

if for each m

$$\left| f(t,\rho) - \sum_{k=1}^{N} e^{q_k(t,\rho)} \sum_{j=0}^{m} c_{kj}(t)\rho^{-j} \right| \leqq K_m \sum_{k=1}^{N} \frac{|e^{q_k(t,\rho)}|}{|\rho|^{m+1}}$$

If the relationship (3.1) is valid for $m \leqq M$, where M is an integer, then this is denoted by

$$f(t,\rho) \underset{M}{\sim} e^{q(t,\rho)} \sum_{j=0}^{\infty} c_j(t)\rho^{-j}$$

It will be seen later that the two boundary arcs of S have the property that, as $|\rho| \to \infty$ on each of them, arg ρ tends to a definite limiting value.

Let the element in the kth row and column of the diagonal matrix Q of Theorem 2.1 be denoted by q_k. Then $q_k'(t,\rho) = \rho^r\lambda_k(t) + \cdots$.

Hypothesis H. *Let there exist a region S of the complex ρ plane bounded by two arcs tending to ∞ such that for each i and j, $1 \leqq i, j \leqq n$, one of the inequalities*

$$\Re[q_i'(t,\rho) - q_j'(t,\rho)] \geqq 0 \tag{3.2}$$

$$\Re[q_i'(t,\rho) - q_j'(t,\rho)] \leqq 0 \tag{3.3}$$

is satisfied for all sufficiently large $|\rho|$, ρ in S, and $a \leqq t \leqq b$.

It is not necessarily the case that S must exist. However, if the interval $[a,b]$ is replaced by $[a,c]$, where c is close enough to a, then S does exist. Indeed, since $\lambda_i(t) - \lambda_j(t) \neq 0$, $i \neq j$, it follows that

$$\arg \rho^r[\lambda_i(t) - \lambda_j(t)] = r\theta + \varphi_{ij}(t)$$

where $\arg \rho = \theta$ and $\varphi_{ij}(t) = \arg [\lambda_i(t) - \lambda_j(t)]$. Clearly

$$\Re\rho^r[\lambda_i(t) - \lambda_j(t)] \neq 0 \tag{3.4}$$

if $\cos [r\theta + \varphi_{ij}(t)] \neq 0$. Thus if c is near enough to a, $\varphi_{ij}(t)$ is near enough to the constant $\varphi_{ij}(a)$ so that a range of θ can be found in which $\cos [r\theta + \varphi_{ij}(t)] \neq 0$ for all $i \neq j$. From this follows the existence of a sector in the ρ plane in which (3.4) holds for all $i \neq j$. Since $\rho^r(\lambda_i - \lambda_j)$ is the dominant term of $q_i' - q_j'$ for large $|\rho|$, it follows readily now that S exists. Thus, if the interval in t is short enough, S always exists.

The fact that the boundary arcs of S tend to definite limits in $\arg \rho = \theta$ as $|\rho| \to \infty$ follows readily from the fact that $\Re[q_i' - q_j']$ is a polynomial in ρ and thus for large $|\rho|$ has its behavior determined mainly by the term of highest power in ρ that appears in it. Let the columns of P of Theorem 2.1 be denoted by $p^{(j)}$, $j = 1, \ldots, n$. Then $p^{(j)}e^{q_j}$ are each formal solutions of (1.1) for $j = 1, 2, \ldots, n$.

Theorem 3.1. *If Hypothesis H holds, then for each fixed integer $m > 0$ and for each integer k, $1 \leqq k \leqq n$, there exists a solution $\varphi_m^{(k)}$ of (1.1) such that*

$$\varphi_m^{(k)} \underset{m}{\sim} p^{(k)}e^{q_k}$$

for ρ in S and $a \leqq t \leqq b$.

REMARK: In case the A_k are of class C^N for some $N > 0$, then the above theorem is valid for $m \leqq N$.

Proof of Theorem 3.1. The proof has very much in common with that of Theorem 4.1, Chap. 5. The truncated series $\hat{P}_m$, as well as $\hat{p}_m^{(k)}$, are defined as being identical in the terms up to ρ^{-m} with those of the matrix P and the vector $p^{(k)}$, respectively, and terminating with the terms of power ρ^{-m}. Thus

$$\hat{P}_m(t,\rho) = \sum_{j=0}^{m} \rho^{-j}P_j(t)$$

Since Pe^Q is a formal solution of (1.1), it follows that

$$(P' + PQ')P^{-1} = \rho^r A$$

The existence of the formal series P^{-1} follows easily with the same argument as in Lemma 4.1, Chap. 5. Consider now $B_m(t,\rho)$ defined by

$$(\hat{P}'_m + \hat{P}_m Q')(\hat{P}_m)^{-1} = \rho^r B_m$$

The identity of B_m and A for terms up to and including those of order $\rho^{-(m+r)}$ follows easily. That is, there exists a constant C_1 depending on m such that

$$|A(t,\rho) - B_m(t,\rho)| \leqq \frac{C_1}{|\rho|^{m+r+1}} \tag{3.5}$$

for large $|\rho|$ and $a \leqq t \leqq b$. Moreover, $\hat{P}_m e^Q$ is actually a fundamental solution of

$$x' = \rho^r B_m(t,\rho)x \tag{3.6}$$

Let k, $1 \leqq k \leqq n$, be fixed in this proof. Let ρ be in S and let

$$\hat{P}_m = V_m^{(1)} + V_m^{(2)}$$

where $V_m^{(1)}$ has its column of index j equal to the same column of $\hat{P}_m$ if for ρ in S and $a \leqq t \leqq b$

$$\Re(q'_j(t,\rho) - q'_k(t,\rho)) \geqq 0 \tag{3.7}$$

The columns of index j in $V_m^{(1)}$ for which (3.7) does not hold are taken as zero. This determines $V_m^{(2)}$ also.

Let (1.1) be written as

$$x' = \rho^r B_m x + \rho^r(A - B_m)x$$

Clearly $\hat{p}_m^{(k)}e^{q_k}$ is a solution of (3.6). If the integral equation

$$\begin{aligned}\varphi(t,\rho) &= \hat{p}_m^{(k)}(t,\rho)e^{q_k(t,\rho)} \\ &\quad - \rho^r V_m^{(1)}(t,\rho)e^{Q(t,\rho)} \int_t^b e^{-Q(\tau,\rho)}(\hat{P}_m(\tau,\rho))^{-1}(A(\tau,\rho) - B_m(\tau,\rho))\varphi(\tau,\rho)\,d\tau \\ &\quad + \rho^r V_m^{(2)}(t,\rho)e^{Q(t,\rho)} \int_a^t e^{-Q(\tau,\rho)}(\hat{P}_m(\tau,\rho))^{-1}(A(\tau,\rho) - B_m(\tau,\rho))\varphi(\tau,\rho)\,d\tau \end{aligned} \tag{3.8}$$

has a continuous solution φ, it follows easily that φ is a solution of (1.1).

Because of the vanishing columns of $V_m^{(1)}$, it is the case that the only exponential terms appearing in the elements of $V_m^{(1)}e^{Q(t,\rho)-Q(\tau,\rho)}$ are $e^{q_j(t,\rho)-q_j(\tau,\rho)}$, where j satisfies (3.7). A similar result holds for

$$V_m^{(2)}e^{Q(t,\rho)-Q(\tau,\rho)}$$

Let

$$K_m^{(1)}(t,\tau,\rho) = V_m^{(1)}(t,\rho)e^{Q(t,\rho)-Q(\tau,\rho)}(\hat{P}_m(\tau,\rho))^{-1}(A(\tau,\rho) - B_m(\tau,\rho))$$

and let $K_m^{(2)}$ be defined similarly with $V_m^{(1)}$ replaced by $V_m^{(2)}$. For $t \leqq \tau \leqq b$ and ρ in S it follows for any j satisfying (3.7) that

$$|e^{q_j(t,\rho)-q_j(\tau,\rho)-q_k(t,\rho)+q_k(\tau,\rho)}| = \exp\left\{-\int_t^\tau \Re[q_j'(\sigma,\rho) - q_k'(\sigma,\rho)]\,d\sigma\right\} \leqq 1$$

Thus for $t \leqq \tau \leqq b$, ρ in S and $|\rho|$ large there exists a constant C_2, depending on m, such that

$$|\rho|^{r+m+1}|K_m^{(1)}(t,\tau,\rho)e^{q_k(\tau,\rho)-q_k(t,\rho)}| \leqq C_2 \tag{3.9}$$

Similarly for $a \leqq \tau \leqq t$,

$$|\rho|^{r+m+1}|K_m^{(2)}(t,\tau,\rho)e^{q_k(\tau,\rho)-q_k(t,\rho)}| \leqq C_3 \tag{3.10}$$

for some constant C_3.

The integral equation (3.8) will be shown to have a solution by use of a successive-approximations procedure. Let $\varphi_{(0)}(t,\rho) = 0$ and let for $l \geqq 0$

$$\varphi_{(l+1)}(t,\rho) = \hat{p}_m^{(k)}(t,\rho)e^{q_k(t,\rho)} - \rho^r \int_t^b K_m^{(1)}(t,\tau,\rho)\varphi_{(l)}(\tau,\rho)\,d\tau + \rho^r \int_a^t K_m^{(2)}(t,\tau,\rho)\varphi_{(l)}(\tau,\rho)\,d\tau$$

Clearly for large $|\rho|$, $\rho \in S$, it follows from (3.9) and (3.10) that

$$|(\varphi_{(l+1)} - \varphi_{(l)})e^{-q_k}| \leqq \frac{(b-a)(C_2+C_3)}{|\rho|^{m+1}} \max |(\varphi_{(l)} - \varphi_{(l-1)})e^{-q_k}|$$

where the max is taken over $a \leqq t \leqq b$. If $|\hat{p}_m^{(k)}(t,\rho)| \leqq C_0$ and if $|\rho|$ is so large that $(b-a)(C_2+C_3) \leqq \frac{1}{2}|\rho|^{m+1}$, then it follows easily that

$$|\varphi_{(l+1)}(t,\rho) - \varphi_{(l)}(t,\rho)|\,|e^{-q_k(t,\rho)}| \leqq \frac{C_0}{2^l}$$

From this follows the uniform convergence of $\{\varphi_{(l)}\}$ to a limit which may be denoted by φ and which is a solution of the integral equation. Moreover, it is also clear that

$$|\varphi_{(l)}(t,\rho)e^{-q_k(t,\rho)}| \leqq 2C_0$$

From the integral equation follows readily that

$$|\varphi(t,\rho)e^{-q_k(t,\rho)} - \hat{p}_m^{(k)}(t,\rho)| \leqq \frac{2C_0(b-a)(C_2+C_3)}{|\rho|^{m+1}}$$

which proves the theorem.

4. The Case of Equal Characteristic Roots

The case where two or more characteristic roots become equal at an isolated point of $[a,b]$ will not be considered here. However, a case of great interest arises where several of the roots $\lambda_i(t)$ are identical over $[a,b]$. That is, for any given i or j, either $\lambda_i(t) = \lambda_j(t)$ over $[a,b]$ or else $\lambda_i(t) \neq \lambda_j(t)$ for every t in $[a,b]$. In this case it can be shown that, on certain subintervals of $[a,b]$, formal solutions exist, but now, instead of involving polynomials in ρ and series in $1/\rho$, the solutions involve $\rho^{1/k}$, where k is some positive integer. Thus the $q_i(t,\rho)$ are polynomials in $\rho^{1/k}$ and the series $P(t,\rho)$ are in powers of $\rho^{-1/k}$.

The proof† that the formal solutions exist is much more complicated than in the case considered in Sec. 2. However, the proof that there exist actual solutions asymptotic to the formal solutions in appropriate sectors is very similar to that given in Sec. 3.

A trivial example is given by

$$x_1' = x_2 \qquad x_2' = \rho x_1$$

Here $r = 1$ and $\lambda_1 = \lambda_2 = 0$. On the other hand, it is easily seen that actual solutions $\varphi = (\varphi_1,\varphi_2)$ are given by

$$\begin{aligned}\varphi_1(t,\rho) &= c_1e^{\rho^{\frac{1}{2}}t} + c_2e^{-\rho^{\frac{1}{2}}t}\\ \varphi_2(t,\rho) &= c_1\rho^{\frac{1}{2}}e^{\rho^{\frac{1}{2}}t} - c_2\rho^{\frac{1}{2}}e^{-\rho^{\frac{1}{2}}t}\end{aligned}$$

where c_1 and c_2 are constants.

5. The nth-order Equation

Consider the nth-order equation

$$u^{(n)} + \rho^r a_1(t,\rho)u^{(n-1)} + \cdots + \rho^{nr}a_n(t,\rho)u = 0 \tag{5.1}$$

over $a \leqq t \leqq b$, where

$$a_j(t,\rho) = \sum_{k=0}^{\infty} a_{jk}(t)\rho^{-k}$$

If $u = x_1$ and if

$$x_1' = \rho^r x_2, \qquad x_2' = \rho^r x_3, \quad . \ . \ . \ , \qquad x_{n-1}' = \rho^r x_n \tag{5.2}$$

† H. L. Turritin, *Asymptotic expansions of solutions of systems of ordinary linear differential equations containing a parameter*, Contributions to the theory of nonlinear oscillations, vol. 2, Princeton, 1952.

then

$$u' = \rho^r x_2, \qquad u'' = \rho^{2r} x_3, \quad . \; . \; . \; , \qquad u^{(n-1)} = \rho^{(n-1)r} x_n$$

and (5.1) becomes

$$x_n' = -\rho^r [a_n(t,\rho)x_1 + \cdot \cdot \cdot + a_1(t,\rho)x_n] \tag{5.3}$$

Thus (5.2) and (5.3) form a system of n first-order equations and the theory of the earlier sections can be used.

The $\lambda_i(t)$ are the characteristic roots of the matrix

$$\begin{pmatrix} 0 & 1 & \cdots & 0 \\ 0 & 0 & \cdots & 0 \\ \cdot & \cdot & \cdots & \cdot \\ 0 & 0 & \cdots & 1 \\ -a_{n0}(t) & -a_{n-1\ 0}(t) & \cdots & -a_{10}(t) \end{pmatrix} \tag{5.4}$$

If the $\lambda_i(t)$ are all distinct, then Theorem 3.1 may be used. If the hypothesis of Theorem 3.1 is fulfilled, then the system, (5.2) and (5.3), has n formal solutions $p^{(i)}e^{q_i}$, and for any integer $m > 0$ there exist n actual independent solutions of (5.1) $\psi_i = \psi_i(t,\rho,m)$, $i = 1, \ldots, n$, such that

$$\psi_i(t,\rho,m) \underset{m}{\sim} p_1^{(i)}(t,\rho)e^{q_i(t,\rho)} \qquad (i = 1, \ldots, n)$$

where the 1 denotes the first component of $p^{(i)}$. The derivatives of ψ_i clearly satisfy

$$\psi_i'(t,\rho,m) \underset{m}{\sim} \rho^r p_2^{(i)}(t,\rho)e^{q_i(t,\rho)} \qquad (i = 1, \ldots, n)$$
$$\psi_i''(t,\rho,m) \underset{m}{\sim} \rho^{2r} p_3^{(i)}(t,\rho)e^{q_i(t,\rho)} \qquad (i = 1, \ldots, n)$$

etc.

As an example, the equation

$$y'' + [\rho^2 + q(t)]y = 0$$

will be considered over the interval $0 \leqq t \leqq 1$ on which q is assumed to be C^∞. Setting $y = x_1$ and $y' = \rho x_2$ and using (5.4), there follows $\lambda_1(t) = i$, $\lambda_2(t) = -i$. Thus a formal solution

$$e^{i\rho t}(c_0(t) + c_1(t)\rho^{-1} + \cdot \cdot \cdot)$$

is considered. This leads to

$$(c_0'' + c_1''\rho^{-1} + \cdot \cdot \cdot) + 2i\rho(c_0' + c_1'\rho^{-1} + \cdot \cdot \cdot)$$
$$+ q(c_0 + c_1\rho^{-1} + \cdot \cdot \cdot) = 0$$

Taking $c_0 = 1$, there follows, equating successive powers of ρ^{-1} to zero,

$$2ic_1' + q = 0$$
$$c_1'' + 2ic_2' + qc_1 = 0$$

etc.

The determination of $c_1(t)$, $c_2(t)$, etc., may be made unique by requiring that they vanish at $t = 0$.

REMARK: The formal series may also be obtained from the relationship

$$\varphi(t,\rho) = e^{i\rho t} + \rho^{-1} \int_t^1 \sin \rho(t - \tau) q(\tau) \varphi(\tau,\rho)\, d\tau \tag{5.5}$$

which allows at the same time a rather easy direct proof that there is a solution which it represents asymptotically. Here S may be taken as the upper half plane $\mathfrak{J}\rho \geqq 0$.

Indeed, using successive approximations with $\varphi_0 = 0$ and

$$\varphi_{l+1} = e^{i\rho t} + \rho^{-1} \int_t^1 \sin \rho(t - \tau) q(\tau) \varphi_l(\tau,\rho)\, d\tau$$

there follows easily, if $|q(t)| \leqq M$, that

$$|\varphi_{l+1} - \varphi_l| \leqq \frac{M^l |e^{i\rho t}|}{|\rho|^l}$$

Thus φ_n converges uniformly for large $|\rho|$ in $\mathfrak{J}\rho \geqq 0$, $0 \leqq t \leqq 1$, to a solution φ. Also if $|\rho| \geqq 2M$

$$|\varphi(t,\rho)| \leqq 2|e^{i\rho t}|$$

in S, $0 \leqq t \leqq 1$. Using (5.5), the above gives

$$|\varphi - e^{i\rho t}| \leqq \frac{2M|e^{i\rho t}|}{|\rho|}$$

If this is used in (5.5), there follows

$$\varphi(t,\rho) = e^{i\rho t} + \frac{e^{i\rho t}}{2i\rho} \int_t^1 q(\tau)\, d\tau - \frac{1}{2i\rho} \int_t^1 e^{i\rho(2\tau - t)} q(\tau)\, d\tau + 0\left(\frac{e^{i\rho t}}{|\rho|^2}\right)$$

Integration by parts used on the second integral yields

$$\varphi(t,\rho) = e^{i\rho t} \left[1 + \int_t^1 \frac{q(\tau)}{2i\rho}\, d\tau + 0\left(\frac{1}{|\rho|^2}\right)\right]$$

This process may be continued indefinitely, thereby proving independently the asymptotic formula for φ in S.

PROBLEMS

1. Let

$$f(t,\rho) = e^{g(t,\rho)}[h_0(t) + h_1(t)\rho^{-1} + \cdots]$$

for large $|\rho|$ and $a \leqq t \leqq b$, where

$$g(t,\rho) = \rho^\mu \sum_{j=0}^{\mu-1} g_j(t)\rho^{-j}$$

The h_j and g_j are of class C^∞ and μ is a nonnegative integer. Let

$$A(t,\rho) = A_0(t) + A_1(t)\rho^{-1} + \cdots$$

where the A_j are C^∞ on $[a,b]$. Show the differential equation $x' = \rho^r A(t,\rho)x + f(t,\rho)$ has a formal solution $\psi(t,\rho) = \rho^{-k}e^{g}[c_0(t) + c_1(t)\rho^{-1} + \cdots]$, where $k = r$ if $\mu \leqq r$ and $k = \mu$ if $\mu \geqq r$, provided (1) no characteristic root of $A_0(t)$ vanishes on $[a,b]$ if $\mu < r$, or (2) no characteristic root of $A_0(t)$ is equal to $g_0'(t)$ for any t on $[a,b]$ if $\mu = r$, or (3) $g_0'(t) \neq 0$ for any t on $[a,b]$ if $\mu > r$.

2. Let the hypothesis of Theorem 2.1 be satisfied and let $q_j(t,\rho)$ be defined as in Sec. 3 above (3.2). Let S be a region in the ρ plane, where for each $j = 1, \ldots, n$ either

$$\Re[q_j'(t,\rho) - g'(t,\rho)] \geqq 0$$

or else

$$\Re[q_j'(t,\rho) - g'(t,\rho)] \leqq 0$$

Show that the differential equation of Prob. 1 has for any fixed $m > 0$ a solution $\varphi = \varphi(t,\rho,m)$, where $\varphi(t,\rho,m) \underset{m}{\sim} \psi(t,\rho)$ for $t \in [a,b]$ and $\rho \in S$.

HINT: Let $\hat{\psi}_m(t,\rho)$ be the truncated sum of the first $m + 1$ terms of the sum in ψ. Let $y = x - \hat{\psi}_m$. Then $y' = \rho^r Ay + F_m(t,\rho)$, where

$$e^{-g}F_m(t,\rho) = 0(|\rho|^{l-k-m-1}) \qquad l = \max(r,\mu)$$

Let $\hat{P}_m(t,\rho)$ be as in the proof of Theorem 3.1 and let $\hat{P}_m = U_m^{(1)} + U_m^{(2)}$, where a column occurs in $U_m^{(1)}$ or $U_m^{(2)}$, according to which inequality in Prob. 2 q_j' satisfies.

Show that there is a unique solution $\chi = \chi(t,\rho,m)$ of

$$\chi(t,\rho,m) = \int_a^b G_m(t,\tau,\rho)F_m(\tau,\rho)\,d\tau + \rho^r \int_a^b G_m(t,\tau,\rho)[B_m(\tau,\rho) - A(\tau,\rho)]\chi(\tau,\rho,m)\,d\tau$$

where $G_m(t,\tau,\rho) = U_m^{(1)}(t,\rho)e^{Q(t,\rho)}e^{-Q(\tau,\rho)}\hat{P}_m^{-1}(\tau,\rho)$ for $a \leqq \tau \leqq t$ and G_m satisfies a similar equation for $t < \tau \leqq b$ except that $U_m^{(1)}$ is replaced by $-U_m^{(2)}$. Show that

$$e^{-g}\chi(t,\rho,m) = 0(|\rho|^{-k-m-1})$$

and that $\varphi(t,\rho,m) = \hat{\psi}_m(t,\rho) + \chi(t,\rho,m)$.

CHAPTER 7

SELF-ADJOINT EIGENVALUE PROBLEMS ON A FINITE INTERVAL

1. Introduction

The solution of boundary-value problems for linear partial differential equations may sometimes be reduced to the solution of ordinary differential equations containing a parameter and subject to certain boundary conditions.

A simple example of this situation is the problem of finding the solution of

$$
\begin{aligned}
Lx = -x'' = lx \qquad \left(' = \frac{d}{dt}\right) \\
x(0) = x(1) = 0
\end{aligned}
\tag{1.1}
$$

on the interval $0 \leqq t \leqq 1$. Here l is a complex parameter and x is a scalar. Solutions of $-x'' = lx$ satisfying $x(0) = 0$ are $c \sin l^{\frac{1}{2}}t$, where c is a constant. Thus (1.1) can have a nontrivial solution, that is, one which is not identically zero, if and only if $\sin l^{\frac{1}{2}} = 0$ or only if $l = \pi^2k^2$, where $k = 1, 2, \ldots$. These values of l are called *eigenvalues*. Corresponding solutions are

$$\chi_k(t) = \sqrt{2} \sin k\pi t \qquad (k = 1, 2, \ldots) \tag{1.2}$$

and are called *eigenfunctions*. It is readily seen that

$$\int_0^1 \chi_j \chi_k \, dt = \delta_{jk} \tag{1.3}$$

where δ_{jk} is zero if $j \neq k$ and one if $j = k$. It is an important fact, which will be proved in the general case later, that a wide class of functions can be represented by a series in the χ_k. Indeed, this is the Fourier sine series.

An even simpler example is given by

$$Lx = ix' = lx \qquad x(0) - x(1) = 0 \tag{1.4}$$

Here it is easily seen that the eigenvalues are $l = 2\pi k$, $k = 0, \pm 1, \pm 2, \ldots$, and that $\chi_k(t) = e^{-2\pi ikt}$. The analogue of (1.3) now is

$$\int_0^1 \chi_j \bar{\chi}_k \, dt = \delta_{jk} \tag{1.5}$$

where $\bar{z}$ denotes the complex conjugate of z. It is seen from (1.3) that the functions (1.2) also satisfy (1.5). A sequence of functions $\{\chi_k\}$ satisfying (1.5) is called *orthonormal* on [0,1].

The principal results concerning the problems (1.1) and (1.4) are valid for all functions in $\mathfrak{L}^2(0,1)$, that is, the set of all complex-valued functions f on $0 \leqq t \leqq 1$ which are Lebesgue-measurable there and for which

$$\int_0^1 |f|^2 \, dt < \infty$$

where the integral is the Lebesgue integral. Since this class of functions includes all the continuous, or even piecewise continuous, functions on $0 \leqq t \leqq 1$, the results naturally hold for these. Therefore the reader unfamiliar with the Lebesgue integral can assume throughout that the functions considered are continuous, or piecewise continuous.

If $f,g \in \mathfrak{L}^2(0,1)$, let

$$(f,g) = \int_0^1 f\bar{g} \, dt \qquad \|f\| = (f,f)^{\frac{1}{2}}$$

Then, if $f \in \mathfrak{L}^2(0,1)$, it is the case for (1.2) that

$$\lim_{m \to \infty} \left\| f - \sum_{k=1}^{m} (f,\chi_k)\chi_k \right\| = 0 \tag{1.6}$$

and the same is true for the χ_k associated with (1.4) if the sum in (1.6) is from $-m$ to m. The numbers (f,χ_k) are called the *Fourier coefficients* of f with respect to the sequence $\{\chi_k\}$. The series

$$\sum_{k=1}^{\infty} (f,\chi_k)\chi_k$$

is said to converge to f in the norm of $\mathfrak{L}^2(0,1)$ if (1.6) holds. The result (1.6) is easily seen to be equivalent to

$$\|f\|^2 = \sum_{k=1}^{\infty} |(f,\chi_k)|^2 \tag{1.7}$$

with 1 replaced by $-\infty$ for the case associated with (1.4). The equality (1.7) is known as the *Parseval equality*. It is (1.6) and (1.7) which will be proved under quite general conditions in this chapter.

Consider a slightly more general problem than (1.4), namely,

$$Lx = ix' = lx \qquad x(1) = ax(0) \tag{1.8}$$

where a is a constant. Let $\{\lambda_j\}$ be the eigenvalues and $\{\chi_j\}$ be the corresponding eigenfunctions, which exist but which need not be computed explicitly for present purposes. Clearly

$$(L\chi_j,\chi_k) = \lambda_j(\chi_j,\chi_k)$$

and

$$(\chi_j,L\chi_k) = \bar{\lambda}_k(\chi_j,\chi_k)$$

Also by (1.8)

$$(L\chi_j,\chi_k) - (\chi_j,L\chi_k) = i\int_0^1 (\chi_j'\bar{\chi}_k + \chi_j\bar{\chi}_k')\,dt = i\chi_j\bar{\chi}_k\Big]_0^1$$
$$= i(a\bar{a} - 1)\chi_j(0)\bar{\chi}_k(0) \qquad (1.9)$$

Thus

$$(\lambda_j - \bar{\lambda}_k)(\chi_j,\chi_k) = i(a\bar{a} - 1)\chi_j(0)\bar{\chi}_k(0)$$

and if $a\bar{a} = 1$ it follows, if $j = k$, that λ_k is real and if $j \neq k$ that

$$(\chi_j,\chi_k) = 0$$

Thus, if $a\bar{a} = 1$, the eigenvalues are real and the eigenfunctions may be taken as an orthonormal set. However, if $a\bar{a} \neq 1$, it is readily seen that the eigenvalues need not be real and that $(\chi_j,\chi_k) \neq 0$.

If the right side of (1.9) vanishes, then the eigenfunctions may be taken as orthonormal. Thus, if the problem (1.8) is such that any two functions u and v of class C^1 on $[0,1]$ which satisfy the boundary conditions also satisfy

$$(Lu,v) - (u,Lv) = 0 \qquad (1.10)$$

then the eigenfunctions of (1.8) form an orthonormal set and the eigenvalues are real. The condition (1.10) is of central importance and is known as the *self-adjointness condition.*

It is readily verified that if u and v are of class C^2 on $[0,1]$ and satisfy the boundary conditions of (1.1), then $(Lu,v) = (u,Lv)$ so that (1.1) is a self-adjoint problem. On the other hand, it is readily verified that with the boundary conditions $x(0) = x(1)$, $x'(0) = 2x'(1)$ the problem $-x'' = lx$ is not self-adjoint.

2. Self-adjoint Eigenvalue Problems

Let L be the nth-order operator given by

$$Lx = p_0x^{(n)} + p_1x^{(n-1)} + \cdots + p_nx$$

where the p_j are complex-valued functions of class C^{n-j} on the closed interval $a \leqq t \leqq b$ and $p_0(t) \neq 0$ on $[a,b]$. Let

$$U_jx = \sum_{k=1}^{n} (M_{jk}x^{(k-1)}(a) + N_{jk}x^{(k-1)}(b)) \qquad (j = 1, \ldots, n)$$

where the M_{jk} and N_{jk} are constants. Denote the relationships $U_jx = 0$, $j = 1, \ldots, n$, by $Ux = 0$. The problem

$$\pi: \quad Lx = lx \qquad Ux = 0$$

is called an *eigenvalue problem.* It is said to be *self-adjoint* if

$$(Lu,v) = (u,Lv) \tag{2.1}$$

for all $u,v \in C^n$ on $[a,b]$ which satisfy the boundary conditions

$$Uu = Uv = 0$$

Here, if $f,g \in \mathfrak{L}^2(a,b)$,

$$(f,g) = \int_a^b f\bar{g}\,dt \qquad \|f\| = (f,f)^{\frac{1}{2}}$$

The number (f,g) is called the *inner product* of f with g, and $\|f\|$ is the *norm* of f in $\mathfrak{L}^2(a,b)$. If $(f,g) = 0$, then f and g are said to be *orthogonal.*

It has already been seen from the examples of Sec. 1 that the class of self-adjoint problems is not vacuous. Further examples of such problems are given in Probs. 1, 2, and 3 at the end of the chapter.

It was shown in (6.13), Chap. 3, that to L there corresponds an adjoint L^+, where

$$L^+x = (-1)^n(\bar{p}_0x)^{(n)} + (-1)^{n-1}(\bar{p}_1x)^{(n-1)} + \cdots + \bar{p}_nx$$

such that for any $u,v \in C^n$ on $[a,b]$

$$(Lu,v) - (u,L^+v) = [uv](b) - [uv](a)$$

Clearly if $L^+ = L$ and U is such that $Uu = Uv = 0$ implies

$$[uv](b) - [uv](a) = 0$$

then π will be self-adjoint. The condition that U have this property is given in Theorem 3.2, Chap. 11.

The problem π always has the trivial solution of the identically zero function. If l is such that π has a nontrivial solution, then l is called an *eigenvalue* of π and the nontrivial solutions of π for that l are called *eigenfunctions.* It will be shown in the next section that eigenvalues always exist for a self-adjoint problem.

Theorem 2.1. *Let the problem π be self-adjoint. Then the eigenvalues are real and constitute an at most enumerable set with no finite cluster point. Eigenfunctions corresponding to distinct eigenvalues are orthogonal.*

Proof. Let $l = \lambda$ be an eigenvalue with χ an eigenfunction of π. Then because $L\chi = \lambda\chi$, the relation (2.1) gives $(\lambda - \bar{\lambda})(\chi,\chi) = 0$. Because $(\chi,\chi) > 0$, it follows that $\lambda = \bar{\lambda}$ and thus an eigenvalue must be real.

If λ_1 and λ_2 are distinct eigenvalues with eigenfunctions χ_1 and χ_2, respectively, then

$$(L\chi_1,\chi_2) - (\chi_1,L\chi_2) = (\lambda_1 - \lambda_2)(\chi_1,\chi_2)$$

and by (2.1) this implies that $(\chi_1,\chi_2) = 0$.

Let $\varphi_j = \varphi_j(t,l)$, $j = 1, \ldots, n$, be solutions of $Lx = lx$ which satisfy the initial conditions

$$\varphi_j^{(k-1)}(c,l) = \delta_{jk} \qquad (j, k = 1, \ldots, n) \tag{2.2}$$

for some c in the interval $[a,b]$. By Theorem 8.4 (also Prob. 7), Chap. 1, the functions $\varphi_j^{(k-1)}$ are continuous in (t,l) for $t \in [a,b]$ and all l, and for fixed t are entire functions of l. Since the φ_j are linearly independent, the problem π has l as an eigenvalue if and only if there exist constants c_j, not all zero, such that $x = \sum_{j=1}^{n} c_j\varphi_j$ satisfies $Ux = 0$. This is the case if and only if

$$\sum_{j=1}^{n} c_jU_k\varphi_j = 0 \qquad (k = 1, \ldots, n)$$

has a nontrivial solution. This system of n equations for the c_j has a nontrivial solution if and only if the determinant Δ of the matrix with $U_k\varphi_j$ in the kth row and jth column is zero. Because the $\varphi_j^{(k-1)}$ are entire functions of l for fixed t, in particular at $t = a$ and $t = b$, it follows that Δ is an entire function of l. This function can have only real zeros because π has no nonreal eigenvalues. Thus Δ is an entire function of l which is not identically zero. Its zeros, which are the eigenvalues of π, can therefore cluster only at $l = \infty$. This completes the proof of the theorem.

The nonhomogeneous problem

$$Lx = lx + f \qquad Ux = 0 \tag{2.3}$$

will now be considered, where $f \in C$ on $[a,b]$. This problem can be solved with the aid of the variation-of-constants formula, Theorem 6.4, Chap. 3 (see also Prob. 21, Chap. 3). Let the φ_j again be the solutions of $Lx = lx$ which satisfy (2.2). For $\tau \leqq t$ let

$$K(t,\tau,l) = \frac{1}{p_0(\tau)W(\varphi_1, \ldots, \varphi_n)(\tau)} \begin{vmatrix} \varphi_1(\tau,l) & \cdots & \varphi_n(\tau,l) \\ \varphi_1'(\tau,l) & \cdots & \varphi_n'(\tau,l) \\ \cdot & \cdots & \cdot \\ \varphi_1^{(n-2)}(\tau,l) & \cdots & \varphi_n^{(n-2)}(\tau,l) \\ \varphi_1(t,l) & \cdots & \varphi_n(t,l) \end{vmatrix} \tag{2.4}$$

and for $t < \tau$ let $K(t,\tau,l) = 0$. The Wronskian $W(\varphi_1, \ldots, \varphi_n)$ in the denominator of K in (2.4) is a function of τ only since

$$W(\varphi_1, \ldots, \varphi_n)(\tau) = \exp\left[\int_c^\tau \frac{-p_1(s)}{p_0(s)}\,ds\right]$$

as is clear from (6.5), Chap. 3. Clearly $(\partial^j K/\partial t^j)(\tau + 0, \tau,l) = 0$ for $j = 0, 1, \ldots, n - 2$, since the determinant in (2.4) vanishes when any two rows are identical. Thus $\partial^j K/\partial t^j$, $j = 0, 1, \ldots, n - 2$, is continuous in (t,τ,l) for t,τ on $[a,b]$ and all l, and is an entire function of l for fixed (t,τ). Moreover, for $j = n - 1$ and n, it is continuous in (t,τ,l) for all l and for $a \leqq \tau \leqq t \leqq b$ and for $a \leqq t \leqq \tau \leqq b$. Also

$$\frac{\partial^{n-1}K}{\partial t^{n-1}}(\tau + 0, \tau,l) - \frac{\partial^{n-1}K}{\partial t^{n-1}}(\tau - 0, \tau,l) = \frac{1}{p_0(\tau)}$$

As a function of t, K satisfies $LK = lK$ if $t \neq \tau$. From Theorem 6.4, Chap. 3, or from the above remarks, it follows that the function u defined by

$$u(t,l) = \int_a^b K(t,\tau,l)f(\tau)\,d\tau = \int_a^t K(t,\tau,l)f(\tau)\,d\tau \tag{2.5}$$

is of class C^n in t, entire in l, and $Lu = lu + f$.

The function K will now be modified so that the conditions $Ux = 0$ are also satisfied. Let

$$G(t,\tau,l) = K(t,\tau,l) + \sum_{j=1}^{n} c_j\varphi_j(t,l) \tag{2.6}$$

where the c_j are chosen so that for fixed τ on (a,b), G as a function of t satisfies $UG = 0$. That is,

$$U_kG = U_kK + \sum_{j=1}^{n} c_jU_k\varphi_j = 0 \qquad (k = 1, \ldots, n)$$

or, since U_kK can be extended to be continuous for $a \leqq \tau \leqq b$,

$$\sum_{j=1}^{n} c_jU_k\varphi_j = -U_kK \qquad (k = 1, \ldots, n;\, a \leqq \tau \leqq b) \tag{2.7}$$

The right member in (2.7) is continuous in (τ,l) for $a \leqq \tau \leqq b$ and all l, and is an entire function of l for fixed τ. Since the determinant Δ is an entire function of l with zeros at the eigenvalues of π, it follows that if Δ does not vanish identically (which is the case for self-adjoint problems), (2.7) determines the c_j as functions of (τ,l) continuous for τ on $[a,b]$ and for all l except the eigenvalues of π. Moreover, for fixed τ, the c_j are meromorphic functions of l. Thus G in (2.6) is determined except at eigenvalues of π. From the fact that (2.5) is a solution of $Lx = lx + f$ it is clear from (2.6) that the function u defined by

$$u(t) = \int_a^b G(t,\tau,l)f(\tau)\,d\tau \tag{2.8}$$

is a solution of (2.3) except at the eigenvalues of π. Indeed, the following theorem is true. Note that in this theorem π is not required to be self-adjoint.

Theorem 2.2. *If for at least one value of l the problem π has no solution except the trivial one (which is always true for the self-adjoint case), then there exists a unique function $G = G(t,\tau,l)$ defined for (t,τ) on the square $a \leqq t, \tau \leqq b$ and for all complex l except the eigenvalues of π and having the following properties:*

(i) *$\partial^k G/\partial t^k$ $(k = 0, 1, \ldots, n-2)$ exist and are continuous in (t,τ,l) for (t,τ) on the square $a \leqq t, \tau \leqq b$ and l not at an eigenvalue of π. Moreover, $\partial^k G/\partial t^k$ for $k = n - 1$ and n are continuous in (t,τ,l) for (t,τ) on each of the triangles $a \leqq t \leqq \tau \leqq b$ and $a \leqq \tau \leqq t \leqq b$ and l not at an eigenvalue of π. For fixed (t,τ) these functions are all meromorphic functions of l.*

(ii) $$\frac{\partial^{n-1}G}{\partial t^{n-1}}(\tau + 0, \tau, l) - \frac{\partial^{n-1}G}{\partial t^{n-1}}(\tau - 0, \tau, l) = \frac{1}{p_0(\tau)}$$

(iii) *As a function of t, G satisfies $Lx = lx$ if $t \neq \tau$.*

(iv) *As a function of t, G satisfies the boundary conditions $Ux = 0$ for $a \leqq \tau \leqq b$.*

The solution of (2.3) is given by the function u defined by (2.8).

The function G is known as *Green's function* for π. The theorem has already been proved above except for the uniqueness. If there were two Green's functions, G and $\tilde{G}$, for some l, not an eigenvalue, then as a function of t, $G - \tilde{G}$ is of class C^{n-1} and since $G - \tilde{G}$ also satisfies $Lx = lx$, $t \neq \tau$, it must indeed be of class C^n. However, since l is not an eigenvalue, the problem π has only the trivial solution and thus $G - \tilde{G} = 0$.

It is readily verified that for (1.1)

$$G(t,\tau,l) = \begin{cases} \dfrac{\sin l^{\frac{1}{2}}t \sin l^{\frac{1}{2}}(1-\tau)}{l^{\frac{1}{2}} \sin l^{\frac{1}{2}}} & (0 \leqq t \leqq \tau \leqq 1) \\[2ex] \dfrac{\sin l^{\frac{1}{2}}\tau \sin l^{\frac{1}{2}}(1-t)}{l^{\frac{1}{2}} \sin l^{\frac{1}{2}}} & (0 \leqq \tau \leqq t \leqq 1) \end{cases}$$

and for (1.4)

$$G(t,\tau,l) = \begin{cases} \dfrac{ie^{il(\tau-t)}}{1-e^{il}} & (0 \leqq t \leqq \tau \leqq 1) \\[2ex] \dfrac{ie^{il(1+\tau-t)}}{1-e^{il}} & (0 \leqq \tau \leqq t \leqq 1) \end{cases}$$

Note that the latter G above is double-valued on $t = \tau$. This is true in general for $\partial^{n-1}G/\partial t^{n-1}$ since it is defined to be continuous in $a \leqq t \leqq \tau \leqq 1$ and $a \leqq \tau \leqq t \leqq 1$.

It will be assumed now that $l = 0$ is not an eigenvalue of the self-adjoint problem π. This is no restriction since there must, in any case, exist a real constant c which is not an eigenvalue; thus if $L_1x = Lx - cx$, the problem π_1: $L_1x = lx$, $Ux = 0$, is again a self-adjoint problem because $(cu,v) = (u,cv)$. Moreover, if λ is an eigenvalue of π_1, then $\lambda + c$ is one of π, and conversely, and the eigenfunctions are the same for π and π_1. Since $l = 0$ is not an eigenvalue of π, $G(t,\tau,0)$ exists. In the rest of this chapter this Green's function for $l = 0$ will be denoted by $G = G(t,\tau)$ and it will be assumed that π is self-adjoint.

Corresponding to this Green's function G, let $\mathcal{G}$ be the linear integral operator defined for all $f \varepsilon C$ on $[a,b]$ by

$$\mathcal{G}f(t) = \int_a^b G(t,\tau)f(\tau)\, d\tau$$

If $f,g \varepsilon C$ on $[a,b]$, then (2.1) applied to $u = \mathcal{G}f$, $v = \mathcal{G}g$ yields

$$(f,\mathcal{G}g) = (\mathcal{G}f,g) \tag{2.9}$$

From (2.9) it follows easily that $(\mathcal{G}f,f)$ is real. A further consequence of (2.9) is that

$$G(t,\tau) = \bar{G}(\tau,t)$$

which also is a sufficient condition that π be self-adjoint. Indeed, let $u,v \varepsilon C^n$ on $[a,b]$ and satisfy $Ux = 0$. Let $f = Lu$, $g = Lv$. Then $u - \mathcal{G}f$ and $v - \mathcal{G}g$ are solutions of $Lx = 0$, $Ux = 0$, and therefore are zero. Thus $u = \mathcal{G}f$ and $v = \mathcal{G}g$ and $(Lu,v) = (u,Lv)$ follows from (2.9).

The operator $\mathcal{G}$ is a type of inverse to the operator L in the sense that

$$L\mathcal{G}f = f \qquad \mathcal{G}Lu = u$$

are valid for all $f \varepsilon C$ on $[a,b]$, and $u \varepsilon C^n$ on $[a,b]$ for which $Uu = 0$.

3. The Existence of Eigenvalues†

With G and $\mathcal{G}$ defined as above, it is clear that if λ is an eigenvalue and φ an eigenfunction of π corresponding to λ, then

$$\varphi = \lambda\mathcal{G}\varphi \tag{3.1}$$

Conversely, if $\varphi \varepsilon C$ on $[a,b]$, then $\mathcal{G}\varphi$ is of class C^n and $L\mathcal{G}\varphi = \varphi$ so that (3.1) implies $L\varphi = \lambda\varphi$. Moreover, $U\varphi = 0$ since $UG = 0$.

If there exists a nontrivial $\varphi \varepsilon C$ on $[a,b]$ and a complex number μ such that $\mathcal{G}\varphi = \mu\varphi$, then μ is said to be an eigenvalue of $\mathcal{G}$ and φ an eigen-

† An alternative treatment is given in Probs. 8 and 9.

function. What was proved above is that *the eigenfunctions of* $\mathcal{G}$ *are identical with those of* π *and the eigenvalues of* $\mathcal{G}$ *are reciprocals of those of* π.

The equation (2.9) expresses the fact that $\mathcal{G}$ is self-adjoint. It will be shown that such a self-adjoint operator must possess eigenvalues and in this way the result will follow for π.

Use will be made of the Schwarz inequality

$$|(f,g)| \leqq \|f\| \, \|g\|$$

and its consequence

$$\|f + g\| \leqq \|f\| + \|g\|$$

Lemma 3.1. *The set of all functions* $\{\mathcal{G}u\}$, *where* $u \in C$ *on* $[a,b]$ *and* $\|u\| \leqq 1$, *is a bounded set of equicontinuous functions.*†

Proof. For $n \geqq 2$, G is uniformly continuous on the square $a \leqq t$, $\tau \leqq b$. Hence, given any $\epsilon > 0$, there exists a $\delta > 0$ such that

$$|G(t_1,\tau) - G(t_2,\tau)| < \epsilon \qquad |t_1 - t_2| < \delta$$

From this it follows that if $u \in C$ and $|t_1 - t_2| < \delta$,

$$|\mathcal{G}u(t_1) - \mathcal{G}u(t_2)| \leqq \epsilon \int_a^b |u(\tau)| \, d\tau \leqq \epsilon(b - a)^{\frac{1}{2}}\|u\| \tag{3.2}$$

This proves the equicontinuity of the set $\{\mathcal{G}u\}$. If $|G(t,\tau)| \leqq \gamma$ for $a \leqq t,\tau \leqq b$, then

$$|\mathcal{G}u(t)| \leqq \gamma(b - a)^{\frac{1}{2}}\|u\| \tag{3.3}$$

which implies the uniform boundedness of the set $\{\mathcal{G}u\}$.

For $n = 1$ the equicontinuity is proved by a slightly different argument, since in this case G has a discontinuity at $t = \tau$. On the right side of (3.2) appears the added term $2\gamma|t_2 - t_1|^{\frac{1}{2}}\|u\|$. Indeed, let $t_1 < t_2$. Then

$$\mathcal{G}u(t_2) - \mathcal{G}u(t_1) = \left(\int_a^{t_1} + \int_{t_2}^b\right) (G(t_2,\tau) - G(t_1,\tau))u(\tau) \, d\tau$$
$$+ \int_{t_1}^{t_2} (G(t_2,\tau) - G(t_1,\tau))u(\tau) \, d\tau$$

The first term on the right is again less than the right side of (3.2). The second term is less than $2\gamma|t_2 - t_1|^{\frac{1}{2}}\|u\|$.

The *norm* of $\mathcal{G}$, denoted by $\|\mathcal{G}\|$, is defined by

$$\|\mathcal{G}\| = \sup_{\|u\|=1} \|\mathcal{G}u\| \qquad (u \in C \text{ on } [a,b])$$

From (3.3) it follows that $\|\mathcal{G}u\| \leqq \gamma(b - a)\|u\|$, and hence $\|\mathcal{G}\| < \infty$. Clearly $\|\mathcal{G}u\| \leqq \|\mathcal{G}\| \, \|u\|$ for all $u \in C$ on $[a,b]$. Because $L\mathcal{G}u = u$, $\|\mathcal{G}\| > 0$.

† This proves that $\mathcal{G}$ is a completely continuous operator.

Lemma 3.2. *The norm of $\mathcal{G}$ satisfies*

$$\|\mathcal{G}\| = \sup_{\|u\|=1} |(\mathcal{G}u,u)| \qquad (u \in C \text{ on } [a,b])$$

Proof. By (2.9) $(\mathcal{G}u,u)$ is real. If $\|u\| = 1$,

$$|(\mathcal{G}u,u)| \leqq \|\mathcal{G}u\| \, \|u\| \leqq \|\mathcal{G}\|$$

and hence $\eta = \sup |(\mathcal{G}u,u)| \leqq \|\mathcal{G}\|$. To prove the reverse inequality, note that

$$(\mathcal{G}(u+v),u+v) = (\mathcal{G}u,u) + (\mathcal{G}v,v) + 2\Re(\mathcal{G}u,v) \leqq \eta\|u+v\|^2$$

and similarly

$$(\mathcal{G}(u-v),u-v) = (\mathcal{G}u,u) + (\mathcal{G}v,v) - 2\Re(\mathcal{G}u,v) \geqq -\eta\|u-v\|^2$$

Subtracting, it follows that

$$4\Re(\mathcal{G}u,v) \leqq 2\eta(\|u\|^2 + \|v\|^2) \tag{3.4}$$

$\mathcal{G}u$ is not zero for $u \in C$ unless $u \equiv 0$, for, if it were, $L\mathcal{G}u = u$ would be zero. Letting $v = \mathcal{G}u/\|\mathcal{G}u\|$ in (3.4), where $\|u\| = 1$, gives $\|\mathcal{G}u\| \leqq \eta$, which completes the proof.

Theorem 3.1. *Either $\|\mathcal{G}\|$ or $-\|\mathcal{G}\|$ is an eigenvalue for $\mathcal{G}$.*

REMARK: This not only proves the existence of an eigenvalue for $\mathcal{G}$, and for π, but shows that an eigenfunction φ corresponding to this eigenvalue is a solution of the extremal problem of finding a function $u \in C$ on $[a,b]$ such that

$$(\mathcal{G}u,u) = \int_a^b \left(\int_a^b G(t,\tau)u(\tau)\,d\tau \right) \bar{u}(t)\,dt$$

attains its least upper bound among functions with $\|u\| = 1$ or else its greatest lower bound, depending on which of these is larger in magnitude.

Proof of Theorem 3.1. Suppose $\|\mathcal{G}\| = \sup (\mathcal{G}u,u)$ for $\|u\| = 1, u \in C$ on $[a,b]$. Then there exists a sequence of functions $u_m \in C$ on $[a,b]$, $\|u_m\| = 1$, such that

$$(\mathcal{G}u_m,u_m) \to \|\mathcal{G}\|$$

Let $\mu_0 = \|\mathcal{G}\|$. Since $\{\mathcal{G}u_m\}$ is an equicontinuous, uniformly bounded sequence,† there exists a subsequence, call it $\{\mathcal{G}u_m\}$ also, which is uniformly convergent on $[a,b]$ to a continuous function φ_0. It will be proved φ_0 is an eigenfunction with eigenvalue μ_0.

Since

$$\max_{a \leqq t \leqq b} |\mathcal{G}u_m - \varphi_0| \to 0 \qquad (m \to \infty)$$

it follows that

$$\|\mathcal{G}u_m - \varphi_0\| \to 0 \qquad (m \to \infty) \tag{3.5}$$

† A proof of the Ascoli theorem is given in Sec. 1, Chap. 1.

Also $\|\mathcal{G}u_m\| \to \|\varphi_0\|$. Now

$$\|\mathcal{G}u_m - \mu_0 u_m\|^2 = \|\mathcal{G}u_m\|^2 + \mu_0^2\|u_m\|^2 - 2\mu_0(\mathcal{G}u_m,u_m) \tag{3.6}$$

and the right side tends to $\|\varphi_0\|^2 - \mu_0^2$ as $m \to \infty$. It follows that $\|\varphi_0\|^2 \geqq \mu_0^2 > 0$, and hence φ_0 is not identically zero on $[a,b]$. From (3.6) it also follows that since $\|\mathcal{G}u_m\|^2 \leqq \mu_0^2$,

$$0 \leqq \|\mathcal{G}u_m - \mu_0 u_m\|^2 \leqq 2\mu_0^2 - 2\mu_0(\mathcal{G}u_m,u_m)$$

which tends to zero as $m \to \infty$. Thus

$$\|\mathcal{G}u_m - \mu_0 u_m\| \to 0 \tag{3.7}$$

But

$$0 \leqq \|\mathcal{G}\varphi_0 - \mu_0\varphi_0\| \leqq \|\mathcal{G}\varphi_0 - \mathcal{G}(\mathcal{G}u_m)\| + \|\mathcal{G}(\mathcal{G}u_m) - \mu_0\mathcal{G}u_m\| + \|\mu_0\mathcal{G}u_m - \mu_0\phi_0\|$$

and using $\|\mathcal{G}u\| \leqq \|\mathcal{G}\|\,\|u\|$, (3.5), and (3.7), this yields $\|\mathcal{G}\varphi_0 - \mu_0\varphi_0\| = 0$, which proves $\mathcal{G}\varphi_0 = \mu_0\varphi_0$.

If $-\|\mathcal{G}\| = \inf\,(\mathcal{G}u,u)$, the proof is similar.

Let $\chi_0 = \varphi_0/\|\varphi_0\|$. Then $\|\chi_0\| = 1$ and χ_0 is said to be *normalized*. Let

$$G_1(t,\tau) = G(t,\tau) - \mu_0\chi_0(t)\bar{\chi}_0(\tau)$$

and define $\mathcal{G}_1$ for $u \in C$ on $[a,b]$ by

$$\mathcal{G}_1 u(t) = \int_a^b G_1(t,\tau)u(\tau)\,d\tau$$

Then $\mathcal{G}_1$ has the same properties as $\mathcal{G}$ was shown to have in Lemmas 3.1 and 3.2. In particular, if $\|\mathcal{G}_1\| \neq 0$, and

$$\sup |(\mathcal{G}_1 u,u)| = |\mu_1|$$

where $u \in C$ on $[a,b]$, $\|u\| = 1$, and μ_1 real, then μ_1 is an eigenvalue for $\mathcal{G}_1$, and there exists a nontrivial $\varphi_1 \in C$ on $[a,b]$ satisfying $\mathcal{G}_1\varphi_1 = \mu_1\varphi_1$. Let $\chi_1 = \varphi_1/\|\varphi_1\|$. Since $(\mathcal{G}_1 u,\chi_0) = 0$ for any $u \in C$ on $[a,b]$, it follows that χ_1 is orthogonal to χ_0. Therefore

$$\mathcal{G}\chi_1 = \mathcal{G}_1\chi_1 = \mu_1\chi_1$$

and hence χ_1 is an eigenfunction of $\mathcal{G}$. From the extremal property, $|\mu_1| \leqq |\mu_0|$.

Letting

$$G_2(t,\tau) = G_1(t,\tau) - \mu_1\chi_1(t)\bar{\chi}_1(\tau)$$

and proceeding as above, the existence of χ_2 and μ_2 is established with $|\mu_2| \leqq |\mu_1|$, and χ_2 orthogonal to χ_1 and χ_0. In this way the existence of an *orthonormal*† sequence $\{\chi_k\}$, $k = 0, 1, 2, \ldots$, is established.

† The sequence $\{\chi_k\}$ is orthonormal if $(\chi_j,\chi_k) = \delta_{jk}$, the Kronecker delta.

This process can terminate only if, for some m, $\|\mathcal{G}_m\| = 0$. But for any f of class C

$$L\mathcal{G}_m f = f - \sum_{j=0}^{m-1} \mu_j(f,\chi_j)L\chi_j$$

With $\|\mathcal{G}_m\| = 0$, this implies

$$f = \sum_{j=0}^{m-1} (f,\chi_j)\chi_j \tag{3.8}$$

Since the χ_j are of class C^1 and f can be taken as $|t - \frac{1}{2}(a+b)|$, which is not C^1, (3.8) is impossible. Thus $\|\mathcal{G}_m\| > 0$ for all m and *there are therefore an infinite number of eigenvalues and eigenfunctions.*

4. The Expansion and Completeness Theorems

The expansion in terms of eigenfunctions of π of a function $f \in C^n$ on $[a,b]$ satisfying the boundary conditions $Ux = 0$ will now be proved. From this the Parseval equality and the extensions of these results to any function $f \in \mathfrak{L}^2(a,b)$ will follow easily. First, an important inequality will be deduced.

Lemma. *If $f \in \mathfrak{L}^2(a,b)$ and $\{\chi_k\}$ is an orthonormal sequence for π, then the series*

$$\sum_{k=0}^{\infty} |(f,\chi_k)|^2$$

is convergent, and

$$\sum_{k=0}^{\infty} |(f,\chi_k)|^2 \leqq \|f\|^2 \qquad \textit{(Bessel's inequality)}$$

Proof. For any finite $m \geqq 0$

$$0 \leqq \|f - \sum_{k=0}^{m} (f,\chi_k)\chi_k\|^2 = \|f\|^2 - \sum_{k=0}^{m} |(f,\chi_k)|^2$$

which proves the convergence of the series in question, and Bessel's inequality.

The number (f,χ_k) is called the kth *Fourier coefficient* of f with respect to the orthonormal set $\{\chi_k\}$.

Theorem 4.1. *Let $f \in C^n$ on $[a,b]$ and satisfy the boundary conditions $Uf = 0$. Then on $[a,b]$*

$$f = \sum_{k=0}^{\infty} (f,\chi_k)\chi_k \tag{4.1}$$

where the series converges uniformly on $[a,b]$.

By multiplying (4.1) by $\bar{f}$ and integrating, there results the following:

Corollary. *If f is as in Theorem* 4.1,

$$\|f\|^2 = \sum_{k=0}^{\infty} |(f,\chi_k)|^2 \qquad \textit{(Parseval equality)}$$

This is also called the *completeness relation.*

Proof of Theorem 4.1. From

$$\int_a^b \bar{G}(t,\tau)\bar{\chi}_k(\tau)\,d\tau = \mu_k\bar{\chi}_k(t)$$

it follows that the kth Fourier coefficient of the function g of τ given by $g(\tau) = \bar{G}(t,\tau) = G(\tau,t)$ for fixed t is $\mu_k\bar{\chi}_k(t)$. The Bessel inequality yields

$$\sum_{k=0}^{m} \mu_k^2|\chi_k(t)|^2 \leqq \int_a^b |G(\tau,t)|^2\,d\tau$$

for all m. Integrating in t and letting $m \to \infty$,

$$\sum_{k=0}^{\infty} \mu_k^2 \leqq \gamma^2(b-a)^2$$

where $\gamma = \sup |G(\tau,t)|$ on $a \leqq t, \tau \leqq b$. In particular, $|\mu_k| \to 0$ as $k \to \infty$.

Consider for an integer $m \geqq 1$

$$G_m(t,\tau) = G(t,\tau) - \sum_{k=0}^{m-1} \mu_k\chi_k(t)\bar{\chi}_k(\tau)$$

From the extremal property of $\mathcal{G}_m$ there follows $\|\mathcal{G}_m\| = |\mu_m|$. Thus for any $u \in C$ on $[a,b]$,

$$\|\mathcal{G}_m u\| = \left\| \mathcal{G}u - \sum_{k=0}^{m-1} \mu_k(u,\chi_k)\chi_k \right\| \leqq |\mu_m|\,\|u\|$$

or since $|\mu_m| \to 0$ as $m \to \infty$,

$$\lim_{m\to\infty} \left\| \mathcal{G}u - \sum_{k=0}^{m-1} \mu_k(u,\chi_k)\chi_k \right\| = 0 \tag{4.2}$$

For any $q > p$,

$$\sum_{k=p}^{q} \mu_k(u,\chi_k)\chi_k = \mathcal{G}\left(\sum_{k=p}^{q} (u,\chi_k)\chi_k\right)$$

Since $|\mathcal{G}u| \leqq \gamma(b-a)^{\frac{1}{2}}\|u\|$, it follows that

$$\left|\sum_{k=p}^{q} \mu_k(u,\chi_k)\chi_k\right| \leqq \gamma(b-a)^{\frac{1}{2}}\left(\sum_{k=p}^{q} |(u,\chi_k)|^2\right)^{\frac{1}{2}}$$

By the Bessel inequality, the last sum tends to zero as $p,q \to \infty$. Thus

$$\sum_{k=0}^{\infty} \mu_k(u,\chi_k)\chi_k$$

is uniformly convergent on $[a,b]$, and therefore represents a continuous function there. Since $\mathcal{G}u$ is also continuous, (4.2) implies that

$$\mathcal{G}u = \sum_{k=0}^{\infty} \mu_k(u,\chi_k)\chi_k \tag{4.3}$$

Given any $f \in C^n$ on $[a,b]$ satisfying $Ux = 0$, then $u = Lf \in C$ on $[a,b]$ and $f = \mathcal{G}u$. From (4.3) the expansion result (4.1) follows, since

$$\mu_k(u,\chi_k) = (u,\mu_k\chi_k) = (u,\mathcal{G}\chi_k) = (\mathcal{G}u,\chi_k) = (f,\chi_k)$$

REMARK: The fact that $\|\mathcal{G}_m\| \to 0$ as $m \to \infty$ suggests the possibility that, since $\mu_k = 1/\lambda_k$,

$$G(t,\tau) = \sum_{k=0}^{\infty} \frac{\chi_k(t)\bar{\chi}_k(\tau)}{\lambda_k}$$

This is, in fact, correct but will not be proved here. It will follow from results of Chap. 12 where much less restrictive results than Theorem 4.1 will be proved.

The expansion theorem and completeness relation will now be extended to the whole space $\mathfrak{L}^2(a,b)$.

Theorem 4.2. *If $f \in \mathfrak{L}^2(a,b)$, then*

$$f = \sum_{k=0}^{\infty} (f,\chi_k)\chi_k$$

where the equality is meant in the sense

$$\lim_{m\to\infty} \left\| f - \sum_{k=0}^{m} (f,\chi_k)\chi_k \right\| = 0 \tag{4.4}$$

Further, Parseval's equality holds:

$$\|f\|^2 = \sum_{k=0}^{\infty} |(f,\chi_k)|^2$$

Proof. The proof depends on the fact that the set of functions of class C^n on $[a,b]$ which satisfy $Ux = 0$ is dense in the space $\mathfrak{L}^2(a,b)$, that is, given any $\epsilon > 0$, there exists such a function $\tilde{f}$ satisfying

$$\|f - \tilde{f}\| < \epsilon \tag{4.5}$$

Now

$$\left\| f - \sum_{k=0}^{m} (f,\chi_k)\chi_k \right\| \leqq \|f - \tilde{f}\| + \left\| \tilde{f} - \sum_{k=0}^{m} (\tilde{f},\chi_k)\chi_k \right\| + \left\| \sum_{k=0}^{m} ((\tilde{f} - f),\chi_k)\chi_k \right\| \quad (4.6)$$

and the last term is easily computed to be

$$\left(\sum_{k=0}^{m} |((\tilde{f} - f),\chi_k)|^2 \right)^{\frac{1}{2}}$$

which, by the Bessel inequality, is less than or equal to

$$\|\tilde{f} - f\|$$

Using Theorem 4.1 for $\tilde{f}$, there exists an integer M, depending on ϵ, such that

$$\left\| \tilde{f} - \sum_{k=0}^{m} (\tilde{f},\chi_k)\chi_k \right\| < \epsilon \qquad (m > M)$$

and hence (4.5), (4.6) yield

$$\left\| f - \sum_{k=0}^{m} (f,\chi_k)\chi_k \right\| < 3\epsilon \qquad (m > M)$$

proving the expansion result (4.4).

Parseval's equality follows directly from (4.4) since

$$\left\| f - \sum_{k=0}^{m} (f,\chi_k)\chi_k \right\|^2 = \|f\|^2 - \sum_{k=0}^{m} |(f,\chi_k)|^2$$

The Parseval equality has the consequence that if $(f,\chi_k) = 0$ for $k = 0, 1, 2, \ldots$, then f is zero almost everywhere, and in particular, if f is continuous, it is the zero function on $[a,b]$. A set of functions $\{\psi\}$ is said to be *closed* in $\mathfrak{L}^2(a,b)$ if, for every $f \in \mathfrak{L}^2(a,b)$, $(f,\psi) = 0$ implies f is zero almost everywhere. Thus the set $\{\chi_k\}$ is closed in $\mathfrak{L}^2(a,b)$. This implies that if $f \in \mathfrak{L}(a,b)$ and $(f,\chi_k) = 0$ for $k = 0, 1, 2, \ldots$, then f is zero almost everywhere; that is, *the set* $\{\chi_k\}$ *is closed in* $\mathfrak{L}(a,b)$, the set of Lebesgue-integrable functions on (a,b). This can be seen in the following way. Suppose $f \in \mathfrak{L}(a,b)$ and $(f,\chi_k) = 0$, $k = 0, 1, 2, \ldots$. Then there exists a continuous function φ satisfying

$$L\varphi = f \qquad U\varphi = 0$$

since zero is not an eigenvalue for π. The proof consists in verifying that, for $f \in \mathfrak{L}(a,b)$, $\varphi = \mathcal{G}f$ is of class C^{n-1} for $a \leqq t \leqq b$ and that $\varphi^{(n-1)}$ is absolutely continuous so that $L\varphi = f$ almost everywhere. Clearly

$$(\varphi,\chi_k) = \lambda_k^{-1}(\varphi,L\chi_k) = \lambda_k^{-1}(L\varphi,\chi_k) = \lambda_k^{-1}(f,\chi_k) = 0$$

which proves φ is orthogonal to all eigenfunctions. But since φ is continuous, φ is the zero function, and this implies, by $L\varphi = f$, that f is zero almost everywhere.

Corresponding to any $f \in \mathfrak{L}^2(a,b)$ there is a unique sequence of complex numbers $c = \{c_k\}$, where

$$c_k = (f,\chi_k)$$

Define the *norm* of c, denoted by $\|c\|$, by

$$\|c\| = \left(\sum_{k=0}^{\infty} |c_k|^2\right)^{\frac{1}{2}}$$

Then the Parseval equality may be written as $\|f\| = \|c\|$. It is an important fact that the correspondence $f \to c$ actually uses up *all* sequences c of complex numbers such that $\|c\| < \infty$. This is the Riesz-Fischer theorem, the proof of which has nothing to do with differential equations, and so will be omitted.†

Riesz-Fischer Theorem. *Let $c = \{c_k\}$ be a sequence of complex numbers such that $\|c\| < \infty$. Then there exists a function $f \in \mathfrak{L}^2(a,b)$ for which $c_k = (f,\chi_k)$, and*

$$\|f\| = \|c\|$$

PROBLEMS

1. Let $Lx = -(px')' + qx$, where p is of class C^1 and q of class C on $[a,b]$ and $p \neq 0$ on $[a,b]$. Let $Ux = 0$ be given by

$$\alpha x(a) + \beta x'(a) = 0 \qquad \gamma x(b) + \delta x'(b) = 0$$

Show the problem π is self-adjoint if and only if p and q are real, $\gamma\bar{\delta} = \bar{\gamma}\delta$ and $\alpha\bar{\beta} = \bar{\alpha}\beta$, which is equivalent to requiring that α, β, γ, and δ all be real.

2. If $Ux = 0$ above is replaced by

$$x(b) - \alpha x(a) - \beta x'(a) = 0 \qquad x'(b) - \gamma x(a) - \delta x'(a) = 0$$

show that the conditions for self-adjointness become $\alpha = c_1e^{i\theta}$, $\beta = c_2e^{i\theta}$, $\gamma = c_3e^{i\theta}$ $\delta = c_4e^{i\theta}$, where c_j and θ are real and $p(b)(c_1c_4 - c_2c_3) = p(a)$.

3. Let $Lx = \sum_{j=0}^{n} (p_{n-j}x^{(j)})^{(j)}$, where $p_{n-j} \in C^{n-j}$ and are real on $[a,b]$ and $p_0(t) \neq 0$ on $[a,b]$. Let $Ux = 0$ be $x^{(j)}(a) = x^{(j)}(b) = 0$, $j = 0, 1, \ldots, n-1$. Prove π is self-adjoint.

† For a proof, see W. Rudin, *Principles of mathematical analysis*, New York, 1953.

4. Let π be a self-adjoint problem with orthonormal eigenfunctions $\{\chi_k\}$. If the norm $\|F\|$ of a function $F = F(t,\tau)$ of class C for $t,\tau \in [a,b]$ is defined by

$$\|F\| = \left(\int_a^b \int_a^b |F(t,\tau)|^2 \, dt \, d\tau \right)^{\frac{1}{2}}$$

then it is possible to approximate $F(t,\tau)$ in this norm by finite sums of the form $\Sigma a_{ij} f_i(t) g_j(\tau)$, where f_i and g_j are of class C^n and satisfy $Uf_i = Ug_j = 0$. Show that the Parseval equality holds for F in the sense that if

$$c_{ij} = \int_a^b \int_a^b F(t,\tau) \bar{\chi}_i(t) \chi_j(\tau) \, dt \, d\tau$$

then

$$\|F\|^2 = \sum_{i,j=0}^{\infty} |c_{ij}|^2$$

Show that this leads to

$$\lim_{m \to \infty} \left\| G(t,\tau,l) - \sum_{k=0}^{m} \frac{\chi_k(t) \bar{\chi}_k(\tau)}{\lambda_k - l} \right\| = 0$$

REMARK: Actually the formula

$$G(t,\tau,l) = \sum_{k=0}^{\infty} \frac{\chi_k(t) \bar{\chi}_k(\tau)}{\lambda_k - l}$$

is valid, as will follow from considerations in Chap. 12.

5. Let $[a,b]$ be $[0,1]$ and let $Lx = -((1 - t^2)x')'$ and $Ux = 0$ be $x(0) = 0$ and $(1 - t^2)x'(t) \to 0$ as $t \to 1 - 0$. Show that by taking

$$G(t,\tau) = \begin{cases} \frac{1}{2} \log \frac{1 + \tau}{1 - \tau} & (0 \leqq \tau \leqq t \leqq 1) \\ \frac{1}{2} \log \frac{1 + t}{1 - t} & (0 \leqq t \leqq \tau \leqq 1) \end{cases}$$

the reasoning of the chapter can be modified to obtain the existence of a complete orthonormal family of eigenfunctions.

HINT: Instead of Lemma 3.1, show that $\mathcal{G}$ is completely continuous in the norm of $\mathfrak{L}^2(0,1)$. This is true for any G with

$$\int_a^b \int_a^b |G(t,\tau)|^2 \, dt \, d\tau < \infty$$

6. Show that if π is self-adjoint the Green's function $G(t,\tau,l)$ satisfies

$$G(t,\tau,l) = \bar{G}(\tau,t,\bar{l})$$

thereby generalizing the result $G(t,\tau,0) = \bar{G}(\tau,t,0)$ already shown when $l = 0$ is not an eigenvalue.

7. Show that if π is self-adjoint the poles of $G(t,\tau,l)$ are simple poles.

HINT: Consider $g(t,l) = \int_a^b G(t,\tau,l) f(\tau) \, d\tau$ for f of class C. Let g have a pole of order $m > 1$ at $l = \lambda_k$. Then near $l = \lambda_k$

$$g(t,l) = \frac{g_m(t)}{(l - \lambda_k)^m} + \frac{g_{m-1}(t)}{(l - \lambda_k)^{m-1}} + \cdots$$

Since $(L - \lambda_k)g = (l - \lambda_k)g + f$ and $Ug = 0$, it follows that $(L - \lambda_k)g_m = 0$, $(L - \lambda_k)g_{m-1} = g_m, \ldots,$ and $U(g_j) = 0, j = m, m - 1, \ldots.$ Since

$$(g_m,g_m) = (g_m,(L - \lambda_k)g_{m-1}) = ((L - \lambda_k)g_m,g_{m-1}) = 0$$

it follows that $g_m = 0$. Since this holds for all f, G has at most a simple pole at λ_k.

8. If π is self-adjoint and $f \varepsilon C$ on $[a,b]$ and $(f,\chi_k) = 0$ for all the eigenfunctions of π, prove that f is zero by making use of $g(t,l)$ of Prob. 7.

HINT: Using the method of Prob. 7, show that because $(f,\chi_k) = 0$, g has no poles and is therefore an entire function of l, $\sum_{j=0}^{\infty} a_j(t)l^j$. Since $Ug = 0$ and $Lg = lg + f$: show $La_0 = f$, $La_1 = a_0$, $La_2 = a_1, \ldots, Ua_j = 0$. Show $(a_{j-1},a_k) = (a_j,a_{k-1})$ and thus that $W_{j+k} = (a_j,a_k)$ depends only on $j + k$. Show that

$$h(l) = (g,a_0) = W_0 + W_1 l + \cdots$$

is an entire function of l as is $\Gamma(l) = W_0 + W_2 l^2 + W_4 l^4 + \cdots$. Show

$$W_{2j}^2 = (a_{j-1},a_{j+1})^2 \leqq W_{2j-2}W_{2j+2}$$

Thus, if $W_2 \neq 0$,

$$\frac{W_{2j}}{W_{2j-2}} \leqq \frac{W_{2j+2}}{W_{2j}} \qquad j = 1, 2, \ldots$$

meaning Γ is not entire. Thus $W_2 = 0$, and so $a_1 = 0$. $La_1 = a_0 = 0$, $La_0 = f = 0$.

REMARK: Note that Prob. 8 gives an independent proof of the closure of the eigenfunctions of π.

9. Using the result of Prob. 8, prove Theorem 4.1.

HINT: Let $u = Lf$. Then, as was shown beginning just below (4.2),

$$g = \sum_{0}^{\infty} \mu_k(u,\chi_k)\chi_k$$

is uniformly convergent on $[a,b]$. Clearly $f - g$ is orthogonal to all χ_k. Thus $f - g$ is zero by Prob. 8, which proves Theorem 4.1.

10. If Green's function for (1.1) is expanded in the eigenfunctions, the series

$$G(t,\tau,l) = \sum_{k=1}^{\infty} \frac{2 \sin k\pi t \sin k\pi\tau}{k^2 - l}$$

is obtained. From the nature of $G(t,\tau,l)$ as a function of t and the convergence properties of Fourier sine series, show the series is convergent for $0 \leqq t \leqq 1$, for all τ, and l not an eigenvalue. Show that the series for $\partial G/\partial t$ also converges for all t. (Note that $\partial G/\partial t$ is of bounded variation as a function of t.)

11. Let L and U be such that π is self-adjoint. Consider now instead of π the problem $Lx = lrx$, $Ux = 0$ on $[a,b]$, where the function $r \varepsilon C$ and $r(t) > 0$ on $[a,b]$. Show that the eigenfunctions $\{\psi_k\}$ can be chosen so that $\{r^{\frac{1}{2}}\psi_k\}$ form an orthonormal sequence. Show that this sequence is complete.

HINT: Let $H(t,\tau) = r^{\frac{1}{2}}(t)r^{\frac{1}{2}}(\tau)G(t,\tau)$ and show that the operator $\mathcal{H}$ defined for all $f \varepsilon C$ on $[a,b]$ by $\mathcal{H}f(t) = \int_a^b H(t,\tau)f(\tau)\,d\tau$ is self-adjoint.

12. Show that $G(t,\tau,l)$ is given by

$$G(t,\tau,l) = \frac{g(t,\tau,l)}{\Delta(l)}$$

where $\Delta(l) = \det(U_j\varphi_k)$,

$$g(t,\tau,l) = \det\begin{pmatrix} K(t,\tau,l) & \varphi_1(t,l) & \cdots & \varphi_n(t,l) \\ U_1K & U_1\varphi_1 & \cdots & U_1\varphi_n \\ \cdot & \cdot & \cdots & \cdot \\ U_nK & U_n\varphi_1 & \cdots & U_n\varphi_n \end{pmatrix}$$

and K is the function defined by (2.4).

13. Let $Lx = p_0x^{(n)} + \cdots + p_nx$ on $[a,b]$ and $p_j \in C^{n-j}[a,b]$ and be real. Then $L = L^+$ if and only if n is even $(n = 2r)$ and Lx can be written as

$$Lx = (q_0x^{(r)})^{(r)} + (q_1x^{(r-1)})^{(r-1)} + \cdots + q_rx$$

where $q_j \in C^{r-j}[a,b]$ and is real.

HINT: Show that, if $L = L^+$ and p_0 is real, then $n = 2r$ and $p_1 = rp_0'$. Thus if $q_0 = p_0$ then

$$Lx = (q_0x^{(r)})^{(r)} + L_1x$$

where L_1 must be of order $n - 2$. Show that $L = L^+$ now implies $L_1 = L_1^+$ and thus establish the result by induction.

14. Let Lx be as above but now let $p_j(t)$ be complex-valued. Show that if $L = L^+$ then

$$Lx = i^nq_0(\cdots(q_0(q_0x)')'\cdots)' + i^{n-1}q_1(\cdots(q_1(q_1x)')'\cdots)' + \cdots + i^2q_{n-2}(q_{n-2}(q_{n-2}x)')' + iq_{n-1}(q_{n-1}x)' + q_nx$$

where the $q_j \in C^{n-j}$ and $(q_j)^{n+1-j}$ are real and $i^nq_0^{n+1} = p_0$.

HINT: Use induction. Show by direct consideration of

$$\int_a^b i^nq_0(\cdots(q_0(q_0u)')'\cdots)'\bar{v}\,dt$$

that the first term of L above is self-adjoint.

15. Show that if $L = L^+$ there exists at least one set of boundary conditions $Ux = 0$ which makes the problem $Lx = lx$, $Ux = 0$, self-adjoint.

HINT: If n is even $(n = 2r)$, then take

$$x(a) = x(b) = x'(a) = x'(b) = \cdots = x^{(r-1)}(a) = x^{(r-1)}(b) = 0$$

If n is odd $(n = 2r + 1)$, take the above conditions and the added condition

$$C_1x^{(r)}(a) = C_2x^{(r)}(b)$$

where $iC_1\bar{C}_1 = p_0^{(r)}(a)$, $iC_2\bar{C}_2 = p_0^{(r)}(b)$.

16. Let A be a square matrix and f a vector both of which are continuous functions of t for $a \leq t \leq b$. Consider the problem

$$x' - A(t)x = f \qquad Ux = Mx(a) + Nx(b) = 0$$

where M and N are constant square matrices. Let the problem with $f = 0$ have only the null solution. Show that there exists a matrix $G(t,\tau)$ continuous for $a \leq t \leq$

$\tau \leqq b$ and for $a \leqq \tau \leqq t \leqq b$ such that

$$\int_a^b G(t,\tau)f(\tau)\,d\tau$$

is the unique solution of the problem.

HINT: Let Φ be a fundamental matrix for $x' = A(t)x$. Let

$$G(t,\tau) = \begin{cases} \Phi(t)\Phi^{-1}(\tau) + \Phi(t)J(\tau) & (\tau < t) \\ \Phi(t)J(\tau) & (\tau > t) \end{cases}$$

where the matrix $J \in C[a,b]$. To satisfy $Ux = 0$

$$M\Phi(a)J(\tau) + N\Phi(b)\Phi^{-1}(\tau) + N\Phi(b)J(\tau) = 0$$

so that

$$J(\tau) = -(M\Phi(a) + N\Phi(b))^{-1}N\Phi(b)\Phi^{-1}(\tau)$$

17. Let the r-by-r matrices P_0 and P_1 be continuous for $a \leqq t \leqq b$. Moreover, let P_0' be continuous and $\det P_0(t) \neq 0$. Let x be a vector with r components and let

$$Lx = P_0x' + P_1x$$

Let P^* denote the adjoint of P, that is, the transposed conjugate, and let

$$L^+x = -(P_0^*x)' + P_1^*x$$

If u and v are vectors with components u_j, v_j, let

$$u \cdot v = u_1\bar{v}_1 + \cdots + u_r\bar{v}_r$$

Show that if $u,v \in C^1[a,b]$

$$Lu \cdot v - u \cdot L^+v = (P_0u \cdot v)'$$

Let $L = L^+$, that is, $P_0 + P_0^* = 0$, $P_0' = P_1 - P_1^*$. Let M and N be r-by-r constant matrices and let $Ux = Mx(a) + Nx(b)$. Suppose M and N are such that for any $u,v \in C^1[a,b]$ and satisfying $Uu = Uv = 0$

$$\int_a^b Lu \cdot v\,dt = \int_a^b u \cdot Lv\,dt$$

Prove the eigenfunctions $\{\psi_{(j)}\}$ of the self-adjoint problem

$$Lx = lx \qquad Ux = 0$$

form a complete orthonormal set. Thus if $\int_a^b u \cdot v\,dt$ is denoted by $(u \cdot v)$ show that for any vector f satisfying $(f \cdot f) < \infty$

$$(f \cdot f) = \sum_{j=-\infty}^{\infty} |(f \cdot \psi_{(j)})|^2$$

and in $\mathfrak{L}^2(a,b)$

$$f(t) = \sum_{j=-\infty}^{\infty} (f \cdot \psi_{(j)})\psi_{(j)}(t)$$

REMARK The cases $P_0 = iE$ and, for even r,

$$P_0 = \begin{pmatrix} 0 & E_{\frac{1}{2}r} \\ -E_{\frac{1}{2}r} & 0 \end{pmatrix}$$

are of special interest.

HINT: The Green's function is an r-by-r matrix and the method of Sec. 4 or Prob. 8 can be used.

18. Let x be a vector with r components and let P_j, $j = 0, \ldots, n$, be r-by-r matrices of class $C^{n-j}[a,b]$ and let $\det P_0(t) \neq 0$. Let

$$Lx = P_0x^{(n)} + \cdots + P_nx$$

and

$$L^+x = (-1)^n(P_0^*x)^{(n)} + (-1)^{n-1}(P_1^*x)^{(n-1)} + \cdots + P_n^*x$$

Show that if u,v are vectors of class $C^n[a,b]$

$$Lu \cdot v - u \cdot L^+v = \frac{d}{dt}[uv]$$

where

$$[uv](t) = \sum_{m=1}^{n} \sum_{j+k=m+1} (-1)^j u^{(k-1)} \cdot (P_{n-m}^* v)^{(j-1)}$$

Let $L = L^+$. Let

$$Ux = M_1x^{(n-1)}(a) + \cdots + M_nx(a) + N_1x^{(n-1)}(b) + \cdots + N_nx(b)$$

where M_j, N_j are matrices of nr rows and r columns and suppose that for all u,v of class $C^n[a,b]$ satisfying $Uu = Uv = 0$

$$\int_a^b Lu \cdot v\, dt = \int_a^b u \cdot Lv\, dt$$

Show that the self-adjoint problem

$$Lx = lx \qquad Ux = 0$$

has a complete orthonormal set of eigenfunctions $\{\psi_{(j)}\}$.

HINT: Let ξ be the vector with nr components $(x, x', \ldots, x^{(n-1)})$. Then

$$Lx - lx = 0$$

can be replaced by a first-order equation in ξ with nr independent solutions with fundamental matrix Ξ, nr-by-nr. Let the first r rows of Ξ be denoted by Φ. Then $L\Phi - l\Phi = 0$ and any solution φ of $Lx - lx = 0$ is given by Φc, where c is a constant column vector with nr rows. The Green's function $G(t,\tau,l)$, an r-by-r matrix, can be constructed.

19. Let L be as in Prob. 13, that is,

$$Lx = (q_0x^{(r)})^{(r)} + (q_1x^{(r-1)})^{(r-1)} + \cdots + q_rx$$

where $q_j \in C^{r-j}[a,b]$, q_j real, and $q_0(t) \neq 0$ on $[a,b]$. Let φ be a solution of $Lx = 0$ and let $\hat{\varphi}$ be the vector with components φ_j, where $\varphi_j = \varphi^{(j-1)}$ $(j = 1, \ldots, r)$,

$$\varphi_{r+j} = (-1)^j\{(q_{r-j}\varphi^{(j)}) + (q_{r-j-1}\varphi^{(j+1)})' + \cdots + (q_0\varphi^{(r)})^{(r-j)}\}$$

$(j = 1, \ldots, r)$. Show that $\hat{\varphi}$ satisfies the formally self-adjoint system

$$P_0\hat{\varphi}' + P_1\hat{\varphi} = 0$$

where $P_0 = -P_0^*$, $P_1 = P_1^*$, are the matrices

$$P_0 = \begin{pmatrix} 0_r & E_r \\ -E_r & 0_r \end{pmatrix} \qquad P_1 = \begin{pmatrix} A & B \\ B^* & C \end{pmatrix}$$

where 0_r, E_r are the zero and unit matrices of r dimensions, and A, B, C are r-by-r matrices given by

$$A = \begin{pmatrix} -q_r & & & & \\ & q_{r-1} & & & \\ & & \cdot & & \\ & & & \cdot & \\ & & & & (-1)^r q_1 \end{pmatrix} \qquad B = \begin{pmatrix} 0 & & & & \\ 1 & \cdot & & & \\ & \cdot & \cdot & & \\ & & \cdot & \cdot & \\ & & & 1 & 0 \end{pmatrix}$$

$$C = \begin{pmatrix} 0 & & & & \\ & \cdot & & & \\ & & \cdot & & \\ & & & 0 & \\ & & & & (-1)^r/q_0 \end{pmatrix}$$

the elements not shown being zero.

CHAPTER 8

OSCILLATION AND COMPARISON THEOREMS FOR SECOND-ORDER LINEAR EQUATIONS AND APPLICATIONS

1. Comparison Theorems

The location of the zeros of the solutions of real second-order differential equations will be considered here. The equation will be assumed to have the form

$$Lx = (p(t)x')' + g(t)x = 0 \qquad (a < t < b) \tag{1.1}$$

Note that (1.1) is considered on the open interval (a,b) rather than on $[a,b]$. The equation $x'' + f(t)x' + h(t)x = 0$ can be put in the form of (1.1) by multiplying it by $\exp\left(\int^t f(t)\, dt\right)$. It will be assumed in what follows that $p(t) > 0$ and that p, p' and g are continuous on (a,b). (The continuity requirement can be relaxed. Indeed, it suffices for g to be integrable and p absolutely continuous.)

A zero of a nontrivial solution of (1.1) is isolated. Indeed, let the solution φ vanish at t_0. Then $\varphi'(t_0) \neq 0$, for otherwise $\varphi(t) \equiv 0$. This proves that t_0 is an isolated zero.

Theorem 1.1. *Suppose φ is a real solution on (a,b) of*

$$(px')' + g_1 x = 0 \tag{1.2}$$

and ψ a real solution on (a,b) of

$$(px')' + g_2 x = 0 \tag{1.3}$$

Let $g_2(t) > g_1(t)$ on (a,b). If t_1 and t_2 are successive zeros of φ on (a,b), then ψ must vanish at some point of (t_1,t_2).

Proof. Suppose ψ does not vanish in (t_1,t_2). Then with no restriction it can be assumed that $\psi(t) > 0$ and also $\varphi(t) > 0$ over (t_1,t_2). Multiplying (1.2) by ψ and (1.3) by φ and subtracting,

$$(p\varphi')'\psi - (p\psi')'\varphi - (g_2 - g_1)\varphi\psi = 0$$

Integrating the above,

$$\int_{t_1}^{t_2} [(p\varphi')'\psi - (p\psi')'\varphi]\, dt > 0$$

Since the bracket above is the derivative of $p(\varphi'\psi - \varphi\psi')$ and since φ vanishes at t_1 and t_2,

$$p(t_2)\varphi'(t_2)\psi(t_2) - p(t_1)\varphi'(t_1)\psi(t_1) > 0 \tag{1.4}$$

Since $\varphi(t_2) = 0$ and $\varphi(t) > 0$ immediately to the left of t_2, $\varphi'(t_2) < 0$. Similarly $\varphi'(t_1) > 0$. Thus the first term on the left is nonpositive, as is the second, which shows that (1.4) is impossible. Thus ψ vanishes at a point inside the open interval (t_1,t_2).

In case $g_1 \equiv g_2$ over (a,b), then φ and ψ are solutions of the same equation. If φ and ψ are independent, then (1.4) is valid with the inequality replaced by an equality, and the above argument shows that ψ vanishes between successive zeros of φ. Since now φ and ψ are interchangeable, it also shows that φ vanishes between successive zeros of ψ. Thus the zeros of two real linearly independent solutions of a real second-order linear differential equation separate one another.

The above method can be further exploited, but the following procedure is simpler:

Let $p(t)x' = y$. Thus (1.1) becomes

$$x' = \frac{y}{p(t)} \qquad y' = -g(t)x \tag{1.5}$$

Let

$$x = r \sin \theta \qquad y = r \cos \theta \tag{1.6}$$

Differentiating the equations (1.6) with respect to t, replacing x' and y by use of (1.5), and then solving for r' and θ', there results

$$r' = \left(\frac{1}{p} - g\right) r \sin \theta \cos \theta \tag{1.7}$$

$$\theta' = \frac{1}{p} \cos^2 \theta + g \sin^2 \theta \tag{1.8}$$

For a solution φ of (1.1) there is the solution $r = \rho(t)$ and $\theta = \omega(t)$ of (1.7), (1.8), where from (1.5) and (1.6)

$$\rho^2 = (p\varphi')^2 + \varphi^2 \qquad \omega = \tan^{-1}\left(\frac{\varphi}{p\varphi'}\right)$$

Since φ and φ' do not vanish simultaneously, it follows that $\rho^2(t) > 0$ on (a,b) and thus with no restriction it can be assumed that $\rho(t) > 0$. A consequence of this is that $\varphi(t) = \rho(t) \sin \omega(t)$ can vanish only where $\omega(t)$ is an integer multiple of π.

Since $\cos^2 \theta$ and $\sin^2 \theta$ are uniformly bounded, the equation (1.8) has a solution over any interval on which $p > 0$ and p and g are piecewise continuous. Indeed, it suffices for $1/p$ and g to be integrable. Because the right side of (1.8) is differentiable in θ, it follows that the solution is unique in the usual sense.

From (1.6) and (1.5) it follows that

$$x(t) \cos \theta - p(t)x'(t) \sin \theta = 0 \tag{1.9}$$

In boundary-value problems a common condition at an end point of an interval $t = a$ is

$$x(a) \cos \alpha - p(a)x'(a) \sin \alpha = 0 \tag{1.10}$$

From (1.9) it is clear that such a condition is equivalent to the simpler condition $\theta(a) = \alpha (\mathrm{mod}\ \pi)$. It is easy to see that (1.10) cannot hold for a solution $x = \varphi(t)$ for two different values of α unless they differ by a multiple of π or unless $\varphi^2(a) + (p\varphi'(a))^2 = \rho^2(a) = 0$. The behavior of the solutions for two equations of the form (1.1) will now be compared. The subscripts 1 and 2 will be used to distinguish between the two equations, that is, $L_i x = (p_i x')' + g_i x = 0$, $i = 1, 2$.

Theorem 1.2. *Let p_i' and g_i be piecewise continuous on $[a,b]$, and let*

$$0 < p_2(t) \leqq p_1(t) \qquad g_2(t) \geqq g_1(t)$$

on $[a,b]$. Let $L_1\varphi_1 = 0$ and $L_2\varphi_2 = 0$ and let $\omega_2(a) \geqq \omega_1(a)$. Then

$$\omega_2(t) \geqq \omega_1(t) \qquad (a \leqq t \leqq b) \tag{1.11}$$

Moreover, if $g_2 > g_1$ on (a,b), then

$$\omega_2(t) > \omega_1(t) \qquad (a < t \leqq b) \tag{1.12}$$

Proof. To prove (1.11), the equations

$$\omega_i' = \frac{1}{p_i} \cos^2 \omega_i + g_i \sin^2 \omega_i \qquad (i = 1, 2) \tag{1.13}$$

are subtracted to yield

$$(\omega_2 - \omega_1)' = \left(g_1 - \frac{1}{p_1}\right) (\sin^2 \omega_2 - \sin^2 \omega_1) + h \tag{1.14}$$

where

$$h = \left(\frac{1}{p_2} - \frac{1}{p_1}\right) \cos^2 \omega_2 + (g_2 - g_1) \sin^2 \omega_2$$

Clearly $h \geqq 0$. If $\omega_2 - \omega_1 = u$, then (1.14) yields

$$u' = fu + h \tag{1.15}$$

where

$$f = \left(g_1 - \frac{1}{p_1}\right) (\sin \omega_2 + \sin \omega_1) \left(\frac{\sin \omega_2 - \sin \omega_1}{\omega_2 - \omega_1}\right)$$

Thus f is piecewise continuous and uniformly bounded. Since $h \geqq 0$, (1.15) yields

$$u' - fu \geqq 0$$

If $F(t) = \int_t^b f(s)\,ds$, then multiplying the inequality above by e^F there results

$$e^F u' + F' e^F u \geqq 0$$

Integrating this over (a,t) gives

$$e^{F(t)} u(t) \geqq e^{F(a)} u(a) \geqq 0 \tag{1.16}$$

which proves (1.11).

If (1.12) fails to hold, then there must be some $c > a$ such that

$$\omega_2(t) = \omega_1(t) \qquad (a \leqq t \leqq c) \tag{1.17}$$

Indeed, suppose this is not the case. Then by (1.11) there must exist a sequence of points $\{t_j\}$ with a as a cluster point such that $\omega_2(t_j) > \omega_1(t_j)$. But if (1.16) is used with a replaced by t_j, it follows that for $t > t_j$ there results $\omega_2(t) > \omega_1(t)$. With t_j arbitrarily near a, this implies (1.12). Thus (1.17) must hold.

Using (1.17), then (1.14) is possible with $g_2 > g_1$ only if

$$\omega_1 = \omega_2 = 0 (\text{mod } \pi)$$

and if $p_1 \equiv p_2$ over (a,c). However, in (1.13) the case

$$\omega_1 = \omega_2 = 0 (\text{mod } \pi)$$

over (a,c) is clearly impossible. This proves (1.12) if $g_2 > g_1$.

2. Existence of Eigenvalues

Application will now be made to the equation

$$(px')' + (\lambda r - q)x = 0 \tag{2.1}$$

where λ is a real parameter and p', r, and q are real and continuous (or piecewise continuous) over $[a,b]$ and $p > 0$, $r > 0$ over $[a,b]$. [By modifying the proofs that follow slightly it is possible for r to vanish at a and at b as well as at isolated points in (a,b).]

Given real α and β, the values of λ for which (2.1) has a solution not identically zero and satisfying

$$x(a) \cos \alpha - p(a)x'(a) \sin \alpha = 0 \tag{2.2}$$

$$x(b) \cos \beta - p(b)x'(b) \sin \beta = 0 \tag{2.3}$$

are called *eigenvalues*. Either one of the conditions (2.2) or (2.3) determines a solution of (2.1) uniquely except for a multiplicative constant.

A nontrivial solution satisfying (2.1), (2.2), and (2.3) for an eigenvalue is called an *eigenfunction.*

Theorem 2.1. *There are an infinite number of eigenvalues $\lambda_0, \lambda_1, \lambda_2, \ldots$ forming a monotone increasing sequence with $\lambda_n \to \infty$ as $n \to \infty$. Moreover, the eigenfunction corresponding to λ_n has exactly n zeros on (a,b).*

Proof. There is no restriction in assuming that $0 \leqq \alpha < \pi$ and that $0 < \beta \leqq \pi$. The solution of (2.1) $\varphi = \varphi(t,\lambda)$ determined by

$$\varphi(a,\lambda) = \sin \alpha \qquad p(a)\varphi'(a,\lambda) = \cos \alpha$$

clearly satisfies (2.2). The eigenvalues are those values of λ for which φ satisfies (2.3). For $x = \varphi(t,\lambda)$ it is clearly the case that ω can be determined so that $\theta = \omega(t,\lambda)$ satisfies $\omega(a,\lambda) = \alpha$.

By Theorem 1.2, $\theta = \omega(t,\lambda)$ is, for fixed t, $a < t \leqq b$, a monotone increasing function of λ. Where $\omega = 0 (\text{mod } \pi)$, φ has a zero. From (1.8)

$$\theta' = \frac{1}{p} \cos^2 \theta + (\lambda r - q) \sin^2 \theta \tag{2.4}$$

it is clear that when $\omega = 0(\text{mod } \pi)$, $\omega' > 0$. This means that ω is an increasing function of t when $\omega = 0(\text{mod } \pi)$. Thus if for some t_k on (a,b), $\omega(t_k,\lambda) = k\pi$, then $\omega(t,\lambda) > k\pi$ for $t > t_k$ and $\omega(t,\lambda) < k\pi$ for $t < t_k$. Moreover, since ω is monotone in λ, it now follows that the zeros of φ, if any, move to the left toward $t = a$ as λ increases. Since ω is continuous in t and λ, and $\omega' > 0$ when $\omega = 0(\text{mod } \pi)$, it follows that the *location of the kth zero of φ on (a,b) at $t_k = t_k(\lambda)$ is a continuous and monotone decreasing function of λ.* Indeed

$$\omega'(t_k,\lambda) \frac{dt_k}{d\lambda} + \frac{\partial \omega}{\partial \lambda} (t_k,\lambda) = 0$$

It is the case that for any fixed $t = c$ in $(a,b]$,

$$\omega(c,\lambda) \to \infty \text{ as } \lambda \to \infty \tag{2.5}$$

and also

$$\omega(c,\lambda) \to 0 \text{ as } \lambda \to -\infty \tag{2.6}$$

The proof of (2.5) will be given first. Since $\alpha \geqq 0$, it follows that $\omega(t,\lambda) \geqq 0$ since $\omega' > 0$ for $\omega = 0(\text{mod } \pi)$. Thus it suffices to show that for some t_0, $a < t_0 < c$, $\omega(c,\lambda) - \omega(t_0,\lambda) \to \infty$ as $\lambda \to \infty$. Let $t_0 = (a + c)/2$. Let P, Q, and R be constants such that over (t_0,c)

$$p(t) \leqq P \qquad r(t) \geqq R > 0 \qquad q(t) \leqq Q$$

Then the equation

$$Px'' + (\lambda R - Q)x = 0 \tag{2.7}$$

with solution $\tilde{\varphi}$ satisfying $\tilde{\varphi}(t_0,\lambda) = \varphi(t_0,\lambda)$, $P\tilde{\varphi}'(t_0,\lambda) = p(t_0)\varphi'(t_0,\lambda)$ has $\tilde{\omega}(t_0,\lambda) = \omega(t_0,\lambda)$ and thus by Theorem 1.2

$$\omega(c,\lambda) - \omega(t_0,\lambda) \geqq \tilde{\omega}(c,\lambda) - \tilde{\omega}(t_0,\lambda) \tag{2.8}$$

The successive zeros of $\tilde{\varphi}$ have spacing $\pi[P/(\lambda R - Q)]^{\frac{1}{2}}$. This tends to zero as $\lambda \to \infty$. Therefore $\tilde{\omega} \equiv 0(\text{mod } \pi)$ for arbitrarily many values of t, and since $\tilde{\omega}' > 0$ at $\tilde{\omega} \equiv 0(\text{mod } \pi)$, $\tilde{\omega} \to \infty$. Thus the right side of (2.8) tends to infinity as $\lambda \to \infty$, which proves the left side must do the same. This completes the proof of (2.5).

To prove (2.6) the equation (2.4) is used. Choose $\delta > 0$ small enough so that $\alpha < \pi - \delta$. If $\delta \leqq \omega \leqq \pi - \delta$ and $\lambda < 0$ and if $0 < P \leqq p$, $0 < R \leqq r$, and $Q \geqq |q|$

$$\omega' < \frac{1}{P} - |\lambda| R \sin^2 \delta + Q$$

Thus $\omega' < 0$ for $\omega = \delta$ if $-\lambda$ is large enough. Moreover,

$$\omega' < \frac{-10}{c - a}$$

for $\delta \leqq \omega \leqq \pi - \delta$. Thus $\omega(c,\lambda) \leqq \delta$ for $-\lambda$ large enough. Since δ is arbitrary, (2.6) follows.

As $\lambda \to -\infty$, $\omega(b,\lambda) \to 0$. Since $\beta > 0$ and since $\omega(b,\lambda)$ is monotone increasing in λ, it follows there is a value of λ, λ_0, for which $\omega(b,\lambda_0) = \beta$. Since $0 \leqq \alpha < \pi$ and $\beta \leqq \pi$, it follows that $0 < \omega(t,\lambda_0) < \pi$ in (a,b) so that the solution $\varphi(t,\lambda_0)$ satisfies (2.3) and does not vanish in (a,b). Letting λ increase beyond λ_0, there is a unique value λ_1 for which

$$\omega(b,\lambda_1) = \beta + \pi$$

Clearly $\varphi(t,\lambda_1)$ satisfies (2.3) and has exactly one zero in (a,b). The nth eigenvalue is determined by $\omega(b,\lambda_n) = \beta + n\pi$. This completes the proof.

3. Periodic Boundary Conditions

The equation (2.1) is subject to the restrictions $p > 0$, $r > 0$, on $[a,b]$, and r, p', and q piecewise continuous. It will be assumed that $p(a) = p(b)$. With no restriction, it can be assumed that $a = 0$, $b = 1$, and $p(0) = p(1) = 1$. The boundary condition (see Prob. 4)

$$x(0) = x(1) \qquad x'(0) = x'(1) \tag{3.1}$$

will be considered, as will the condition

$$x(0) = -x(1) \qquad x'(0) = -x'(1) \tag{3.2}$$

Theorem 3.1. *The eigenvalues for* (2.1) *with* (3.1), λ_i, $i \geqq 0$, *and for* (2.1) *with* (3.2), $\tilde{\lambda}_i$, $i \geqq 1$, *form sequences such that*

$$-\infty < \lambda_0 < \tilde{\lambda}_1 \leqq \tilde{\lambda}_2 < \lambda_1 \leqq \lambda_2 < \tilde{\lambda}_3 \leqq \tilde{\lambda}_4 < \lambda_3 \leqq \lambda_4 < \cdots \tag{3.3}$$

For $\lambda = \lambda_0$ *there exists a unique eigenfunction,* φ_0. *If* $\lambda_{2i+1} < \lambda_{2i+2}$ *for some* $i \geqq 0$, *then there is a unique eigenfunction* φ_{2i+1} *at* $\lambda = \lambda_{2i+1}$ *and a unique eigenfunction* φ_{2i+2} *at* $\lambda = \lambda_{2i+2}$. *If, however,* $\lambda_{2i+1} = \lambda_{2i+2}$, *then there are two independent eigenfunctions* φ_{2i+1}, φ_{2i+2} *at* $\lambda = \lambda_{2i+1} = \lambda_{2i+2}$. *Similar results hold for the cases* $\tilde{\lambda}_{2i+1} < \tilde{\lambda}_{2i+2}$ *and* $\tilde{\lambda}_{2i+1} = \tilde{\lambda}_{2i+2}$, *where the eigenfunctions are denoted by* $\tilde{\varphi}_{2i+1}$ *and* $\tilde{\varphi}_{2i+2}$. *Furthermore,* φ_0 *has no zeros in* $[0,1]$; φ_{2i+1} *and* φ_{2i+2}, $i \geqq 0$, *each have exactly* $2i + 2$ *zeros in* $[0,1)$; *and* $\tilde{\varphi}_{2i+1}$ *and* $\tilde{\varphi}_{2i+2}$ *each have exactly* $2i + 1$ *zeros in* $[0,1)$.

Proof. Let φ and ψ be the solutions of (2.1) satisfying

$$\varphi(0,\lambda) = \psi'(0,\lambda) = 1 \qquad \varphi'(0,\lambda) = \psi(0,\lambda) = 0 \tag{3.4}$$

From (6.5), Chap. 3, [or from (2.1)]

$$p(t)[\varphi(t,\lambda)\psi'(t,\lambda) - \varphi'(t,\lambda)\psi(t,\lambda)] = 1 \tag{3.5}$$

For (3.1) to hold, it is necessary and sufficient that there exist constants C_1 and C_2 not both zero, such that $C_1\varphi + C_2\psi$ satisfies (3.1), which yields

$$\begin{aligned} [\varphi(1,\lambda) - 1]C_1 + \psi(1,\lambda)C_2 &= 0 \\ \varphi'(1,\lambda)C_1 + [\psi'(1,\lambda) - 1]C_2 &= 0 \end{aligned} \tag{3.6}$$

A necessary and sufficient condition for two independent solutions to satisfy (3.1) is

$$\varphi(1,\lambda) = \psi'(1,\lambda) = 1 \qquad \psi(1,\lambda) = \varphi'(1,\lambda) = 0 \tag{3.7}$$

A necessary and sufficient condition for (3.6) to have a nontrivial solution is that the determinant of the coefficients should vanish, which yields, with the use of (3.5) at $t = 1$,

$$f(\lambda) = 2 \tag{3.8}$$

where

$$f(\lambda) = \varphi(1,\lambda) + \psi'(1,\lambda) \tag{3.9}$$

The corresponding condition for (3.2) is

$$f(\lambda) = -2 \tag{3.10}$$

The values of λ satisfying (3.8) are the eigenvalues for (3.1) and similarly (3.10) for (3.2). If (3.8), but not (3.7), holds for a value of λ, then there is exactly one eigenfunction of (3.1) for this value of λ, and λ is called a *simple* eigenvalue. Not only is f a continuous function of λ but it is actually an entire function, as may be seen from Theorem 8.4 (or Prob. 7), Chap. 1.

Let the eigenvalues of (2.1) with the boundary condition

$$x(0) = x(1) = 0 \tag{3.11}$$

be μ_i, $i = 0, 1, 2, \ldots$. Then the following result will be proved after the completion of the proof of Theorem 3.1.

Lemma 3.1. *With μ_i the eigenvalues of (3.11), there exists a ν_0 such that*

$$\nu_0 < \mu_0 < \mu_1 < \cdots$$

and

$$f(\nu_0) \geqq 2 \qquad f(\mu_{2i}) \leqq -2 \qquad f(\mu_{2i+1}) \geqq 2 \qquad (i = 0, 1, \ldots) \tag{3.12}$$

If $f(\hat{\lambda}) = 2$ or -2 for some $\hat{\lambda} \neq \mu_i$, then such a $\hat{\lambda}$ is a simple eigenvalue for (3.1) or (3.2) and for such a $\hat{\lambda}$,

$$\frac{df}{d\lambda} < 0 \qquad (\hat{\lambda} < \mu_0);\ (-1)^i \frac{df}{d\lambda} > 0 \qquad (\mu_i < \hat{\lambda} < \mu_{i+1}) \qquad (i = 0, 1, \ldots) \tag{3.13}$$

If $f(\mu_{2i+1}) = 2$ and $df/d\lambda \neq 0$ at $\lambda = \mu_{2i+1}$, then μ_{2i+1} is a simple eigenvalue for (3.1). If $f(\mu_{2i+1}) = 2$ and if $df/d\lambda = 0$ at $\lambda = \mu_{2i+1}$, then at μ_{2i+1} there are two independent eigenfunctions for (3.1). Moreover, in this case,

$$\frac{d^2f}{d\lambda^2}(\mu_{2i+1}) < 0 \tag{3.14}$$

A similar result holds for (3.2) if $f(\mu_{2i}) = -2$. In this case the sign of the analogue of (3.14) is reversed.

An immediate consequence of this lemma is the existence of $\{\lambda_i\}$ and $\{\tilde{\lambda}_i\}$ satisfying (3.3) and the existence of the corresponding eigenfunctions. Indeed, clearly

$$\lambda_0 < \tilde{\lambda}_1 \leqq \mu_0 \leqq \tilde{\lambda}_2 < \lambda_1 \leqq \mu_1 \leqq \lambda_2 < \tilde{\lambda}_3 \leqq \mu_2 \leqq \tilde{\lambda}_4 < \lambda_3 \leqq \cdots \tag{3.15}$$

To show that the eigenfunctions have the specified number of zeros, the oscillation result, Theorem 1.1, is used. By the condition (3.1) it follows that the eigenfunctions φ_i have an even number of zeros in $[0,1)$, and by (3.2) the $\tilde{\varphi}_i$ an odd number. The eigenfunctions of (3.11) are $\psi(t,\mu_i)$ with i zeros in $(0,1)$ (by Theorem 2.1). Since $\lambda_0 < \mu_0$, it follows that φ_0 cannot have two zeros in $[0,1)$. Since φ_0 has an even number of zeros, the number must be zero. Since $\mu_{2i} < \lambda_{2i+1} \leqq \lambda_{2i+2} < \mu_{2i+2}$, $i \geqq 0$, it follows that φ_{2i+1} and φ_{2i+2} have more than $2i + 1$ zeros in $[0,1)$ and less than $2i + 4$ and thus exactly $2i + 2$.

Since $\tilde{\lambda}_1 \leqq \tilde{\lambda}_2 < u_1$, it follows that $\tilde{\varphi}_1$ and $\tilde{\varphi}_2$ have less than three zeros in $[0,1)$ and by (3.2) have at least one zero. Since the total number is odd, there must be exactly one. For $\tilde{\varphi}_{2i+1}$, $\tilde{\varphi}_{2i+2}$, $i \geqq 1$, using $\mu_{2i-1} < \tilde{\lambda}_{2i+1} \leqq \tilde{\lambda}_{2i+2} < \mu_{2i+1}$, it follows readily that there are exactly $2i + 1$ zeros in $[0,1)$. Thus there remains only the lemma to prove.

Proof of Lemma 3.1. With μ_i the eigenvalues of (2.1) with (3.11), it follows that $\psi(t,\mu_i)$ are the eigenfunctions. That is, $\psi(1,\mu_i) = 0$ and $\psi(t,\mu_i)$ has i zeros in (0,1). Thus $\psi'(1,\mu_i) > 0$ for i odd and < 0 for i even. From (3.5) follows $\varphi(1,\mu_i)\psi'(1,\mu_i) = 1$ so that

$$f(\mu_i) = \psi'(1,\mu_i) + \frac{1}{\psi'(1,\mu_i)}$$

Since for real $x > 0$, $x + 1/x \geqq 2$, the results (3.12) for μ_i now follow.

If ν_0 is the least eigenvalue of (2.1) with $x'(0) = x'(1) = 0$, then $\varphi(t,\nu_0)$ is the eigenfunction and it has no zeros in [0,1]. Thus $\nu_0 < \mu_0$ and $\varphi(1,\nu_0) > 0$. Since $\varphi'(1,\nu_0) = 0$ it follows from (3.5) that

$$\varphi(1,\nu_0)\psi'(1,\nu_0) = 1$$

Thus

$$f(\nu_0) = \varphi(1,\nu_0) + \frac{1}{\varphi(1,\nu_0)} \geqq 2$$

which completes the proof of (3.12).

In order to consider $df/d\lambda$, where $f = 2$ or -2, the function $u = \partial\varphi/\partial\lambda$ is considered. Clearly $u(0,\lambda) = u'(0,\lambda) = 0$ and from (2.1)

$$(pu')' + (\lambda r - q)u = -r\varphi$$

Thus from the variation-of-constants formula (or as can be verified directly),

$$u(t,\lambda) = \int_0^t [\varphi(t,\lambda)\psi(\tau,\lambda) - \varphi(\tau,\lambda)\psi(t,\lambda)]r(\tau)\varphi(\tau,\lambda)\,d\tau$$

Thus

$$\frac{\partial\varphi}{\partial\lambda}(1,\lambda) = \int_0^1 [\varphi(1,\lambda)\psi(\tau,\lambda) - \varphi(\tau,\lambda)\psi(1,\lambda)]r(\tau)\varphi(\tau,\lambda)\,d\tau \quad (3.16)$$

and in the same way

$$\frac{\partial\psi'}{\partial\lambda}(1,\lambda) = \int_0^1 [\varphi'(1,\lambda)\psi(\tau,\lambda) - \varphi(\tau,\lambda)\psi'(1,\lambda)]r(\tau)\psi(\tau,\lambda)\,d\tau$$

Thus, not indicating λ explicitly,

$$\frac{df}{d\lambda} = \int_0^1 [\psi^2(\tau)\varphi'(1) + \psi(\tau)\varphi(\tau)(\varphi(1) - \psi'(1)) - \varphi^2(\tau)\psi(1)]r(\tau)\,d\tau \quad (3.17)$$

The bracket in (3.17) regarded as a quadratic form in $\psi(\tau)$, $\varphi(\tau)$ does not change sign if $(\varphi(1) - \psi'(1))^2 + 4\varphi'(1)\psi(1) \leqq 0$. Using (3.5), this becomes

$$[\varphi(1,\lambda) + \psi'(1,\lambda)]^2 \leqq 4$$

Thus, if $-2 \leqq f(\lambda) \leqq 2$, the bracket in (3.17) has a fixed sign. If $f(\lambda) = 2$ or -2 then, except possibly for a factor -1, the bracket is a perfect square and $df/d\lambda$ cannot vanish unless the bracket is identically zero in τ. Because $\psi(\tau)$ and $\varphi(\tau)$ are independent, the bracket is identically zero if and only if all the coefficients vanish, which together with (3.5) is the condition (3.7) if $f = 2$ and the corresponding one if $f = -2$. Thus $df/d\lambda = 0$, where $f = 2$ or -2, if and only if the eigenvalue is not simple.

If $\lambda < \mu_0$ or if $\mu_i < \lambda < \mu_{i+1}$, then $\psi(1,\lambda) \neq 0$ and thus, if $f = 2$ or -2, the bracket in (3.17) is not identically zero. Being a perfect square, $df/d\lambda$ has the same sign as $-\psi(1,\lambda)$, which proves (3.13).

There remains only the proof of (3.14). At $\lambda = \mu_{2i+1}$ then, $f = 2$ and $df/d\lambda = 0$ so that (3.7) holds. Thus

$$\psi(1,\mu_{2i+1}) = \varphi'(1,\mu_{2i+1}) = 0 \qquad \psi'(1,\mu_{2i+1}) = \varphi(1,\mu_{2i+1}) = 1 \tag{3.18}$$

Using the notation

$$\varphi_\lambda = \frac{\partial \varphi}{\partial \lambda}(1,\lambda) \qquad \psi_\lambda = \frac{\partial \psi}{\partial \lambda}(1,\lambda)$$

and similarly for φ_λ', ψ_λ', then

$$\frac{d^2 f}{d\lambda^2} = \varphi_{\lambda\lambda} + \psi_{\lambda\lambda}' \tag{3.19}$$

From (3.5), differentiating with respect to λ gives

$$\psi_\lambda \varphi' + \psi \varphi_\lambda' - \psi_\lambda' \varphi - \psi' \varphi_\lambda = 0 \tag{3.20}$$

Taking account of (3.18), there results

$$\psi_\lambda'(1,\mu_{2i+1}) = -\varphi_\lambda(1,\mu_{2i+1}) \tag{3.21}$$

Differentiating (3.20) again and using (3.18) and (3.21),

$$2\psi_\lambda \varphi_\lambda' + 2\varphi_\lambda^2 - \psi_{\lambda\lambda}' - \varphi_{\lambda\lambda} = 0 \qquad (\lambda = \mu_{2i+1})$$

In (3.19) this yields

$$\frac{d^2 f}{d\lambda^2}(1,\mu_{2i+1}) = 2[\varphi_\lambda^2(1,\mu_{2i+1}) + \psi_\lambda(1,\mu_{2i+1})\varphi_\lambda'(1,\mu_{2i+1})] \tag{3.22}$$

Using (3.18) again, it follows from (3.16) that

$$\varphi_\lambda(1,\mu_{2i+1}) = \int_0^1 \psi(\tau,\mu_{2i+1})\varphi(\tau,\mu_{2i+1})r(\tau)\, d\tau$$

and in the same way

$$\begin{aligned} \psi_\lambda(1,\mu_{2i+1}) &= \int_0^1 \psi^2(\tau,\mu_{2i+1})r(\tau)\,d\tau \\ \varphi'_\lambda(1,\mu_{2i+1}) &= -\int_0^1 \varphi^2(\tau,\mu_{2i+1})r(\tau)\,d\tau \end{aligned} \tag{3.23}$$

Since ψ and φ are independent, the above relations and the Schwarz inequality imply that the right member of (3.22) is negative, which proves (3.14) and completes the proof of the lemma.

4. Stability Regions of Second-order Equations with Periodic Coefficients

Here the real equation

$$(p(t)x')' + [ar(t) + bq(t)]x = 0 \tag{4.1}$$

will be considered with a and b constant and with $p > 0$, $r > 0$, and p, p', r, and q continuous over $0 \leqq t \leqq 1$ and also periodic of period 1. That is, $r(0) = r(1)$, $q(0) = q(1)$ and, as can be assumed with no restriction, $p(0) = p(1) = 1$. In Sec. 5, Chap. 3, the existence of characteristic exponents and multipliers was proved. If $x = \psi(t,a,b)$ and $x = \varphi(t,a,b)$ are solutions of (4.1) with

$$\psi(0,a,b) = \varphi'(0,a,b) = 0 \qquad \psi'(0,a,b) = \varphi(0,a,b) = 1$$

then for fixed t, ψ, ψ', φ, and φ' are entire functions of (a,b) for all a and b. To determine the multipliers, the solution $x = C_1\varphi + C_2\psi$ is considered which satisfies

$$\begin{aligned} C_1\varphi(1,a,b) + C_2\psi(1,a,b) &= \sigma C_1 \\ C_1\varphi'(1,a,b) + C_2\psi'(1,a,b) &= \sigma C_2 \end{aligned}$$

for some σ. For a nontrivial solution (C_1,C_2) to exist, the determinant of the coefficients must vanish, which gives the characteristic equation

$$\sigma^2 - \sigma[\varphi(1,a,b) + \psi'(1,a,b)] + 1 = 0 \tag{4.2}$$

where use is made of (3.5). If

$$f(a,b) = \varphi(1,a,b) + \psi'(1,a,b)$$

then the roots σ_1 and σ_2 of (4.2) are distinct complex conjugates of magnitude 1 if

$$f^2(a,b) < 4 \tag{4.3}$$

while the roots are real and distinct if

$$f^2(a,b) > 4 \tag{4.4}$$

Because $\sigma_1\sigma_2 = 1$, in this latter case one root is always larger than 1 in magnitude and the other less than 1.

If the roots are distinct, two independent solutions exist, $x = u_1(t)e^{\alpha_1 t}$ and $x = u_2(t)e^{\alpha_2 t}$, where u_1 and u_2 are periodic of period 1 and $e^{\alpha_i} = \sigma_i$, $i = 1, 2$. Thus, in case (4.3) prevails, all solutions of (4.1) are uniformly bounded over $-\infty < t < \infty$. If (4.4) prevails, this is certainly not the case, even over $(-\infty,0)$ or over $(0,\infty)$. Therefore, in this section the values of (a,b) for which (4.3) holds will be called *stable* while those for which (4.4) holds will be called *unstable*. From the continuity of f, it follows that the stable regions and the unstable regions of the (a,b) plane have their boundaries made up of points where $f^2(a,b) = 4$ or, in other words, of points where either

$$f(a,b) = 2 \tag{4.5}$$

or

$$f(a,b) = -2 \tag{4.6}$$

For any fixed b the equation (4.1) is of the form (2.1) with a in place of λ. Thus the conditions (4.5) and (4.6) are precisely those already considered in connection with the eigenvalues of (3.1) and of (3.2). When (4.5) is satisfied, the equation (4.1) has a solution of period 1 while (4.6) corresponds to a solution which satisfies $x(0) = -x(1)$, $x'(0) = -x'(1)$ and thus has period 2. It will be designated as having half-period 1. From (3.15) it follows that for any fixed b the values of a, $a_i(b)$, $i = 0, 1, 2, \ldots$, at which (4.5) is satisfied and $\tilde{a}_i(b)$, $i = 1, 2, \ldots$, where (4.6) is satisfied, are related by

$$-\infty < a_0(b) < \tilde{a}_1(b) \leqq \mu_0(b) \leqq \tilde{a}_2(b) < a_1(b) \leqq \mu_1(b) \leqq a_2(b) < \tilde{a}_3(b) \\ \leqq \mu_2(b) \leqq \tilde{a}_4(b) < a_3(b) \leqq \cdots \tag{4.7}$$

where $\mu_i(b)$ are the eigenvalues of (4.1) for fixed b with the conditions $x(0) = x(1) = 0$. That the μ_i are continuous functions of b for each i follows from the fact that $(\partial\psi/\partial a)(1,a,b) \neq 0$, where $\psi(1,a,b) = 0$, since, much as in (3.23), under these conditions

$$\frac{\partial\psi}{\partial a}(1,a,b) = \varphi(1,a,b) \int_0^1 r(\tau)\psi^2(\tau)\,d\tau$$

It remains now to show that a_i and $\tilde{a}_i$ are continuous single-valued functions of b, $-\infty < b < \infty$. Before showing this, the consequences of this fact will be considered. For fixed b and all a satisfying $a_{2i+1}(b) < a < a_{2i+2}(b)$ it follows from (3.12) and (3.15) that $f(a,b) > 2$ and thus that (4.1) is unstable. In the same way, $\tilde{a}_{2i+1} < a < \tilde{a}_{2i+2}$ is unstable. *The stable regions are given by* $a_{2i}(b) < a < \tilde{a}_{2i+1}(b)$, $i \geqq 0$, *and by* $\tilde{a}_{2i+2}(b) < a < a_{2i+1}(b)$, $i \geqq 0$.

The $a_i(b)$ are the unique solutions of $f(a_i(b),b) = 2$ and similarly for $\tilde{a}_i(b)$. Thus that $a = a_i(b)$ [or $a = \tilde{a}_i(b)$] determines a continuous curve (with a as a function of b) in the (a,b) plane follows at once wherever

$\partial f\,(a_i(b),b)/\partial a \neq 0$. Since $f(a_i(b),b) = 2$, it follows from Lemma 3.1 that $\partial f/\partial a \neq 0$ can fail only where i is one of a pair $2j + 1$, $2j + 2$ which satisfies

$$a_{2j+1}(b) = \mu_{2j+1}(b) = a_{2j+2}(b) \tag{4.8}$$

Let $b = \beta$ denote a value of b, where (4.8) holds. Then $a_{2j+1}(\beta - 0)$ and $a_{2j+2}(\beta - 0)$ must both be $\mu_{2j+1}(\beta)$, for if

$$\mu_{2j}(\beta) \leqq \liminf_{b \to \beta - 0} a_{2j+1}(b) = \alpha < \mu_{2j+1}(\beta)$$

then by continuity $f(\alpha,\beta) = 2$ so that $\mu_{2j}(\beta) \leqq a_{2j+1}(\beta) = \alpha < \mu_{2j+1}(\beta)$, which contradicts (4.8). A similar procedure works for a_{2j+2} and for $\beta + 0$. Thus a_i is a continuous function of b for all b. In the same way, $\tilde{a}_i$ is a continuous function of b.

Important familiar examples of (4.1) occur with $p \equiv r \equiv 1$. In the *Mathieu equation* $q(t) = \cos 2\pi t$ and in the *Hill equation* any q, aside from being periodic, has average value zero. In the equation (*Meissner*) where $q(t) = 1$, $0 < t < \frac{1}{2}$, and $q(t) = -1$, $\frac{1}{2} < t < 1$, explicit calculations can readily be made for $f(a,b)$.

PROBLEMS

1. If φ and ψ are independent real solutions of (1.1), show by considering $(\varphi/\psi)'$ that the zeros of φ and ψ separate each other.

2. Let the hypothesis of Theorem 1.1 be satisfied and let there exist some $\delta > 0$ such that $\varphi(t) > 0$, $\psi(t) > 0$ on $(a, a + \delta)$. Let

$$\lim_{t \to a+0} p(t)[\varphi'\psi - \varphi\psi'] \geqq 0$$

Prove that if $\varphi(t_1) = 0$ for some t_1 on (a,b) then there is a t_2 on (a,t_1) such that $\psi(t_2) = 0$.

3. Consider (2.1), (2.2), and (2.3) over the interval $[a,b]$ but now let $q < 0$ on (a,b) and suppose r changes its sign on (a,b) while, as before, $p > 0$ on $[a,b]$. Show that the eigenvalues have $\lambda = +\infty$ and $\lambda = -\infty$ as cluster points.

HINT: For the case $\lambda > 0$ consider

$$\left(\frac{p}{\lambda}x'\right)' + \left(r - \frac{q}{\lambda}\right)x = 0$$

Then as λ increases, p/λ decreases and $r - q/\lambda$ increases. Let $\varphi(a,\lambda) = \sin \alpha$, $p(a)\varphi'(a,\lambda) = \cos \alpha$. Then let $\theta = \omega(t,\lambda)$ satisfy $\omega(a,\lambda) = \alpha$. Clearly $\omega(b,\lambda)$ is an increasing function of λ, and $\omega(b,0)$ is bounded. To show $\omega(b,\lambda) \to \infty$ as $\lambda \to \infty$, consider an interval $[t_1,t_2]$ on which $r(t) > R > 0$, where R is a constant. Let $p(t) < P$ on (t_1,t_2). Then for large λ, the solutions of $\frac{1}{\lambda}Px'' + \frac{1}{2}Rx = 0$ have zeros spaced $\pi\left(\frac{2P}{\lambda R}\right)^{\frac{1}{2}}$ on $[t_1,t_2]$. Thus $\omega(b,\lambda) > c\lambda^{\frac{1}{2}}$ for some constant c. A similar argument applies as $\lambda \to -\infty$.

4. Consider (2.1) p', q, and r continuous and $p > 0$, $r > 0$ on $[0,1]$. Develop the conclusion analogous to that of Theorem 3.1 for the boundary conditions

$$\begin{aligned} x(0) &= ax(1) + bx'(1) \\ x'(0) &= cx(1) + dx'(1) \end{aligned}$$

where a, b, c, and d are real constants and where $(ad - bc)p(0) = p(1)$.

HINT: Show that the eigenvalues occur at the roots of $f(\lambda) = 2$, where

$$f(\lambda) = a\varphi(1,\lambda) + b\varphi'(1,\lambda) + c\psi(1,\lambda) + d\psi'(1,\lambda)$$

Show that if μ_i are the eigenvalues of $x(0) = 0$, $ax(1) + bx'(1) = 0$, then $f^2(\mu_i) \geqq 4$ and $f(\mu_i)$ has alternating signs.

5. In Theorem 3.1 let $\lambda_{2i+1} < \lambda_{2i+2}$. Show that (2.1) with $\lambda = \lambda_{2i+1}$ has a solution ψ_{2i+1} independent of φ_{2i+1} such that

$$\psi_{2i+1}(t) = p_{2i+1}(t) + t\varphi_{2i+1}(t)$$

where p_{2i+1} is a periodic function with period 1. Show that similar results hold for λ_{2i+2}, $\bar{\lambda}_{2i+1}$, and $\bar{\lambda}_{2i+2}$.

6. Using the notation of Sec. 4, can an $\tilde{a}_j(b)$ and $a_i(b)$ ever intersect? What is the significance of Prob. 10, Chap. 3, in the terminology of Sec. 4? Sketch possible configurations of $a_j(b)$ and $\tilde{a}_j(b)$ showing stable and unstable domains in the (a,b) plane for (4.1) with $p(t) \equiv r(t) \equiv 1$.

7. In (4.1) let $a > 0$ and $\int_0^1 q(t)\,dt = 0$. Show that if u is a real solution satisfying $u(t + 1) = \lambda_0 u(t)$, where λ_0 is a constant, then u must vanish at at least one point in the interval $[0,1]$.

HINT: If not,

$$\int_0^1 \frac{pu'^2}{u^2}\,dt + a\int_0^1 r\,dt = 0$$

Show that u vanishes at two points t_1, t_2, where $|t_2 - t_1| \leqq 1$.

8. Let f be real and of class C^2 on $[a,b]$, and let $f(a) = f(b) = 0$ and $f > 0$ on (a,b). Prove that

$$(b - a)\int_a^b \frac{|f''(t)|}{f(t)}\,dt > 4$$

HINT: Let f attain its maximum at c. Then for some τ_1 and τ_2

$$\begin{aligned} \frac{1}{c-a} + \frac{1}{b-c} &= \frac{1}{f(c)}\left[\frac{f(c) - f(a)}{c-a} - \frac{f(b) - f(c)}{b-c}\right] \\ &= \frac{f'(\tau_1) - f'(\tau_2)}{f(c)} \leq \int_{\tau_1}^{\tau_2} \frac{|f''(t)|\,dt}{f(c)} \end{aligned}$$

9. Let r be nonnegative, continuous, and of period 1. If

$$\int_0^1 r(t)\,dt \leqq 4$$

show that $x'' + r(t)x = 0$ has stable solutions on $(-\infty, \infty)$.

HINT: Use Probs. 7 and 8.

CHAPTER 9

SINGULAR SELF-ADJOINT BOUNDARY-VALUE PROBLEMS FOR SECOND-ORDER EQUATIONS

1. Introduction

The treatment of Chap. 7 fails to apply in case the finite interval (a,b) becomes infinite or in case the coefficients in the differential operator have a sufficiently singular behavior at a or b. These cases are all regarded as singular, and the second-order singular case will be treated in this chapter.

As a preliminary example, the problem

$$-x'' = lx \qquad x(0) = 0 \qquad \left({}' = \frac{d}{dt}\right)$$

for $0 \leqq t < \infty$ will be considered as a limiting case of the problem on the finite interval $0 \leqq t \leqq b$ with the condition

$$x(b) = 0$$

added, and then letting $b \to \infty$. The finite interval problem of course gives rise to the orthonormal system $\{\psi_k\}$, where

$$\psi_k(t) = \left(\frac{2}{b}\right)^{\frac{1}{2}} \sin \frac{k\pi t}{b} \qquad (k = 1, 2, \ldots)$$

Any function f which is continuous over $0 \leqq t \leqq c$ and vanishes for $t \geqq c$ satisfies the completeness relationship

$$\int_0^c |f(t)|^2 \, dt = \frac{2}{b} \sum_{k=1}^{\infty} \left| \int_0^c \sin \frac{k\pi t}{b} f(t) \, dt \right|^2$$

in the event that $b > c$. If

$$g(s) = \int_0^c \sin st \, f(t) \, dt \tag{1.1}$$

then the completeness relationship becomes

$$\int_0^\infty |f(t)|^2 \, dt = \frac{2}{b} \sum_{k=1}^{\infty} \left| g\left(\frac{k\pi}{b}\right) \right|^2 \tag{1.2}$$

where the c in the left side has been replaced by ∞ since f vanishes for $t > c$. Let ρ_b be a nondecreasing step function of s which increases by $2/b$ when s passes through $k\pi/b$, $(k = 1, 2, \ldots)$, and is otherwise constant. Assume also that ρ_b has been normalized so that $\rho_b(0) = 0$. Then (1.2) can be written as

$$\int_0^\infty |f(t)|^2\,dt = \int_0^\infty |g(s)|^2\,d\rho_b(s) \tag{1.3}$$

Clearly, as $b \to \infty$, $\rho_b(s) \to 2s/\pi$. Thus, proceeding without regard to rigor, (1.3) yields as $b \to \infty$

$$\int_0^\infty |f(t)|^2\,dt = \frac{2}{\pi}\int_0^\infty |g(s)|^2\,ds \tag{1.4}$$

This is, of course, the Plancherel equation for the Fourier sine transform for the restricted class of functions under consideration here.

It is easy to give a rigorous proof of (1.4). Suppose that $f(0) = 0$, and that f has a continuous first derivative on $[0,c]$. Then from (1.1) it follows by integrating by parts that

$$|g(s)| \leqq \frac{1}{s}\int_0^c |f'(t)|\,dt = \frac{M}{s} \qquad (s > 0)$$

where M represents the integral. Thus for $s \geqq 1$,

$$|g(s)|^2 \leqq \frac{M^2}{s^2} \tag{1.5}$$

Since g is continuous, it is the case that for any fixed large μ

$$\lim_{b\to\infty}\int_0^\mu |g(s)|^2\,d\rho_b(s) = \frac{2}{\pi}\int_0^\mu |g(s)|^2\,ds \tag{1.6}$$

From (1.5) and integrating by parts, it follows that

$$\int_\mu^\infty |g(s)|^2\,d\rho_b(s) \leqq M^2\int_\mu^\infty s^{-2}\,d\rho_b(s) \leqq \frac{4M^2}{\pi}\int_\mu^\infty s^{-2}\,ds = \frac{4M^2}{\pi\mu} \tag{1.7}$$

Similarly

$$\int_\mu^\infty |g(s)|^2\,ds \leqq \frac{M^2}{\mu} \tag{1.8}$$

Thus using (1.6), (1.7), and (1.8) and letting $\mu \to \infty$, formula (1.4) is validated for f restricted as indicated. Such f are dense in the space $\mathfrak{L}^2(0,\infty)$, and, using standard theorems from Lebesgue integration, (1.4) can be proved for any $f \in \mathfrak{L}^2(0,\infty)$. Hence the analogue of the completeness theorem of Chap. 7, Sec. 4, is valid in the case of the simple example just considered. It might be of interest to the reader to parallel the above argument for the problem

$$-x'' = lx \qquad x'(0) = 0$$

and see how the cosine transform theorem is associated with the same differential operator but with a different boundary condition.

The method of the above example, with necessary major elaborations, provides a means of treating the general second-order problem in the singular case. Throughout the remainder of this chapter L will denote the formally self-adjoint differential operator defined by

$$Lx = -(px')' + qx$$

where it is assumed that p, p', q are real and continuous, $p(t) > 0$, on any real t interval under consideration.†

Of fundamental importance in all that follows is Green's formula which states that if $[t_1,t_2]$ is any interval over which L is defined and f and g are any two functions for which Lf, Lg make sense, then

$$\int_{t_1}^{t_2} (\bar{g}Lf - f\overline{Lg})\, dt = [fg](t_2) - [fg](t_1) \qquad (1.9)$$

where

$$[fg](t) = p(t)(f(t)\bar{g}'(t) - f'(t)\bar{g}(t))$$

In particular, if f and g are solutions of the same equation, $Lx = lx$, where l is a complex number, then Green's formula applied to f and $\bar{g}$ shows that $[f\bar{g}](t)$ is a constant independent of t, and hence can be denoted by just $[f\bar{g}]$.

The case of the semi-infinite interval $0 \leq t < \infty$ will be dealt with first, followed by a treatment of the case of the interval $-\infty < t < \infty$. In the case of a finite open interval (a,b) the treatments of the earlier chapters remain valid if $|p'/p|$, $|q/p|$, and $|1/p|$ have finite integrals over (a,b). If the behavior at one end, say b, is worse than this, the problem may be treated over $[a, \tilde{b}]$, $a < \tilde{b} < b$, and then the nature of the problem as $\tilde{b} \to b - 0$ is considered. This is completely analogous to the treatment of the problem on $[0, \infty)$, and it will be seen that all results obtained on $[0, \infty)$ are valid in the case of an interval $[a,b)$ where the coefficients in L have a singular behavior at b. Similar remarks hold concerning the case where the coefficients are singular at a and b, and the results are the same as for the case $(-\infty, \infty)$.

† Actually, it suffices for p to be absolutely continuous and q integrable over every finite subinterval.

2. The Limit-point and Limit-circle Cases

Here the interval under consideration is $[0, \infty)$. If for a particular complex number l_0 every solution φ of the differential equation

$$Lx = l_0 x$$

satisfies $\int_0^\infty |\varphi|^2 \, dt < \infty$, that is, $\varphi \in \mathfrak{L}^2(0, \infty)$, then L is said to be of the *limit-circle type* at infinity; otherwise L is said to be of the *limit-point type* at infinity.† In order to justify this definition, it must be shown that the classification depends only on L and not on the particular l_0 chosen.

Theorem 2.1. *If every solution of $Lx = l_0 x$ is of class $\mathfrak{L}^2(0, \infty)$ for some complex number l_0, then, for arbitrary complex l, every solution of $Lx = lx$ is of class $\mathfrak{L}^2(0, \infty)$.*

Proof. It is given that two linearly independent solutions φ and ψ of $Lx = l_0 x$ are of class $\mathfrak{L}^2(0, \infty)$. Let χ be any solution of $Lx = lx$, which may be written as

$$Lx = l_0 x + (l - l_0)x$$

By multiplying φ by a constant if necessary (to achieve $[\varphi\bar{\psi}] = 1$) the variation-of-constants formula yields

$$\chi(t) = c_1\varphi(t) + c_2\psi(t) + (l - l_0) \int_c^t (\varphi(t)\psi(\tau) - \varphi(\tau)\psi(t))\chi(\tau) \, d\tau \tag{2.1}$$

where c, c_1, c_2 are constants. If the notation

$$\|\chi\|_c = \left(\int_c^t |\chi|^2 \, dt\right)^{\frac{1}{2}}$$

is used, and if M is such that $\|\varphi\|_c \leqq M$, $\|\psi\|_c \leqq M$ for all $t \geqq c$, then the Schwarz inequality gives

$$\left| \int_c^t (\varphi(t)\psi(\tau) - \varphi(\tau)\psi(t))\chi(\tau) \, d\tau \right| \leqq M(|\varphi(t)| + |\psi(t)|)\|\chi\|_c$$

Using this in (2.1), the Minkowski inequality yields

$$\|\chi\|_c \leqq (|c_1| + |c_2|)M + 2|l - l_0|M^2\|\chi\|_c$$

If c is chosen large enough so that $|l - l_0|M^2 < \frac{1}{4}$, then

$$\|\chi\|_c \leqq 2(|c_1| + |c_2|)M$$

Since the right side of this inequality is independent of t, it follows that $\chi \in \mathfrak{L}^2(0, \infty)$ and the theorem is proved.

† The geometric significance of the terms *limit circle* and *limit point* will be made apparent shortly.

In the limit-point case, clearly, at most one linearly independent solution of $Lx = lx$ is of class $\mathfrak{L}^2(0, \infty)$. It will be shown presently that in this case there is exactly one solution of $Lx = lx$ of class $\mathfrak{L}^2(0, \infty)$ for any l such that $\mathfrak{I}l \neq 0$.

Let φ, ψ be two solutions of $Lx = lx$ satisfying

$$\begin{aligned} \varphi(0,l) &= \sin\alpha & \psi(0,l) &= \cos\alpha \\ p(0)\varphi'(0,l) &= -\cos\alpha & p(0)\psi'(0,l) &= \sin\alpha \end{aligned} \tag{2.2}$$

where $0 \leqq \alpha < \pi$. Then clearly φ, ψ are linearly independent solutions, and from Theorem 8.4 or Prob. 7, Chap. 1, φ, φ', ψ, ψ' are entire functions of l and continuous in (t,l). Moreover, since $[\varphi\bar{\psi}](0) = 1$, one has $[\varphi\bar{\psi}](t) = 1$ for all t. These solutions are real for real l and satisfy the following boundary conditions at 0

$$\begin{aligned} \cos\alpha\,\varphi(0,l) + \sin\alpha\,p(0)\varphi'(0,l) &= 0 \\ \sin\alpha\,\psi(0,l) - \cos\alpha\,p(0)\psi'(0,l) &= 0 \end{aligned}$$

Every solution χ of $Lx = lx$ except ψ is, up to a constant multiple, of the form

$$\chi = \varphi + m\psi \tag{2.3}$$

for some m which will depend on l.

Consider now a real boundary condition at some point b, $0 < b < \infty$, say,

$$\cos\beta\,x(b) + \sin\beta\,p(b)x'(b) = 0 \qquad (0 \leqq \beta < \pi) \tag{2.4}$$

and ask what must m be like in order that the solution χ, (2.3), satisfy (2.4). Clearly m must satisfy

$$m = -\frac{\cot\beta\,\varphi(b,l) + p(b)\varphi'(b,l)}{\cot\beta\,\psi(b,l) + p(b)\psi'(b,l)} \tag{2.5}$$

As l, b, β vary, m becomes a function of these arguments $m = m(l,b,\beta)$, and since φ, φ', ψ, ψ' are entire in l it follows that m is meromorphic in l and real for real l. If $z = \cot\beta$ and if (l,b) are held fixed, (2.5) may be written as

$$m = -\frac{Az + B}{Cz + D} \tag{2.6}$$

with A, B, C, D fixed while z varies over the real line as β varies from 0 to π. From well-known properties of the mapping (2.6), the real axis of the z plane has as its image a circle C_b in the m plane. Thus χ will satisfy (2.4) if and only if m lies on C_b.

From

$$z = -\frac{B + Dm}{A + Cm}$$

the equation of the image of the real axis, $\Im z = 0$, becomes

$$(\bar{A} + \bar{C}\bar{m})(B + Dm) - (A + Cm)(\bar{B} + \bar{D}\bar{m}) = 0$$

which is the equation for C_b. It follows easily that the center of C_b is

$$\tilde{m}_b = \frac{A\bar{D} - B\bar{C}}{\bar{C}D - C\bar{D}}$$

and the radius is

$$r_b = \frac{|AD - BC|}{|\bar{C}D - C\bar{D}|}$$

From the fact that

$$\begin{aligned} A &= \varphi(b,l) \qquad & B &= p(b)\varphi'(b,l) \\ C &= \psi(b,l) \qquad & D &= p(b)\psi'(b,l) \end{aligned}$$

it is readily seen that the equation of C_b is

$$[\chi\chi](b) = 0 \tag{2.7}$$

and that

$$\begin{aligned} A\bar{D} - B\bar{C} &= [\varphi\psi](b) \\ \bar{C}D - C\bar{D} &= -[\psi\psi](b) \\ AD - BC &= [\varphi\bar{\psi}](b) = 1 \end{aligned}$$

so that

$$\tilde{m}_b = -\frac{[\varphi\psi](b)}{[\psi\psi](b)} \qquad r_b = \frac{1}{|[\psi\psi](b)|} \tag{2.8}$$

Since the coefficient of $m\bar{m}$ in (2.7) is $[\psi\psi](b)$, it follows that the interior of C_b in the m plane is given by

$$\frac{[\chi\chi](b)}{[\psi\psi](b)} < 0 \tag{2.9}$$

By Green's formula [(1.9)],

$$[\psi\psi](b) = 2i\Im l \int_0^b |\psi|^2\,dt$$

and

$$[\chi\chi](b) = 2i\Im l \int_0^b |\chi|^2\,dt + [\chi\chi](0)$$

Since $[\chi\chi](0) = -2i\Im m$, (2.9) becomes

$$\int_0^b |\chi|^2\,dt < \frac{\Im m}{\Im l} \qquad (\Im l \neq 0) \tag{2.10}$$

which determines the interior of C_b. Points m are on C_b if and only if

$$\int_0^b |\chi|^2\,dt = \frac{\Im m}{\Im l} \qquad (\Im l \neq 0) \tag{2.11}$$

The radius r_b in (2.8) is given for $\Im l > 0$ by

$$r_b^{-1} = 2\Im l \int_0^b |\psi|^2\,dt \tag{2.12}$$

Now let $0 < a < b < \infty$. Then if m is inside or on C_b

$$\int_0^a |\chi|^2\,dt < \int_0^b |\chi|^2\,dt \leq \frac{\Im m}{\Im l}$$

and therefore m is inside C_a. This means C_a contains C_b in its interior if $a < b$. Thus for a given l, ($\Im l > 0$), as $b \to \infty$ the circles C_b converge either to a circle C_∞ or to a point m_∞. If the C_b converge to a circle, then its radius $r_\infty = \lim r_b$ is positive, and from (2.12) this implies $\psi \in \mathfrak{L}^2(0, \infty)$. If $\hat{m}_\infty$ is any point on C_∞, then $\hat{m}_\infty$ is inside any C_b for $b > 0$. Hence

$$\int_0^b |\varphi + \hat{m}_\infty\psi|^2\,dt < \frac{\Im \hat{m}_\infty}{\Im l}$$

and letting $b \to \infty$ one sees that $\varphi + \hat{m}_\infty\psi \in \mathfrak{L}^2(0, \infty)$. The same argument holds if $\hat{m}_\infty$ reduces to the point m_∞. Therefore, if $\Im l \neq 0$, there always exists a solution of $Lx = lx$ of class $\mathfrak{L}^2(0, \infty)$. In the case $C_b \to C_\infty$, all solutions are of class $\mathfrak{L}^2(0, \infty)$ for $\Im l \neq 0$, since both ψ and $\varphi + \hat{m}_\infty\psi$ are, and this identifies the limit-circle case with existence of the circle C_∞. Correspondingly, the limit-point case is identified with the existence of the point m_∞. In the case $C_b \to m_\infty$ there results $\lim r_b = 0$, and from (2.12) this implies that ψ is not of class $\mathfrak{L}^2(0, \infty)$. Therefore in this situation there is only one linearly independent solution of class $\mathfrak{L}^2(0, \infty)$ for $\Im l \neq 0$.

In the limit-circle case m is on C_b if and only if (2.11) holds. Since $\chi = \varphi(t,l) + m\psi(t,l)$, it follows that m is on C_∞ if and only if

$$\Im l \int_0^\infty |\chi|^2\,dt = \Im m$$

Since $[\chi\chi](0) = -2i\Im m$, it follows from the formula above [(2.10)] that m is on the limit circle if and only if $[\chi\chi](\infty) = 0$. The following theorem has been proved.

Theorem 2.2. *If $\Im l \neq 0$ and φ, ψ are the linearly independent solutions of $Lx = lx$ satisfying (2.2), then the solution $\chi = \varphi + m\psi$ satisfies the real boundary condition (2.4) if and only if m lies on a circle C_b in the complex plane whose equation is*

$$[\chi\chi](b) = 0$$

As $b \to \infty$ either $C_b \to C_\infty$, a limit circle, or $C_b \to m_\infty$, a limit point. All solutions of $Lx = lx$ are $\mathfrak{L}^2(0, \infty)$ in the former case, and if $\Im l \neq 0$ exactly one linearly independent solution is $\mathfrak{L}^2(0, \infty)$ in the latter case. Moreover, *in the limit-circle case, a point is on the limit circle* $C_\infty(l)$ *if and only if* $[\chi\chi](\infty) = 0$.

In the limit-point case, if m is any point on C_b, then $m \to m_\infty$, the limit point, and this holds independent of the choice of β in the boundary condition (2.4). In particular, this will hold when $\beta = 0$, and thus the limit point is given by

$$m_\infty(l) = -\lim_{b\to\infty} \frac{\varphi(b,l)}{\psi(b,l)} \tag{2.13}$$

The Green's function G associated with the boundary-value problem

$$\begin{aligned} Lx &= lx \\ \sin\alpha\, x(0) - \cos\alpha\, p(0)x'(0) &= 0 \\ \cos\beta\, x(b) + \sin\beta\, p(0)x'(b) &= 0 \end{aligned}$$

is clearly given by

$$G(t,\tau,l) = \begin{cases} \psi(t,l)\{\varphi(\tau,l) + m(l,b,\beta)\psi(\tau,l)\} & (t \leqq \tau) \\ \psi(\tau,l)\{\varphi(t,l) + m(l,b,\beta)\psi(t,l)\} & (t > \tau) \end{cases}$$

It follows directly from Theorem 2.2 that in the limit-point case Green's function tends to a unique limit as $b \to \infty$ given by the same formula but with m replaced by m_∞. In the limit-circle case there are an infinite number of limit functions to which Green's function may tend, depending on how β varies as b increases. In any case, the limit function to which G tends is of class $\mathfrak{L}^2(0,\infty)$ as a function of t.

Theorem 2.3. *In the limit-point case the limit point* m_∞ *is an analytic function of* l *for* $\Im l > 0$ *(and* $\Im l < 0$*).* $\Im m_\infty > 0$ *for* $\Im l > 0$ *and if* m_∞ *has zeros or poles on the real axis they are all simple.*

Proof. From (2.8) it follows that the center and radius of the circle C_1 are continuous functions of l for $\Im l > 0$. Thus, since C_b is interior to C_1 for $b > 1$, it follows that if l is restricted to a closed bounded subset Λ of $\Im l > 0$, then the points $m = m(l,b,\beta)$ on C_b are uniformly bounded as $b \to \infty$. The functions m_b, where $m_b(l,\beta) = m(l,b,\beta)$, being meromorphic and bounded on Λ, are analytic there. Hence, by Cauchy's theorem, the functions m_b constitute an equicontinuous set on Λ, and m_b converges uniformly to m_∞. Being the uniform limit of analytic functions, m_∞ itself is analytic on Λ, and hence on $\Im l > 0$.

Since m_∞ is inside of C_b, it follows from (2.10) that $\Im m_\infty > 0$ for $\Im l > 0$. This proves that if m_∞ has zeros or poles on the real axis they are simple and that the poles have negative residue. These remarks apply also, of course, to the meromorphic function of l, m_b.

It is important to know whether a given operator L is in the limit-point or limit-circle case. A useful sufficient condition for L to be in the limit-point case is the following:

Theorem 2.4. *Let* M *be a positive differentiable function, and* k_1 *and* k_2

two positive constants such that for large t

$$q(t) \geqq -k_1 M(t) \qquad \int_t^\infty (pM)^{-\frac{1}{2}}\, dt = \infty$$
$$|p^{\frac{1}{2}}(t) M'(t) M^{-\frac{3}{2}}(t)| \leqq k_2 \tag{2.14}$$

Then L is in the limit-point case at infinity.

Proof. It will be shown that $Lx = 0$ does not have two linearly independent solutions of class $\mathfrak{L}^2(0, \infty)$. Suppose χ is a real solution of $Lx = 0$, and assume $\chi \in \mathfrak{L}^2(0, \infty)$. From $(p\chi')' = q\chi$ follows for some $c > 0$

$$\int_c^t \frac{(p\chi')'\chi}{M}\, dt = \int_c^t \frac{q}{M}\chi^2\, dt \geqq -k_1 \int_c^t \chi^2\, dt$$

Integrating by parts and using the fact that $\chi \in \mathfrak{L}^2(0, \infty)$ there exists a constant k_3 such that

$$\frac{-p\chi'\chi}{M} + \int_c^t \frac{p(\chi')^2}{M}\, dt - \int_c^t \frac{p\chi'\chi M'}{M^2}\, dt < k_3 \tag{2.15}$$

Let

$$H(t) = \int_c^t \frac{p(\chi')^2}{M}\, dt$$

Then using (2.14) and then the Schwarz inequality

$$\left| \int_c^t \frac{p\chi'\chi M'}{M^2}\, dt \right|^2 \leqq k_2^2 \left(\int_c^t \frac{p^{\frac{1}{2}}|\chi'\chi|}{M^{\frac{1}{2}}}\, dt \right)^2 \leqq k_2^2 H(t) \int_c^t \chi^2\, dt$$

Thus by (2.14) and (2.15) there exists a constant k_4 such that

$$-\frac{p\chi'\chi}{M} + H - k_4 H^{\frac{1}{2}} < k_3 \tag{2.16}$$

If $H(t) \to \infty$, as $t \to \infty$, then (2.16) implies that for all large t, $p\chi'\chi/M > H/2$. This means χ and χ' have the same sign for all large t, which contradicts $\chi \in \mathfrak{L}^2(0, \infty)$. Thus H remains finite so that

$$\int_c^\infty \frac{p(\chi')^2}{M}\, dt < \infty \tag{2.17}$$

Now suppose φ and ψ are two linearly independent solutions of $Lx = 0$ which are of class $\mathfrak{L}^2(0, \infty)$, that is, suppose L is in the limit-circle case. It can be assumed that these solutions are real and

$$[\varphi\bar{\psi}] = p(\varphi\psi' - \psi\varphi') = 1$$

This implies

$$\varphi \frac{p^{\frac{1}{2}}\psi'}{M^{\frac{1}{2}}} - \psi \frac{p^{\frac{1}{2}}\varphi'}{M^{\frac{1}{2}}} = \frac{1}{(pM)^{\frac{1}{2}}}$$

By (2.17) and the Schwarz inequality, the left side of the above equation is integrable over (c, ∞). By hypothesis (2.14), the right side is not. Thus the limit-circle case is ruled out.

In the case $M(t) = 1$ for $0 \leqq t < \infty$ the following corollary results.

Corollary 1. *If $q(t) \geqq -k$, where k is a positive constant, and*

$$\int^{\infty} p^{-\frac{1}{2}}\,dt = \infty$$

then L is in the limit-point case at infinity.

Many second-order differential operators of practical interest have $p(t) = 1$ for $0 \leqq t < \infty$ (in fact, a simple transformation can always effect this), and in this situation Theorem 2.4 implies the following simple criterion.

Corollary 2. *If $p(t) = 1$ for $0 \leqq t < \infty$ and $q(t) \geqq -kt^2$ for some positive constant k, then L is in the limit-point case at infinity.*

3. The Completeness and Expansion Theorems in the Limit-point Case at Infinity

As a necessary preliminary, the results for the finite interval $0 \leqq t \leqq b < \infty$ will be given a slightly different formulation. Consider the problem

$$\begin{aligned} Lx &= -(px')' + qx = lx \\ \sin\alpha\, x(0) - \cos\alpha\, p(0)x'(0) &= 0 \\ \cos\beta\, x(b) + \sin\beta\, p(b)x'(b) &= 0 \end{aligned} \tag{3.1}$$

where $0 \leqq \alpha, \beta < \pi$. This is a self-adjoint boundary-value problem on $0 \leqq t \leqq b$, and consequently there exists a sequence $\{\lambda_{bn}\}$, $n = 1, 2, \ldots$, of real eigenvalues and a corresponding complete orthonormal set $\{\theta_{bn}\}$ of eigenfunctions. As in Sec. 2, let φ, ψ be the solutions of $Lx = lx$ satisfying the conditions (2.2). Then ψ satisfies the first boundary condition of (3.1) and no solution of $Lx = lx$ independent of ψ can satisfy this condition. Therefore

$$\theta_{bn}(t) = r_{bn}\psi(t,\lambda_{bn})$$

where r_{bn} is a constant, independent of t. The completeness theorem applied to any continuous function on $0 \leqq t < \infty$ which vanishes outside $0 \leqq t \leqq c$, where $0 < c < b$, yields

$$\int_0^b |f(t)|^2\,dt = \sum_{n=1}^{\infty} |r_{bn}|^2 \left| \int_0^b f(t)\psi(t,\lambda_{bn})\,dt \right|^2 \tag{3.2}$$

Let

$$g(\lambda) = \int_0^{\infty} f(t)\psi(t,\lambda)\,dt \tag{3.3}$$

and let ρ_b be a monotone nondecreasing step function of λ having a jump of $|r_{bn}|^2$ at each eigenvalue λ_{bn}, and otherwise constant. Assume further that $\rho_b(\lambda + 0) = \rho_b(\lambda)$ and $\rho_b(0) = 0$. Then the Parseval equality (3.2) may be written as

$$\int_0^\infty |f(t)|^2 \, dt = \int_{-\infty}^\infty |g(\lambda)|^2 \, d\rho_b(\lambda) \tag{3.4}$$

The function ρ_b will be called the *spectral function* for the problem (3.1).

The fundamental idea behind generalizing the formula (3.4) to the interval $0 \leqq t < \infty$ is to show that as $b \to \infty$ the function ρ_b tends to a monotone nondecreasing function ρ (which need not be a step function) in such a way that (3.4) still remains valid in some sense when ρ_b is replaced by ρ.

If σ is any monotone nondecreasing function on $-\infty < \lambda < \infty$, let $\mathfrak{L}^2(\sigma)$ denote the set of all functions h which are measurable with respect to the Lebesgue-Stieltjes measure defined by σ and such that

$$\int_{-\infty}^\infty |h(\lambda)|^2 \, d\sigma(\lambda) < \infty$$

Theorem 3.1. *Let L be in the limit-point case at* ∞. *Then*

(i) *There exists a monotone nondecreasing function* ρ *on* $-\infty < \lambda < \infty$ *such that*

$$\rho(\lambda) - \rho(\mu) = \lim_{b \to \infty} (\rho_b(\lambda) - \rho_b(\mu)) \tag{3.5}$$

at points of continuity λ,μ *of* ρ.

(ii) *If* $f \in \mathfrak{L}^2(0,\infty)$ *there exists a function* $g \in \mathfrak{L}^2(\rho)$ *such that*

$$\lim_{a \to \infty} \int_{-\infty}^\infty \left| g(\lambda) - \int_0^a f(t)\psi(t,\lambda) \, dt \right|^2 d\rho(\lambda) = 0 \tag{3.6}$$

and

$$\int_0^\infty |f(t)|^2 \, dt = \int_{-\infty}^\infty |g(\lambda)|^2 \, d\rho(\lambda) \tag{3.7}$$

(iii) *The integral*

$$\int_{-\infty}^\infty g(\lambda)\psi(t,\lambda) \, d\rho(\lambda)$$

converges in $\mathfrak{L}^2(0,\infty)$ *to f, that is,*

$$\lim_{(\mu,\nu) \to (-\infty,\infty)} \int_0^\infty \left| f(t) - \int_\mu^\nu g(\lambda)\psi(t,\lambda) \, d\rho(\lambda) \right|^2 dt = 0 \tag{3.8}$$

(iv) *If* m_∞ *is the limit point, considered as a function of l,*

$$\rho(\lambda) - \rho(\mu) = \lim_{\epsilon \to 0+} \frac{1}{\pi} \int_\mu^\lambda \Im m_\infty(\nu + i\epsilon) \, d\nu \tag{3.9}$$

at points of continuity λ,μ *of* ρ *and inversely*

$$m_\infty(l) - m_\infty(l_0) = \int_{-\infty}^{\infty} \left(\frac{1}{\lambda - l} - \frac{1}{\lambda - l_0}\right) d\rho(\lambda) + c(l - l_0) \quad \textbf{(3.10)}$$

where† c *is a nonnegative constant, and* $\Im l_0 \neq 0$.

The function ρ is called the *spectral function* for the problem

$$Lx = lx \qquad \sin \alpha\, x(0) - \cos \alpha\, p(0)x'(0) = 0$$

By

$$\int_0^\infty f(t)\psi(t,\lambda)\, dt$$

will be meant the value at λ of the function g defined by (3.6). Thus

$$g(\lambda) = \int_0^\infty f(t)\psi(t,\lambda)\, dt \quad (3.11)$$

and g may be regarded as a transform of f by means of the function ψ. The relationship (3.7) is the analogue of (3.4) and is called the *completeness theorem*, or *Parseval equality*. It follows that

$$f(t) = \int_{-\infty}^{\infty} g(\lambda)\psi(t,\lambda)\, d\rho(\lambda) \quad (3.12)$$

where the equality is meant in the sense of (3.8), and (3.12) is called the *expansion theorem*. If f, for example, is continuous and vanishes for all large t, then the integral in (3.11) exists in the ordinary sense.

The correspondence $f \to g$ maps the space $\mathfrak{L}^2(0,\infty)$ into $\mathfrak{L}^2(\rho)$, and it is an important additional fact that this mapping is onto, that is, all of $L^2(\rho)$ is used up in the process.‡

Theorem 3.2. *If* $g \in \mathfrak{L}^2(\rho)$, *there exists an* $f \in \mathfrak{L}^2(0,\infty)$ *satisfying* (3.8), *and by means of this* f *the function* g *may be represented in the form* (3.11).

Before proving Theorem 3.1, the following results required in the proof will be stated. The first is a form of the Helly theorem on sequences of monotone functions adapted to the infinite interval, whereas the second is an integration theorem.

Selection Theorem. *Let* $\{h_n\}$, $n = 1, 2, \ldots$, *be a sequence of real nondecreasing functions on* $-\infty < \lambda < \infty$, *and let* H *be a continuous nonnegative function on the same interval. If*

$$|h_n(\lambda)| \leqq H(\lambda) \qquad (n = 1, 2, \ldots ; -\infty < \lambda < \infty)$$

† By appraising $\rho_b(\lambda)$ more precisely than will be done here it can be shown that $c = 0$ in (3.10).

‡ The spaces $\mathfrak{L}^2(0,\infty)$ and $\mathfrak{L}^2(\rho)$ are Hilbert spaces and the content of Theorems 3.1 and 3.2 may be summed up by saying that the correspondence $f \longleftrightarrow g$ is a one-to-one norm preserving mapping of $\mathfrak{L}^2(0,\infty)$ onto $\mathfrak{L}^2(\rho)$, that is, a unitary mapping.

then there exist a subsequence $\{h_{n_k}\}$ and a nondecreasing function h such that

$$|h(\lambda)| \leqq H(\lambda) \qquad (-\infty < \lambda < \infty)$$

and

$$\lim_{k\to\infty} h_{n_k}(\lambda) = h(\lambda)$$

Integration Theorem. *Suppose $\{h_n\}$ is a real, uniformly bounded, sequence of nondecreasing functions on a finite interval $a \leqq \lambda \leqq c$, and assume*

$$\lim_{n\to\infty} h_n(\lambda) = h(\lambda) \qquad (a \leqq \lambda \leqq c)$$

If f is any continuous function on $a \leqq \lambda \leqq c$ then

$$\lim_{n\to\infty} \int_a^c f(\lambda)\, dh_n(\lambda) = \int_a^c f(\lambda)\, dh(\lambda)$$

Proof of Theorem 3.1. Let $m_b(l)$ be a point on the circle C_b, where $\Im l > 0$. Then the completeness theorem (3.2) applied to the solution $\chi_b = \varphi + m_b\psi$ of $Lx = lx$ yields

$$\int_0^b |\chi_b(t)|^2\, dt = \sum_{n=1}^{\infty} |r_{bn}|^2 \left| \int_0^b \chi_b(t)\psi(t,\lambda_{bn})\, dt \right|^2 \tag{3.13}$$

Green's formula applied to χ_b and ψ results in

$$(l - \lambda_{bn}) \int_0^b \chi_b(t)\psi(t,\lambda_{bn})\, dt = [\chi_b\psi_{bn}](b) - [\chi_b\psi_{bn}](0)$$

where $\psi_{bn}(t) = \psi(t,\lambda_{bn})$. A simple calculation shows that $[\chi_b\psi_{bn}](b) = 0$ since both χ_b and ψ_{bn} satisfy the same boundary condition at b, and $[\chi_b\psi_{bn}](0) = 1$. Therefore (3.13) implies

$$\int_0^b |\chi_b(t)|^2\, dt = \int_{-\infty}^{\infty} \frac{d\rho_b(\lambda)}{|\lambda - l|^2}$$

taking into account the definition of the monotone function ρ_b. The fact that m_b is on C_b yields, by (2.11),

$$\int_0^b |\chi_b(t)|^2\, dt = \frac{\Im m_b(l)}{\Im l}$$

From these equations follows the important equality

$$\int_{-\infty}^{\infty} \frac{d\rho_b(\lambda)}{|\lambda - l|^2} = \frac{\Im m_b(l)}{\Im l} \qquad (\Im l > 0) \tag{3.14}$$

Let $l = i$ in (3.14). Since C_b is in C_1 for $b > 1$, there exists a constant

k such that $\mathfrak{I}m_b(i)$ is less than k for $b > 1$. Thus (3.14) yields

$$\int_{-\infty}^{\infty} \frac{d\rho_b(\lambda)}{1 + \lambda^2} < k \tag{3.15}$$

or, for $\nu > 0$,

$$\int_{-\nu}^{\nu} d\rho_b(\lambda) < k(1 + \nu^2)$$

This, together with $\rho_b(0) = 0$, gives

$$|\rho_b(\lambda)| < k(1 + \lambda^2) \qquad (-\infty < \lambda < \infty)$$

Choosing a sequence $b_n \to \infty$, and choosing a β, $0 \leqq \beta < \pi$, for each b_n, it follows from the selection theorem that a subsequence of the sequence $\{\rho_{b_n}\}$ exists, converging to a limit function ρ which is monotone nondecreasing and satisfies

$$|\rho(\lambda)| \leqq k(1 + \lambda^2)$$

Now let f have a continuous second derivative on $0 \leqq t < \infty$, and vanish near $t = 0$ and for all large t. Then applying (3.4) to Lf for large enough b results in

$$\int_0^{\infty} |(Lf)(t)|^2 \, dt = \int_{-\infty}^{\infty} \left| \int_0^{\infty} (Lf)(t)\psi(t,\lambda) \, dt \right|^2 d\rho_b(\lambda) \tag{3.16}$$

Using Green's formula for f and ψ,

$$\int_0^{\infty} (Lf)(t)\psi(t,\lambda) \, dt = \lambda g(\lambda)$$

and hence (3.16) becomes

$$\int_0^{\infty} |(Lf)(t)|^2 \, dt = \int_{-\infty}^{\infty} \lambda^2 |g(\lambda)|^2 \, d\rho_b(\lambda)$$

From this it follows that for $\mu > 0$,

$$\begin{aligned} \int_{\Delta} |g(\lambda)|^2 \, d\rho_b(\lambda) &\leqq \mu^{-2} \int_{\Delta} \lambda^2 |g(\lambda)|^2 \, d\rho_b(\lambda) \\ &\leqq \mu^{-2} \int_0^{\infty} |(Lf)(t)|^2 \, dt \end{aligned} \tag{3.17}$$

where $\Delta = (-\infty, \infty) - (-\mu, \mu]$. Applying the completeness relationship to f itself

$$\int_0^{\infty} |f(t)|^2 \, dt = \left(\int_{-\mu}^{\mu} + \int_{\Delta} \right) |g(\lambda)|^2 \, d\rho_b(\lambda)$$

Letting b tend to infinity through the subsequence found above, it follows, using (3.17) and the integration theorem, that

$$\left| \int_0^{\infty} |f(t)|^2 \, dt - \int_{-\mu}^{\mu} |g(\lambda)|^2 \, d\rho(\lambda) \right| \leqq \mu^{-2} \int_0^{\infty} |(Lf)(t)|^2 \, dt$$

Now allowing $\mu \to \infty$, there results the Parseval equality

$$\int_0^\infty |f(t)|^2\,dt = \int_{-\infty}^\infty |g(\lambda)|^2\,d\rho(\lambda) \tag{3.18}$$

for any f restricted as above.

Rather standard arguments now suffice to show that the Parseval equality holds for any $f \in \mathfrak{L}^2(0,\infty)$. First suppose $f \in \mathfrak{L}^2(0,\infty)$ and vanishes for large $t > 0$. Then there exists a sequence of functions $f_n \in \mathfrak{L}^2(0,\infty)$ possessing continuous second derivatives and vanishing near $t = 0$ and for all large t such that

$$\lim_{n\to\infty} \int_0^\infty |f_n - f|^2\,dt = 0 \tag{3.19}$$

and from (3.18) applied to $f_n - f_m$

$$\int_0^\infty |f_n - f_m|^2\,dt = \int_{-\infty}^\infty |g_n(\lambda) - g_m(\lambda)|^2\,d\rho(\lambda) \tag{3.20}$$

where

$$g_n(\lambda) = \int_0^\infty f_n(t)\psi(t,\lambda)\,dt \tag{3.21}$$

Since the left side of (3.20) tends to zero as $n, m \to \infty$, it follows that the sequence $\{g_n\}$ converges in the mean in $\mathfrak{L}^2(\rho)$, and since the latter space is complete there exists a $g \in \mathfrak{L}^2(\rho)$ which is the limit in the mean of this sequence. From (3.21) it is clear that g is the continuous function given by

$$g(\lambda) = \int_0^\infty f(t)\psi(t,\lambda)\,dt$$

Using (3.18) again

$$\begin{aligned}\int_0^\infty |f(t)|^2\,dt &= \lim_{n\to\infty} \int_0^\infty |f_n(t)|^2\,dt = \lim_{n\to\infty} \int_{-\infty}^\infty |g_n(\lambda)|^2\,d\rho(\lambda)\\ &= \int_{-\infty}^\infty |g(\lambda)|^2\,d\rho(\lambda)\end{aligned}$$

which proves the Parseval relation for any $f \in \mathfrak{L}^2(0,\infty)$ vanishing for all large $t > 0$. Suppose f is any function of class $\mathfrak{L}^2(0,\infty)$ and define

$$\begin{aligned} f_a(t) &= f(t) \qquad (0 \leqq t \leqq a)\\ &= 0 \qquad (a < t)\end{aligned}$$

and

$$g_a(\lambda) = \int_0^\infty f_a(t)\psi(t,\lambda)\,dt = \int_0^a f(t)\psi(t,\lambda)\,dt$$

Since

$$\int_{-\infty}^\infty |g_a(\lambda) - g_d(\lambda)|^2\,d\rho(\lambda) = \int_a^d |f(t)|^2\,dt \qquad (a < d)$$

the set $\{g_a\}$ converges (as $a \to \infty$) in the mean in $\mathfrak{L}^2(\rho)$ to a function $g \in \mathfrak{L}^2(\rho)$. From letting $a \to \infty$ in

$$\int_{-\infty}^{\infty} |g_a(\lambda)|^2 \, d\rho(\lambda) = \int_0^a |f(t)|^2 \, dt$$

there now follows the Parseval equality (3.7) for any $f \in \mathfrak{L}^2(0, \infty)$.

The proof of the expansion theorem (3.12) [in the sense of (3.8)] will now be given. Let $\Delta = (\mu,\lambda]$ and

$$f_\Delta(t) = \int_\Delta g(\lambda)\psi(t,\lambda) \, d\rho(\lambda) \tag{3.22}$$

where g is the function appearing in (3.6). The relation (3.7) implies that, if $f_1, f_2 \in \mathfrak{L}^2(0, \infty)$ and g_1, g_2 are the corresponding transforms, then†

$$\int_0^\infty f_1 \bar{f}_2 \, dt = \int_{-\infty}^{\infty} g_1 \bar{g}_2 \, d\rho(\lambda) \tag{3.23}$$

Let $P \in \mathfrak{L}^2(0, \infty)$ and vanish for $t > a > 0$, and let the transform of P be Q. Then from (3.22) follows, on multiplying by $\bar{P}$ and integrating,

$$\begin{aligned} \int_0^a f_\Delta(t)\bar{P}(t) \, dt &= \int_0^a \left(\int_\Delta g(\lambda)\psi(t,\lambda) \, d\rho(\lambda) \right) \bar{P}(t) \, dt \\ &= \int_\Delta g(\lambda) \left(\int_0^a \bar{P}(t)\psi(t,\lambda) \, dt \right) d\rho(\lambda) \\ &= \int_\Delta g(\lambda)\bar{Q}(\lambda) \, d\rho(\lambda) \end{aligned}$$

From (3.23) for $f_1 = f$ and $f_2 = P$,

$$\int_0^\infty f\bar{P} \, dt = \int_{-\infty}^{\infty} g(\lambda)\bar{Q}(\lambda) \, d\rho(\lambda)$$

Subtracting the above and letting $\Delta^c = (-\infty, \infty) - \Delta$

$$\int_0^\infty (f - f_\Delta)\bar{P} \, dt = \int_{\Delta^c} g(\lambda)\bar{Q}(\lambda) \, d\rho(\lambda)$$

and using the Schwarz inequality

$$\begin{aligned} \left| \int_0^\infty (f - f_\Delta)\bar{P} \, dt \right|^2 &\leq \int_{\Delta^c} |g(\lambda)|^2 \, d\rho(\lambda) \int_{-\infty}^{\infty} |Q(\lambda)|^2 \, d\rho(\lambda) \\ &= \int_{\Delta^c} |g(\lambda)|^2 \, d\rho(\lambda) \int_0^\infty |P|^2 \, dt \end{aligned}$$

Applying this inequality to the function P given by

$$\begin{aligned} P(t) &= f(t) - f_\Delta(t) \qquad (0 \leq t \leq a) \\ &= 0 \qquad (a < t) \end{aligned}$$

† $4f_1\bar{f}_2 = |f_1 + f_2|^2 - |f_1 - f_2|^2 + i|f_1 + if_2|^2 - i|f_1 - if_2|^2$

one obtains

$$\int_0^a |f - f_\Delta|^2 \, dt \leqq \int_{\Delta^c} |g(\lambda)|^2 \, d\rho(\lambda) \tag{3.24}$$

Since the right side does not depend on a, the above holds with $a \to \infty$. Letting $\Delta \to (-\infty, \infty)$ yields the expansion result (3.8).

Clearly the uniqueness relation (3.5) is an immediate consequence of (3.9). Thus it remains to prove the relations (3.9) and (3.10). From (3.14) it follows that, for any fixed l, $\Im l > 0$, there exists a constant k such that

$$\int_{-\mu}^{\mu} \frac{d\rho_b(\lambda)}{|\lambda - l|^2} \leqq k$$

for $b > 1$ and all $\mu \geqq 0$. Letting $b \to \infty$ through the subsequence chosen below (3.15), it follows that the above is true with ρ in place of ρ_b. Since it holds for all μ,

$$\int_{-\infty}^{\infty} \frac{d\rho(\lambda)}{|\lambda - l|^2} < \infty$$

From (3.15), for $\mu > 1$, there exists a constant k such that

$$\int_{\mu}^{\infty} \frac{d\rho_b(\lambda)}{\lambda^3} < \frac{k}{\mu}$$

for $b > 1$ and similarly over $(-\infty, -\mu)$. If $\Im l \neq 0$, $\Im l_0 \neq 0$ and

$$\int_{-\infty}^{\infty} \left(\frac{1}{|\lambda - l|^2} - \frac{1}{|\lambda - l_0|^2} \right) d\rho_b(\lambda) \tag{3.25}$$

is considered over $(-\infty, -\mu)$, $(-\mu, \mu)$ and (μ, ∞), it follows that, if $b \to \infty$ through a chosen subsequence and if then $\mu \to \infty$, (3.25) must tend to

$$\int_{-\infty}^{\infty} \left(\frac{1}{|\lambda - l|^2} - \frac{1}{|\lambda - l_0|^2} \right) d\rho(\lambda)$$

But (3.25) is just

$$\frac{\Im m_b(l)}{\Im l} - \frac{\Im m_b(l_0)}{\Im l_0}$$

which tends, as $b \to \infty$, to

$$\frac{\Im m_\infty(l)}{\Im l} - \frac{\Im m_\infty(l_0)}{\Im l_0}$$

Therefore

$$\frac{\Im m_\infty(l)}{\Im l} = \int_{-\infty}^{\infty} \frac{d\rho(\lambda)}{|\lambda - l|^2} + c \tag{3.26}$$

where c is a constant independent of l, provided $\Im l > 0$ (or $\Im l < 0$).

From Theorem 2.3 $\Im m_\infty(l)/\Im l > 0$. From (3.26), letting $\Re l = 0$, $\Im l \to \infty$, it readily follows that $c \geqq 0$.† From (3.26) results

$$\Im(m_\infty(l) - m_\infty(l_0)) = \Im\left(\int_{-\infty}^{\infty}\left(\frac{1}{\lambda - l} - \frac{1}{\lambda - l_0}\right) d\rho(\lambda)\right) + c\Im(l - l_0) \tag{3.27}$$

Since m_∞ is analytic in l for $\Im l > 0$ (or $\Im l < 0$), $\Im m_\infty$ determines $\Re m_\infty$ to within an additive constant. The imaginary part of the analytic function of l (for fixed $l_0, \Im l_0 \neq 0$) defined by the right side of (3.10) is $\Im(m_\infty(l) - m_\infty(l_0))$ by virtue of (3.27). Thus, this must be $m_\infty(l) - m_\infty(l_0)$, which proves (3.10).

Let λ, μ be points of continuity for ρ. Then from (3.26)

$$\begin{aligned}\lim_{\epsilon\to+0}\int_\mu^\lambda \Im m_\infty(\nu + i\epsilon)d\nu &= \lim_{\epsilon\to+0}\int_\mu^\lambda\int_{-\infty}^{\infty}\frac{\epsilon\, d\rho(\sigma)}{(\sigma-\nu)^2+\epsilon^2}d\nu\\ &= \lim_{\epsilon\to+0}\int_{-\infty}^{\infty}\left[\tan^{-1}\left(\frac{\lambda-\sigma}{\epsilon}\right) - \tan^{-1}\left(\frac{\mu-\sigma}{\epsilon}\right)\right]d\rho(\sigma)\\ &= \pi(\rho(\lambda) - \rho(\mu))\end{aligned}$$

yielding (3.9), and thus completing the proof of Theorem 3.1.

In proving Theorem 3.2 the following will be required.

Lemma 3.1. *Let* $g \in \mathfrak{L}^2(\rho)$ *and*

$$f_\Delta(t) = \int_\Delta g(\lambda)\psi(t,\lambda)\, d\rho(\lambda)$$

where Δ *is a finite* λ*-interval. Then, as* $\Delta \to (-\infty, \infty)$, f_Δ *is convergent in* $\mathfrak{L}^2(0, \infty)$, *and thus tends in the mean to a function* $f \in \mathfrak{L}^2(0, \infty)$.

Proof. Let $\Delta_1 \supset \Delta_2$ and let $P \in \mathfrak{L}^2(0, \infty)$ be a function vanishing for large t. Then if Q is the transform of P,

$$\int_0^\infty (f_{\Delta_1} - f_{\Delta_2})\bar{P}\, dt = \int_{\Delta_1 - \Delta_2} g(\lambda)\bar{Q}(\lambda)\, d\rho(\lambda)$$

Using the Schwarz inequality much as above (3.24), and letting

$$P = f_{\Delta_1} - f_{\Delta_2}$$

for $0 \leqq t \leqq a$, and $P = 0$ for $t > a$, there follows

$$\int_0^a |f_{\Delta_1} - f_{\Delta_2}|^2\, dt \leqq \int_{\Delta_1 - \Delta_2} |g(\lambda)|^2\, d\rho(\lambda)$$

Since the right side does not depend on a, this inequality holds with $a = \infty$. Letting Δ_1 and $\Delta_2 \to (-\infty, \infty)$, the proof is completed.

REMARK: At this point the self-adjointness of L in the limit-point case can be proved. This is done in Prob. 13.

† As has already been remarked by appraising ρ_b, it can be shown that, in fact, $c = 0$.

Proof of Theorem 3.2. In view of Lemma 3.1, it remains to show that g comes from f by (3.11), that is,

$$g(\lambda) = \int_0^\infty f(t)\psi(t,\lambda)\,dt \tag{3.11}$$

where the equality is meant in the mean in $\mathfrak{L}^2(\rho)$. From Theorem 3.1 there exists a $\tilde{g} \in \mathfrak{L}^2(\rho)$ such that (3.11) holds with g replaced by $\tilde{g}$. Therefore the problem reduces to showing

$$\int_{-\infty}^{\infty} |g(\lambda) - \tilde{g}(\lambda)|^2\,d\rho(\lambda) = 0$$

Let $r = g - \tilde{g}$; clearly $r \in \mathfrak{L}^2(\rho)$. Using the fact that f is the limit in the mean of $\tilde{f}_\Delta$, where

$$\tilde{f}_\Delta(t) = \int_\Delta \tilde{g}(\sigma)\psi(t,\sigma)\,d\rho(\sigma) \qquad \Delta = (\mu,\lambda]$$

it follows that the function $h_\Delta = f_\Delta - \tilde{f}_\Delta$,

$$h_\Delta(t) = \int_\Delta r(\lambda)\psi(t,\lambda)\,d\rho(\lambda)$$

tends in the mean to zero, that is,

$$\lim_{\Delta \to (-\infty,\infty)} \int_0^\infty |h_\Delta(t)|^2\,dt = 0 \tag{3.28}$$

It will be shown that h_Δ is the zero function.

Let l be a complex number with $\Im l > 0$, and put

$$H_\Delta(t,l) = \int_\Delta \frac{r(\lambda)}{\lambda - l}\psi(t,\lambda)\,d\rho(\lambda) \tag{3.29}$$

Then, since $L\psi = \lambda\psi$,

$$LH_\Delta = lH_\Delta + h_\Delta \tag{3.30}$$

From (3.29) follows easily that H_Δ satisfies the same boundary condition at zero as does ψ, namely,

$$\sin\alpha\, H_\Delta(0,l) - \cos\alpha\, p(0)H'_\Delta(0,l) = 0$$

By the variation-of-constants formula, (3.30) yields

$$H_\Delta(t,l) = \int_0^t [\varphi(t,l)\psi(\tau,l) - \varphi(\tau,l)\psi(t,l)]h_\Delta(\tau)\,d\tau + c_\Delta\psi(t,l) \tag{3.31}$$

where c_Δ is a constant (which may depend on Δ). From the fact that $r \in \mathfrak{L}^2(\rho)$, $r/(\lambda - l) \in \mathfrak{L}^2(\rho)$ and hence H_Δ, as a function of t for fixed l, $\Im l > 0$, converges in the mean in $\mathfrak{L}^2(0,\infty)$ to a function H as $\Delta \to (-\infty,\infty)$. Using this in (3.31), it follows from (3.28) that there exists a constant c so that

$$H(t,l) = c\psi(t,l)$$

Since ψ is not of class $\mathfrak{L}^2(0,\infty)$ for $\mathfrak{I}l > 0$, the constant c must be zero, and hence $H(t,l) = 0$ for $\mathfrak{I}l > 0$. Thus H_Δ tends to the zero function in $\mathfrak{L}^2(0,\infty)$ for $\mathfrak{I}l > 0$.

Let

$$\Gamma_s(\lambda) = \int_0^s \psi(t,\lambda)\,dt$$

that is, Γ_s is the transform of the function which is one for $0 \leqq t \leqq s$ and zero for $t > s$. Thus $\Gamma_s \in \mathfrak{L}^2(\rho)$. Integrating (3.29) with respect to t and using $H_\Delta \to 0$ in $\mathfrak{L}^2(0,\infty)$ there results

$$\int_{-\infty}^{\infty} \frac{r(\lambda)}{\lambda - l}\,\Gamma_s(\lambda)\,d\rho(\lambda) = 0 \tag{3.32}$$

Since $\Gamma_s \in \mathfrak{L}^2(\rho)$ the Schwarz inequality shows that the integral on the left of (3.32) is absolutely convergent. Indeed

$$\int_{-\infty}^{\infty} |r(\lambda)\Gamma_s(\lambda)|\,d\rho(\lambda) < \infty \tag{3.33}$$

Formula (3.32) can be inverted much as in the proof of (3.9) for ρ. For this purpose it will be assumed that g, and hence r, is real. This is no restriction since every g is a sum $g_1 + ig_2$, where g_1 and g_2 are real. Taking the imaginary part of (3.32) and integrating, one obtains

$$\begin{aligned}\lim_{\epsilon\to+0} \int_\mu^\lambda \int_{-\infty}^{\infty} \frac{\epsilon}{(\sigma-\nu)^2+\epsilon^2}\, r(\sigma)\Gamma_s(\sigma)\,d\rho(\sigma)\,d\nu \\ = \int_{-\infty}^{\infty} r(\sigma)\Gamma_s(\sigma) \lim_{\epsilon\to+0}\left[\tan^{-1}\left(\frac{\lambda-\sigma}{\epsilon}\right) - \tan^{-1}\left(\frac{\mu-\sigma}{\epsilon}\right)\right] d\rho(\sigma) \\ = \pi \int_\mu^\lambda r(\sigma)\Gamma_s(\sigma)\,d\rho(\sigma) = 0\end{aligned} \tag{3.34}$$

Since for $\Delta = (\mu,\lambda]$,

$$\int_\Delta r(\sigma)\Gamma_s(\sigma)\,d\rho(\sigma) = \int_\Delta r(\sigma)\int_0^s \psi(t,\sigma)\,dt\,d\rho(\sigma) = \int_0^s\left(\int_\Delta r(\sigma)\psi(t,\sigma)\,d\rho(\sigma)\right)dt$$

it follows that the function k given by

$$k(s) = \int_\Delta r(\sigma)\Gamma_s(\sigma)\,d\rho(\sigma)$$

has a continuous derivative, and from (3.34) this must vanish. Thus

$$h_\Delta(t) = \int_\Delta r(\lambda)\psi(t,\lambda)\,d\rho(\lambda) = 0 \tag{3.35}$$

Since $\psi(0,\lambda) = \cos\alpha$, it follows from (3.35) with $t = 0$ that if $\cos\alpha \neq 0$

$$\int_\Delta r(\lambda)\,d\rho(\lambda) = 0 \tag{3.36}$$

If $\cos\alpha = 0$, then $\sin\alpha \neq 0$ and, by differentiating (3.35) with respect to t first and then setting $t = 0$, (3.36) follows. Thus (3.36) is always valid. Because of the arbitrary nature of Δ, (3.36) implies that for any $a > 0$

$$\int_{-a}^{a} \gamma(\lambda)r(\lambda)\,d\rho(\lambda) = 0 \tag{3.37}$$

for any step function γ. The step functions are dense in $\mathfrak{L}^2(\rho)$, and since $r \in \mathfrak{L}^2(\rho)$ it follows that γ can be chosen so that the left side of (3.37) is arbitrarily close to

$$\int_{-a}^{a} |r(\lambda)|^2\,d\rho(\lambda)$$

which must therefore vanish. Since this holds for all a, the theorem is proved.

4. The Limit-circle Case at Infinity

If L is in the limit-circle case at infinity, the circles $C_b(l)$ converge to $C_\infty(l)$ as $b \to \infty$ for each l, $\mathfrak{I}l \neq 0$. Each circle $C_b(l)$ is traced by points $m = m(l,b,\beta)$ as β ranges over $0 \leq \beta < \pi$ for fixed b and l. [In case $\mathfrak{I}l = 0$, the circle $C_b(l)$ becomes a straight line.] Let l_0 be *fixed*, $\mathfrak{I}l_0 \neq 0$. A point $\hat{m}_\infty(l_0)$ on the circle $C_\infty(l_0)$ is the limit point of a sequence $m(l_0,b_j,\beta_j)$, $j = 1, 2, \ldots$, with $b_j \to \infty$ as $j \to \infty$.

Let m_j denote the function of l given by $m_j(l) = m(l,b_j,\beta_j)$. These are meromorphic functions of l, real for real l and, as was seen from $\mathfrak{I}m/\mathfrak{I}l > 0$ for $\mathfrak{I}l \neq 0$, the poles and zeros of these functions can lie on the real axis only and are simple. Let ρ_j denote the step function ρ_b associated with the condition β_j at b_j. Then the following theorem will be proved.

Theorem 4.1. *Let $\hat{m}_\infty(l_0)$ be a point on $C_\infty(l_0)$ and (b_j,β_j) a sequence such that $m(l_0,b_j,\beta_j) = m_j(l_0)$ tends to $\hat{m}_\infty(l_0)$ as $j \to \infty$. Then for all l*

$$\lim_{j\to\infty} m_j(l) = \hat{m}_\infty(l) \tag{4.1}$$

and [in the sense of (3.5)]

$$\lim_{j\to\infty} \rho_j(\lambda) = \hat{\rho}(\lambda) \tag{4.2}$$

where $\hat{m}_\infty$ is a meromorphic function of l, real for real l, and with poles and zeros that are real and simple. Moreover, $\hat{\rho}$ is a step function discontinuous at the poles, $l = \lambda_k$, $k = 1, 2, \ldots$, of $\hat{m}_\infty$ only and with a jump at λ_k equal to minus the residue of $\hat{m}_\infty$ at λ_k. The functions ψ_k, where $\psi_k(t) = \psi(t,\lambda_k)$, *form a complete orthogonal family in* $\mathfrak{L}^2(0, \infty)$. *If* $\hat{\chi}_\infty$ *is the function defined by* $\hat{\chi}_\infty(t) = \varphi(t,l_0) + \hat{m}_\infty(l_0)\psi(t,l_0)$, *then*

$$[\psi_k \hat{\chi}_\infty](\infty) = 0 \tag{4.3}$$

for all k. *On the other hand, for* ψ_l, *where* $\psi_l(t) = \psi(t,l)$,

$$[\psi_l \hat{\chi}_\infty](\infty) \neq 0 \qquad (l \neq \lambda_k,\ k = 1, 2, \ldots)$$

The condition (4.3) is actually a boundary condition satisfied by the ψ_k at $t = \infty$. With each point on the limit circle $C_\infty(l_0)$ is associated such a boundary condition. In the course of proving Theorem 4.1, two other theorems will be proved which will make clearer the nature of this boundary condition.

Proof of Theorem 4.1. Let

$$\chi_j(t,l) = \varphi(t,l) + m_j(l)\psi(t,l) \tag{4.4}$$

Apply Green's formula to $\chi_j(t,l)$ and $\bar{\chi}_j(t,l_0)$. Since both functions satisfy the same condition at b_j, it follows that

$$m_j(l) - m_j(l_0) = (l - l_0) \int_0^{b_j} \chi_j(t,l)\chi_j(t,l_0)\, dt \tag{4.5}$$

Using (4.4) in (4.5),

$$m_j(l) = \frac{m_j(l_0) + (l - l_0) \int_0^{b_j} \varphi(t,l)\chi_j(t,l_0)\, dt}{1 - (l - l_0) \int_0^{b_j} \psi(t,l)\chi_j(t,l_0)\, dt} \tag{4.6}$$

In the limit-circle case all solutions of $Lx = lx$ are $\mathfrak{L}^2(0, \infty)$. Therefore, as $j \to \infty$, the entire function of l whose value at l is given by

$$\int_0^{b_j} \varphi(t,l)\chi_j(t,l_0)\, dt = \int_0^{b_j} \varphi(t,l)[\varphi(t,l_0) + m_j(l_0)\psi(t,l_0)]\, dt$$

which appears in the numerator of (4.6), tends to the limit

$$\int_0^{\infty} \varphi(t,l)[\varphi(t,l_0) + \hat{m}_\infty(l_0)\psi(t,l_0)]\, dt \tag{4.7}$$

If l is restricted to some finite part Λ of the l plane, then it was shown in the course of the proof of Theorem 2.1 that φ and ψ have norms in $\mathfrak{L}^2(0, \infty)$ which are uniformly bounded in Λ. Thus, by the Schwarz inequality, the integrals in (4.7) are uniformly convergent in l over any finite part of the l plane. This implies that (4.7) defines an entire function of l. The same is true for the integral in the denominator of (4.6). Thus, as $j \to \infty$, the meromorphic function m_j tends to a limit $\hat{m}_\infty$ which is also a meromorphic function, and this proves (4.1). The properties of $\hat{m}_\infty$

follow from $\Im \hat{m}_\infty(l)/\Im l > 0$ for $\Im l \neq 0$. It follows from (4.6) that

$$\hat{m}_\infty(l) = \frac{\hat{m}_\infty(l_0) + (l - l_0)\int_0^\infty \varphi(t,l)\hat{\chi}_\infty(t,l_0)\,dt}{1 - (l - l_0)\int_0^\infty \psi(t,l)\hat{\chi}_\infty(t,l_0)\,dt} \tag{4.8}$$

As in the proof of Theorem 3.1, the Helly selection theorem shows that a subsequence of $\{\rho_j\}$ exists which converges to a limit $\hat{\rho}$. Since (3.14) is valid, the argument of Theorem 3.1 then shows that (3.9) is valid with ρ replaced by $\hat{\rho}$ and m_∞ by $\hat{m}_\infty$. This proves that $\hat{\rho}$ is independent of the choice of sequence and thus that (4.2) is valid. Since $\hat{m}_\infty$ is real on the real axis, (3.9) also proves that $\hat{\rho}$ is a step function discontinuous only at the λ_k and with a jump at λ_k equal to minus the residue of the pole of $\hat{m}_\infty$ there.

The completeness of the set $\{\psi_k\}$ follows from (3.6), (3.7), and (3.8) with ρ replaced by $\hat{\rho}$. The orthogonality will be proved after Theorems 4.2 and 4.3.

From (4.8) it follows that at any pole λ_k of $\hat{m}_\infty$ the denominator must vanish, that is,

$$(\lambda_k - l_0)\int_0^\infty \psi(t,\lambda_k)\hat{\chi}_\infty(t,l_0)\,dt = 1 \tag{4.9}$$

By Green's formula it is easily seen, since $\psi(t,\lambda_k)$ is real, that

$$(\lambda_k - l_0)\int_0^b \psi(t,\lambda_k)\hat{\chi}_\infty(t,l_0)\,dt = 1 - [\hat{\chi}_\infty\psi_k](b)$$

Using (4.9) and letting $b \to \infty$, it follows that (4.3) is valid.

Let $\hat{\mathfrak{D}}$ denote the set of all functions u such that

(i) u is differentiable and u' is absolutely continuous on $0 \leq t \leq b$ for all $b < \infty$,
(ii) u and $Lu \in \mathfrak{L}^2(0, \infty)$,
(iii) $\sin\alpha\, u(0) - \cos\alpha\, p(0)u'(0) = 0$,
(iv) $[u\hat{\chi}_\infty](\infty) = 0$

Let

$$\hat{G}(t,\tau,l_0) = \begin{cases} \psi(t,l_0)\hat{\chi}_\infty(\tau,l_0) & (t \leq \tau) \\ \psi(\tau,l_0)\hat{\chi}_\infty(t,l_0) & (t > \tau) \end{cases} \tag{4.10}$$

and for any $f \in \mathfrak{L}^2(0, \infty)$ let

$$\hat{\mathfrak{G}}(l_0)f(t) = \int_0^\infty \hat{G}(t,\tau,l_0)f(\tau)\,d\tau$$

The integral is absolutely convergent since f and $\hat{\chi}_\infty$ are both of class $\mathfrak{L}^2(0, \infty)$.

Theorem 4.2. *For any* $f \in \mathfrak{L}^2(0, \infty)$ *the function* $u = \hat{\mathfrak{G}}(l_0)f \in \hat{\mathfrak{D}}$ *and*

$(L - l_0)u = f$. Conversely, if $u \in \hat{\mathfrak{D}}$, then $f = (L - l_0)u \in \mathfrak{L}^2(0, \infty)$ and $u = \hat{\mathcal{G}}(l_0)f$.†

Proof. It has been shown in Theorem 2.2 that

$$[\hat{\chi}_\infty \hat{\chi}_\infty](\infty) = 0 \tag{4.11}$$

Let φ, ψ denote the functions given by $\varphi(t) = \varphi(t, l_0)$, $\psi(t) = \psi(t, l_0)$. From (2.8) it follows easily that, if $\tilde{m}_\infty$ is the center of the circle $C_\infty(l_0)$, then

$$[\varphi\psi](\infty) + \tilde{m}_\infty[\psi\psi](\infty) = 0 \tag{4.12}$$

and the reciprocal of the radius of $C_\infty(l_0)$ is

$$|[\psi\psi](\infty)| > 0$$

Clearly, since $\hat{\chi}_\infty = \varphi + \hat{m}_\infty \psi$, (4.12) yields

$$[\psi\hat{\chi}_\infty](\infty) = (\bar{\hat{m}}_\infty - \tilde{m}_\infty)[\psi\psi](\infty)$$

so that

$$[\psi\hat{\chi}_\infty](\infty) \neq 0 \tag{4.13}$$

The proof of the first half of the theorem follows from the use of (4.10). Indeed, if $u = \hat{\mathcal{G}}(l_0)f$, then

$$u'(t) = \hat{\chi}'_\infty(t, l_0) \int_0^t \psi(\tau, l_0)f(\tau)\, d\tau + \psi'(t, l_0) \int_t^\infty \hat{\chi}_\infty(\tau, l_0)f(\tau)\, d\tau$$

From this it follows that (i) is satisfied, and

$$u''(t) = \hat{\chi}''_\infty(t, l_0) \int_0^t \psi(\tau, l_0)f(\tau)\, d\tau + \psi''(t, l_0) \int_t^\infty \hat{\chi}_\infty(\tau, l_0)f(\tau)\, d\tau + \frac{[\psi\bar{\hat{\chi}}_\infty](t)f(t)}{p(t)}$$

Since $[\psi\bar{\hat{\chi}}_\infty](t) = [\psi\bar{\hat{\chi}}_\infty](0) = -1$, one sees easily that $Lu = l_0 u + f$. That $u \in \mathfrak{L}^2(0, \infty)$ follows from the fact that ψ and $\hat{\chi}_\infty$ are $\mathfrak{L}^2(0, \infty)$ and the use of the Schwarz and Minkowski inequalities. Since $Lu = l_0 u + f$, $Lu \in \mathfrak{L}^2(0, \infty)$, and (ii) holds. Condition (iii) is valid since ψ satisfies (iii), and (iv) follows with the aid of (4.11).

The second half of the theorem will now be proved. Let $f = Lu - l_0 u$. Then $f \in \mathfrak{L}^2(0, \infty)$ and, from the first part of the theorem, $\hat{\mathcal{G}}(l_0)f$ is of class $\hat{\mathfrak{D}}$. Thus $w = u - \hat{\mathcal{G}}(l_0)f$ is of class $\hat{\mathfrak{D}}$ and satisfies $Lw - l_0 w = 0$. Hence $w = c_1\psi + c_2\hat{\chi}_\infty$ for some constants c_1, c_2. Used in (iii) and (iv), it follows that c_2 and c_1, respectively, must be zero because $\hat{\chi}_\infty$ cannot satisfy (iii) and, by (4.13), ψ cannot satisfy (iv). This completes the proof.

This result can now be used to prove the following theorem.

† The statement of Theorem 4.2 is just the statement that $\hat{\mathcal{G}}(l_0)$ is the inverse of the operator $L - l_0$ with domain $\hat{\mathfrak{D}}$.

Theorem 4.3. *The boundary-value problem*

$$Lx = lx \qquad \sin\alpha\, x(0) - \cos\alpha\, p(0)x'(0) = 0 \qquad [x\mathfrak{X}_\infty](\infty) = 0 \quad (4.14)$$

is self-adjoint; that is, for all u and v of class $\hat{\mathfrak{D}}$

$$\int_0^\infty (Lu)\bar{v}\,dt = \int_0^\infty u(\overline{Lv})\,dt \tag{4.15}$$

Proof. From Green's formula and the fact that $[uv](0) = 0$, (4.15) is equivalent to

$$[uv](\infty) = 0 \tag{4.16}$$

From Theorem 4.2 there exist $f,g \in \mathfrak{L}^2(0,\infty)$ such that $u = \hat{\mathfrak{G}}(l_0)f$ and $v = \hat{\mathfrak{G}}(l_0)g$. Expressing $[uv](b)$ in terms of the integrals involving the Green's function (4.10) and f and g and letting $b \to \infty$, (4.16) follows from (4.11), and the theorem is proved.

It is a consequence of (4.3) that ψ_k is of class $\hat{\mathfrak{D}}$. From (4.15) the orthogonality of the ψ_k is immediate. No ψ_l, $l \neq \lambda_k$ for all k, can satisfy $[x\mathfrak{X}_\infty](\infty) = 0$, for if it did ψ_l would be of class $\hat{\mathfrak{D}}$ and thus would be orthogonal to all ψ_k. This is impossible since the ψ_k are complete.

5. Singular Behavior at Both Ends of an Interval

The cases where the coefficients in L have singular behavior at both ends of an interval, or singular behavior at one end and a semi-infinite interval, or an interval extending over the whole t axis, are all handled similarly. Here the case where L is defined over $-\infty < t < \infty$ will be treated. Recall that

$$Lx = -(px')' + qx$$

where now it is assumed that p, p', q are real and continuous, $p(t) > 0$, on $-\infty < t < \infty$. (These restrictions on p,q can be relaxed somewhat.) Let φ_1,φ_2 be solutions of $Lx = lx$, real for real l, satisfying the initial conditions

$$\begin{aligned} \varphi_1(0,l) &= 1 & \varphi_2(0,l) &= 0 \\ p(0)\varphi_1'(0,l) &= 0 & p(0)\varphi_2'(0,l) &= 1 \end{aligned}$$

Then φ_1,φ_2 are entire functions of l for fixed t.

Let δ: $a \leqq t \leqq b$ be any finite interval containing zero, and consider the self-adjoint boundary-value problem on δ:

$$\begin{aligned} Lx &= lx \\ \cos\alpha\, x(a) + \sin\alpha\, p(a)x'(a) &= 0 \\ \cos\beta\, x(b) + \sin\beta\, p(b)x'(b) &= 0 \end{aligned} \tag{5.1}$$

where $0 \leqq \alpha, \beta < \pi$. There exists a sequence of real eigenvalues $\{\lambda_{\delta n}\}$, $n = 1, 2, \ldots$, and a complete orthonormal set of eigenfunctions $\{h_{\delta n}\}$.

In terms of these, the Parseval equality holds for any $f \in \mathfrak{L}^2(\delta)$,

$$\int_\delta |f(t)|^2\,dt = \sum_{n=1}^{\infty} \left| \int_\delta f(t)\bar{h}_{\delta n}(t)\,dt \right|^2 \tag{5.2}$$

If $f_1,f_2 \in \mathfrak{L}^2(\delta)$, then

$$\int_\delta f_1(t)\bar{f}_2(t)\,dt = \sum_{n=1}^{\infty} \int_\delta f_1(t)\bar{h}_{\delta n}(t)\,dt\,\overline{\int_\delta f_2(t)\bar{h}_{\delta n}(t)\,dt} \tag{5.3}$$

Since φ_1,φ_2 form a basis for the solutions of $Lx = lx$, it follows that

$$h_{\delta n}(t) = r_{\delta n1}\varphi_1(t,\lambda_{\delta n}) + r_{\delta n2}\varphi_2(t,\lambda_{\delta n}) \tag{5.4}$$

where $r_{\delta n1}$ and $r_{\delta n2}$ are complex constants. Placing (5.4) into (5.2), the latter may be rewritten as

$$\int_\delta |f(t)|^2\,dt = \int_{-\infty}^{\infty} \sum_{j,k=1}^{2} \bar{g}_{\delta j}(\lambda)g_{\delta k}(\lambda)\,d\rho_{\delta jk}(\lambda) \tag{5.5}$$

where

$$g_{\delta j}(\lambda) = \int_\delta f(t)\varphi_j(t,\lambda)\,dt$$

and the matrix $\rho_\delta = (\rho_{\delta jk})$, called the *spectral matrix* associated with the self-adjoint problem (5.1), consists of step functions with jumps at the eigenvalues $\lambda_{\delta n}$ given by

$$\rho_{\delta jk}(\lambda_{\delta n} + 0) - \rho_{\delta jk}(\lambda_{\delta n} - 0) = \sum r_{\delta mj}\bar{r}_{\delta mk}$$

where the sum is taken over all m such that $\lambda_{\delta m} = \lambda_{\delta n}$. Let

$$\rho_\delta(\lambda + 0) = \rho_\delta(\lambda)$$

and $\rho_\delta(0)$ be the zero matrix. Clearly ρ_δ possesses the properties:

(i) ρ_δ is Hermitian $\quad (\rho_{\delta jk} = \bar{\rho}_{\delta kj})$
(ii) $\rho_\delta(\Delta) = \rho_\delta(\lambda) - \rho_\delta(\mu)$ is positive semidefinite if $\lambda > \mu$ $(\Delta = (\mu,\lambda])$
(iii) The total variation of $\rho_{\delta jk}$ is finite on every finite λ interval.

Any matrix ρ_δ satisfying (ii) is said to be *nondecreasing*. The matrix ρ_δ is the counterpart on δ of the nondecreasing spectral function ρ_b for the problem (3.1).

Applying the Parseval equality (5.5) to any continuous function f on $-\infty < t < \infty$ which vanishes outside some interval δ_1 contained properly in δ, one obtains

$$\int_{-\infty}^{\infty} |f(t)|^2\,dt = \int_{-\infty}^{\infty} \sum_{j,k=1}^{2} \bar{g}_j(\lambda)g_k(\lambda)\,d\rho_{\delta jk}(\lambda) \tag{5.6}$$

where

$$g_j(\lambda) = \int_{-\infty}^{\infty} f(t)\varphi_j(t,\lambda)\,dt$$

As $\delta \to (-\infty,\infty)$ (that is, $a \to -\infty, b \to \infty$), it can be shown that, if L is in the limit-point case at $-\infty$ and ∞, there exists a matrix ρ having the properties (i) through (iii) such that $\rho_\delta \to \rho$, and (5.6) is valid for any $f \in \mathfrak{L}^2(-\infty,\infty)$. If L is in the limit-circle case at one, or both, of the points $-\infty$ or ∞, limiting matrices ρ still exist such that (5.6) holds, but there is the usual nonuniqueness.

The key to proving the existence of a limiting matrix ρ is an equality for ρ_δ which replaces the equality (3.14) for ρ_b. Let $\chi_a = \varphi_1 + m_a\varphi_2$ be a solution of $Lx = lx$ ($\Im l \neq 0$) satisfying the boundary condition

$$\cos\alpha\, x(a) + \sin\alpha\, p(a)x'(a) = 0$$

and similarly let $\chi_b = \varphi_1 + m_b\varphi_2$ be a solution of the same equation satisfying

$$\cos\beta\, x(b) + \sin\beta\, p(b)x'(b) = 0$$

Then, as has been shown in Theorem 2.2, m_a and m_b lie on circles C_a and C_b in the complex m plane whose equations are, respectively,

$$[\chi_a\chi_a](a) = 0 \qquad [\chi_b\chi_b](b) = 0 \tag{5.7}$$

It is easily seen that Green's function G_δ for the problem (5.1) exists, provided $\Im l \neq 0$, and is given by

$$G_\delta(t,\tau,l) = \begin{cases} \dfrac{\chi_a(t,l)\chi_b(\tau,l)}{m_a(l) - m_b(l)} & (t \leqq \tau) \\[2ex] \dfrac{\chi_a(\tau,l)\chi_b(t,l)}{m_a(l) - m_b(l)} & (t > \tau) \end{cases}$$

The completeness relationship in the form (5.3) is now applied to the functions

$$f_1(t) = \frac{\partial^j G_\delta}{\partial\tau^j}(t,0,l), \qquad f_2(t) = \frac{\partial^k G_\delta}{\partial\tau^k}(t,0,l) \qquad (j,\ k = 0,\ 1)$$

yielding

$$\int_\delta \frac{\partial^j G_\delta}{\partial\tau^j}(t,0,l)\,\frac{\partial^k \bar{G}_\delta}{\partial\tau^k}(t,0,l)\,dt$$

$$= \sum_{n=1}^{\infty} \int_\delta \frac{\partial^j G_\delta}{\partial\tau^j}(t,0,l)\bar{h}_{\delta n}(t)\,dt\ \overline{\int_\delta \frac{\partial^k G_\delta}{\partial\tau^k}(t,0,l)\bar{h}_{\delta n}(t)\,dt} \tag{5.8}$$

From the definition of G_δ it follows that

$$\begin{aligned} G_\delta(t,0,l) &= \frac{\chi_a(t,l)}{m_a(l) - m_b(l)} \qquad (t \leqq 0) \\ &= \frac{\chi_b(t,l)}{m_a(l) - m_b(l)} \qquad (t > 0) \end{aligned} \tag{5.9}$$

and

$$\begin{aligned} \frac{\partial G_\delta}{\partial \tau}(t,0,l) &= \frac{m_b(l)\chi_a(t,l)}{p(0)(m_a(l) - m_b(l))} \qquad (t \leqq 0) \\ &= \frac{m_a(l)\chi_b(t,l)}{p(0)(m_a(l) - m_b(l))} \qquad (t > 0) \end{aligned} \tag{5.10}$$

Using (5.9) and (5.10) and Green's formula, the integrals in (5.8) can be evaluated. For example,

$$\begin{aligned} 2i\Im l \int_\delta |G_\delta(t,0,l)|^2\,dt &= 2i\Im l|m_a(l) - m_b(l)|^{-2}\left\{\int_a^0 |\chi_a(t,l)|^2\,dt \right. \\ &\qquad \left. + \int_0^b |\chi_b(t,l)|^2\,dt\right\} \\ &= |m_a(l) - m_b(l)|^{-2}\{[\chi_a\chi_a](0) - [\chi_b\chi_b](0)\} \\ &= 2i\Im(m_b(l) - m_a(l))|m_a(l) - m_b(l)|^{-2} \end{aligned}$$

making use of (5.7). Therefore

$$\int_\delta |G_\delta(t,0,l)|^2\,dt = \frac{\Im(m_a(l) - m_b(l))^{-1}}{\Im l} \tag{5.11}$$

Similarly,

$$\begin{aligned} (m_a(l) - m_b(l))(l - \lambda_{\delta n}) &\int_\delta G_\delta(t,0,l)\bar{h}_{\delta n}(t)\,dt \\ &= [\chi_b h_{\delta n}](b) - [\chi_b h_{\delta n}](0) + [\chi_a h_{\delta n}](0) - [\chi_a h_{\delta n}](a) \\ &= [(\chi_a - \chi_b)h_{\delta n}](0) \\ &= (m_a(l) - m_b(l))[\varphi_2 h_{\delta n}](0) \\ &= (m_b(l) - m_a(l))\bar{r}_{\delta n1} \end{aligned}$$

and hence

$$\int_\delta G_\delta(t,0,l)\bar{h}_{\delta n}(t)\,dt = \frac{\bar{r}_{\delta n1}}{\lambda_{\delta n} - l} \tag{5.12}$$

Here use has been made of the fact that $[\chi_b h_{\delta n}](b) = 0$ which follows since both χ_b and $h_{\delta n}$ satisfy the same boundary condition at b; similarly $[\chi_a h_{\delta n}](a) = 0$. Now (5.8), (5.11), and (5.12) yield for $j = k = 0$,

$$\int_{-\infty}^{\infty} \frac{d\rho_{\delta 11}(\lambda)}{|\lambda - l|^2} = \frac{\Im M_{\delta 11}(l)}{\Im l}$$

where

$$M_{\delta 11}(l) = (m_a(l) - m_b(l))^{-1} \tag{5.13}$$

Further similar calculation shows that

$$\int_{-\infty}^{\infty} \frac{d\rho_{\delta jk}(\lambda)}{|\lambda - l|^2} = \frac{\Im M_{\delta jk}(l)}{\Im l} \tag{5.14}$$

where $M_{\delta 11}$ is given by (5.13), and

$$\begin{aligned} M_{\delta 12}(l) &= M_{\delta 21}(l) = \tfrac{1}{2}(m_a(l) + m_b(l))(m_a(l) - m_b(l))^{-1} \\ M_{\delta 22}(l) &= m_a(l)m_b(l)(m_a(l) - m_b(l))^{-1} \end{aligned}$$

Formula (5.14) replaces (3.14) for ρ_b.

Since

$$\Im l \int_a^0 |\chi_a(t,l)|^2 \, dt = -\Im m_a(l) \qquad \Im l \int_0^b |\chi_b(t,l)|^2 \, dt = \Im m_b(l)$$

for a fixed l, $\Im l \neq 0$, $m_a(l)$ and $m_b(l)$ are in opposite half planes. Suppose $l = i$ in (5.14). Then points $m_a(i)$ lie on a circle C_a which is in C_{-1} for $a < -1$, whereas points $m_b(i)$ lie on C_b which is in C_1 for $b > 1$. Thus there is a constant $c > 0$ such that $|m_a(i) - m_b(i)| > c$ for $a < -1$, $b > 1$. Since $m_a(i)$and $m_b(i)$ are uniformly bounded for $a < -1$, $b > 1$, it follows from (5.14), and the definition of the $M_{\delta jk}(l)$, that

$$\int_{-\infty}^{\infty} \frac{d\rho_{\delta jj}(\lambda)}{1 + \lambda^2} < K \qquad (j = 1, 2)$$

for some constant K. Since

$$2|r_{\delta nj}\bar{r}_{\delta nk}| \leqq |r_{\delta nj}|^2 + |r_{\delta nk}|^2$$

it follows that

$$\int_{-\infty}^{\infty} \frac{|d\rho_{\delta jk}(\lambda)|}{1 + \lambda^2} < K$$

holds for $j \neq k$ also.

Using the Helly selection theorem, which applies equally well to functions with dominated total variation, much as in the proof of Theorem 3.1, it follows that there exists a sequence of intervals $\delta_n = [a_n, b_n]$, $\delta_n \to (-\infty, \infty)$, and corresponding boundary conditions prescribed by α_n, β_n, such that $\rho_{\delta_n jk}(\lambda)$ tends to a limit $\rho_{jk}(\lambda)$ as $n \to \infty$. The matrix $\rho = (\rho_{jk})$ possesses the same properties (i) through (iii) as ρ_δ.

If L is in the limit-point case at $-\infty$ and ∞, ρ is unique since in this situation both m_a and m_b tend to points $m_{-\infty}$, m_∞, respectively, and as in the proof of Theorem 3.1 the formula

$$\rho_{jk}(\lambda) - \rho_{jk}(\mu) = \lim_{\epsilon \to +0} \frac{1}{\pi} \int_\mu^\lambda \Im M_{jk}(\nu + i\epsilon) \, d\nu$$

can be proved. Here $M_{jk}(l)$ is the limit (which exists) of $M_{\delta jk}(l)$ as $\delta \to (-\infty, \infty)$.

There is also a corresponding expansion and completeness theorem, the proof of which parallels that of Theorem 3.1, and will be omitted.

Theorem 5.1. *Let L be in the limit-point case at $-\infty$ and ∞. There exists a nondecreasing Hermitian matrix $\rho = (\rho_{jk})$ whose elements are of bounded variation on every finite λ interval, and which is essentially unique in the sense that*

$$\rho_{\delta jk}(\lambda) - \rho_{\delta jk}(\mu) \to \rho_{jk}(\lambda) - \rho_{jk}(\mu) \qquad (\delta \to (-\infty, \infty))$$

at points of continuity λ,μ of ρ_{jk}. Further

$$\rho_{jk}(\lambda) - \rho_{jk}(\mu) = \lim_{\epsilon \to +0} \frac{1}{\pi} \int_{\mu}^{\lambda} \Im M_{jk}(\nu + i\epsilon)\, d\nu \tag{5.15}$$

where

$$\begin{aligned} M_{11}(l) &= (m_{-\infty}(l) - m_{\infty}(l))^{-1} \\ M_{12}(l) &= M_{21}(l) = \tfrac{1}{2}(m_{-\infty}(l) + m_{\infty}(l))(m_{-\infty}(l) - m_{\infty}(l))^{-1} \\ M_{22}(l) &= m_{-\infty}(l)m_{\infty}(l)(m_{-\infty}(l) - m_{\infty}(l))^{-1} \end{aligned} \tag{5.16}$$

Analogous results hold if L is in the limit-circle case at either or both of the end points of the interval. In order to obtain a unique spectral matrix, boundary conditions must be added at the end point where L is in the limit-circle case, as was done in Sec. 4. If both ends are in the limit-circle case, then m_{∞} and $m_{-\infty}$ are meromorphic and thus so are M_{11}, M_{12}, and M_{22}.

For any limiting matrix ρ let $\mathfrak{L}^2(\rho)$ denote the set of all vectors g with components g_1,g_2 functions of λ such that

$$\|g\|^2 = \int_{-\infty}^{\infty} \sum_{j,k=1}^{2} \bar{g}_j(\lambda) g_k(\lambda)\, d\rho_{jk}(\lambda) < \infty$$

The nondecreasing nature of ρ guarantees that the integral above is nonnegative.

Theorem 5.2. *Let ρ be any limit matrix of the set $\{\rho_\delta\}$. If $f \in \mathfrak{L}^2(-\infty,\infty)$ the vector $g = (g_1,g_2)$, where*

$$g_j(\lambda) = \int_{-\infty}^{\infty} f(t)\varphi_j(t,\lambda)\, dt$$

converges in $\mathfrak{L}^2(\rho)$, that is, there exists a $g \in \mathfrak{L}^2(\rho)$ such that

$$\|g - g_{cd}\| \to 0 \qquad (c \to -\infty, d \to \infty)$$

where

$$g_{cdj}(\lambda) = \int_{c}^{d} f(t)\varphi_j(t,\lambda)\, dt \qquad (-\infty < c < d < \infty)$$

In terms of this g, the Parseval equality

$$\int_{-\infty}^{\infty} |f(t)|^2 \, dt = \int_{-\infty}^{\infty} \sum_{j,k=1}^{2} \bar{g}_j(\lambda) g_k(\lambda) \, d\rho_{jk}(\lambda)$$

and the expansion

$$f(t) = \int_{-\infty}^{\infty} \sum_{j,k=1}^{2} \varphi_j(t,\lambda) g_k(\lambda) \, d\rho_{jk}(\lambda)$$

are valid, the latter integral converging in the mean in $\mathfrak{L}^2(-\infty,\infty)$.

The *spectrum* associated with a problem for which ρ is uniquely determined is the set of nonconstancy points of ρ, that is, the set of all nonconstancy points of all elements ρ_{jk} of ρ. Since ρ is Hermitian and nondecreasing, it follows that

$$|\rho_{jk}(\Delta)|^2 \leqq \rho_{jj}(\Delta)\rho_{kk}(\Delta)$$

where

$$\rho_{jk}(\Delta) = \rho_{jk}(\lambda) - \rho_{jk}(\mu) \qquad \Delta = (\mu,\lambda]$$

Hence the set of nonconstancy points of all elements of ρ is the same as the set of nonconstancy points of all *diagonal* elements ρ_{jj} of ρ. Clearly the spectrum is a closed set. The *point spectrum* is the set of all discontinuity points of ρ, and the *continuous spectrum* is the set of continuity points of ρ which are in the spectrum. Points in the point spectrum are also called *eigenvalues* and solutions of the problem for such points are called *eigenfunctions.*†

Example 1. Perhaps the simplest case is where

$$Lx = -x''$$

Here

$$\varphi_1(t,l) = \cos\sqrt{l}\,t \qquad \varphi_2(t,l) = \frac{\sin\sqrt{l}\,t}{\sqrt{l}}$$

and it is obvious that L is in the limit-point case at both $-\infty$ and ∞. For $\mathfrak{I}l > 0$ the solution $e^{-i\sqrt{l}t} \in \mathfrak{L}^2(-\infty,0)$ and hence

$$\cos\sqrt{l}\,t + m_{-\infty}(l)\frac{\sin\sqrt{l}\,t}{\sqrt{l}} = c(\cos\sqrt{l}\,t - i\sin\sqrt{l}\,t)$$

for some constant c. Putting $t = 0$, there results $c = 1$, and therefore

$$m_{-\infty}(l) = -i\sqrt{l}$$

Similarly

$$m_{\infty}(l) = i\sqrt{l}$$

† Every eigenfunction is of class $\mathfrak{L}^2(-\infty,\infty)$. See Probs. 6 and 7.

Thus

$$M_{11}(l) = \frac{i}{2\sqrt{l}} \qquad M_{22}(l) = \frac{i\sqrt{l}}{2} \qquad M_{12}(l) = M_{21}(l) = 0$$

and consequently, from (5.15),

$$\begin{aligned} d\rho_{11}(\lambda) &= \frac{1}{2\pi}\frac{d\lambda}{\sqrt{\lambda}} && (\lambda > 0) \\ &= 0 && (\lambda < 0) \\ d\rho_{22}(\lambda) &= \frac{1}{2\pi}\sqrt{\lambda}\,d\lambda && (\lambda > 0) \\ &= 0 && (\lambda < 0) \\ d\rho_{12}(\lambda) &= d\rho_{21}(\lambda) = 0 \end{aligned}$$

The expansion theorem becomes for any $f \varepsilon \mathfrak{L}^2(-\infty, \infty)$

$$f(t) = \frac{1}{2\pi}\int_0^\infty \frac{\cos\sqrt{\lambda}\,t}{\sqrt{\lambda}} g_1(\lambda)\,d\lambda + \frac{1}{2\pi}\int_0^\infty \sin\sqrt{\lambda}\,t\, g_2(\lambda)\,d\lambda$$

where

$$g_1(\lambda) = \int_{-\infty}^{\infty} f(t)\cos\sqrt{\lambda}\,t\,dt$$

$$g_2(\lambda) = \int_{-\infty}^{\infty} f(t)\frac{\sin\sqrt{\lambda}\,t}{\sqrt{\lambda}}\,dt$$

and the interpretation on the integrals is the same as in Theorem 5.2. This is precisely the Fourier integral formula for functions $f \varepsilon \mathfrak{L}^2(-\infty, \infty)$. (The spectrum of the problem $Lx = -x'' = lx$ on $-\infty < t < \infty$ is the set $0 \leqq \lambda < \infty$.)

Example 2. Another interesting case is the *Hermite operator* L given by

$$Lx = -x'' + t^2x \qquad (-\infty < t < \infty)$$

From Corollary 2 of Theorem 2.4 it follows that since $q(t) = t^2 \to \infty$ as $t \to \pm\infty$ the limit-point case prevails at $-\infty$ and ∞. Considered over $-\infty < t \leqq 0$ and $0 \leqq t < \infty$, the spectrum of any boundary-value problem with a boundary condition at zero is discrete; see Prob. 1. This implies that the functions $m_{-\infty}$, m_∞ are meromorphic with simple poles on the real axis. For $\mathfrak{I}l \neq 0$, $m_{-\infty}(l)$ and $m_\infty(l)$ are in opposite half planes; this implies $m_{-\infty}(l) - m_\infty(l)$ has only isolated zeros. From (5.15) and (5.16) it then follows that the ρ_{jk} are step functions with discontinuities at an enumerable set $\{\lambda_n\}$, $n = 1, 2, \ldots$. The orthonormal set of eigenfunctions $\{h_n\}$ which are solutions of $Lx = \lambda_n x$ of class $\mathfrak{L}^2(-\infty, \infty)$ are, except for constant factors, $e^{-\frac{1}{2}t^2}H_n(t)$, where the H_n are the Hermite polynomials. Indeed, let $x = e^{-\frac{1}{2}t^2}y$. Then $Lx = lx$ becomes

$$-y'' + 2ty' + y = ly$$

If $u = \sum_{k=0}^{\infty} a_k t^k$ is a solution of this equation, then

$$\frac{a_{k+2}}{a_k} = \frac{2k + 1 - l}{(k + 1)(k + 2)} \tag{5.17}$$

Thus there are two solutions, u_1 even, and u_2 odd, and if l is not an odd positive integer, (5.17) implies that

$$\lim_{|t|\to\infty} \frac{\log |u_j|}{t^2} = 1 \qquad (j = 1, 2)$$

so that $(c_1u_1 + c_2u_2)e^{-\frac{1}{2}t^2}$ is unbounded for all c_1,c_2 not 0. Thus $ue^{-\frac{1}{2}t^2}$ is $\mathfrak{L}^2(-\infty, \infty)$ if and only if l is an odd positive integer and u is a polynomial in t.

PROBLEMS

1. Let $Lx = -x'' + qx$, where q is real and continuous on $[0, \infty)$ and $q(t) \to \infty$ as $t \to \infty$. Show that L is in the limit-point case. Show the spectral function ρ of $Lx = lx$, $\sin\alpha\, x(0) - \cos\alpha\, x'(0) = 0$, is a step function with discontinuities at $\{\lambda_k\}$, $k = 0, 1, \ldots$, where $\lambda_0 < \lambda_1 < \cdots$. Show that the eigenfunction ψ_k, where

$$\psi_k(t) = \psi(t,\lambda_k)$$

has exactly k zeros on $0 < t < \infty$.

HINT: Make use of Probs. 1(*a*), (*b*), (*c*), (*d*), and (*e*).

(*a*) For any given real λ show that ψ_λ, where $\psi_\lambda(t) = \psi(t,\lambda)$, does one of three things as $t \to \infty$: ψ_λ and $\psi_\lambda' \to \infty, -\infty$, or 0.

HINT: Let $t_0 = t_0(\lambda)$ be chosen so that $q(t) - \lambda > 1$ for $t > t_0$. Use $\psi_\lambda'' = (q - \lambda)\psi_\lambda$ and show ψ_λ and ψ_λ' can have at most one zero for $t > t_0$.

(*b*) Show that there exists a monotone increasing sequence $\lambda_0, \lambda_1, \ldots$ with $\lambda_n \to \infty$ as $n \to \infty$ such that ψ_λ has exactly n zeros on $0 < t < \infty$ for $\lambda_{n-1} < \lambda \leqq \lambda_n$, where $\lambda_{-1} = -\infty$.

HINT: By Prob. 1(*a*), ψ_λ has a finite number of zeros on $(0, \infty)$. By the proof of Theorem 2.1, Chap. 8, ψ_λ and ψ_λ' are of the same sign as $\lambda \to -\infty$ for any given $\bar{t} > 0$ and thus by Prob. 1(*a*), ψ_λ has no zeros for λ near $-\infty$ [and $|\psi(t,\lambda)| \to \infty$ as $t \to \infty$ so that L is in limit-point case]. By the same theorem, ψ_λ has zeros on $(0, \infty)$ if λ is large enough and the zeros move continuously to the left as λ increases and to the right as λ decreases. Thus there exists a λ_n such that for $\lambda < \lambda_n$, ψ_λ has at most n zeros on $(0, \infty)$ and for $\lambda > \lambda_n$ at least $n + 1$ zeros on $(0, \infty)$. If ϵ is small enough, then $\psi_{\lambda_n+\epsilon}$ has exactly $n + 1$ zeros, for if it has $n + j$, then as $\epsilon \to +0$ the last j zeros must move toward $t = \infty$. Thus, when the $(n + 1)$st zero is large enough, the method of Prob. 1(*a*) shows there are no further zeros and also that the nth zero is not to the right of $t_0(\lambda_n)$. Thus ψ_n has at least n zeros. On the other hand, if ψ_n has exactly k zeros, so does $\psi_{\lambda_n-\epsilon}$ for small enough ϵ. Thus ψ_n has exactly n zeros.

(*c*) The spectral function ρ is a step function which can be discontinuous only at $\lambda_0, \lambda_1, \ldots$.

HINT: Consider the problem on $(0,b)$ with $x(b) = 0$ and eigenvalues λ_{bj}. Then since ψ_{bj}, where $\psi_{bj}(t) = \psi(t,\lambda_{bj})$, has exactly $j + 1$ zeros on $(0,b]$, it follows that

$\lambda_{bj} > \lambda_j$. On the other hand, given any $\epsilon > 0$ and n, $\lambda_{bn} < \lambda_n + \epsilon$ for b large enough since $\psi_{\lambda_n+\epsilon}$ has $n + 1$ zeros on $(0, \infty)$. Use $\rho_b(\lambda) \to \rho(\lambda)$ as $b \to \infty$.

(*d*) Show that the ψ_k are $\mathfrak{L}^2(0, \infty)$ and orthogonal.

HINT: By Prob. 1(*a*), $\psi_k \to 0$ as $t \to \infty$ since, if $|\psi_k| \to \infty$, the number of zeros of ψ_λ on $(0, \infty)$ cannot change with a small change of λ from λ_k. Since ψ_k and $\psi_k' \to 0$ monotonically, show that $\int^\infty |\psi_k''|\, dt < \infty$ and therefore $\int^\infty |\psi_k|\, dt < \infty$ and thus $\psi_k \in \mathfrak{L}^2(0, \infty)$. For orthogonality, use Green's formula and $\psi_j\psi_k' - \psi_j'\psi_k \to 0$ as $t \to \infty$.

(*e*) The spectral function is discontinuous at all λ_k, $k \geqq 0$.

HINT: If ρ does not have a jump at λ_n, then since ψ_n is orthogonal to all ψ_j, $j \neq n$, the Parseval equality fails.

2. Let $p = 1$ and q be real and continuous on $[0, \infty)$ and

$$\liminf_{t \to \infty} q(t) = \mu$$

for some μ, $-\infty < \mu \leqq \infty$. The spectral function ρ of $Lx = lx$,

$$\sin \alpha\, x(0) - \cos \alpha\, p(0)x'(0) = 0$$

is a step function for $\lambda < \mu$ with jumps only on an increasing sequence $\lambda_0 < \lambda_1 < \cdots < \mu$. The number of λ_j (if finite) is the number of zeros of ψ_μ on $0 < t < \infty$. If the sequence is not empty, then ψ_n has exactly n zeros on $0 < t < \infty$.

HINT: See Prob. 1.

3. Show that if n is the number of zeros of the solution of $-x'' + qx = 0$, $x(0) = 0$, on $(0, \infty)$, where q is real, then

$$n < \int_0^\infty t|q(t)|\, dt$$

HINT: Let $|q(t)| = h(t)$. If α and β are successive zeros of a solution φ of

$$x'' + hx = 0$$

then

$$\varphi(t) = a(t - \alpha) - \int_\alpha^t (t - s)h(s)\varphi(s)\, ds$$

Hence show that $|\varphi(t)| \leqq |a|(t - \alpha)$ for $\alpha \leqq t \leqq \beta$ and

$$1 \leqq \int_\alpha^\beta (s - \alpha)h(s)\, ds$$

4. Let q be real and integrable on $(0, \infty)$, that is,

$$\int_0^\infty |q(t)|\, dt < \infty$$

Let $Lx = -x'' + qx$ and consider $Lx = lx$, $x(0) = 0$. Show that L is in the limit-point case and that for $\lambda > 0$, the spectral function is of class C^1. Indeed, show that if $l = s^2$, $\psi(t,s)$, where $\psi(0,s) = 0$, $\psi'(0,s) = 1$, satisfies

$$\psi(t,s) - \frac{A(s)}{s} \sin (st - \alpha(s)) \to 0$$

as $t \to \infty$, where A is continuous and positive, and α is continuous and real. Prove that for $\lambda > 0$

$$\frac{d\rho}{d\lambda}(\lambda) = \frac{\lambda^{\frac{1}{2}}}{\pi A^2(\lambda^{\frac{1}{2}})}$$

HINT: For $s > 0$ show, by successive approximations, that

$$\varphi_1(t,s) = e^{its} - \int_t^\infty \frac{\sin s(t-\tau)}{s} q(\tau)\varphi_1(\tau,s)\, d\tau$$

has a bounded solution and hence that $\varphi_1(t,s) - e^{its} \to 0$ as $t \to \infty$. Show $L\varphi_1 = s^2\varphi_1$. (Note that φ_1 exists for $\mathfrak{I}s \geqq 0$.) Show similarly there is a φ_2 related to e^{-its}. Neither φ_1 nor φ_2 is $\mathfrak{L}^2(0,\infty)$. Show, by Prob. 1, Chap. 1, from

$$\psi(t,s) = \frac{\sin st}{s} + \int_0^t \frac{\sin s(t-\tau)}{s} \psi(\tau,s)q(\tau)\, d\tau$$

that $|\psi|$ is bounded as $t \to \infty$ and from this that as $t \to \infty$

$$\psi(t,s) - \frac{A(s)}{s} \sin (st - \alpha(s)) \to 0$$

where

$$A(s)e^{i\alpha} = 1 + \int_0^\infty e^{is\tau}q(\tau)\psi(\tau,s)\, d\tau$$

and the integral converges uniformly. Since $\psi = c_1\varphi_1 + c_2\varphi_2$, show by letting $t \to \infty$ that $c_1 = Ae^{-i\alpha}/(2is)$ and $c_2 = \bar{c}_1$. Hence $A(\tilde{s}) = 0$ implies that $\psi(t,\tilde{s}) = 0$ for all t. Consider the problem $Lx = lx$, $x(0) = x(b) = 0$. Then the eigenvalues occur at those values of s for which $\psi(b,s) = 0$. For b large there occur $(s_2 - s_1)b/\pi + c$ eigenvalues in (s_1,s_2), where $|c| < 4$ and

$$\frac{1}{b}\int_0^b \psi^2\, dt = \frac{A^2(s)}{2s^2} + \epsilon$$

where $\epsilon \to 0$ as $b \to \infty$. Thus for large b

$$\bar{\rho}_b(s_2) - \bar{\rho}_b(s_1) = \frac{2(s_2 - s_1)s^2}{\pi(A^2 + 2\epsilon s^2)}$$

where s is in (s_1,s_2). As $b \to \infty$

$$\bar{\rho}(s_2) - \bar{\rho}(s_1) = \frac{2(s_2 - s_1)\tilde{s}^2}{\pi A^2(\tilde{s})}$$

for some $\tilde{s}$ in (s_1,s_2). Thus

$$\frac{d\bar{\rho}}{ds}(s) = \frac{2s^2}{\pi A^2(s)}$$

where $\bar{\rho}(s) = \rho(s^2)$.

5. In Prob. 4 let s be complex and show that the function F defined by

$$F(s) = A(s)e^{i\alpha(s)}$$

is analytic in s for $\mathfrak{I}s > 0$. Show that F can vanish for $\mathfrak{I}s > 0$ only if $\mathfrak{R}s = 0$. Show that where $F(s) = 0$, $\psi(t,s) = 0(e^{-|s|t})$ and hence an eigenvalue occurs.

6. In the case of a boundary-value problem on $0 \leqq t < \infty$ which has a spectral function ρ show that if $\bar{\lambda}$ is in the point spectrum then $\bar{\psi}$, where $\bar{\psi}(t) = \psi(t,\bar{\lambda})$, is of class $\mathfrak{L}^2(0,\infty)$.

HINT: Let $f = \bar{\psi}$ for $t < a$ and $f = 0$ for $t > a$. Let

$$\int_0^a f(t)\psi(t,\lambda)\, dt = g_a(\lambda)$$

Then

$$\int_0^a |f|^2\, dt = \int_{-\infty}^\infty |g_a(\lambda)|^2\, d\rho(\lambda)$$

Let $\rho(\bar\lambda + 0) - \rho(\bar\lambda - 0) = \tilde r$. Show

$$\int_0^a |\psi|^2\, dt < \frac{1}{\tilde r}$$

7. For $Lx = lx$, $-\infty < t < \infty$, prove that a discontinuity in the matrix ρ at $\lambda = \bar\lambda$ implies that at least one solution of $Lx = \bar\lambda x$ is of class $\mathfrak{L}^2(-\infty, \infty)$.

HINT: Let the discontinuities of ρ_{11}, ρ_{12}, and ρ_{22} at $\bar\lambda$ be r_{11}, r_{12}, and r_{22}, respectively. If $f \in \mathfrak{L}^2(-\infty, \infty)$ and is real,

$$\int_{-a}^a f^2\, dt \geqq r_{11}\left(\int_{-a}^a f\varphi_1\, dt\right)^2 - 2r_{12}\left(\int_{-a}^a f\varphi_1\, dt\right)\left(\int_{-a}^a f\varphi_2\, dt\right) + r_{22}\left(\int_{-a}^a f\varphi_2\, dt\right)_2$$

Taking $f = r_{11}^{\frac{1}{2}}\varphi_1 + b\varphi_2$, where $b = r_{12}/r_{11}^{\frac{1}{2}}$, prove $f \in \mathfrak{L}^2(-\infty, \infty)$. If $r_{11} = 0$, prove $\varphi_2 \in \mathfrak{L}^2(-\infty, \infty)$.

REMARK: If $r_{12}^2 < r_{11}r_{22}$, there are two independent solutions in $\mathfrak{L}^2(-\infty, \infty)$.

8. For a boundary-value problem on $0 \leqq t < \infty$ which is in the limit-point case at ∞, the spectrum depends on α in the boundary condition at $t = 0$ and will be denoted by $S(\alpha)$. Let the set of cluster points of $S(\alpha)$ be denoted by $S'(\alpha)$. Prove that $S'(\alpha)$ does not depend on α.

HINT: Let $S^*(\alpha)$ denote the complement of $S'(\alpha)$ on $-\infty < \lambda < \infty$. On $S^*(\alpha)$, ρ is constant except for isolated jumps so that, from (3.10) and (3.9), m_∞, where $m_\infty(l) = m_\infty(l,\alpha)$, is a meromorphic function of l and real on $S^*(\alpha)$. For $\alpha = \alpha_1$ then, $m_{\infty 1}$, where $m_{\infty 1}(l) = m_\infty(l,\alpha_1)$, is meromorphic on $S^*(\alpha_1)$. If $\gamma = \alpha_1 - \alpha$,

$$m_\infty(l,\alpha) = \frac{1 + \cot\gamma\, m_\infty(l,\alpha_1)}{\cot\gamma - m_\infty(l,\alpha_1)}$$

Thus m_∞ is meromorphic and real on $S^*(\alpha_1)$. This proves $S^*(\alpha) \supseteqq S^*(\alpha_1)$. Since the roles of α and α_1 are interchangeable, the theorem is proved.

9. S^*, the complement of S', defined above, is open and thus consists of the union of intervals. From the formula for $m_\infty(l,\alpha)$ above, it is clear that each point λ of an interval is a pole of m_∞ for some choice of α, namely, $\cot(\alpha_1 - \alpha) = m_\infty(\lambda,\alpha_1)$. Thus each point of an interval is in the point spectrum $P(\alpha)$ of the boundary-value problem for some α. On each interval, then, α may be regarded as a function of λ with value $\alpha(\lambda)$ being such that $\lambda \in P(\alpha)$. Prove that this function of λ is regular on an interval of S^* and is monotone increasing.

HINT: From (3.10), $(dm_\infty/d\lambda)(\lambda,\alpha_1) > 0$. From $\cot(\alpha_1 - \alpha) = m_\infty(\lambda,\alpha_1)$ follows

$$\frac{d\alpha}{d\lambda}(\lambda) = \frac{1}{1 + m_\infty^2(\lambda,\alpha_1)}\frac{dm_\infty}{d\lambda}(\lambda,\alpha_1)$$

10. Let $Lx = -x'' + q(t)x$ on $0 \leqq t < \infty$, where q is real, continuous, and periodic of period 1. Prove that S' is the union of the closure of the intervals on $-\infty < \lambda < \infty$ on which $x'' + (\lambda - q)x = 0$ is stable; that is, prove, in the notation of Sec. 3, Chap. 8, that S' consists of those points for which $f^2(\lambda) \leqq 4$.

HINT: L is in the limit-point case [by Theorem 2.4 or from the fact that on a stable interval no solution is $\mathfrak{L}^2(0, \infty)$]. Thus for $\mathfrak{I}l \neq 0$ one characteristic multiplier $\gamma(l)$ satisfies $|\gamma(l)| < 1$. Multipliers are roots of $\sigma^2 - \sigma f(l) + 1 = 0$, where

$$f(l) = \varphi'(1,l) + \psi(1,l)$$

in the notation of Chap. 8, and $\gamma(l)$ is one of the roots $(f \pm \sqrt{f^2 - 4})/2$. For the problem with $x(0) = 0$, $\varphi + m_\infty\psi \in \mathfrak{L}^2(0, \infty)$ and is therefore a solution with multiplier

γ. Thus $\varphi(1,l) + m_\infty(l)\psi(1,l) = \gamma$ and therefore $m_\infty(l) = (\gamma - \varphi(1,l))/\psi(1,l)$. On the interior of stable intervals $f^2 < 4$, and $\gamma(l)$ tends to a nonreal number of magnitude 1 as $\Im l \to 0$. Thus $\Im m_\infty \neq 0$ on stable intervals. On unstable intervals $\Im m_\infty \to 0$ except at points where the entire function $\psi(1,l) = 0$.

11. Let L be as above but now take the interval $-\infty < t < \infty$. Show that here the spectrum S is identical with S' of Prob. 10 and S is continuous.

HINT: $M_{11} = \psi(1,l)/(\gamma^{-1} - \gamma)$ and $M_{22} = (1 - \varphi f + \varphi^2)/((\gamma^{-1} - \gamma)\psi)$. Since no solution of $Lx = \lambda x$ is $\mathfrak{L}^2(-\infty, \infty)$, it follows from Prob. 7 that S is continuous.

12. Let $f \in \mathfrak{L}^2(0, \infty)$ and let L be in the limit-point case on $0 \leqq t < \infty$ as in Theorem 3.1. If l is not in the spectrum (in particular, if $\Im l \neq 0$), the problem $Lx = lx + f$, $\sin\alpha\, x(0) - \cos\alpha\, p(0)x'(0) = 0$, has a unique solution h in $\mathfrak{L}^2(0, \infty)$. If

$$f(t) = \int_0^\infty g(\lambda)\psi(t,\lambda)\, d\rho(\lambda)$$

then

$$h(t) = \int_{-\infty}^\infty \frac{g(\lambda)\psi(t,\lambda)\, d\rho(\lambda)}{\lambda - l}$$

where the integral converges in $\mathfrak{L}^2(0, \infty)$. (Thus $L - l$ has an inverse if l is not in the spectrum.)

HINT: Let $h_a = \int_{-a}^a g\psi\, d\rho/(\lambda - l)$. By Lemma 3.1 (or by Theorem 3.2), h_a converges in $\mathfrak{L}^2(0, \infty)$ to a limit h as $a \to \infty$. Clearly h_a satisfies the boundary condition at $t = 0$. Also $Lh_a = lh_a + f_a$, where $f_a = \int_{-a}^a g\psi\, d\rho$ and from this

$$h_a = \int_0^t [\varphi(t,l)\psi(\tau,l) - \psi(t,l)\varphi(\tau,l)]f_a(\tau)\, d\tau + c(a)\psi$$

where $c(a)$ is a constant. Thus, as $a \to \infty$,

$$h = \int_0^t [\varphi(t,l)\psi(\tau,l) - \psi(t,l)\varphi(\tau,l)]f(\tau)\, d\tau + c(\infty)\psi$$

where $c(\infty)$ must exist because $h_a \to h$, $f_a \to f$. Thus $Lh = lh + f$, and h satisfies the condition at $t = 0$. It remains to prove h is unique. For $\Im l \neq 0$ this follows from the fact that ψ is not $\mathfrak{L}^2(0, \infty)$ for $\Im l \neq 0$. For $\Im l = 0$, let $l = \bar{\lambda}$. Clearly $\bar{\psi}$, where $\bar{\psi}(t) = \psi(t,\bar{\lambda})$, is an eigenfunction corresponding to some ρ_b for some condition β at b since $\bar{\psi}$ is real. If $\bar{\psi} \in \mathfrak{L}^2(0, \infty)$, the jump of ρ_b at $\bar{\lambda}$ does not tend to zero as $b \to \infty$, and $\bar{\lambda}$ is in the spectrum of ρ. Thus, if $\bar{\lambda}$ is not in the spectrum, $\bar{\psi}$ is not $\mathfrak{L}^2(0, \infty)$ and so h is unique.

13. Let $\mathfrak{D}$ denote the class of functions satisfying (i), (ii), and (iii) below (4.9). The boundary-value problem $Lx = lx$, $\sin\alpha\, x(0) - \cos\alpha\, p(0)x'(0) = 0$, is said to be *self-adjoint* if for any $u,v \in \mathfrak{D}$

$$(Lu,v) = \int_0^\infty (Lu)\bar{v}\, dt = \int_0^\infty u(\overline{Lv})\, dt = (u,Lv)$$

Show that if L is in the limit-point case the problem is self-adjoint.

HINT: Let $\Im l \neq 0$, $Lu - lu = f$, and $Lv - \bar{l}v = p$. Then $f,p \in \mathfrak{L}^2(0, \infty)$. Let $g = \int_0^\infty f\psi\, dt$ and q be similarly related to p. Let $u_a = \int_{-a}^a \psi g\, d\rho/(\lambda - l)$. Then by Prob. 12, $u_a \to u$ in $\mathfrak{L}^2(0, \infty)$. Let v_a be similarly defined.

$$\int_0^\mu f\, dt \int_{-a}^a \frac{\bar{q}\psi\, d\rho}{\lambda - l} = \int_{-a}^a \frac{\bar{q}\, d\rho}{\lambda - l} \int_0^\mu f\psi\, dt$$

Letting $\mu \to \infty$

$$(f,v_a) = \int_0^\infty f\bar{v}_a\,dt = \int_{-a}^{a} \frac{g\bar{q}}{\lambda - l}\,d\rho$$

But (u_a,p) is also given by the integral on the right. Thus $(f,v_a) = (u_a,p)$. Letting $a \to \infty$, $(Lu,v) = (u,Lv)$.

14. It is shown in Theorem 3.1 that ρ is unique in the sense that all $\rho_b \to \rho$ as $b \to \infty$. Prove the following stronger result: If (3.6) and (3.8) hold with ρ replaced by some $\tilde{\rho}$, then $\tilde{\rho}$ must be ρ given by (3.9).

HINT: Let $f \in \mathfrak{L}^2(0,\infty)$. Then the unique solution h of Prob. 12 can be represented two ways, so that for $\mathfrak{I}l > 0$

$$\int_{-a}^{a} \frac{g(\lambda)\psi(t,\lambda)}{\lambda - l}\,d(\rho - \tilde{\rho})(\lambda) = F_a(t,l)$$

converges in $\mathfrak{L}^2(0,\infty)$ to 0 as $a \to \infty$. Let $\Gamma(\lambda,s) = \int_0^s \psi(t,\lambda)\,dt$. Clearly $\Gamma \in \mathfrak{L}^2(\rho)$ and $\mathfrak{L}^2(\tilde{\rho})$, being of the form (3.11). Taking $\int_0^s F_a\,dt$ and letting $a \to \infty$

$$\int_{-\infty}^{\infty} \frac{g(\lambda)\Gamma(\lambda,s)}{\lambda - l}\,d(\rho - \tilde{\rho})(\lambda) = 0$$

The same holds with l replaced by $\bar{l}$. If $l = \mu + i\nu$, then

$$(\lambda - l)^{-1} - (\lambda - \bar{l})^{-1} = \frac{2i\nu}{|\lambda - l|^2}$$

Thus

$$\nu \int_{-\infty}^{\infty} \frac{g(\lambda)\Gamma(\lambda,s)}{(\lambda - \mu)^2 + \nu^2}\,d(\rho - \tilde{\rho})(\lambda) = 0$$

Integrating with respect to μ from μ_1 to μ_2 and letting $\nu \to 0$,

$$\int_{\mu_1}^{\mu_2} g(\lambda)\Gamma(\lambda,s)\,d(\rho - \tilde{\rho})(\lambda) = 0$$

for all μ_1 and μ_2. Differentiate the above with respect to s. Then

$$\int_{\mu_1}^{\mu_2} g(\lambda)\psi(s,\lambda)\,d(\rho - \tilde{\rho})(\lambda) = 0$$

Let $f(t) = 1$ for $0 < t < \tau$ and $f(t) = 0$ for $t > \tau$. Then $g(\lambda)$ is $\Gamma(\lambda,\tau)$ and differentiating with respect to τ

$$\int_{\mu_1}^{\mu_2} \psi(\tau,\lambda)\psi(s,\lambda)\,d(\rho - \tilde{\rho})(\lambda) = 0$$

If $\alpha \neq \pi/2$, let $s = \tau = 0$ and use $\psi(0,\lambda) = \cos\alpha$. If $\alpha = \pi/2$, differentiate with respect to τ and s and then let $\tau = s = 0$. In any case,

$$\int_{\mu_1}^{\mu_2} d(\rho - \tilde{\rho})(\lambda) = 0$$

proving the result.

15. In the limit-circle case, Theorem 4.1, show that $\Sigma'\lambda_k^{-2} < \infty$, where the prime denotes omission of a vanishing eigenvalue.

HINT: $\psi_k = (\lambda_k - l_0)\hat{\mathcal{G}}(l_0)\psi_k$. Thus, by the Bessel inequality applied to $\hat{G}(t,\tau,l_0)$ for fixed τ,

$$\int_0^\infty |\hat{G}(t,\tau,l_0)|^2\,dt \geq \sum_{1}^{N} \frac{|r_k\psi(\tau,\lambda_k)|^2}{|\lambda_k - l_0|^2}$$

for all N, where r_k is the normalizing factor. Integrating in τ, the result follows.

16. At zero, $Lx = -x'' + [(r^2 - \frac{1}{4})/t^2]x$, $0 < r < 1$, is in the limit-circle case. If $l = s^2$, solutions of $Lx = lx$ are given by $t^{\frac{1}{2}}J_r(st)$ and $t^{\frac{1}{2}}J_{-r}(st)$, where J_r and J_{-r} are Bessel functions. Let $x'(1) = 0$ and consider a boundary condition at $t = a < 1$. Determine $m_a(l)$ and show that

$$m_0(l) = -\frac{1}{s}\frac{cs^{-r}J_r(s) - s^rJ_{-r}(s)}{cs^{-r}J_r'(s) - s^rJ_{-r}'(s)}$$

for any real choice of the constant c.

17. If L is in the limit-point case at ∞ and $-\infty$, show that the problem $Lx = lx$ is self-adjoint in the sense that

$$\int_{-\infty}^{\infty} (Lu)\bar{v}\,dt = \int_{-\infty}^{\infty} u(\overline{Lv})\,dt$$

for all functions $u,v \in \mathfrak{L}^2(-\infty,\infty)$ which have absolutely continuous first derivatives in every closed subinterval of $(-\infty,\infty)$ and such that $Lu, Lv \in \mathfrak{L}^2(-\infty,\infty)$.

HINT: Modify the method of Prob. 13.

18. State and prove the analogue of Theorem 3.2 for the case $(-\infty,\infty)$, where L is in the limit-point case at both $-\infty$ and ∞.

19. Expand on the remark below (5.16) and formulate and prove precise theorems.

20. Let x be a vector with components x_1 and x_2. Let Lx be the vector with components

$$L_1x = p^{\frac{1}{2}}(p^{\frac{1}{2}}x_2)' + rx_2 + q_1x_1 \qquad \left(' = \frac{d}{dt}\right)$$

$$L_2x = -p^{\frac{1}{2}}(p^{\frac{1}{2}}x_1)' + rx_1 + q_2x_2$$

where p and p' are continuous, $p > 0$, and r, q_1, and q_2 real and continuous for $0 \leqq t < \infty$. Let u and v be vectors of class C^1 on $0 \leqq t < \infty$. Let $u \cdot v$ denote $u_1\bar{v}_1 + u_2\bar{v}_2$. Show that

$$\int_{t_1}^{t_2} (Lu \cdot v - u \cdot Lv)\,dt = [uv](t_2) - [uv](t_1)$$

where $[uv](t) = p(t)(u_2(t)\bar{v}_1(t) - u_1(t)\bar{v}_2(t))$.

Let φ and ψ be solutions of $Lx = lx$ which satisfy

$$p^{\frac{1}{2}}(0)\psi_1(0,l) = \sin\alpha \qquad p^{\frac{1}{2}}(0)\psi_2(0,l) = \cos\alpha$$
$$p^{\frac{1}{2}}(0)\varphi_1(0,l) = -\cos\alpha \qquad p^{\frac{1}{2}}(0)\varphi_2(0,l) = \sin\alpha$$

(*a*) Develop the theory of Sec. 2 for $Lx = lx$, $x_1(0)\cos\alpha - x_2(0)\sin\alpha = 0$.

(*b*) Develop the theory of Sec. 3 for the above problem in the limit-point case (define the latter notion). In particular, show that if the vector f is $\mathfrak{L}^2(0,\infty)$, that is, $\int_0^\infty f \cdot f\,dt < \infty$, there exists a ρ such that

$$g(\lambda) = \int_0^\infty f(t) \cdot \psi(t,\lambda)\,dt$$

exists and

$$f(t) = \int_{-\infty}^{\infty} \psi(t,\lambda)g(\lambda)\,d\rho(\lambda)$$

(*c*) Develop the theory of Sec. 4 for this system.

(*d*) Formulate the results of Sec. 5 for this problem.

CHAPTER 10

SINGULAR SELF-ADJOINT BOUNDARY-VALUE PROBLEMS FOR nTH-ORDER EQUATIONS

1. Introduction

In this chapter the theory of Chap. 9 will be extended to the nth-order case. Here the formal differential operator L is defined by

$$Lx = p_0 x^{(n)} + p_1 x^{(n-1)} + \cdots + p_n x$$

It is assumed that the p_k are complex-valued functions with $n - k$ continuous derivatives on an open interval $a < t < b$, where the cases $a = -\infty$, $b = \infty$, or both are allowed. Further, $p_0(t) \neq 0$ on $a < t < b$, and L coincides with its Lagrange adjoint L^+ given by

$$L^+x = (-1)^n(\bar{p}_0 x)^{(n)} + (-1)^{n-1}(\bar{p}_1 x)^{(n-1)} + \cdots + \bar{p}_n x$$

Notice that n may be odd and that for $n = 2$ complex-valued coefficients may now occur.

First, a general expansion result and Parseval equality will be proved. and then the inverse transform theorem will be demonstrated for two important cases. The existence of Green's functions and the relationship of these with the spectral matrices will then be considered.

The simplest example of the present theory occurs with $Lx = ix'$ on the interval $-\infty < t < \infty$. Here $ix' = lx$ has e^{-ilt} as a solution and the expansion theorem for this case is just the Plancherel theorem.

As before, the method used will depend on setting up self-adjoint boundary-value problems on closed intervals δ: $\tilde{a} \leqq t \leqq \tilde{b}$, where $a < \tilde{a} < \tilde{b} < b$. Thus on δ it is assumed that n linearly independent boundary conditions

$$U_{\delta j}x = \sum_{k=1}^{n} (M_{\delta jk}x^{(k-1)}(\tilde{a}) + N_{\delta jk}x^{(k-1)}(\tilde{b})) = 0 \qquad (j = 1, \ldots, n) \quad (1.1)$$

which are self-adjoint are given. (That such conditions always exist when $L = L^+$ is shown in Prob. 15, Chap. 7.) The conditions (1.1), determined by the matrices $M_\delta = (M_{\delta jk})$ and $N_\delta = (N_{\delta jk})$, are abbreviated by $U_\delta x = 0$. The boundary-value problem on δ

$$Lx = lx \qquad U_\delta x = 0 \tag{1.2}$$

is self-adjoint, and therefore there exists a complete orthonormal set of eigenfunctions $\{\chi_{\delta k}\}$ and corresponding eigenvalues $\{\lambda_{\delta k}\}$.

For functions u,v in the space $\mathfrak{L}^2(\delta)$ the inner product and norm are defined by

$$(u,v)_\delta = \int_\delta u\bar{v}\,dt \qquad \|u\|_\delta = (u,u)^{\frac{1}{2}}$$

For functions in $\mathfrak{L}^2(a,b)$ these will be denoted by (u,v) and $\|u\|$, that is,

$$(u,v) = \int_a^b u\bar{v}\,dt \qquad \|u\| = (u,u)^{\frac{1}{2}}$$

2. The Expansion Theorem and Parseval Equality

The Parseval equality for the self-adjoint problem (1.2) is

$$\|u\|_\delta^2 = \sum_{k=1}^{\infty} |(u,\chi_{\delta k})_\delta|^2 \tag{2.1}$$

where $u \in \mathfrak{L}^2(\delta)$. Let $\varphi_j = \varphi_j(t,l)$, $(j = 1, \ldots, n)$, be solutions of $Lx = lx$ which for some fixed c, $a < c < b$, satisfy

$$\varphi_j^{(k-1)}(c,l) = \delta_{jk} \qquad (j, k = 1, \ldots, n) \tag{2.2}$$

where δ_{jk} is the Kronecker delta. The $\varphi_j^{(k-1)}$ are entire functions of l for fixed t. Since the φ_j are independent solutions

$$\chi_{\delta k}(t) = \sum_{j=1}^{n} r_{\delta kj}\varphi_j(t,\lambda_{\delta k}) \tag{2.3}$$

where the $r_{\delta kj}$ are complex constants. Using (2.3) in (2.1), the Parseval equation can be written as

$$\|u\|_\delta^2 = \int_{-\infty}^{\infty} \sum_{j,k=1}^{n} \bar{g}_{\delta j}(\lambda) g_{\delta k}(\lambda)\, d\rho_{\delta jk}(\lambda) \tag{2.4}$$

where

$$g_{\delta j}(\lambda) = \int_\delta \bar{\varphi}_j(t,\lambda)u(t)\,dt \tag{2.5}$$

and the matrix $\rho_\delta = (\rho_{\delta jk})$ consists of step functions with discontinuities at the eigenvalues. The jumps at the eigenvalues are given by

$$\rho_{\delta jk}(\lambda_p + 0) - \rho_{\delta jk}(\lambda_p - 0) = \sum r_{\delta mj}\bar{r}_{\delta mk} \tag{2.6}$$

where the sum is taken over all m such that $\lambda_{\delta m} = \lambda_{\delta p}$ (note that several

$\chi_{\delta m}$ can correspond to one eigenvalue $\lambda_{\delta p}$). Clearly the matrix has the properties:

(i) ρ_δ is Hermitian
(ii) $\rho_\delta(\Delta) = \rho_\delta(\lambda) - \rho_\delta(\mu)$ is positive semidefinite if $\lambda > \mu$, $\Delta = (\mu,\lambda]$
(iii) The total variation of $\rho_{\delta jk}$ is finite on every finite λ interval.

It is further assumed that $\rho_\delta(0)$ is the zero matrix and $\rho_\delta(\lambda + 0) = \rho_\delta(\lambda)$. Because of (ii), ρ_δ is said to be nondecreasing. The matrix ρ_δ is called the *spectral matrix*. Note that it depends not only on the problem (1.2) but also on the choice of the independent solutions of $Lx = lx$.

In this section it will be shown that the relation (2.4) is valid with δ replaced everywhere by (a,b). The following theorem gives the existence of at least one limiting matrix ρ as $\delta \to (a,b)$.

Theorem 2.1. *Let $\{\delta\}$ be a set of intervals tending to (a,b) and $\{U_\delta x = 0\}$ a corresponding set of self-adjoint boundary conditions. Then $\{\delta\}$ contains a sequence $\{\delta_j\}$ tending to (a,b) as $j \to \infty$ such that*

$$\rho(\lambda) = \lim_{j\to\infty} \rho_{\delta_j}(\lambda) \tag{2.7}$$

exists on $-\infty < \lambda < \infty$. Moreover, the limit matrix ρ satisfies (i), (ii), *and* (iii) *above.*

Proof. The proof is an immediate consequence of the Helly selection theorem and the following fact: *given any $\mu > 0$ there exists an $M(\mu) < \infty$, not depending on δ or U_δ, such that for $|\lambda| \leqq \mu$*

$$|\rho_{\delta jk}(\lambda)| \leqq M(\mu) \tag{2.8}$$

To prove (2.8) it is sufficient to take $j = k$, since from (2.6)

$$2\int_{-\mu}^{\mu} |d\rho_{\delta jk}(\lambda)| \leqq \int_{-\mu}^{\mu} d\rho_{\delta jj}(\lambda) + \int_{-\mu}^{\mu} d\rho_{\delta kk}(\lambda)$$

The functions $\varphi_j^{(k-1)}$ are continuous in (t,λ) and at $t = c$ are equal to δ_{jk}. Thus, given μ, there is an $h > 0$ such that

$$|\varphi_j^{(k-1)}(t,\lambda) - \delta_{jk}| < \frac{1}{6n^2} \tag{2.9}$$

for $c \leqq t \leqq c + h$ and $|\lambda| \leqq \mu$. Let $\tilde{f}$ be a nonnegative function of class C^n on (a,b) vanishing along with its first $n - 1$ derivatives, outside of $(c, c + h)$ and normalized so that

$$\int_c^{c+h} \tilde{f}(t)\, dt = 1 \tag{2.10}$$

The Bessel inequality applied to $(-1)^{m-1}\tilde{f}^{(m-1)}$ for some fixed m, $(m = 1,$

. . . , n), gives

$$\int_c^{c+h} |\tilde{f}^{(m-1)}|^2 \, dt \geqq \int_{-\mu}^{\mu} \sum_{j,k=1}^{n} \bar{g}_j(\lambda) g_k(\lambda) \, d\rho_{\delta jk}(\lambda) \tag{2.11}$$

where here

$$g_k(\lambda) = (-1)^{m-1} \int_c^{c+h} \bar{\varphi}_k(t,\lambda) \tilde{f}^{(m-1)}(t) \, dt$$
$$= \int_c^{c+h} \bar{\varphi}_k^{(m-1)}(t,\lambda) \tilde{f}(t) \, dt$$

By (2.9) and (2.10),

$$|g_k(\lambda) - \delta_{km}| < \frac{1}{6n^2}$$

which, with (2.11), yields

$$\int_c^{c+h} |\tilde{f}^{(m-1)}|^2 \, dt \geqq \frac{1}{2} \int_{-\mu}^{\mu} d\rho_{\delta mm}(\lambda) - \frac{1}{4n} \int_{-\mu}^{\mu} \sum_{j=1}^{n} d\rho_{\delta jj}(\lambda)$$

Summing the above from $m = 1$ to $m = n$, the result follows, and hence Theorem 2.1.

For any such limiting ρ, the space $\mathfrak{L}^2(\rho)$ is defined as the set of all vector functions $g = (g_j)$, $j = 1, \ldots, n$, which are measurable with respect to ρ and such that

$$\|g\|^2 = \int_{-\infty}^{\infty} \sum_{j,k=1}^{n} \bar{g}_j(\lambda) g_k(\lambda) \, d\rho_{jk}(\lambda) < \infty$$

Theorem 2.2. *Let ρ be any limit matrix given by Theorem* 2.1. *If $f \in \mathfrak{L}^2(a,b)$ there exists a vector $g \in \mathfrak{L}^2(\rho)$ such that if*

$$g_{\delta j}(\lambda) = \int_\delta \bar{\varphi}_j(t,\lambda) f(t) \, dt \qquad (\delta \subset (a,b)) \tag{2.12}$$

then

$$\|g - g_\delta\| \to 0 \qquad (\delta \to (a,b)) \tag{2.13}$$

In terms of this g, the Parseval equality

$$\|f\| = \|g\| \tag{2.14}$$

and expansion

$$f(t) = \int_{-\infty}^{\infty} \sum_{j,k=1}^{n} \varphi_j(t,\lambda) g_k(\lambda) \, d\rho_{jk}(\lambda) \tag{2.15}$$

are valid, where the latter integral converges to f in the norm of $\mathfrak{L}^2(a,b)$.

Proof. The proof follows the same development as that of Theorem 3.1, Chap. 9, Eqs. (3.16) to (3.24).

As in Chap. 9, by

$$\int_a^b \bar{\varphi}_j(t,\lambda)f(t)\,dt \qquad (j = 1, \ldots, n)$$

will be meant the jth component of the vector g whose existence is proved in Theorem 2.2.

The analogues of the remaining parts of Theorems 3.1 and 3.2 in Chap. 9 will be proved in Secs. 3 and 4 below.

3. The Inverse-transform Theorem and the Uniqueness of the Spectral Matrix

The inverse-transform theorem states that every vector $g \in \mathfrak{L}^2(\rho)$ arises from a function $f \in \mathfrak{L}^2(a,b)$, as in Theorem 2.2. This is true under certain additional assumptions on L, which correspond in the case $n = 2$ of Chap. 9 to the situation where L is in the limit-point case at b, as in Sec. 3, or in the limit-point case at both a and b, as in Sec. 5. In principle, the proof is like that of Theorem 3.2, Chap. 9. As in that proof, the following lemma is required.

Lemma 3.1. *Let $g \in \mathfrak{L}^2(\rho)$ and*

$$f_\Delta(t) = \int_\Delta \sum_{j,k=1}^{n} \varphi_j(t,\lambda)g_k(\lambda)\,d\rho_{jk}(\lambda)$$

where Δ is a finite λ interval. Then, as $\Delta \to (-\infty,\infty)$, f_Δ converges in $\mathfrak{L}^2(a,b)$ to a function $f \in \mathfrak{L}^2(a,b)$.

Proof. The proof is the same as for Lemma 3.1, Chap. 9.

Another fact used in the proof is the explicit representation of the inverse of the operator $L - l$ for $\Im l \neq 0$.

Lemma 3.2. *Suppose for some l, $\Im l \neq 0$, that $Lx = lx$ has no nontrivial solution in $\mathfrak{L}^2(a,b)$. If $f \in \mathfrak{L}^2(a,b)$ and g is any vector in $\mathfrak{L}^2(\rho)$ such that*

$$f(t) = \int_{-\infty}^{\infty} \sum_{j,k=1}^{n} \varphi_j(t,\lambda)g_k(\lambda)\,d\rho_{jk}(\lambda)$$

where the integral converges in $\mathfrak{L}^2(a,b)$ to f, then the equation $(L - l)x = f$ has a unique solution F in $\mathfrak{L}^2(a,b)$ given by

$$F(t) = \int_{-\infty}^{\infty} \sum_{j,k=1}^{n} (\lambda - l)^{-1}\varphi_j(t,\lambda)g_k(\lambda)\,d\rho_{jk}(\lambda)$$

where the latter integral converges in $\mathfrak{L}^2(a,b)$ to F.

Proof. If for any finite λ interval Δ,

$$f_\Delta(t) = \int_\Delta \sum_{j,k=1}^{n} \varphi_j(t,\lambda)g_k(\lambda)\, d\rho_{jk}(\lambda)$$

and

$$F_\Delta(t) = \int_\Delta \sum_{j,k=1}^{n} (\lambda - l)^{-1}\varphi_j(t,\lambda)g_k(\lambda)\, d\rho_{jk}(\lambda)$$

then $(L - l)F_\Delta = f_\Delta$. By Lemma 3.1 there exists an $F \in \mathfrak{L}^2(a,b)$ such that $\|F_\Delta - F\| \to 0$ as $\Delta \to (-\infty, \infty)$. Also, by hypothesis, $\|f_\Delta - f\| \to 0$ as $\Delta \to (-\infty, \infty)$. From the variation-of-constants formula there exist continuous functions θ_j such that

$$F_\Delta(t) = \sum_{j=1}^{n} \varphi_j(t,l) \int_c^t \theta_j(\tau) f_\Delta(\tau)\, d\tau + \sum_{j=1}^{n} c_j(\Delta)\varphi_j(t,l)$$

where the $c_j(\Delta)$ are constants, and $a < c < b$. Because F_Δ and f_Δ converge in $\mathfrak{L}^2(a,b)$ and because the φ_j are independent, it follows that

$$c_j(\infty) = \lim_{\Delta \to (-\infty,\infty)} c_j(\Delta)$$

exists (where the limit is taken through a suitable sequence Δ_k) and

$$F(t) = \sum_{j=1}^{n} \varphi_j(t,l) \int_c^t \theta_j(\tau) f(\tau)\, d\tau + \sum_{j=1}^{n} c_j(\infty)\varphi_j(t,l)$$

This is the variation-of-constants formula and shows that F is a solution of $(L - l)x = f$. Since $F \in \mathfrak{L}^2(a,b)$ and since $Lx = lx$ has no nontrivial solution in $\mathfrak{L}^2(a,b)$, F is unique.

Theorem 3.1. *Suppose neither equation $(L \pm i)x = 0$ has a nontrivial solution in $\mathfrak{L}^2(a,b)$. If $g \in \mathfrak{L}^2(\rho)$, there exists an $f \in \mathfrak{L}^2(a,b)$ such that*

$$f(t) = \int_{-\infty}^{\infty} \sum_{j,k=1}^{n} \varphi_j(t,\lambda)g_k(\lambda)\, d\rho_{jk}(\lambda)$$

where the integral converges to f in $\mathfrak{L}^2(a,b)$, and

$$g_k(\lambda) = \int_a^b \bar{\varphi}_k(t,\lambda) f(t)\, dt \tag{3.1}$$

REMARKS: If $n = 2$ and the p_k are real, the condition on the solutions of $(L \pm i)x = 0$ is equivalent to the statement that L is in the limit-point case at a and b. It can be shown that if neither $(L \pm i)x = 0$ has a nontrivial solution in $\mathfrak{L}^2(a,b)$, then $(L - l)x = 0$ has no nontrivial solution in $\mathfrak{L}^2(a,b)$ for any l, $\mathfrak{I}l \neq 0$; see Sec. 4.

The interpretation to be put on (3.1) is the usual one, namely, if

$$g_{\delta k}(\lambda) = \int_\delta \bar{\varphi}_k(t,\lambda)f(t)\,dt \qquad (\delta \subset (a,b))$$

then $\|g - g_\delta\| \to 0$ as $\delta \to (a,b)$.

Proof of Theorem 3.1. The existence of f is provided by Lemma 3.1. Using this f, there exists by Theorem 2.2 a $\tilde{g} \in \mathfrak{L}^2(\rho)$ such that

$$\tilde{g}_k(\lambda) = \int_a^b \bar{\varphi}_k(t,\lambda)f(t)\,dt$$

and if

$$\tilde{f}_\Delta(t) = \int_\Delta \sum_{j,k=1}^{n} \varphi_j(t,\lambda)\tilde{g}_k(\lambda)\,d\rho_{jk}(\lambda) \tag{3.2}$$

then $\|f - \tilde{f}_\Delta\| \to 0$ as $\Delta \to (-\infty,\infty)$. The problem is to show

$$\|g - \tilde{g}\| = 0$$

If f_Δ is defined as in (3.2) with $\tilde{g}$ replaced by g, then $\|f - f_\Delta\| \to 0$ as $\Delta \to (-\infty,\infty)$. Using Lemma 3.2, it follows that the unique solution F of $(L - i)x = f$ which is in $\mathfrak{L}^2(a,b)$ is given by

$$\begin{aligned} F(t) &= \int_{-\infty}^{\infty} \sum_{j,k=1}^{n} (\lambda - i)^{-1}\varphi_j(t,\lambda)g_k(\lambda)\,d\rho_{jk}(\lambda) \\ &= \int_{-\infty}^{\infty} \sum_{j,k=1}^{n} (\lambda - i)^{-1}\varphi_j(t,\lambda)\tilde{g}_k(\lambda)\,d\rho_{jk}(\lambda) \end{aligned}$$

Similarly, the unique solution of $(L - i)x = F$ which is in $\mathfrak{L}^2(a,b)$ is

$$\begin{aligned} F_1(t) &= \int_{-\infty}^{\infty} \sum_{j,k=1}^{n} (\lambda - i)^{-2}\varphi_j(t,\lambda)g_k(\lambda)\,d\rho_{jk}(\lambda) \\ &= \int_{-\infty}^{\infty} \sum_{j,k=1}^{n} (\lambda - i)^{-2}\varphi_j(t,\lambda)\tilde{g}_k(\lambda)\,d\rho_{jk}(\lambda) \end{aligned}$$

This may be repeated to obtain for $p = 1, 2, \ldots,$

$$\int_{-\infty}^{\infty} \sum_{j,k=1}^{n} (\lambda - i)^{-p}\varphi_j(t,\lambda)r_k(\lambda)\,d\rho_{jk}(\lambda) = 0 \tag{3.3}$$

where $r = g - \tilde{g}$.

Let Γ be the transform of the function which is one on $(c, c + s)$ and zero elsewhere, that is,

$$\Gamma_j(\lambda,s) = \int_c^{c+s} \bar{\varphi}_j(t,\lambda)\,dt$$

Integrating (3.3) with respect to t from c to $c + s$, one obtains

$$\int_{-\infty}^{\infty} \sum_{j,k=1}^{n} (\lambda - i)^{-p}\bar{\Gamma}_j(\lambda,s)r_k(\lambda)\, d\rho_{jk}(\lambda) = 0 \tag{3.4}$$

Since Γ and r are in $\mathfrak{L}^2(\rho)$, the function H of l defined by

$$H(l) = \int_{-\infty}^{\infty} \sum_{j,k=1}^{n} (\lambda - l)^{-1}\bar{\Gamma}_j(\lambda,s)r_k(\lambda)\, d\rho_{jk}(\lambda)$$

is analytic for $\Im l > 0$. By (3.4) H and all its derivatives vanish for $l = i$. Thus $H(l) = 0$ for $\Im l > 0$. A similar result holds for the lower half plane, that is, with l replaced by $\bar{l}$. If $l = \mu + i\nu$, then

$$\frac{1}{\lambda - l} - \frac{1}{\lambda - \bar{l}} = \frac{2i\nu}{(\lambda - \mu)^2 + \nu^2}$$

Thus, using $H(l) - H(\bar{l}) = 0$,

$$\nu \int_{-\infty}^{\infty} \sum_{j,k=1}^{n} \frac{\bar{\Gamma}_j(\lambda,s)r_k(\lambda)\, d\rho_{jk}(\lambda)}{(\lambda - \mu)^2 + \nu^2} = 0 \tag{3.5}$$

If λ_1 and λ_2 are points of continuity of ρ and (3.5) integrated with respect to μ from λ_1 to λ_2, holding ν fixed, and then letting $\nu \to 0$, it follows that

$$\int_{\lambda_1}^{\lambda_2} \sum_{j,k=1}^{n} \bar{\Gamma}_j(\lambda,s)r_k(\lambda)\, d\rho_{jk}(\lambda) = 0$$

If this is differentiated with respect to s,

$$\int_{\lambda_1}^{\lambda_2} \sum_{j,k=1}^{n} \varphi_j(c + s, \lambda)r_k(\lambda)\, d\rho_{jk}(\lambda) = 0$$

and if derivatives are taken with respect to s and evaluated at $s = 0$, there follows by (2.2)

$$\int_{\lambda_1}^{\lambda_2} \sum_{k=1}^{n} r_k(\lambda)\, d\rho_{jk}(\lambda) = 0 \qquad (j = 1, \ldots, n) \tag{3.6}$$

Because of the arbitrary character of λ_1 and λ_2, if γ_j are any step functions which vanish for large $|\lambda|$, then

$$\int_{-\infty}^{\infty} \sum_{j,k=1}^{n} \gamma_j(\lambda)r_k(\lambda)\, d\rho_{jk}(\lambda) = 0$$

This implies $\|r\| = 0$, which was to be proved.

Theorem 3.2. *Let neither equation* $(L \pm i)x = 0$ *have a nontrivial solution in* $\mathfrak{L}^2(a,b)$. *Then the matrix* ρ *is unique in the sense that if* $\tilde{\rho}$ *is any other matrix for which Theorem 2.2 is valid then*

$$\tilde{\rho}(\lambda) - \tilde{\rho}(\mu) = \rho(\lambda) - \rho(\mu)$$

at points of continuity λ, μ *of* ρ *and* $\tilde{\rho}$.

REMARK. An immediate consequence of Theorem 3.2 is that for any λ, μ which are points of continuity of ρ

$$\rho_\delta(\lambda) - \rho_\delta(\mu) \to \rho(\lambda) - \rho(\mu)$$

as $\delta \to (a,b)$, irrespective of how U_δ varies with δ. Thus if neither equation $(L \pm i)x = 0$ has a solution $(\not\equiv 0)$ in $\mathfrak{L}^2(a,b)$, ρ is called *the spectral matrix* of the problem $Lx = lx$ on (a,b). The spectrum, point spectrum, and continuous spectrum for this problem are defined in terms of ρ, as in Sec. 5, Chap. 9.

Proof of Theorem 3.2. Let $f(t) = 1$, $c \leqq t \leqq c + \tau$ and $f(t) = 0$ otherwise, and let

$$g_k(\lambda,\tau) = \int_c^{c+\tau} \bar{\varphi}_k(t,\lambda)\, dt$$

Then, much as in the proof of Theorem 3.1, F, the unique solution of $Lx = ix + f$, is given by

$$\begin{aligned} F(t) &= \int_{-\infty}^{\infty} \sum_{j,k=1}^{n} (\lambda - i)^{-1}\varphi_j(t,\lambda) g_k(\lambda,\tau)\, d\rho_{jk}(\lambda) \\ &= \int_{-\infty}^{\infty} \sum_{j,k=1}^{n} (\lambda - i)^{-1}\varphi_j(t,\lambda) g_k(\lambda,\tau)\, d\tilde{\rho}_{jk}(\lambda) \end{aligned}$$

Also, if $\sigma_{jk} = \rho_{jk} - \tilde{\rho}_{jk}$,

$$H(l) = \int_{-\infty}^{\infty} \sum_{j,k=1}^{n} (\lambda - l)^{-1}\bar{\Gamma}_j(\lambda,s) g_k(\lambda,\tau)\, d\sigma_{jk}(\lambda)$$

must vanish for $\mathfrak{J}l > 0$ and this leads to

$$\int_{\lambda_1}^{\lambda_2} \sum_{k=1}^{n} g_k(\lambda,\tau)\, d\sigma_{jk}(\lambda) = 0$$

in place of (3.6). Taking the derivative with respect to τ of the left side above, there results

$$\int_{\lambda_1}^{\lambda_2} \sum_{k=1}^{n} \bar{\varphi}_k(c + \tau,\lambda)\, d\sigma_{jk}(\lambda) = 0$$

Differentiating $m - 1$ times with respect to τ and setting $\tau = 0$, there results

$$\int_{\lambda_1}^{\lambda_2} d\sigma_{jm}(\lambda) = 0$$

which proves the theorem.

The other case where the inverse-transform theorem, and also the uniqueness of ρ, will be proved is one for which the open interval (a,b) is replaced by $[a,b)$, $-\infty < a \leq t < b$, and where the interval $\delta = [\tilde{a},\tilde{b}]$ is now $[a,\tilde{b}]$. It is assumed that $p_k \in C^{n-k}$ on $[a,b)$ and $p_0(t) \neq 0$ on $[a,b)$. The boundary conditions $U_\delta x = 0$ are assumed to contain m linearly independent conditions of the form

$$\sum_{j=1}^{n} M_{ij} x^{(j-1)}(a) = 0 \qquad (i = 1, \ldots, m) \tag{3.7}$$

where the M_{ij} are constants not depending on δ, that is, on $\tilde{b}$. The conditions (3.7) will be denoted by $U^{(1)}x = 0$ and the remaining conditions of $U_\delta x = 0$, which may depend on δ, will be denoted by $U_\delta^{(2)}x = 0$. The case $n = 2m$ will be considered and it will be assumed that $U^{(1)}$ is such that $U_\delta^{(2)}$ can be determined so that $U_\delta x = 0$ are self-adjoint conditions.

In this case it is convenient to take $c = a$. Since the rank of the matrix $M = (M_{ij})$ is m, there exist exactly $n - m = m$ linearly independent solutions $\psi_1, \ldots, \psi_m$ of $Lx = lx$ satisfying (3.7). Let $\psi_{m+1}, \ldots, \psi_n$ be m solutions of $Lx = lx$ such that $\psi_1, \ldots, \psi_n$ form a fundamental set with initial conditions at a independent of l. The eigenfunctions $\{\chi_{\delta k}\}$ of the problem $Lx = lx$, $U_\delta x = 0$ are of the form

$$\chi_{\delta k}(t) = \sum_{j=1}^{m} \hat{r}_{\delta kj} \psi_j(t,\lambda_{\delta k})$$

since only $\psi_1, \ldots, \psi_m$ satisfy the conditions (3.7), and in terms of $\psi_1, \ldots, \psi_m$ the Parseval equality becomes for $u \in \mathfrak{L}^2(\delta)$

$$\|u\|_\delta^2 = \int_{-\infty}^{\infty} \sum_{j,k=1}^{m} \bar{g}_{\delta j}(\lambda) g_{\delta k}(\lambda) \, d\hat{\rho}_{\delta jk}(\lambda)$$

where

$$g_{\delta j}(\lambda) = \int_\delta \bar{\psi}_j(t,\lambda) u(t) \, dt$$

Since

$$\varphi_j = \sum_{k=1}^{n} c_{kj} \psi_k \qquad (j = 1, \ldots, n)$$

for some constant nonsingular matrix $C = (c_{jk})$, it follows that ρ_δ is related to the matrix $\hat{\rho}_\delta$ by $\hat{\rho}_\delta = C\rho_\delta C^*$. Here $\hat{\rho}_\delta = (\hat{\rho}_{\delta jk})$ is such that $\hat{\rho}_{\delta jk} = 0$ for $j,k > m$, and $\hat{\rho}_\delta$ has all the properties that ρ_δ has. If ρ is any limit matrix obtained by letting $\delta \to [a,b)$ through a sequence of intervals, then clearly $\hat{\rho} = C\rho C^*$ is a matrix with elements having row or column index exceeding m equal to zero. In terms of $\hat{\rho}$, the Parseval equality and expansion theorem become for any $f \in \mathfrak{L}^2(a,b)$

$$\|f\|^2 = \int_{-\infty}^{\infty} \sum_{j,k=1}^{m} \bar{g}_j(\lambda) g_k(\lambda)\, d\hat{\rho}_{jk}(\lambda) \tag{3.8}$$

where

$$g_j(\lambda) = \int_a^b \bar{\psi}_j(t,\lambda) f(t)\, dt \tag{3.9}$$

and

$$f(t) = \int_{-\infty}^{\infty} \sum_{j,k=1}^{m} \psi_j(t,\lambda) g_k(\lambda)\, d\hat{\rho}_{jk}(\lambda) \tag{3.10}$$

where the integral on the right of (3.9) converges in the norm of $\mathfrak{L}^2(\hat{\rho})$ and that of (3.10) in $\mathfrak{L}^2(a,b)$.

The analogue of Lemma 3.1 is now valid, that is, if $g \in \mathfrak{L}^2(\hat{\rho})$ there exists an $f \in \mathfrak{L}^2(a,b)$ for which (3.10) is valid. Similarly, the representation of f in terms of $\hat{\rho}$ gives the analogue of Lemma 3.2 as follows:

Lemma 3.3. *Suppose for some l, $\mathfrak{I}l \neq 0$, that the problem*

$$Lx = lx \qquad U^{(1)}x = 0 \tag{3.11}$$

has no nontrivial solution in $\mathfrak{L}^2(a,b)$. If $f \in \mathfrak{L}^2(a,b)$ and g is any vector in $\mathfrak{L}^2(\hat{\rho})$ such that (3.10) *holds, then the problem*

$$(L - l)x = f \qquad U^{(1)}x = 0$$

has a unique solution $F \in \mathfrak{L}^2(a,b)$ given by

$$F(t) = \int_{-\infty}^{\infty} \sum_{j,k=1}^{m} (\lambda - l)^{-1} \psi_j(t,\lambda) g_k(\lambda)\, d\hat{\rho}_{jk}(\lambda)$$

where the latter integral converges in $\mathfrak{L}^2(a,b)$ to F.

Proof. The proof is the same as for Lemma 3.2 if it is observed that, for any finite λ interval Δ, the function F_Δ defined by

$$F_\Delta(t) = \int_\Delta \sum_{j,k=1}^{m} (\lambda - l)^{-1} \psi_j(t,\lambda) g_k(\lambda)\, d\hat{\rho}_{jk}(\lambda)$$

satisfies $(L - l)F_\Delta = f_\Delta$, and $U^{(1)}F_\Delta = 0$ since the ψ_j satisfy the condition

$U^{(1)}x = 0$. Thus for some constants $c_j(\Delta)$ and continuous functions θ_j

$$F_\Delta(t) = \sum_{j=1}^{n} \psi_j(t,l) \int_c^t \theta_j(\tau)f(\tau)\,d\tau + \sum_{j=1}^{m} c_j(\Delta)\psi_j(t,l)$$

The analogues of Theorems 3.1 and 3.2 are now immediate.

Theorem 3.3. *Suppose neither problem*

$$(L \pm i)x = 0 \qquad U^{(1)}x = 0$$

has a nontrivial solution in $\mathfrak{L}^2(a,b)$. *If* $g \in \mathfrak{L}^2(\hat{\rho})$, *there exists an* $f \in \mathfrak{L}^2(a,b)$ *given by* (3.10), *and in terms of this* f *the vector* g *is represented by* (3.9). *Moreover, the matrix* $\hat{\rho}$ *is unique in the sense that if* $\tilde{\rho}$ *is any other matrix for which* (3.8) *through* (3.10) *is valid, then*

$$\tilde{\rho}(\lambda) - \tilde{\rho}(\mu) = \hat{\rho}(\lambda) - \hat{\rho}(\mu)$$

at continuity points λ,μ *of* $\tilde{\rho}$ *and* $\hat{\rho}$.

Thus if λ,μ are continuity points of $\hat{\rho}$, then

$$\hat{\rho}_\delta(\lambda) - \hat{\rho}_\delta(\mu) \to \hat{\rho}(\lambda) - \hat{\rho}(\mu)$$

as $\delta = [a,\bar{b}] \to [a,b)$, irrespective of how $U_\delta^{(2)}$ varies with δ. The matrix $\hat{\rho}$ is called *the spectral matrix* of the problem (3.11) with respect to the set $\{\psi_j\}$.

The two cases, that of Theorem 3.1 and that of Theorem 3.3, will be referred to as Cases I and II, respectively.

The inverse-transform theorem and the uniqueness of the spectral matrix hold for other cases also if boundary conditions are added in the manner shown in Chap. 9 for the limit-circle case, but this will not be considered here.

Let $\mathfrak{D}$ represent the set of all functions $u \in \mathfrak{L}^2(a,b)$ which are of class C^{n-1} on (a,b), $u^{(n-1)}$ is absolutely continuous on every closed subinterval, and $Lu \in \mathfrak{L}^2(a,b)$. If a is finite, let $\hat{\mathfrak{D}}$ represent the set of all $u \in \mathfrak{D}$ such that $u \in C^{n-1}$ on $[a,b)$ and $U^{(1)}u = 0$. In Case I the problem $Lx = lx$ on (a,b), and in Case II the problem $Lx = lx$, $U^{(1)}x = 0$ on $[a,b)$, are self-adjoint, in the sense that in Case I $(Lu,v) = (u,Lv)$ for all $u,v \in \mathfrak{D}$, and in Case II this is valid for all $u,v \in \hat{\mathfrak{D}}$. The proof in each case can be made to follow that of Prob. 13, Chap. 9.

4. Green's Function

In the treatment of Chap. 9 for $n = 2$, the existence of Green's function in the limit-point and limit-circle cases was an immediate by-product of the method. Here a somewhat different approach will be followed. The existence of Green's function G_δ for the self-adjoint problem $Lx = lx$, $U_\delta x = 0$ on δ was proved in Chap. 7. It will be shown that there exists a sequence of intervals $\delta_m \to (a,b)$ such that the corresponding Green's functions G_{δ_m} tend to a function G which is a Green's function in the Cases I and II discussed in Sec. 3. The relation of Green's function to the spectral matrix shown in Sec. 5, Chap. 9, will be shown to hold in general in the section that follows.

Use will be made of the existence of a function $K = K(t,\tau)$ defined for $a < t, \tau < b$ which serves as a kernel in the variation-of-constants formula for the solution of $Lx = f$. This function is such that

$$\int_\delta K(t,\tau)f(\tau)\, d\tau$$

is a solution of $Lx = f$ on any subinterval δ of (a,b), and K has the same differentiability properties as any Green's function associated with L. In particular, $\partial^{n-1}K/\partial t^{n-1}$ has the same discontinuity at $t = \tau$ as the Green's functions. There are many such K, and the existence of one such is given in (2.4) of Chap. 7 for the case $Lx = lx + f$; see also Prob. 22, Chap. 3.

In the following it will be convenient to denote functions such as K, considered as a function of t for fixed τ, as $K(\ ,\tau)$, and similarly if considered as a function of τ alone it will be denoted by $K(t,\ )$.

Lemma 4.1. *The set of functions* $\{G_\delta\}$ *is uniformly bounded and (for* $n > 1$*) equicontinuous on every compact* (t,τ,l) *region where* $\mathfrak{I}l \neq 0$. *(If* $n = 1$, $t = \tau$ *is excepted.)*

Proof. Let the closed interval δ_0 be contained properly in the closed interval δ_1, which in turn is properly contained in (a,b). Let μ be any real-valued function of class C^n on (a,b) such that $\mu(t) = 1$ on some open interval $\bar{\delta}_0$, $\delta_0 \subset \bar{\delta}_0 \subset \delta_1$, and $\mu(t) = 0$ for t outside δ_1. Then define J by

$$J(t,\tau) = \mu(t)K(t,\tau)$$

Let $\delta \supset \delta_1$ and $\tau \in \delta_0$. The function $u = G_\delta(\ ,\tau,l) - J(\ ,\tau)$ is of class C^n on δ and satisfies the boundary conditions $U_\delta u = 0$. Therefore, since $(L - l)u = -(L_t - l)J$, where L_t denotes L applied to J considered as a function of t,

$$G_\delta(t,\tau,l) = J(t,\tau) - \int_{\delta_1} G_\delta(t,s,l)[L_sJ(s,\tau) - lJ(s,\tau)]\, ds \tag{4.1}$$

Applying the Schwarz inequality to (4.1), there results

$$|G_\delta(t,\tau,l)| \leqq |J(t,\tau)| + \|G_\delta(t,\ ,l)\|_\delta \, \|LJ(\ ,\tau) - lJ(\ ,\tau)\|_{\delta_1} \tag{4.2}$$

The uniform boundedness of the G_δ for $t,\tau \in \delta_0$ and l ranging over some compact set Λ with $\mathfrak{I}l \neq 0$ will follow from (4.2) once it has been shown that $\|G_\delta(t,\ ,l)\|_\delta$ is bounded uniformly for $t \in \delta_0$, $l \in \Lambda$.

However, this follows from the fact that if $u = \mathcal{G}_\delta(l)f$, where

$$\mathcal{G}_\delta(l)f(t) = \int_\delta G_\delta(t,\tau,l)f(\tau)\,d\tau$$

and $f \in \mathfrak{L}^2(\delta)$, then

$$\|u\|_\delta \leqq |\Im l|^{-1}\|f\|_\delta \tag{4.3}$$

Indeed, u satisfies $(L - l)u = f$ and $U_\delta u = 0$. Using Green's formula,

$$\int_\delta (Lu)\bar{u}\,dt - \int_\delta u(\overline{Lu})\,dt = 0$$

or

$$2i\Im l\|u\|_\delta^2 = \int_\delta (\bar{f}u - f\bar{u})\,dt$$

and using the Schwarz inequality, this yields (4.3).

Applying (4.3) to $u = G_\delta(\ ,\tau,l) - J(\ ,\tau)$ for $\tau \in \delta_0$ results in

$$\|G_\delta(\ ,\tau,l)\|_\delta \leqq \|J(\ ,\tau)\|_{\delta_1} + |\Im l|^{-1}\|LJ(\ ,\tau) - lJ(\ ,\tau)\|_{\delta_1} \tag{4.4}$$

Thus $\|G_\delta(\ ,\tau,l)\|_\delta$ is uniformly bounded for $\tau \in \delta_0$, $l \in \Lambda$, and $\delta \supset \delta_1$. From the symmetry relation $G_\delta(t,\tau,l) = \bar{G}_\delta(\tau,t,\bar{l})$ it follows that $\|G_\delta(t,\ ,l)\|_\delta$ is also uniformly bounded, and hence by (4.2) so is $\{G_\delta\}$ for $t,\tau \in \delta_0$, $l \in \Lambda$, and $\delta \supset \delta_1$.

From (4.1) follows (for $n > 1$)

$$\frac{\partial G_\delta}{\partial \tau}(t,\tau,l) = \frac{\partial J}{\partial \tau}(t,\tau) - \int_{\delta_1} G_\delta(t,s,l)\frac{\partial}{\partial \tau}[L_sJ(s,\tau) - lJ(s,\tau)]\,ds \tag{4.5}$$

and, using the Schwarz inequality on the integral, the uniform boundedness of the set $\{\partial G_\delta/\partial\tau\}$ for $t,\tau \in \delta_0$ and $l \in \Lambda$ results. {If $n = 1$, the integral in (4.1) is taken as a sum of an integral from a_1 to τ and an integral from τ to b_1, where $\delta_1 = [a_1,b_1]$.} The symmetry of G_δ implies the uniform boundedness of $\{\partial G_\delta/\partial t\}$ also. The uniform boundedness of $\{\partial G_\delta/\partial l\}$ follows from the analyticity of G_δ in l and the uniform boundedness of $\{G_\delta\}$. The uniform boundedness of all first partial derivatives of G_δ implies the equicontinuity of the set $\{G_\delta\}$. (If $n = 1$, the set $\{G_\delta - J\}$ is equicontinuous.) This completes the proof of the lemma.

This lemma, together with the Ascoli lemma, proves that there exists a sequence of intervals $\delta_m \subset (a,b)$ $(m = 2, 3, \ldots)$, where $\delta_m \to (a,b)$ as $m \to \infty$, such that the corresponding Green's functions $G_m = G_{\delta_m}$ tend uniformly on any fixed compact subset Λ_1 of $a < t$, $\tau < b$, $\Im l \neq 0$, to a limit function. A subsequence will tend to a limit function uniformly on a compact subset $\Lambda_2 \supset \Lambda_1$. By taking a sequence $\{\Lambda_i\}$ tending to the set $a < t$, $\tau < b$, $\Im l \neq 0$, and using the diagonal process, there exists a sequence of Green's functions which tend uniformly on any compact subset of $a < t$, $\tau < b$, $\Im l \neq 0$, to a limit function G. This G is defined for $a < t$, $\tau < b$, $\Im l \neq 0$, and being the uniform limit of continuous functions, is continuous. Since the G_m are analytic in l for $\Im l \neq 0$, the same holds for G. The relation $G_m(t,\tau,l) = \bar{G}_m(\tau,t,\bar{l})$ implies that $G(t,\tau,l) = \bar{G}(\tau,t,\bar{l})$.

Theorem 4.1. *Let G be the limit of any convergent sequence $\{G_m\}$ of the set $\{G_\delta\}$ of Green's functions associated with given self-adjoint boundary-value problems*

$$Lx = lx \qquad U_\delta x = 0$$

on closed bounded subintervals δ of (a,b). Then G is continuous for $a < t$, $\tau < b$, $(t \neq \tau$ for $n = 1)$, $\Im l \neq 0$, analytic in l, and possesses the properties:

(i) $\partial^k G/\partial t^k$ $(k = 0, 1, \ldots, n-2)$ *exist, are continuous on $a < t, \tau < b$, and $\partial^{n-1}G/\partial t^{n-1}$, $\partial^n G/\partial t^n$ are continuous on each of the regions $t \leqq \tau$, $\tau \leqq t$.*

(ii) $$\frac{\partial^{n-1}G}{\partial t^{n-1}}(\tau + 0,\tau,l) - \frac{\partial^{n-1}G}{\partial t^{n-1}}(\tau - 0,\tau,l) = \frac{1}{p_0(\tau)} \qquad (a < \tau < b)$$

(iii) *As a function of t, G satisfies $Lx = lx$ if $t \neq \tau$.*

(iv) $$\frac{\partial^{j+k}G_m}{\partial t^j \partial \tau^k} \to \frac{\partial^{j+k}G}{\partial t^j \partial \tau^k} \qquad (j, k = 0, 1, \ldots, n-1)$$

uniformly on any compact (t,τ,l) region, where $\Im l \neq 0$, and $t \neq \tau$ if j or k is $n - 1$.

(v) $G(t,\tau,l) = \bar{G}(\tau,t,\bar{l})$

(vi) $G(t,\ ,l) \in \mathfrak{L}^2(a,b) \qquad (a < t < b)$

(vii) *If $f \in \mathfrak{L}^2(a,b)$ the function v defined by*

$$v(t) = \int_a^b G(t,\tau,l)f(\tau)\,d\tau \qquad (\Im l \neq 0)$$

is an element of $\mathfrak{D}$ and

$$Lv = lv + f$$

Proof. The representation (4.1) gives for $t,\tau \in \delta_0$, $\Im l \neq 0$, and $\delta_0 \subset \delta_1 \subset \delta$,

$$\frac{\partial^j G_\delta}{\partial \tau^j}(t,\tau,l) = \frac{\partial^j K}{\partial \tau^j}(t,\tau) + \int_{\delta_1} G_\delta(t,s,l)\frac{\partial^j}{\partial \tau^j}[lJ(s,\tau) - L_s J(s,\tau)]\,ds \tag{4.6}$$

for $j = 0, 1, \ldots, n-1$. Recall that since $\partial^{n-1}G_\delta/\partial\tau^{n-1}$ and $\partial^{n-1}K/\partial\tau^{n-1}$ have the same discontinuity at $t = \tau$ their difference is continuous there. Moreover, from (4.6), if $t \neq \tau$,

$$\frac{\partial^n G_\delta}{\partial \tau^n}(t,\tau,l) = \frac{\partial^n K}{\partial \tau^n}(t,\tau) + \frac{(-1)^n l G_\delta(t,\tau,l)}{\bar{p}_0(\tau)} + \int_{\delta_1} G_\delta(t,s,l)\frac{\partial^n}{\partial \tau^n}[lJ(s,\tau) - L_s J(s,\tau)]\,ds \tag{4.7}$$

Observing (4.1) with $\delta = \delta_m$, and letting $m \to \infty$, yields

$$G(t,\tau,l) = K(t,\tau) + \int_{\delta_1} G(t,s,l)[lJ(s,\tau) - L_sJ(s,\tau)]\, ds \qquad (4.8)$$

and therefore the partial derivatives $\partial^j G/\partial\tau^j$ exist and

$$\frac{\partial^j G}{\partial\tau^j}(t,\tau,l) = \frac{\partial^j K}{\partial\tau^j}(t,\tau) + \int_{\delta_1} G(t,s,l)\frac{\partial^j}{\partial\tau^j}[lJ(s,\tau) - L_sJ(s,\tau)]\, ds \qquad (4.9)$$

for $j = 0, 1, \ldots, n-1$, $a_0 \leqq t,\tau \leqq b_0$, $\Im l \neq 0$. Moreover,

$$\frac{\partial^n G}{\partial\tau^n}(t,\tau,l) = \frac{\partial^n K}{\partial\tau^n}(t,\tau) + \frac{(-1)^n lG(t,\tau,l)}{\bar{p}_0(\tau)} + \int_{\delta} G(t,s,l)\frac{\partial^n}{\partial\tau^n}[lJ(s,\tau) - L_sJ(s,\tau)]\, ds \qquad (4.10)$$

for $t \neq \tau$. Since (v) has been proved, the relations (4.9) and (4.10) prove (i).

Clearly $\partial^{n-1}G/\partial t^{n-1}$ has the same jump at $t = \tau$ as $\partial^{n-1}K/\partial t^{n-1}$, proving (ii). From (4.8) it follows that, as a function of t, G satisfies $Lx = lx$, provided $t \neq \tau$, proving (iii). Since the right sides of (4.6) and (4.7) with $\delta = \delta_m$ tend, as $m \to \infty$, to the right sides in (4.9) and (4.10), it is seen that

$$\frac{\partial^j G_m}{\partial\tau^j} \to \frac{\partial^j G}{\partial\tau^j} \qquad (j = 0, 1, \ldots, n) \qquad (4.11)$$

uniformly on any compact (t,τ,l) region, where $\Im l \neq 0$, and provided $t \neq \tau$ when $j = n-1, n$. The symmetry relations imply that

$$\frac{\partial^j G_m}{\partial t^j} \to \frac{\partial^j G}{\partial t^j} \qquad (j = 0, 1, \ldots, n)$$

under the same conditions that (4.11) is valid. Returning to (4.6) through (4.10), it is easy to see that the mixed derivatives $\partial^{j+k}G/\partial t^j\partial\tau^k$ ($j, k = 0, 1, \ldots, n-1$) exist and satisfy (iv). Relation (v) has been proved.

The proof of (vi) is based on (4.4). From that inequality there exists a constant c_1 (depending on δ_0 and δ_1 only) such that

$$\|G_\delta(\ ,\tau,l)\|_\delta \leqq c_1|\Im l|^{-1}(|l| + 1) + c_1 \qquad (\tau \in \delta_0)$$

But $\|G_\delta(\ ,\tau,l)\|_{\bar\delta} \leqq \|G_\delta(\ ,\tau,l)\|_\delta$ for $\bar\delta \subset \delta$, and letting first $\delta \to (a,b)$ through the sequence δ_m, and then $\bar\delta \to (a,b)$, it follows that, for any fixed (τ,l), $\Im l \neq 0$, $G(\ ,\tau,l) \in \mathfrak{L}^2(a,b)$. This also gives for fixed (t,l), $\Im l \neq 0$, $G(t,\ ,l) \in \mathfrak{L}^2(a,b)$.

It remains to prove (vii). If $f \in \mathfrak{L}^2(a,b)$, then the integral

$$\int_a^b G(t,\tau,l)f(\tau)\,d\tau \qquad (a < t < b,\ \Im l \neq 0)$$

converges absolutely by (vi) [and uniformly for t in any finite subinterval of (a,b)]. It defines a function v, and, using the properties developed above for G, it is not difficult to see that v has continuous derivatives up to order $n - 1$, $v^{(n-1)}$ is absolutely continuous on every closed subinterval of (a,b), and

$$Lv = lv + f$$

For example, to prove the existence and continuity of v', one first shows by means of

$$u'(t) = \int_\delta \frac{\partial G_\delta}{\partial t}(t,\tau,l)f(\tau)\,d\tau$$

and a treatment of (4.5) similar to that of (4.1) in deriving (4.4) that $\|\partial G(t,\ \ ,l)/\partial t\|$ is bounded for fixed l, $\Im l \neq 0$, uniformly for t on any finite subinterval of (a,b). Thus the integral

$$\int_a^b \frac{\partial G}{\partial t}(t,\tau,l)f(\tau)\,d\tau$$

converges uniformly for t on any finite subinterval of (a,b), and hence represents a continuous function on (a,b) which is easily verified to be v'. From (4.3) for $\delta = \delta_m$, letting $m \to \infty$, it follows that

$$\|v\| \leqq |\Im l|^{-1}\|f\| \tag{4.12}$$

which proves $v \in \mathfrak{L}^2(a,b)$. Since $Lv = lv + f$,

$$\|Lv\| \leqq \left(1 + \frac{|l|}{|\Im l|}\right)\|f\|$$

yielding $Lv \in \mathfrak{L}^2(a,b)$, and completing the proof that $v \in \mathfrak{D}$.

For any $f \in \mathfrak{L}^2(a,b)$ let $\mathcal{G}(l)f$ denote the function given by (vii) in Theorem 4.1.

The functions G arising in Cases I and II will now be considered. In Case II suppose G is obtained by taking for δ_0, δ_1 and all δ the interval closed at a. Thus $\delta = [a,\bar{b}]$. The boundary conditions $U_\delta x = 0$ which determine G_δ are assumed in this case to include $U^{(1)}x = 0$. Thus in Case II G_δ as a function of t satisfies $U^{(1)}x = 0$. All the convergence properties of Theorem 4.1 can be shown to hold uniformly over $a \leqq t$, $\tau \leqq b_0$ for any $b_0 < b$. Thus the limiting G also satisfies $U^{(1)}x = 0$ and thus for such G, v of (vii) is an element of $\hat{\mathfrak{D}}$.

It will now be shown that in Cases I and II the hypothesis that

$$Lx \pm ix = 0$$

has no nontrivial solutions of class $\mathfrak{D}$ and $\hat{\mathfrak{D}}$, respectively, implies that for any l, $\Im l \neq 0$, $Lx - lx = 0$ has no nontrivial solution of class $\mathfrak{D}$ or $\hat{\mathfrak{D}}$, respectively. The argument is essentially the same for both cases and will be proved for Case I.

Suppose for some l_0, $\Im l_0 > 0$, that $(L - l_0)x = 0$ has only the trivial solution in $\mathfrak{L}^2(a,b)$. Let $|l - l_0| < \Im l_0$ and suppose $v \in \mathfrak{D}$ and $(L - l)v = 0$. If

$$u = v - (l - l_0)\mathcal{G}(l_0)v \tag{4.13}$$

then $u \in \mathfrak{D}$. Clearly $(L - l_0)u = (L - l)v = 0$. Thus $u = 0$ and (4.13) implies, with the use of (4.12), that

$$\|v\| \leqq \frac{|l - l_0|}{|\Im l_0|} \|v\| < \|v\|$$

if $\|v\| \neq 0$. Thus $v = 0$ and the result is proved for $|l - l_0| < \Im l_0$, which implies the result for $\Im l > 0$. It is proved similarly for $\Im l < 0$.

Now suppose G for $\Im l \neq 0$ is not unique in Cases I and II and let $\tilde{G}$ have the same properties as G. Then as a function of t, $G - \tilde{G}$ is of class $C^n(a,b)$ and is thus of class $\mathfrak{D}$ or $\hat{\mathfrak{D}}$, respectively, and a solution of $(L - l)x = 0$. This is impossible, proving the following theorem:

Theorem 4.2. *In Cases* I *and* II

$$G_\delta \to G \qquad [\delta \to (a,b) \text{ in Case I}]$$
$$G_\delta \to G \qquad [\delta \to [a,b) \text{ in Case II and } U^{(1)}G_\delta = 0]$$

uniformly on any compact (t,τ,l) region where $\Im l \neq 0$ independent of the choice of boundary conditions U_δ in Case I *and $U_\delta^{(2)}$ in Case* II. *The function G is unique in that it is the only function with the properties listed in Theorem* 4.1 *(and in Case* II *satisfying $U^{(1)}x = 0$).*

The function G is called *Green's function* for the problem $Lx = lx$ on (a,b) in Case I, and for the problem $Lx = lx$, $U^{(1)}x = 0$ on $[a,b)$ in Case II. In either case, let $\mathcal{G}(l)$ denote the operator defined for all $f \in \mathfrak{L}^2(a,b)$ by

$$\mathcal{G}(l)f(t) = \int_a^b G(t,\tau,l)f(\tau)\,d\tau \qquad (\Im l \neq 0)$$

Then Theorem 4.2 yields the following: *The operator $\mathcal{G}(l)$ is the inverse of $L - l$ with domain $\mathfrak{D}$ in Case* I *and with domain $\hat{\mathfrak{D}}$ in Case* II.

5. Representation of the Spectral Matrix by Green's Function

Here the existence of the spectral matrix ρ will be established, independent of the treatment of Sec. 2, using methods related to those of Chap. 9. In particular, the spectral matrix and Green's functions will be related by formulas analogous to (3.9), (3.10), Chap. 9.

First, the nonsingular case is treated. Let G_δ be Green's function associated with the self-adjoint boundary-value problem $Lx = lx$,

$U_\delta x = 0$ on δ. Let

$$H_\delta(t,\tau,l) = G_\delta(t,\tau,l) - G_\delta(t,\tau,\bar{l})$$

Theorem 5.1. *The spectral matrix* ρ_δ *described by* (2.6) *satisfies*

$$2i\Im l \int_{-\infty}^{\infty} \frac{d\rho_{\delta jk}(\lambda)}{|\lambda - l|^2} = \frac{\partial^{j+k-2}H_\delta}{\partial t^{j-1}\partial\tau^{k-1}}(c,c,l) \tag{5.1}$$

where $\Im l \neq 0$ *and* $j,k = 1, \ldots, n$.

Proof. If $\Im m \neq 0$, $\Im l \neq 0$, then

$$(l - m) \int_\delta G_\delta(t,s,l)G_\delta(s,\tau,m)\,ds = G_\delta(t,\tau,l) - G_\delta(t,\tau,m) \tag{5.2}$$

To show this, let u be the solution of

$$(L - l)u = G_\delta(\ ,\tau,m) \qquad U_\delta u = 0$$

Then clearly

$$u(t) = \int_\delta G_\delta(t,s,l)G_\delta(s,\tau,m)\,ds$$

Let $v = G_\delta(\ ,\tau,l) - G_\delta(\ ,\tau,m)$. Then $v \in C^n(\delta)$ and $U_\delta v = 0$. Moreover, clearly $Lv = lv + (l - m)G_\delta(\ ,\tau,m)$. Thus $v = (l - m)u$ and this proves (5.2). In (5.2) take $m = \bar{l}$. Then it becomes

$$2i\Im l \int_\delta G_\delta(t,s,l)G_\delta(s,\tau,\bar{l})\,ds = H_\delta(t,\tau,l)$$

Since $G_\delta(s,\tau,\bar{l}) = \bar{G}_\delta(\tau,s,l)$, this yields

$$2i\Im l \int_\delta \frac{\partial^j G_\delta}{\partial t^j}(t,s,l)\frac{\partial^k \bar{G}_\delta}{\partial\tau^k}(\tau,s,l)\,ds = \frac{\partial^{j+k}H_\delta}{\partial t^j\partial\tau^k}(t,\tau,l) \tag{5.3}$$

for $j, k = 0, 1, \ldots, n - 1$.

Letting $\{\chi_{\delta k}\}$ and $\{\lambda_{\delta k}\}$ be the eigenfunctions and eigenvalues for the problem $Lx = lx$, $U_\delta x = 0$, it follows from

$$L\chi_{\delta m} = l\chi_{\delta m} + (\lambda_{\delta m} - l)\chi_{\delta m}$$

that

$$\chi_{\delta m}(t) = (\lambda_{\delta m} - l)\int_\delta G_\delta(t,s,l)\chi_{\delta m}(s)\,ds$$

and thus for $k = 0, \ldots, n - 1$

$$\chi_{\delta m}^{(k)}(t) = (\lambda_{\delta m} - l)\int_\delta \frac{\partial^k G_\delta}{\partial t^k}(t,s,l)\chi_{\delta m}(s)\,ds$$

Thus the mth Fourier coefficient of $\partial^k\bar{G}_\delta(t,\ ,l)/\partial t^k$ with respect to $\chi_{\delta m}$ is

$\bar{\chi}_{\delta m}^{(k)}(t)/(\lambda_{\delta m} - \bar{l})$. Using the Parseval equation on the left side of (5.3) yields

$$2i\Im l \int_\delta \frac{\partial^j G_\delta}{\partial t^j}(t,s,l)\frac{\partial^k \bar{G}_\delta}{\partial \tau^k}(\tau,s,l)\,ds = 2i\Im l \sum_{m=1}^{\infty} \frac{\chi_{\delta m}^{(j)}(t)\bar{\chi}_{\delta m}^{(k)}(\tau)}{|\lambda_{\delta m} - l|^2} = \frac{\partial^{j+k} H_\delta}{\partial t^j \partial \tau^k}(t,\tau,l)$$

Using the definition of the matrix ρ_δ, this can be rewritten as

$$2i\Im l \int_{-\infty}^{\infty} \sum_{p,q=1}^{n} \varphi_p^{(j)}(t,\lambda)\bar{\varphi}_q^{(k)}(\tau,l)|\lambda - l|^{-2}\,d\rho_{\delta pq}(\lambda) = \frac{\partial^{j+k} H_\delta}{\partial t^j \partial \tau^k}(t,\tau,l)$$

Setting $t = \tau = c$ and using (2.2), (5.1) is proved.

Let G be the limit of a convergent sequence $\{G_m\}$ of $\{G_\delta\}$, and let

$$H(t,\tau,l) = G(t,\tau,l) - G(t,\tau,\bar{l})$$

Further, let

$$P_{\delta jk}(l) = \frac{\partial^{j+k-2} H_\delta}{\partial t^{j-1}\partial \tau^{k-1}}(c,c,l)$$

and

$$P_{jk}(l) = \frac{\partial^{j+k-2} H}{\partial t^{j-1}\partial \tau^{k-1}}(c,c,l)$$

for $j, k = 1, \ldots, n$.

Theorem 5.2. *Let $\{G_m\}$ be any convergent sequence of the set $\{G_\delta\}$ and let $\rho_m = (\rho_{mjk})$ be the spectral matrix associated with G_m. Then there exists an Hermitian, nondecreasing matrix $\rho = (\rho_{jk})$ whose elements are of bounded variation on every finite λ interval, such that*

$$\rho_m(\lambda) - \rho_m(\mu) \to \rho(\lambda) - \rho(\mu) \qquad (m \to \infty) \tag{5.4}$$

if λ,μ are continuity points of ρ. Further, at such points λ,μ

$$\rho_{jk}(\lambda) - \rho_{jk}(\mu) = \frac{1}{2\pi i}\lim_{\epsilon \to +0}\int_\mu^\lambda P_{jk}(\nu + i\epsilon)\,d\nu \tag{5.5}$$

REMARKS: Although Theorem 5.2 gives an alternate existence proof for ρ, the new result is given by (5.5). It is clear that in Cases I and II it is not necessary to take a sequence of $\{\delta\}$ but suffices for $\delta \to (a,b)$ and $\delta \to [a,b)$, respectively, for $\rho_\delta \to \rho$. In these cases, G, H, and therefore the P_{jk} are unique, and by (5.5) so is any limiting ρ.

Proof of Theorem 5.2. The proof is much like that of Theorem 3.1, Chap. 9, and will be sketched briefly. From (5.1)

$$2i\Im l \int_{-\infty}^{\infty} \frac{d\rho_{mjk}(\lambda)}{|\lambda - l|^2} = P_{mjk}(l)$$

and for $l = i$

$$\int_{-\mu}^{\mu} \frac{d\rho_{mjj}(\lambda)}{1+\lambda^2} \leq \frac{P_{mjj}(i)}{2i} \tag{5.6}$$

since ρ_{mjj} is monotone. The right side of (5.6) is bounded, since by Theorem 4.1 (iv) $P_{mjj}(i) \to P_{jj}(i)$. Thus

$$\int_{-\infty}^{\infty} \frac{d\rho_{mjj}(\lambda)}{1+\lambda^2} < A \qquad (j = 1, \ldots, n)$$

where A is a constant and

$$|\rho_{mjj}(\lambda)| < A(1+\lambda^2)$$

Since

$$|\rho_{mjk}(\Delta)|^2 \leqq \rho_{mjj}(\Delta)\rho_{mkk}(\Delta)$$

with $\rho_{mjk}(\Delta) = \rho_{mjk}(\lambda) - \rho_{mjk}(\mu)$, $\Delta = (\mu,\lambda]$, it follows that the total variation of ρ_{mjk} on any finite λ interval is bounded independent of m. The Helly selection theorem implies the existence of a subsequence of $\{\rho_m\}$ tending to a ρ having the properties stated in the theorem. Moreover, the argument of (3.25) and (3.26), Chap. 9, leads readily to the fact that

$$\frac{P_{jk}(l)}{2i\Im l} - \int_{-\infty}^{\infty} \frac{d\rho_{jk}(\lambda)}{|\lambda - l|^2}$$

is a constant independent of l for $\Im l \neq 0$. An inversion results in (5.5). Since $P_{mjk} \to P_{jk}$, the relation (5.4) follows from (5.5).

PROBLEMS

1. Let L be the differential operator defined for vectors x of r components by

$$Lx = P_0x^{(n)} + P_1x^{(n-1)} + \cdots + P_nx$$

where the P_k are r-by-r matrices of complex-valued functions of class C^{n-k} on an open interval (a,b). Assume $\det P_0(t) \neq 0$ for $a < t < b$ and that L is formally self-adjoint in that it coincides with its Lagrange adjoint L^+ given by

$$L^+x = (-1)^n(P_0^*x)^{(n)} + (-1)^{n-1}(P_1^*x)^{(n-1)} + \cdots + P_n^*x$$

Formulate and prove the analogues of the expansion theorem, Parseval equality, and the inverse-transform theorem in Cases I and II. Also, prove the existence of a Green's matrix in the two cases, and prove the analogue of Theorem 5.2.

2. Prove that the problem $Lx = ix' + a(t)x = lx$ is self-adjoint on $-\infty < t < \infty$, where a is a real continuous function on this interval. Show that the spectrum is the λ axis, $-\infty < \lambda < \infty$. Give the expansion theorem for this case.

3. Consider the operator L defined for vector functions with r components

$$Lx = ix' + A(t)x$$

where A is an r-by-r matrix of continuous functions on $-\infty < t < \infty$ such that $A^*(t) = A(t)$. Then L is formally self-adjoint. Prove that the problem $Lx = lx$ is self-adjoint on $(-\infty, \infty)$, no boundary conditions required.

HINT: Prove that every solution φ of $Lx = lx$ is of the form $\varphi(t,l) = e^{-ilt}\psi(t)$, where ψ is a solution of the equation $x' = -iA(t)x$. Let $f \cdot g = \sum_{j=1}^{r} f_j \bar{g}_j$ for vectors f,g. Show that $(\psi \cdot \psi)' = 0$ so that $\psi(t) \cdot \psi(t) = c$, a constant, and

$$\varphi(t,l) \cdot \varphi(t,l) = e^{2\Im lt}\psi(t) \cdot \psi(t) = ce^{2\Im lt}$$

Therefore there are no nontrivial solutions of $Lx = lx$ for any complex l which are of class $\mathfrak{L}_r^2(-\infty, \infty)$, that is, for which $\int_{-\infty}^{\infty} \varphi \cdot \varphi \, dt < \infty$.

Compute the spectral matrix ρ in this case and prove that the spectrum is the entire λ axis, $-\infty < \lambda < \infty$. This problem is the generalization to systems of Prob. 2.

4. Let L be the operator defined for vector functions of $r = 2s$ $(s \geqq 1)$ components by

$$Lx = Ix' + Ax$$

where I is the skew-symmetric matrix $I = -I^{\prime} = -I^{-1}$ given by

$$I = \begin{pmatrix} 0_s & E_s \\ -E_s & 0_s \end{pmatrix}$$

E_s, 0_s being the identity and zero s-by-s matrices, respectively, and A is a real constant matrix such that $A = A^{\prime}$. Thus L is formally self-adjoint. Prove that there are no nontrivial solutions of the equation $Lx = lx$ for *any* complex l which are of class $\mathfrak{L}_r^2(-\infty, \infty)$. Prove that there are exactly s linearly independent solutions of class $\mathfrak{L}_r^2(0, \infty)$ for $\Im l \neq 0$.

HINT: The equation $Lx = lx$ is one with constant coefficients, and a fundamental matrix is given by

$$\Phi(t,l) = \exp\,[tI(A - lE)]$$

which shows there are no solutions $\mathfrak{L}_r^2(-\infty, \infty)$. The nature of the solutions depends on the characteristic roots of the matrix $I(A - lE)$. The characteristic polynomial of this matrix is

$$f(\mu,l) = \det\,(\mu E - I(A - lE))$$

so that $f(\mu,l) = \bar{f}(\bar{\mu},\bar{l})$. Since

$$f(\mu,l) = \det\,(I^{-1}(\mu E - I(A - lE))I)$$

show $f(\mu,l) = f(-\mu,l)$. Because $A + \mu I$ is Hermitian if $\Re\mu = 0$, it follows that for $\Im l \neq 0$ none of the characteristic roots of $I(A - lE)$ can have a zero real part.

The problem $Ix' + Ax = lx$ on $(-\infty, \infty)$ is thus self-adjoint. Prove that for $0 \leqq t < \infty$ the problem $Ix' + Ax = lx$, together with an appropriate set of s homogeneous boundary conditions at $t = 0$, yields a self-adjoint problem. Investigate the nature of the spectrum in these two cases.

5. Show that an alternative way to obtain a self-adjoint problem in Prob. 4 on $0 \leqq t < \infty$ is by use of

$$(\varphi^* I\varphi)' = (l - \bar{l})\varphi^*\varphi$$

for any solution φ of $Ix' + Ax = lx$.

HINT: Adjoin s conditions $U^{(1)}x = 0$ at $t = 0$ so that for any φ satisfying $U^{(1)}\varphi = 0$ it is the case that $\varphi^* I\varphi = 0$ at $t = 0$.

6. Let L be the operator given by $Lx = Ix' + A(t)x$, where I is the matrix defined in Prob. 4 and A is an r-by-r ($r = 2s$, $s \geq 1$) matrix of continuous real functions on $-\infty < t < \infty$ which are periodic of period 1, i.e., $A(t + 1) = A(t)$, and $A = A'$. Prove that the problem $Lx = lx$ on $-\infty < t < \infty$ is self-adjoint with no boundary conditions required. Consider self-adjoint problems on $[0, \infty)$.

HINT: From Sec. 5, Chap. 3, the fundamental matrix Φ of $Lx = lx$ satisfying $\Phi(0,l) = E$, the identity matrix, is of the form $\Phi(t,l) = P(t,l)e^{tR(l)}$, where

$$P(t + 1, l) = P(t,l)$$

and $R(l)$ is a constant matrix for fixed l. This shows that there are no solutions $\mathfrak{L}_r^2(-\infty, \infty)$ and no point spectrum. The problem on $[0, \infty)$ can be treated by the device of Prob. 5.

7. If (a,b) is an open real interval, the space $\mathfrak{L}^2(a,b)$ is a Hilbert space with inner product

$$(u,v) = \int_a^b u\bar{v}\, dt$$

Let $\mathfrak{D}$ be the set of all $u \in \mathfrak{L}^2(a,b)$ of class C^{n-1} on (a,b), $u^{(n-1)}$ absolutely continuous on every closed subinterval of (a,b), and $Lu \in \mathfrak{L}^2(a,b)$, where L is the formally self-adjoint differential operator defined in Sec. 1. Let T be the operator in $\mathfrak{L}^2(a,b)$ with domain $\mathfrak{D}$ and $Tu = Lu$ for $u \in \mathfrak{D}$. Let $\mathfrak{D}_S$ denote the set of all $u \in \mathfrak{D}$ such that $u = 0$ outside some closed bounded subinterval of (a,b), the interval may depend on u, and let S be the operator in $\mathfrak{L}^2(a,b)$ with domain $\mathfrak{D}_S$ defined by $Su = Lu$ for $u \in \mathfrak{D}_S$. Prove that S is a symmetric operator and that its closure $\tilde{S}$ is the adjoint T^* of T.

HINT: Use the variation-of-constants formula.

8. Suppose $T = T^*$, that is, T is self-adjoint. Let, for any closed subinterval δ of (a,b), the set $\mathfrak{D}_\delta$ be all $u \in \mathfrak{L}^2(a,b)$ such that $u \in C^{n-1}$ on δ, $u^{(n-1)}$ is absolutely continuous on δ, and $Lu \in \mathfrak{L}^2(\delta)$, and u satisfies a set of self-adjoint boundary conditions $U_\delta u = 0$. Define T_δ to be the operator with domain D_δ given by $T_\delta u(t) = Lu(t)$, $t \in \delta$, and $T_\delta u(t) = 0$ for t not in δ. Show that T_δ is self-adjoint. Let the spectral resolutions of T and T_δ be given by

$$T = \int_{-\infty}^{\infty} \lambda\, dE(\lambda) \qquad T_\delta = \int_{-\infty}^{\infty} \lambda\, dE_\delta(\lambda)$$

respectively. Prove that $\|E_\delta(\lambda)u - E(\lambda)u\| \to 0$, as $\delta \to (a,b)$, for every $u \in \mathfrak{L}^2(a,b)$ if λ is not an eigenvalue of T.

HINT: Let T_∞ be the pointwise limit of T_δ. Prove that the closure $\tilde{T}_\infty$ of T_∞ is just T. Apply a result due to Rellich [see B. v. Sz. Nagy, *Spektraldarstellung linearer Transformationen des Hilbertschen Raumes*, Ergeb. Math. vol. 5 (1942) p. 56.]

9. Using the result of Prob. 8 and Theorem 2.1, prove that

$$E(\Delta)f(t) = \int_\Delta \sum_{j,k=1}^{n} \varphi_j(t,\lambda)g_k(\lambda)\, d\rho_{jk}(\lambda)$$

where $f \in \mathfrak{L}^2(a,b)$, $\Delta = (\mu,\lambda]$, $E(\Delta) = E(\lambda) - E(\mu)$, and the φ_j, g_k are as defined in Theorem 2.2. From this, prove the Parseval equality and expansion theorem.

10. Formulate and prove results corresponding to Probs. 8 and 9 which are analogues of Case II, Chap. 10.

CHAPTER 11

ALGEBRAIC PROPERTIES OF LINEAR BOUNDARY-VALUE PROBLEMS ON A FINITE INTERVAL

1. Introduction†

Let $a \leqq t \leqq b$ be a closed bounded interval and let L be the linear differential operator of order n ($n \geqq 1$) defined by

$$Lx = p_0 x^{(n)} + p_1 x^{(n-1)} + \cdots + p_{n-1}x' + p_n x$$

where the p_k are complex-valued functions of class C^{n-k} on $[a,b]$, and $p_0(t) \neq 0$ on $[a,b]$. The principal concern of this chapter will be with boundary-value problems such as investigating the solutions of $Lx = 0$ on $[a,b]$ which satisfy a set of homogeneous boundary conditions of the type

$$\sum_{k=1}^{n} (M_{jk}x^{(k-1)}(a) + N_{jk}x^{(k-1)}(b)) = 0 \qquad (j = 1, \ldots, m) \tag{1.1}$$

where the M_{jk} and N_{jk} are complex constants. Corresponding to any homogeneous boundary-value problem is a well-defined "adjoint" problem which is associated with the Lagrange adjoint L^+ of L given by

$$L^+x = (-1)^n(\bar{p}_0 x)^{(n)} + (-1)^{n-1}(\bar{p}_1 x)^{(n-1)} + \cdots + \bar{p}_n x$$

and a set of boundary conditions "complementary" in a sense to those for the problem associated with L. It will be shown that many properties of the original problem are mirrored into "complementary" properties of the adjoint problem. This chapter is largely algebraic.

The fundamental results follow from two important formulas, that of Green and the boundary-form formula. The latter will be discussed in Sec. 2. Recall that Green's formula says that if, for example, $u,v \in C^n$ on $a \leqq t \leqq b$, then

$$\int_{t_1}^{t_2} (Lu)\bar{v}\, dt - \int_{t_1}^{t_2} u(\overline{L^+v})\, dt = [uv](t_2) - [uv](t_1) \tag{1.2}$$

† Only Sec. 5 of Chap. 12 requires Chap. 11, and for this purpose Theorems 3.1, 3.2, and 4.1 can be omitted.

where $a \leqq t_1 < t_2 \leqq b$, and $[uv](t)$ is the form in $(u, u', \ldots, u^{(n-1)})$ and $(v, v', \ldots, v^{(n-1)})$ given by

$$[uv](t) = \sum_{m=1}^{n} \sum_{j+k=m-1} (-1)^j u^{(k)}(t)(p_{n-m}\bar{v})^{(j)}(t)$$

If the form $[uv](t)$ is written as

$$[uv](t) = \sum_{j,k=1}^{n} B_{jk}(t)u^{(k-1)}(t)\bar{v}^{(j-1)}(t)$$

then it is clear that $B_{jk}(t) = 0$ for $j + k > n + 1$, and

$$B_{jk}(t) = (-1)^{j-1}p_0(t)$$

for $j + k = n + 1$. Therefore the matrix $B(t)$ with elements $B_{jk}(t)$ is a triangular one of the form

$$B(t) = \begin{pmatrix} B_{11} & B_{12} & \cdots & \cdot & p_0(t) \\ \cdot & \cdot & & -p_0(t) & 0 \\ \cdot & \cdot & & \cdot & \cdot \\ \cdot & \cdot & & \cdot & \cdot \\ (-1)^{n-1}p_0(t) & 0 & \cdots & 0 & 0 \end{pmatrix}$$

Thus det $B(t) = (p_0(t))^n$ and hence $B(t)$ is nonsingular for $a \leqq t \leqq b$.

It will be convenient now to introduce the notion of a *semibilinear* form. Such a form is a complex-valued function $\mathcal{S}$ defined for pairs of vectors f,g with k components satisfying

$$\mathcal{S}(\alpha f + \beta g,h) = \alpha\mathcal{S}(f,h) + \beta\mathcal{S}(g,h)$$
$$\mathcal{S}(f,\alpha g + \beta h) = \bar{\alpha}\mathcal{S}(f,g) + \bar{\beta}\mathcal{S}(f,h)$$

for any complex numbers α, β, and vectors f,g,h. If $f = (f_1, \ldots, f_k)$ and $g = (g_1, \ldots, g_k)$ the product $f \cdot g$ is defined as

$$f \cdot g = \sum_{i=1}^{k} f_i\bar{g}_i$$

If S is a k-by-k matrix, with elements s_{ij}, $Sf \cdot g$ is a semibilinear form

$$\mathcal{S}(f,g) = Sf \cdot g = \sum_{i,j=1}^{k} s_{ij}f_j\bar{g}_i \tag{1.3}$$

It is clear that $[uv](t)$ is a semibilinear form with matrix $B(t)$. The right side of Green's formula (1.2) may also be considered as a form in $(u(t_1), u'(t_1), \ldots, u^{(n-1)}(t_1), u(t_2), \ldots, u^{(n-1)}(t_2))$ and $(v(t_1), v'(t_1),$

$\ldots, v^{(n-1)}(t_1), v(t_2), \ldots, v^{(n-1)}(t_2))$. The $2n$-by-$2n$ matrix $\hat{B}$ corresponding for this form is given by

$$\hat{B} = \begin{pmatrix} -B(t_1) & 0_n \\ 0_n & B(t_2) \end{pmatrix} \tag{1.4}$$

where 0_n is the zero matrix. Clearly $\det \hat{B} = (-1)^n \det B(t_1) \det B(t_2)$. Consequently $\hat{B}$ is nonsingular for all t_1, t_2 in $[a,b]$.

2. The Boundary-form Formula

Given any set of $2mn$ complex constants M_{ij}, N_{ij} ($i = 1, \ldots, m$; $j = 1, \ldots, n$), define m *boundary operators* $U_1, \ldots, U_m$ for functions x on $a \leqq t \leqq b$, for which $x^{(j)}$ ($j = 1, \ldots, n-1$) exists at a and b, by

$$U_i x = \sum_{j=1}^{n} (M_{ij} x^{(j-1)}(a) + N_{ij} x^{(j-1)}(b)) \qquad (i = 1, \ldots, m) \tag{2.1}$$

Clearly, if α and β are complex numbers and $x_1, x_2 \in C^{n-1}$ on $[a,b]$, then $U_i(\alpha x_1 + \beta x_2) = \alpha U_i x_1 + \beta U_i x_2$, that is, the U_i are linear operators. The operators U_i will also be called *boundary forms*. They are said to be *linearly independent* if the only set of complex constants $c_1, \ldots, c_m$ for which

$$\sum_{i=1}^{m} c_i U_i x = 0$$

for all $x \in C^{n-1}$ on $[a,b]$ is the set $c_1 = c_2 = \cdots = c_m = 0$.

The forms (2.1) may be described more briefly if ξ is defined to be the vector associated with x with components $x, x', \ldots, x^{(n-1)}$, and M, N are the m-by-n matrices with elements M_{ij}, N_{ij}, respectively. Also, let U denote the *vector boundary form* with components $U_1, \ldots, U_m$. In these notations (2.1) becomes simply

$$Ux = M\xi(a) + N\xi(b) \tag{2.2}$$

If $(M:N)$ denotes the matrix with m rows and $2n$ columns

$$(M:N) = \begin{pmatrix} M_{11} & \cdots & M_{1n} & N_{11} & \cdots & N_{1n} \\ \cdot & & \cdot & \cdot & & \cdot \\ \cdot & & \cdot & \cdot & & \cdot \\ \cdot & & \cdot & \cdot & & \cdot \\ M_{m1} & \cdots & M_{mn} & N_{m1} & \cdots & N_{mn} \end{pmatrix}$$

then it is easy to see that $U_1, \ldots, U_m$ are linearly independent if and only if rank $(M:N) = m$. If the latter condition holds in (2.2), U will be said to have rank m. It will always be assumed that for any vector form U the number of components is equal to its rank.

To any m linearly independent boundary forms $U_1, \ldots, U_m$ it is always possible (in many ways) to adjoin $2n - m$ linearly independent forms $U_{m+1}, \ldots, U_{2n}$ such that the combined system $U_1, \ldots, U_{2n}$ constitute $2n$ linearly independent boundary forms. This is equivalent to imbedding the matrix $(M:N)$ in a $2n$-by-$2n$ nonsingular matrix. Let U_c be the vector form with the components $U_{m+1}, \ldots, U_{2n}$. If U is any form of rank m and U_c any form of rank $2n - m$ such that the vector form with components $U_1, \ldots, U_{2n}$ has rank $2n$, then U and U_c are said to be *complementary boundary forms.*

REMARK. The only application of the results of this chapter will be for the case $m = n$ and therefore the reader may restrict himself to this case in what follows, if he desires.

The boundary-form formula will show how the form on the right side of Green's formula (1.2) may be considered as a linear combination of a boundary form and a complementary form. In order to prove this, two remarks concerning the semibilinear form (1.3) will be required. It will be recalled that the adjoint of a matrix $A = (a_{ij})$ is the transposed complex conjugate matrix $A^* = (\bar{a}_{ji})$. Thus

$$Sf \cdot g = f \cdot S^*g$$

Now let $\mathcal{S}$ be the semibilinear form associated with a nonsingular matrix S, and suppose $\tilde{f} = Ff$, where F is a nonsingular matrix. *There exists a unique nonsingular matrix G such that if $\tilde{g} = Gg$, then $\mathcal{S}(f,g) = \tilde{f} \cdot \tilde{g}$ for all f and g.* To see this, note that $\mathcal{S}(f,g) = Sf \cdot g = SF^{-1}\tilde{f} \cdot g$. Hence the matrix $G = (SF^{-1})^*$, which is clearly nonsingular, will satisfy the requirement and is uniquely determined.

Suppose $\mathcal{S}$ has the unit matrix E; i.e., $\mathcal{S}(f,g) = f \cdot g$. Let F be a nonsingular matrix such that the first j $(1 \leqq j < k)$ components of $\tilde{f} = Ff$ are the same as those of f. Then the unique nonsingular matrix G such that $\tilde{g} = Gg$ and $\tilde{f} \cdot \tilde{g} = f \cdot g$ is such that *the last $k - j$ components of $\tilde{g}$ are linear combinations of the last $k - j$ components of g with nonsingular coefficient matrix.* To prove this, note that F must have the form

$$F = \begin{pmatrix} E_j & 0_+ \\ F_+ & F_{k-j} \end{pmatrix} \tag{2.3}$$

where 0_+ is the zero matrix with j rows and $k - j$ columns. Let

$$G = \begin{pmatrix} G_j & G_- \\ G_- & G_{k-j} \end{pmatrix} \tag{2.4}$$

where G_j is a matrix with j rows and columns. Clearly

$$f \cdot g = Ff \cdot Gg = G^*Ff \cdot g$$

must be an identity in f and g. Thus $G^*F = E_k$, and this means from (2.3) and (2.4) that $G^*_{=}F_{k-j} = 0_+$, and since $\det F_{k-j} = \det F \neq 0$, $G^*_{=} = 0_+$ or $G_{=} = 0_{=}$, the $(k - j)$-by-j zero matrix. From this follows that G_{k-j} is nonsingular because G is nonsingular, and this completes the proof.

Theorem 2.1 (Boundary-form Formula). *Given any boundary form U of rank m, and any complementary form U_c, there exist unique boundary forms U_c^+, U^+ of rank m and $2n - m$, respectively, such that*

$$[xy](b) - [xy](a) = Ux \cdot U_c^+y + U_cx \cdot U^+y \tag{2.5}$$

If $\tilde{U}_c$ is any other complementary form to U, and $\tilde{U}_c^+$, $\tilde{U}^+$ the corresponding forms of rank m and $2n - m$, then

$$\tilde{U}^+y = C^*U^+y$$

for some nonsingular matrix C.

Proof. The left side of (2.5) may be considered as a semibilinear form $\mathcal{S}$ for vectors f with components $(x(a), \ldots, x^{(n-1)}(a), x(b), \ldots, x^{(n-1)}(b))$, and g with components $(y(a), \ldots, y^{(n-1)}(a), y(b), \ldots, y^{(n-1)}(b))$ with nonsingular matrix $\hat{B}$. Thus if

$$Ux = M\xi(a) + N\xi(b)$$

then $Ux = (M:N)f$. Also $U_cx = (\tilde{M}:\tilde{N})f$ for two appropriate matrices $\tilde{M},\tilde{N}$ for which

$$H = \begin{pmatrix} M & N \\ \tilde{M} & \tilde{N} \end{pmatrix}$$

is of rank $2n$. Thus

$$\begin{pmatrix} Ux \\ U_cx \end{pmatrix} = Hf$$

and, by the sentence in italics in the middle of page 287, there exists a unique $2n$-by-$2n$ nonsingular matrix J such that $\mathcal{S}(f,g) = Hf \cdot Jg$. If

$$Jg = \begin{pmatrix} U_c^+y \\ U^+y \end{pmatrix}$$

then (2.5) holds.

The second statement in the theorem follows from the remark in italics made on page 287 above (2.3), but here Hf and Jg correspond to f and g in that remark.

3. Homogeneous Boundary-value Problems and Adjoint Problems

For any boundary form U of rank m there is associated the *homogeneous boundary condition*

$$Ux = 0 \tag{3.1}$$

for functions $x \in C^{n-1}$ on $[a,b]$, and if U^+ is any boundary form of rank $2n - m$ determined as in Theorem 2.1, then the homogeneous boundary condition

$$U^+x = 0 \tag{3.2}$$

is called an *adjoint boundary condition* to (3.1). It follows from Green's formula and the boundary-form formula that with $(u,v) = \int_a^b u\bar{v}\,dt$,

$$(Lu,v) = (u,L^+v)$$

for all $u \in C^n$ on $[a,b]$ satisfying (3.1), and all $v \in C^n$ on $[a,b]$ satisfying (3.2).

Let D and D^+ be the linear subsets of the set of all functions $u \in C^n$ which satisfy (3.1) and (3.2), respectively. Then Theorem 2.1 shows that D^+ is uniquely determined by U, although U^+ is not. If the theorem is applied to U^+ instead of U, it follows that (3.1) is an adjoint boundary condition to (3.2) and D is uniquely determined by U^+.

Associated with U^+ are two matrices of complex constants P and Q, each with n rows and $2n - m$ columns, such that $(P^*:Q^*)$ has rank $2n - m$ and

$$U^+x = P^*\xi(a) + Q^*\xi(b)$$

It is of interest to characterize the adjoint condition (3.2) directly in terms of the matrices M, N, P, Q.

Theorem 3.1. *The boundary condition $U^+x = 0$ is adjoint to $Ux = 0$ if and only if*

$$MB^{-1}(a)P = NB^{-1}(b)Q \tag{3.3}$$

where $B(t)$ is the matrix associated with the form $[xy](t)$.

Proof. Let η be the vector with components $(y, y', \ldots, y^{(n-1)})$. Then $[xy](t) = B(t)\xi(t) \cdot \eta(t)$.

First, suppose $U^+x = 0$ is an adjoint boundary condition to $Ux = 0$. From Theorem 2.1 there exist forms U_c, U_c^+ of rank $2n - m$ and m, respectively, such that the boundary-form formula holds. Put

$$U_cx = M_c\xi(a) + N_c\xi(b) \qquad \text{rank}\,(M_c:N_c) = 2n - m$$
$$U_c^+y = P_c^*\eta(a) + Q_c^*\eta(b) \qquad \text{rank}\,(P_c^*:Q_c^*) = m$$

Writing out the boundary-form formula, there results the following identity in $\xi(a)$, $\xi(b)$, $\eta(a)$, $\eta(b)$:

$$\begin{aligned}(P_cM + PM_c)\xi(a) \cdot \eta(a) + (Q_cM + QM_c)\xi(a) \cdot \eta(b) \\ + (P_cN + PN_c)\xi(b) \cdot \eta(a) + (Q_cN + QN_c)\xi(b) \cdot \eta(b) \\ = B(b)\xi(b) \cdot \eta(b) - B(a)\xi(a) \cdot \eta(a)\end{aligned}$$

Thus

$$P_cM + PM_c = -B(a) \qquad P_cN + PN_c = 0_n$$
$$Q_cM + QM_c = 0_n \qquad Q_cN + QN_c = B(b)$$

and, since $B(a)$, $B(b)$ are nonsingular, this implies that

$$\begin{pmatrix} -B^{-1}(a)P_c & -B^{-1}(a)P \\ B^{-1}(b)Q_c & B^{-1}(b)Q \end{pmatrix} \begin{pmatrix} M & N \\ M_c & N_c \end{pmatrix} = \begin{pmatrix} E_n & 0_n \\ 0_n & E_n \end{pmatrix} \tag{3.4}$$

The matrix involving $M, \ldots, N_c$ is nonsingular, and hence so is the matrix on the left. Since the latter is a left reciprocal for the matrix involving $M, \ldots, N_c$, it is also a right reciprocal. Thus

$$\begin{pmatrix} M & N \\ M_c & N_c \end{pmatrix} \begin{pmatrix} -B^{-1}(a)P_c & -B^{-1}(a)P \\ B^{-1}(b)Q_c & B^{-1}(b)Q \end{pmatrix} = \begin{pmatrix} E_m & 0_+ \\ 0_- & E_{2n-m} \end{pmatrix}$$

where 0_+, 0_- are matrices with all elements zero. Therefore

$$-MB^{-1}(a)P + NB^{-1}(b)Q = 0_+$$

which is the relation (3.3).

Conversely, suppose U_1^+ is a form of rank $2n - m$, where

$$U_1^+ y = P_1^* \eta(a) + Q_1^* \eta(b)$$

and the relation

$$MB^{-1}(a)P_1 = NB^{-1}(b)Q_1 \tag{3.5}$$

holds. Since rank $(M:N) = m$, it follows that there exist exactly $2n - m$ linearly independent $2n$-rowed vector solutions of the linear system $(M:N)u = 0$. From (3.5) one sees that the $2n - m$ columns of the matrix

$$H_1 = \begin{pmatrix} B^{-1}(a)P_1 \\ -B^{-1}(b)Q_1 \end{pmatrix} \tag{3.6}$$

are solutions of this system. Since rank $(P_1^* : Q_1^*) = 2n - m$,

$$\text{rank} \begin{pmatrix} P_1 \\ Q_1 \end{pmatrix} = 2n - m$$

and because $B(a)$, $B(b)$ are nonsingular, the rank of the matrix H_1 in (3.6) is $2n - m$.

If $U^+x = P^*\xi(a) + Q^*\xi(b) = 0$ is a boundary condition adjoint to $Ux = 0$, then it follows that the matrix on the left in (3.4) is nonsingular, and this implies that if

$$H = \begin{pmatrix} B^{-1}(a)P \\ -B^{-1}(b)Q \end{pmatrix} \tag{3.7}$$

then the rank of H must be $2n - m$. Therefore by (3.3) the columns of H also form $2n - m$ linearly independent solutions of $(M:N)u = 0$. Hence there exists a nonsingular $(2n - m)$-by-$(2n - m)$ matrix A such that $H_1 = HA$, implying that $B^{-1}(a)P_1 = B^{-1}(a)PA$, $B^{-1}(b)Q_1 =$

$B^{-1}(b)QA$, or $P_1 = PA$, $Q_1 = QA$. Consequently $U_1^+ y = A^* U^+ y$ and this proves $U_1^+ y = 0$ is an adjoint boundary condition to $Ux = 0$.

If U is a boundary form of rank m, the problem of finding solutions of

$$\pi_m: \qquad Lx = 0 \qquad Ux = 0$$

on $[a,b]$ is called a *homogeneous boundary-value problem of rank* m. The problem

$$\pi_{2n-m}^+: \qquad L^+x = 0 \qquad U^+x = 0$$

on $[a,b]$ is defined to be the *adjoint boundary-value problem* to π_m. Clearly π_m is the adjoint problem to π_{2n-m}^+. The identically zero function on $[a,b]$ is a solution of both π_m and π_{2n-m}^+, and this will be referred to as the *trivial solution.*

An immediate consequence of Theorem 3.1 is the following:

Theorem 3.2. *If $m = n$, the boundary condition $Ux = 0$ is adjoint to itself if and only if*

$$MB^{-1}(a)M^* = NB^{-1}(b)N^*$$

Thus if the above holds and $L^+ = L$, the boundary problem π_n is self-adjoint; that is, if $u,v \in C^n$ on $[a,b]$ and satisfy $Ux = 0$, then

$$(Lu,v) = (u,Lv)$$

This last equation follows from Green's formula and the boundary-form formula.

If $\varphi_1, \ldots, \varphi_n$ is a fundamental set for $Lx = 0$, let Φ denote the nonsingular matrix

$$\Phi = \begin{pmatrix} \varphi_1 & \cdots & \varphi_n \\ \varphi_1' & \cdots & \varphi_n' \\ \cdot & \cdots & \cdot \\ \varphi_1^{(n-1)} & \cdots & \varphi_n^{(n-1)} \end{pmatrix}$$

It is called a *fundamental matrix associated with* $Lx = 0$. Similarly, if $\psi_1, \ldots, \psi_n$ is a fundamental set for $L^+x = 0$, let Ψ denote the matrix

$$\Psi = \begin{pmatrix} \psi_1 & \cdots & \psi_n \\ \cdot & \cdots & \cdot \\ \psi_1^{(n-1)} & \cdots & \psi_n^{(n-1)} \end{pmatrix}$$

The meanings of U and U^+ are extended to matrices by defining

$$U\Phi = M\Phi(a) + N\Phi(b)$$
$$U^+\Psi = P^*\Psi(a) + Q^*\Psi(b)$$

Theorem 3.3. *The problem π_m has exactly k $(0 \leqq k \leqq n)$ linearly independent solutions if and only if $U\Phi$ has rank $n - k$, where Φ is any fundamental matrix associated with $Lx = 0$.*

Proof. The function φ satisfies $Lx = 0$ if and only if the corresponding vector $\hat{\varphi}$ with components $\varphi, \varphi', \ldots, \varphi^{(n-1)}$ is of the form $\hat{\varphi} = \Phi c$, where c is a constant vector. Thus $U\varphi = 0$ if and only if

$$U(\Phi c) = (U\Phi)c = 0$$

The number of linearly independent vectors c satisfying $(U\Phi)c = 0$ is $n - \text{rank}\ (U\Phi)$.

If Φ_1 is any other fundamental matrix associated with $Lx = 0$, then $\Phi_1 = \Phi C$, where C is a nonsingular constant matrix. Therefore

$$\text{rank}\ (U\Phi_1) = \text{rank}\ (U\Phi)$$

completing the proof.

If π_m has exactly k linearly independent solutions, say $\varphi_1, \ldots, \varphi_k$, then any linear combination $\sum_{i=1}^{k} c_i\varphi_i$, where the c_i are complex numbers, is again a solution. Moreover, if φ is any solution of π_m, then $\varphi = \sum_{i=1}^{k} c_i\varphi_i$ for some constants c_i. Thus the solutions of π_m form a vector space over the complex numbers of dimension k.

There exists a certain duality between the number of nontrivial solutions of π_m and π^+_{2n-m}.

Theorem 3.4. *If π_m has exactly k linearly independent solutions, then π^+_{2n-m} has exactly $k + m - n$ linearly independent solutions.*

Proof. Let $\varphi_1, \ldots, \varphi_k$ be k linearly independent solutions of π_m. Suppose U_c, where

$$U_cx = M_c\xi(a) + N_c\xi(b)$$

is a boundary form of rank $2n - m$ complementary to U. It will first be proved that the vectors $U_c\varphi_i$ $(i = 1, \ldots, k)$ are linearly independent. Suppose they are not. Then for some constants, $\alpha_1, \ldots, \alpha_k$, not all zero,

$$\sum_{i=1}^{k} \alpha_i U_c\varphi_i = 0$$

and this implies

$$U_c\left(\sum_{i=1}^{k} \alpha_i\varphi_i\right) = 0 \tag{3.8}$$

However.

$$U\left(\sum_{i=1}^{k} \alpha_i\varphi_i\right) = 0 \tag{3.9}$$

for each of the φ_i satisfies $Ux = 0$. Thus if $\bar{\varphi} = \sum_{i=1}^{k} \alpha_i\varphi_i$ and the corresponding ξ vector is $\bar{\xi}$, then (3.8) and (3.9) together give

$$M\bar{\xi}(a) + N\bar{\xi}(b) = 0$$
$$M_c\bar{\xi}(a) + N_c\bar{\xi}(b) = 0$$

Since

$$\operatorname{rank}\begin{pmatrix} M & N \\ M_c & N_c \end{pmatrix} = 2n$$

it follows that $\bar{\xi}(a) = \bar{\xi}(b) = 0$. But $L\bar{\varphi} = 0$, and thus by uniqueness $\bar{\varphi}(t) = 0$ for $a \leqq t \leqq b$. This contradicts the definition of $\bar{\varphi}$ as a nontrivial linear combination of $\varphi_1, \ldots, \varphi_k$. Hence

$$\alpha_1 = \alpha_2 = \cdots = \alpha_k = 0$$

Let $\psi_1, \ldots, \psi_n$ be n linearly independent solutions of $L^+x = 0$, and suppose Ψ is the corresponding fundamental matrix. Green's formula gives

$$0 = (L\varphi_i,\psi_j) - (\varphi_i,L^+\psi_j) = [\varphi_i\psi_j](b) - [\varphi_i\psi_j](a)$$

for $i = 1, \ldots, k$, $j = 1, \ldots, n$. By the boundary-form formula this is equal to

$$U\varphi_i \cdot U_c^+\psi_j + U_c\varphi_i \cdot U^+\psi_j$$

and since $U\varphi_i = 0$ $(i = 1, \ldots, k)$ there results

$$U_c\varphi_i \cdot U^+\psi_j = 0$$

or, since $f \cdot g = g^*f$ for any column vectors f and g,

$$(U^+\Psi)^*U_c\varphi_i = 0 \qquad (i = 1, \ldots, k)$$

Hence the system $(U^+\Psi)^*v = 0$ has the k linearly independent $2n - m$ dimensional vectors $U_c\varphi_1, \ldots, U_c\varphi_k$ as solutions. Therefore

$$\operatorname{rank}(U^+\Psi) = \operatorname{rank}(U^+\Psi)^* \leqq (2n - m) - k$$

Suppose rank $(U^+\Psi) = r < (2n - m) - k$. Then it can be shown by similar reasoning that, if Φ is any fundamental matrix associated with $Lx = 0$, rank $(U\Phi) \leqq m - (n - r) < n - k$, which is a contradiction. Therefore rank $(U^+\Psi) = 2n - m - k$, and this implies that there exist exactly $k + m - n$ linearly independent solutions of π_{2n-m}^+, by Theorem 3.3.

In particular, π_n *and* π_n^+ *have the same number of independent solutions.*

4. Nonhomogeneous Boundary-value Problems and Green's Function

A *nonhomogeneous boundary-value problem* associated with π_m is a problem of the form

$$Lx = f \qquad Ux = \gamma \tag{4.1}$$

on $a \leqq t \leqq b$, where f is a complex-valued continuous function on $[a,b]$ and γ is a complex constant vector such that either f is not the zero function or $\gamma \neq 0$. Here it will be assumed that U is a boundary form of rank m. Clearly if φ and $\tilde{\varphi}$ are two solutions of (4.1), the difference $\varphi - \tilde{\varphi}$ is a solution of π_m. Hence, if π_m has k linearly independent solutions $\varphi_1, \ldots, \varphi_k$, then $\varphi = \tilde{\varphi} + \sum_{i=1}^{k} c_i\varphi_i$ for some constants $c_1, \ldots, c_k$.

The problem (4.1) does not always possess solutions; the theorem below gives a necessary and sufficient condition for the existence of a solution. The following result will be used. Let A be a matrix and b a vector. Then $Ax = b$ has a solution if and only if $b \cdot u = 0$ for every solution u of $A^*x = 0$.

Theorem 4.1. *The nonhomogeneous problem* (4.1) *has a solution if and only if*

$$(f,\psi) = \gamma \cdot U_c^+\psi \tag{4.2}$$

holds for every solution ψ *of the adjoint homogeneous problem* π_{2n-m}^+.

If $\gamma = 0$, then the condition (4.2) says that f must be orthogonal to all solutions ψ of π_{2n-m}^+.

Proof of Theorem 4.1. If φ is a solution of (4.1) and ψ satisfies π_{2n-m}^+, then Green's formula and the boundary-form formula yield

$$(L\varphi,\psi) - (\varphi,L^+\psi) = U\varphi \cdot U_c^+\psi + U_c\varphi \cdot U^+\psi$$

and (4.2) results immediately.

Conversely, suppose (4.2) holds for all ψ satisfying π_{2n-m}^+. Every solution φ of $Lx = f$ is of the form

$$\varphi = \sum_{i=1}^{n} c_i\varphi_i + \tilde{\varphi}$$

where $\varphi_1, \ldots, \varphi_n$ is a fundamental set for $Lx = 0$, c_i are constants, and $\tilde{\varphi}$ a particular solution of $Lx = f$. Thus (4.1) has a solution if and only if there exist constants $c_1, \ldots, c_n$ such that

$$\sum_{i=1}^{n} c_iU\varphi_i + U\tilde{\varphi} = \gamma$$

or

$$(U\Phi)c = \gamma - U\tilde{\varphi} \tag{4.3}$$

where Φ is the fundamental matrix corresponding to $\varphi_1, \ldots, \varphi_n$, and c is the constant vector with components $c_1, \ldots, c_n$. Now (4.3) has a solution c if and only if $\gamma - U\bar{\varphi}$ is orthogonal to every solution u of the corresponding adjoint homogeneous system

$$(U\Phi)^*u = 0 \tag{4.4}$$

that is,

$$(\gamma - U\bar{\varphi}) \cdot u = 0 \tag{4.5}$$

Let π^+_{2n-m} have exactly k^+ linearly independent solutions, $\psi_1, \ldots, \psi_{k^+}$. By precisely the same argument used in the proof of Theorem 3.4, it can be shown that the k^+ vectors $U^+_c\psi_1, \ldots, U^+_c\psi_{k^+}$ are linearly independent m-dimensional vectors which are solutions of (4.4). The number of linearly independent solutions of (4.4) is $m - \text{rank}\,(U\Phi) = m - (n - k)$, where k is the number of linearly independent solutions of π_m. But from Theorem 3.4, $k^+ = m - n + k$, and hence (4.5) holds for every u satisfying (4.4) if and only if

$$(\gamma - U\bar{\varphi}) \cdot U^+_c\psi_i = 0 \qquad (i = 1, \ldots, k^+) \tag{4.6}$$

Applying Green's and the boundary-form formulas to $\bar{\varphi}$ and ψ_i, one obtains

$$(f,\psi_i) = U\bar{\varphi} \cdot U^+_c\psi_i \tag{4.7}$$

and together with (4.2) this yields (4.6). Therefore there exists a constant vector c such that (4.3) obtains, proving the existence of a solution of (4.1).

Corollary. *The problem* (4.1) *has a unique solution if* $m = n$ *and the only solution of* π_n *is the trivial one.*

Proof. By Theorem 3.4, π^+_n has only the trivial solution so that only $\psi = 0$ enters (4.2). A more direct proof is that $U\Phi$ must have rank n in (4.3) which leads to the above at once.

Suppose $m = n$. By Theorem 3.4 this implies that π_n and π^+_n have the same number k of linearly independent solutions. If $k = 0$, it is possible to solve the nonhomogeneous problem (4.1) with $\gamma = 0$ explicitly in terms of the Green's function.

The existence of the Green's function $G(t,\tau,l)$ for the problem

$$Lx - lx = f \qquad Ux = 0$$

was established in Chap. 7. It was there shown that, if the homogeneous problem had no solution at l, then G existed. (It was also shown that, if for one value of l the homogeneous problem had no solution, then it would have solutions only for a set of l which are the zeros of an entire function.) Here l will be taken as zero and it will be assumed π_n has only the trivial solution. $G(t,\tau,0)$ will be denoted by $G(t,\tau)$. The unique solution of (4.1) with $\gamma = 0$ is given by $\mathcal{G}f$, where

$$\mathcal{G}f(t) = \int_a^b G(t,\tau)f(\tau)\, d\tau$$

If π_n has only the trivial solution, then clearly π_n^+ has only the trivial solution, and hence Green's function G^+ for π_n^+ exists and is unique.

Theorem 4.2. *If π_n has only the trivial solution, Green's function G^+ for π_n^+ is given by*

$$G^+(t,\tau) = \bar{G}(\tau,t) \tag{4.8}$$

Proof. Let $a < \tau_1 < \tau_2 < b$ and consider the functions G_1 and G_2^+ given by $G_1(t) = G(t,\tau_1)$, $G_2^+(t) = G^+(t,\tau_2)$. Then applying Green's formula to each of the intervals $[a, \tau_1 - 0]$, $[\tau_1 + 0, \tau_2 - 0]$, $[\tau_2 + 0, b]$, there results

$$[G_1G_2^+](\tau_1 - 0) - [G_1G_2^+](a) + [G_1G_2^+](\tau_2 - 0) - [G_1G_2^+](\tau_1 + 0) + [G_1G_2^+](b) - [G_1G_2^+](\tau_2 + 0) = 0 \tag{4.9}$$

By the boundary-form formula

$$[G_1G_2^+](b) - [G_1G_2^+](a) = 0 \tag{4.10}$$

From the form of $[xy](t)$ it follows that the only terms of interest in (4.9) are those involving the $(n-1)$st derivatives, and these are

$$p_0(t)[(-1)^{n-1}x(t)\bar{y}^{(n-1)}(t) + x^{(n-1)}(t)\bar{y}(t)] \tag{4.11}$$

Now G satisfies

$$\frac{\partial^{n-1}G}{\partial t^{n-1}}(\tau + 0, \tau) - \frac{\partial^{n-1}G}{\partial t^{n-1}}(\tau - 0, \tau) = \frac{1}{p_0(\tau)} \tag{4.12}$$

and similarly for G^+

$$\frac{\partial^{n-1}G^+}{\partial t^{n-1}}(\tau + 0, \tau) - \frac{\partial^{n-1}G^+}{\partial t^{n-1}}(\tau - 0, \tau) = \frac{1}{(-1)^n\bar{p}_0(\tau)} \tag{4.13}$$

It follows easily from (4.9) through (4.13) that $\bar{G}^+(\tau_1,\tau_2) - G(\tau_2,\tau_1) = 0$. Similar reasoning shows this is true for $\tau_1 > \tau_2$, proving the theorem.

To consider $G(t,\tau,l)$, the differential operation $(L - l)$ is considered instead of L. Let $L_1 = L - l$ and consider the problem

$$L_1x = 0 \qquad Ux = 0 \tag{4.14}$$

The adjoint problem is given in terms of $L_1^+ = L^+ - \bar{l}$ and U^+. Applying Theorem 4.2 to (4.14), it follows that

$$G^+(t,\tau,l) = \bar{G}(\tau,t,\bar{l}) \tag{4.15}$$

For the self-adjoint problem where $L^+ = L$ and U and U^+ are equivalent, it follows that

$$G(t,\tau,l) = \bar{G}(\tau,t,\bar{l})$$

which was proved in Chap. 7.

PROBLEMS

1. Let $Lx = -(px')' + qx$, where p is positive and p' is absolutely continuous on $[a,b]$ and q is continuous and real. Let

$$Ux = \begin{pmatrix} m_{11} & m_{12} \\ m_{21} & m_{22} \end{pmatrix} \begin{pmatrix} x(a) \\ x'(a) \end{pmatrix} + \begin{pmatrix} n_{11} & n_{12} \\ n_{21} & n_{22} \end{pmatrix} \begin{pmatrix} x(b) \\ x'(b) \end{pmatrix}$$

Show that U is its own adjoint if and only if

$$\frac{\bar{m}_{11}m_{12} - \bar{m}_{12}m_{11}}{p(a)} = \frac{\bar{n}_{11}n_{12} - \bar{n}_{12}n_{11}}{p(b)}$$

$$\frac{\bar{m}_{21}m_{22} - m_{21}\bar{m}_{22}}{p(a)} = \frac{\bar{n}_{21}n_{22} - n_{21}\bar{n}_{22}}{p(b)}$$

$$\frac{\bar{m}_{11}m_{22} - m_{21}\bar{m}_{12}}{p(a)} = \frac{\bar{n}_{11}n_{22} - n_{21}\bar{n}_{12}}{p(b)}$$

If M and N are real, notice that only the last condition is required.

2. Prove (4.15) using $(Lu,v) = (u,L^+v)$ for $u,v \in C^n[a,b]$ and satisfying $Uu = 0$, $U^+v = 0$.

HINT: For $f,g \in C[a,b]$, let $u(t) = \int_a^b G(t,\tau,\bar{l})f(\tau)\,d\tau$, $v(t) = \int_a^b G^+(t,\tau,l)g(\tau)\,d\tau$. Then $Lu = \bar{l}u + f$, $L^+v = lv + g$, $Uu = 0$. $U^+v = 0$.

3. Let the adjoint of L be L^+; let the boundary form U have rank n and adjoint U^+. Suppose for all $u,v \in C^n[a,b]$ and satisfying $Ux = 0$

$$(Lu,v) = (u,Lv)$$

Then show that $L^+ = L$ and that U is its own adjoint, that is, $U^+x = 0$ is equivalent to $Ux = 0$.

HINT: For all $u,v \in C^n[a,b]$ and vanishing identically near a and b, $(u,Lv) = (u,L^+v)$. This implies $(L - L^+)v = 0$ for all such v and thus $L = L^+$, since a homogeneous linear differential equation cannot have solutions vanishing over an interval and not identically zero. If U^+ is adjoint to U,

$$(Lu,v) - (u,Lv) = Uu \cdot U_c^+v + U_cu \cdot U^+v$$

so that for all u,v satisfying $Ux = 0$, $U_cu \cdot U^+v = 0$. From this it follows that $U^+v = 0$ for all v satisfying $Uv = 0$. Since the ranks of the matrices associated with U and U^+ are both n, this implies the result.

4. Let L, U, and L^+ be defined as in Prob. 18, Chap. 7, but now it is no longer assumed that self-adjointness prevails. Show that U^+ is determined and an adjoint problem associated with $Lx = lx$, $Ux = 0$.

5. Since U is assumed to have nr components above, show that the problem above and its adjoint have the same number of linearly independent solutions.

6. Show that if L and U are as in Prob. 4, and U has nr components, if the r-by-r matrix $G(t,\tau,l)$ is the Green's function for solving $Lx = lx + f$, $Ux = 0$ and $G^+(t,\tau,l)$ is that associated with the adjoint problem, then

$$G^+(t,\tau,l) = G^*(\tau,t,\bar{l})$$

CHAPTER 12

NON-SELF-ADJOINT BOUNDARY-VALUE PROBLEMS

1. Introduction

In the case where a boundary-value problem on a finite interval is not necessarily self-adjoint, the methods of Chap. 7 are no longer adequate, and a new approach is required in order to obtain an expansion theorem. Such an approach is furnished by the *Cauchy integral method.* The method is valid for the self-adjoint problem already treated in Chap. 7 and yields complete information about the convergence of the expansion for any integrable function.

The essence of this method can be easily seen by looking at the expansion theorem in the self-adjoint case in a slightly different light. Let L denote an nth-order ordinary differential operator which is formally self-adjoint, and consider a self-adjoint boundary-value problem

$$\pi: \qquad Lx = lx \qquad Ux = 0$$

on a closed bounded interval $a \leqq t \leqq b$. Then there exists a complete orthonormal set of eigenfunctions $\{\chi_k\}$, and a Green's function

$$G = G(t,\tau,l)$$

for the equation $Lx = lx$, provided l is not one of the eigenvalues λ_k, $k = 1, 2, \ldots$. The expansion theorem of Chap. 7 states that for any function $f \in \mathfrak{L}^2(a,b)$

$$f = \sum_{k=1}^{\infty} (f,\chi_k)\chi_k$$

where

$$(f,\chi_k) = \int_a^b f(t)\bar{\chi}_k(t)\,dt$$

and the series converges in the mean to f. Let

$$\mathcal{G}(l)f(t) = \int_a^b G(t,\tau,l)f(\tau)\,d\tau$$

Then since $L\chi_k = \lambda_k\chi_k = \bar{l}\chi_k + (\lambda_k - \bar{l})\chi_k$, it follows that

$$(\mathcal{G}(l)f,\chi_k) = (f,\mathcal{G}(\bar{l})\chi_k) = (\lambda_k - l)^{-1}(f,\chi_k)$$

and hence the Fourier series for $\mathcal{G}(l)f$ is given by

$$\mathcal{G}(l)f = \sum_{k=1}^{\infty} (\lambda_k - l)^{-1}(f,\chi_k)\chi_k$$

Using the residue theorem with C_m a simple closed curve in the l plane which encircles each of the poles $\lambda_1, \ldots, \lambda_m$ exactly once in the counterclockwise direction, this leads formally to

$$\sum_{k=1}^{\infty} (f,\chi_k)\chi_k = -\frac{1}{2\pi i} \lim_{m\to\infty} \int_{C_m} \mathcal{G}(l)f\,dl$$

the *Cauchy integral formula* for the series expansion of f.

In the non-self-adjoint case it is possible, as was shown in Chap. 7, to define Green's function $G(t,\tau,l)$ whenever the entire function $\Delta = \Delta(l)$ does not vanish identically. Further, it makes sense to speak of the integral

$$\int_{C_m} \mathcal{G}(l)f\,dl$$

where C_m is a simple closed curve encircling eigenvalues $\lambda_1, \ldots, \lambda_m$, which may be complex. Then the expansion theorem results by showing that for suitably restricted functions f,

$$f = -\frac{1}{2\pi i} \lim_{m\to\infty} \int_{C_m} \mathcal{G}(l)f\,dl \tag{1.1}$$

Thus f is represented as minus the sum of the residues of $\mathcal{G}(l)f$.

In this chapter the second-order problem

$$\begin{aligned} Lx &= -x'' + q(t)x = lx \\ U_ix &= a_{i1}x(0) + a_{i2}x'(0) + a_{i3}x(\pi) + a_{i4}x'(\pi) = 0 \qquad (i = 1, 2) \end{aligned} \tag{1.2}$$

will be first considered. Here q is a continuous† complex-valued function on $0 \leqq t \leqq \pi$, and the a_{ij} are complex constants. By a transformation of the form $\bar{t} = \alpha t + \beta$, any closed bounded interval $a \leqq t \leqq b$ can be carried into $0 \leqq t \leqq \pi$, so that it is no restriction to limit attention to the latter interval. The generalization to the case where L is of the nth order is straightforward, and will be outlined in Sec. 4.

The matrix $A = (a_{ij})$ of two rows and four columns specifies the boundary conditions. It will be assumed that it has rank 2; otherwise there would exist only one linearly independent boundary condition.

† It will be seen from the proofs that the continuity restriction on q can be relaxed greatly. In fact, all results hold if q is only required to be integrable on $0 \leqq t \leqq \pi$.

Let φ_1, φ_2 be linearly independent solutions of $Lx = lx$ satisfying

$$\varphi_j^{(k-1)}(0,l) = \delta_{jk}$$

for $j, k = 1, 2$. These functions are continuous in (t,l) for $0 \leqq t \leqq \pi$ and all l, and for fixed t are entire functions of l. A nontrivial solution of (1.2) will exist for a given l if and only if the determinant

$$\Delta = \begin{vmatrix} U_1\varphi_1 & U_1\varphi_2 \\ U_2\varphi_1 & U_2\varphi_2 \end{vmatrix}$$

vanishes. As a function of l, Δ is entire, and its zeros are the eigenvalues for (1.2).

In the self-adjoint case it was seen that Δ could not vanish identically, for the eigenvalues are all real. Moreover, Δ had to vanish at an enumerable set of points on the real axis. For the non-self-adjoint case the situation can be quite different. For example, if q is the zero function on $0 \leqq t \leqq \pi$, and

$$A = \begin{pmatrix} 1 & 0 & 1 & 0 \\ 0 & 1 & 0 & -1 \end{pmatrix}$$

then elementary calculations show that Δ vanishes for all l. On the other hand, if

$$A = \begin{pmatrix} 1 & 0 & 2 & 0 \\ 0 & 1 & 0 & -2 \end{pmatrix}$$

for the same q, then Δ is a constant, not zero, and therefore there are no eigenvalues. Clearly it is then necessary to give sufficient conditions which will insure that these degenerate cases will not occur.

The method of this chapter will be to show that in a large number of cases the general problem (1.2) can be reduced to the study of the same problem when $q(t) = 0$ for $0 \leqq t \leqq \pi$. In the latter case, the function Δ of l can be given explicitly, as can the Green's function.

Where the above method fails, the problem (1.2) is handled by dealing directly with (1.2) and by using the results of Chap. 6 to get the asymptotic behavior of its Green's function as $|l| \to \infty$, as is indicated at the end of Sec. 3.

2. Green's Function and the Expansion Theorem for the Case $Lx = -x''$

Since later interest will center on the problem (1.2) with the corresponding Green's function G, it will be convenient to denote by Λ the special operator given by

$$\Lambda x = -x''$$

and by Γ, Green's function for the problem on $0 \leqq t \leqq \pi$,

$$\Lambda x = lx \qquad U_1x = 0 \qquad U_2x = 0 \tag{2.1}$$

when it exists. Setting $l = \rho^2$, it follows that, except when $\rho = 0$, $e^{i\rho t}$ and $e^{-i\rho t}$ are independent solutions of $\Lambda x = \rho^2 x$. The zeros of the function ω of ρ defined by

$$\omega(\rho) = \begin{vmatrix} U_1 e^{i\rho t} & U_1 e^{-i\rho t} \\ U_2 e^{i\rho t} & U_2 e^{-i\rho t} \end{vmatrix} \tag{2.2}$$

are those values of ρ for which (2.1) has a nontrivial solution for $l = \rho^2$, except at $\rho = 0$. The Green's function for (2.1) exists for those ρ for which $\omega(\rho) \neq 0$, and is given for $\rho \neq 0$ by

$$\Gamma(t,\tau,\rho^2) = \frac{M(t,\tau,\rho)}{\omega(\rho)} \tag{2.3}$$

where M is defined by

$$M(t,\tau,\rho) = \begin{vmatrix} \gamma(t,\tau,\rho) & e^{i\rho t} & e^{-i\rho t} \\ U_1\gamma & U_1 e^{i\rho t} & U_1 e^{-i\rho t} \\ U_2\gamma & U_2 e^{i\rho t} & U_2 e^{-i\rho t} \end{vmatrix} \tag{2.4}$$

and the function γ is a fundamental solution of the equation $\Lambda x = \rho^2 x$. It can be defined as that solution of this equation [when considered as a function of t for fixed $\tau(0 < \tau < \pi)$] on $0 < \tau < t \leqq \pi$ which satisfies $\gamma(\tau + 0,\ \tau,\ \rho) = 0$, $(\partial\gamma/\partial t)(\tau + 0,\ \tau,\ \rho) = -1$, and $\gamma(t,\tau,\rho) = 0$ for $0 \leqq t \leqq \tau$. With this definition of γ, a solution φ of $\Lambda x = \rho^2 x + f$, where f is integrable on $0 \leqq t \leqq \pi$, is given by

$$\varphi(t) = \int_0^t \gamma(t,\tau,\rho) f(\tau)\, d\tau$$

Formula (2.3) may be verified directly from the definition of Γ or by reference to Prob. 12, Chap. 7. Since Γ is a meromorphic function of l and therefore of ρ, it must be given by (2.3) for $\rho = 0$ also.

The explicit nature of Γ will now be examined. From the definition of ω it follows that

$$\omega(\rho) = -e^{i\rho\pi} \begin{vmatrix} a_{11} - i\rho a_{12} & a_{13} + i\rho a_{14} \\ a_{21} - i\rho a_{22} & a_{23} + i\rho a_{24} \end{vmatrix}$$

$$+ e^{-i\rho\pi} \begin{vmatrix} a_{11} + i\rho a_{12} & a_{13} - i\rho a_{14} \\ a_{21} + i\rho a_{22} & a_{23} - i\rho a_{24} \end{vmatrix} - 2i\rho \begin{vmatrix} a_{11} & a_{12} \\ a_{21} & a_{22} \end{vmatrix} - 2i\rho \begin{vmatrix} a_{13} & a_{14} \\ a_{23} & a_{24} \end{vmatrix}$$

Using the notation

$$A_{jk} = \begin{vmatrix} a_{1j} & a_{1k} \\ a_{2j} & a_{2k} \end{vmatrix}$$

this can be written as

$$\omega(\rho) = -P(\rho)e^{i\rho\pi} + P(-\rho)e^{-i\rho\pi} - 2i(A_{12} + A_{34})\rho \tag{2.5}$$

where

$$P(\rho) = A_{24}\rho^2 + i(A_{14} - A_{23})\rho + A_{13} \tag{2.6}$$

Another form for (2.5) is

$$\omega(\rho) = -2i(A_{24}\rho^2 + A_{13}) \sin \pi\rho + 2i(A_{23} - A_{14})\rho \cos \pi\rho - 2i(A_{12} + A_{34})\rho \quad (2.7)$$

from which it is obvious that ω *is an entire function with an infinite number of zeros unless* $A_{24} = 0$, $A_{13} = 0$, *and* $A_{23} = A_{14}$.

The function γ is readily seen to be

$$\begin{aligned} \gamma(t,\tau,\rho) &= -\rho^{-1} \sin \rho(t - \tau) \qquad &(\tau < t) \\ &= 0 &(t \leqq \tau) \end{aligned}$$

After a little tedious calculation, M is found to satisfy for $t \leqq \tau$.

$$\begin{aligned} -2i\rho M(t,\tau,\rho) = P(\rho)e^{i\rho(\pi-\tau+t)} + P(-\rho)e^{-i\rho(\pi-\tau+t)} + Q(\rho)e^{i\rho(\pi-\tau-t)} \\ + Q(-\rho)e^{-i\rho(\pi-\tau-t)} + 2iA_{34}\rho[e^{i\rho(t-\tau)} - e^{-i\rho(t-\tau)}] \quad (2.8) \end{aligned}$$

where P is given by (2.6) and

$$Q(\rho) = A_{24}\rho^2 - i(A_{14} + A_{23})\rho - A_{13}$$

For $t > \tau$,

$$\begin{aligned} -2i\rho M(t,\tau,\rho) = P(\rho)e^{i\rho(\pi-t+\tau)} + P(-\rho)e^{-i\rho(\pi-t+\tau)} + Q(\rho)e^{i\rho(\pi-t-\tau)} \\ + Q(-\rho)e^{-i\rho(\pi-t-\tau)} + 2iA_{12}\rho[e^{i\rho(\tau-t)} - e^{-i\rho(\tau-t)}] \quad (2.9) \end{aligned}$$

In terms of trigonometric functions, (2.8) becomes for $t \leqq \tau$

$$\begin{aligned} -i\rho M(t,\tau,\rho) &= (A_{24}\rho^2 + A_{13}) \cos \rho(\pi - \tau + t) \\ &- (A_{14} - A_{23})\rho \sin \rho(\pi - \tau + t) + (A_{24}\rho^2 - A_{13}) \cos \rho(\pi - \tau - t) \\ &+ (A_{14} + A_{23})\rho \sin \rho(\pi - \tau - t) - 2\rho A_{34} \sin \rho(t - \tau) \quad (2.10) \end{aligned}$$

A similar formula for $-i\rho M(t,\tau,\rho)$ results for $t > \tau$ by interchanging t and τ, and replacing A_{34} by A_{12}, in the right side of (2.10).

If $A_{24} \neq 0$, it follows easily from (2.7) that the zeros of ω, for large $|\rho|$, are close to $\rho = \pm m$, where m goes through the positive integers. Let each integer $\pm m$ be enclosed by a circle of radius $\frac{1}{4}$ and with that integer as center. Denote the points interior to these circles by E. Then the circles C_m with equations $|\rho| = m + \frac{1}{2}$ do not intersect E (and thus do not intersect any zeros of ω) for large integer m. Hence they may be taken as simple closed curves surrounding the zeros of ω. Since $l = \rho^2$, the circles C_m go into circles $\tilde{C}_m$: $|l| = (m + \frac{1}{2})^2$ in the l-plane.

Let σ_m be given for integrable functions f on $0 \leqq t \leqq \pi$ by

$$\begin{aligned} \sigma_m(t) &= \frac{-1}{2\pi i} \int_{\tilde{C}_m} \left(\int_0^\pi \Gamma(t,\tau,l)f(\tau)\, d\tau \right) dl \\ &= -\frac{1}{2\pi i} \int_{C_m} \left(\int_0^\pi \Gamma(t,\tau,\rho^2)f(\tau)\, d\tau \right) \rho\, d\rho \quad (2.11) \end{aligned}$$

where each circle is traversed once in the positive direction. It will be shown, in the case $A_{24} \neq 0$, that the difference between σ_m and the mth partial sum of the Fourier cosine development of f tends to zero as $m \to \infty$, uniformly on $0 \leqq t \leqq \pi$. The cosine partial sums s_m are given by

$$s_m(t) = \sum_{k=0}^{m} c_k \cos kt$$

where

$$c_0 = \frac{1}{\pi}\int_0^\pi f(\tau)\,d\tau \qquad c_k = \frac{2}{\pi}\int_0^\pi \cos k\tau\, f(\tau)\,d\tau \qquad (k = 1, 2, \ldots)$$

Theorem 2.1. *If $A_{24} \neq 0$, the difference $\sigma_m - s_m$ tends to zero as $m \to \infty$, uniformly on $0 \leqq t \leqq \pi$.*

REMARK: The sequence $\{\sigma_m\}$ is said to be *equiconvergent* with the sequence $\{s_m\}$ on an interval if $\sigma_m - s_m \to 0$ uniformly on the interval as $m \to \infty$.

Proof of Theorem 2.1. Let $\rho = u + iv$, where u,v are real. From (2.7) it follows that

$$\omega(\rho) = -2iA_{24}\rho^2 \sin \pi\rho + 0(|\rho|e^{\pi|v|}) \qquad (|\rho| \to \infty)$$

and, if ρ is outside the set E defined above, $|\sin \pi\rho| >$ const $e^{\pi|v|}$. Thus if ρ is not in E,

$$\frac{1}{\omega(\rho)} = \frac{1}{-2iA_{24}\rho^2 \sin \pi\rho} + 0\left(\frac{e^{-\pi|v|}}{|\rho|^3}\right) \qquad (|\rho| \to \infty)$$

Since $\Gamma(t,\tau,\rho^2) = M(t,\tau,\rho)/\omega(\rho)$, it then follows from (2.10) that for large $|\rho|$, ρ outside E,

$$\Gamma(t,\tau,\rho^2) = -\frac{1}{2}\frac{\cos \rho(\pi - \tau + t) + \cos \rho(\pi - \tau - t)}{\rho \sin \pi\rho} + 0\left(\frac{e^{-|v|(\tau-t)}}{|\rho|^2}\right) + 0\left(\frac{e^{-|v|(\pi-(\tau-t))}}{|\rho|^2}\right) \qquad (t \leqq \tau)$$

or

$$\Gamma(t,\tau,\rho^2) = -\frac{\cos \rho(\pi - \tau) \cos \rho t}{\rho \sin \pi\rho} + 0\left(\frac{e^{-|v|(\tau-t)}}{|\rho|^2}\right) + 0\left(\frac{e^{-|v|(\pi-(\tau-t))}}{|\rho|^2}\right) \qquad (t \leqq \tau) \quad (2.12)$$

For $t > \tau$ a similar estimate holds for Γ, the only change being that t and τ are interchanged in (2.12). Thus from (2.11) σ_m may be thought of as consisting of the sum of two terms $\sigma_m^{(1)}$ and $\sigma_m^{(2)}$, where

$$\sigma_m^{(1)}(t) = \frac{1}{2\pi i}\int_{C_m}\left\{\int_0^t \frac{\cos \rho(\pi - t) \cos \rho\tau}{\sin \pi\rho} f(\tau)\,d\tau + \int_t^\pi \frac{\cos \rho(\pi - \tau) \cos \rho t}{\sin \pi\rho} f(\tau)\,d\tau\right\} d\rho$$

Interchanging the order of integration and using the residue theorem. there results

$$\begin{aligned}\sigma_m^{(1)}(t) &= \frac{1}{\pi}\int_0^\pi \left(1 + 2\sum_{k=1}^{m} \cos kt \cos k\tau\right) f(\tau)\, d\tau \\ &= s_m(t)\end{aligned}$$

the mth partial sum of the Fourier cosine expansion of f.

It only remains to show that $\sigma_m^{(2)}(t)$ tends to zero uniformly in $0 \leqq t \leqq \pi$ as $m \to \infty$. From (2.12) one is led to estimate

$$\int_{C_m}\int_t^\pi \frac{e^{-|v|(\tau-t)}}{|\rho|} |f(\tau)|\, d\tau\, |d\rho| \tag{2.13}$$

If $\delta > 0$, then this term is

$$0\left(\int_{C_m} e^{-|v|\delta}\left|\frac{d\rho}{\rho}\right|\right) + 0\left(\int_t^{t+\delta} |f(\tau)|\, d\tau\right)$$

A simple calculation shows that, since C_m has radius $r_m = m + \frac{1}{2}$,

$$\int_{C_m} e^{-|v|\delta}\left|\frac{d\rho}{\rho}\right| < \frac{16}{\delta r_m} \tag{2.14}$$

By making δ small enough $\int_t^{t+\delta} |f(\tau)|\, d\tau$ can be made arbitrarily small, independent of t, since $\int_0^t |f(\tau)|\, d\tau$ is uniformly continuous on $0 \leqq t \leqq \pi$. Having chosen δ, the integer m can be made large enough so that the integral in (2.14) is made arbitrarily small. Thus (2.13) tends to zero uniformly in t as $m \to \infty$. Entirely similar considerations apply to the last term in (2.12) except that the integral of $|f(\tau)|$ over $(\pi - \delta, \pi)$ enters instead of over $(t, t + \delta)$. The terms for $t > \tau$ are handled similarly, and this proves the theorem.

The case $A_{24} \neq 0$ is not the only one for which a theorem of the type Theorem 2.1 holds. For example, if $A_{24} = 0$, and $A_{14} - A_{23} \neq 0$, then (2.7) and (2.10) imply the existence of circles C_m and a set E as before, where C_m does not intersect E for large m. Similar estimates to (2.12) hold. It would be of value to the student to formulate and prove the analogue of Theorem 2.1 for this case and also the case where $A_{13} \neq 0$ and all other $A_{ij} = 0$.

It can be shown directly that if f is differentiable on $0 \leqq t \leqq \pi$, then σ_m given by (2.11) converges to f as $m \to \infty$ except possibly at 0 and π. Briefly, what is involved is to observe that M is a linear sum of exponentials in τ. Thus

$$\int_t^\pi M(t,\tau,\rho) f(\tau)\, d\tau$$

is a sum of terms of which

$$\int_t^\pi e^{i\rho(\pi-\tau+t)}f(\tau)\,d\tau$$

is typical. This term can be integrated by parts, yielding

$$\frac{f(t)e^{i\rho\pi}}{i\rho} - \frac{f(\pi)e^{i\rho t}}{i\rho} + \frac{1}{i\rho}\int_t^\pi e^{i\rho(\pi-\tau+t)}f'(\tau)\,d\tau$$

The integration in ρ then gives $f(t)$ from terms like the first one above. The second and third terms are disposed of much as in the arguments used to prove Theorem 2.1.

3. Green's Function and the Expansion Theorem for the Case $Lx = -x'' + q(t)x$

Let G denote Green's function for the problem

$$Lx = -x'' + q(t)x = lx \qquad U_1x = 0 \qquad U_2x = 0 \tag{3.1}$$

where U_1, U_2 are given in (1.2), and q is a continuous complex-valued function on $0 \leqq t \leqq \pi$. As in Sec. 2, let Γ be Green's function for the problem

$$\Lambda x = -x'' = lx \qquad U_1x = 0 \qquad U_2x = 0 \tag{3.2}$$

Since $\partial G/\partial t$ and $\partial\Gamma/\partial t$ have the same discontinuity at $t = \tau$, it follows that $G - \Gamma$, considered as a function of t, is of class C^1, and, except possibly at $t = \tau$, is of class C^2. However, since G satisfies $Lx = lx$ (except at $t = \tau$) and Γ satisfies $\Lambda x = lx$ (except at $t = \tau$), $G - \Gamma$ is, in fact, of class C^2 because

$$\Lambda(G - \Gamma) - l(G - \Gamma) = -q(t)G$$

It follows from this equation and $U_i(G - \Gamma) = 0$, $i = 1, 2$, that, except at the poles of G and Γ,

$$G(t,\tau,l) - \Gamma(t,\tau,l) = -\int_0^\pi \Gamma(t,s,l)q(s)G(s,\tau,l)\,ds \tag{3.3}$$

It will be seen that the integral equation (3.3) determines the essential behavior of G in terms of the known behavior of Γ. It will be shown below that when Γ meets certain requirements for large $|l|$, (3.3) has a solution G, and it follows at once from (3.3) that this G has all the desired properties of Green's function for large $|l|$ and therefore must be Green's function. Thus, under these conditions, it will be the case that for *any* continuous q the function G exists and is meromorphic because if G exists for even one value of l, $\Delta \not\equiv 0$.

Under the assumption $A_{24} \neq 0$, it follows from (2.12), and a similar relation for $t > \tau$, that for large $|\rho|$, ρ outside E,

$$|\Gamma(t,\tau,\rho^2)| \leqq k|\rho|^{-1}h_v(t,\tau) \tag{3.4}$$

where k is a constant dependent only on the boundary conditions, and

$$h_v(t,\tau) = e^{-|v||t-\tau|} + e^{-|v|(\pi-|t-\tau|)}$$

The existence of a solution G of (3.3) will be proved under the assumption that Γ satisfies (3.4).

Clearly the case $A_{24} \neq 0$ *is not the only situation where* (3.4) *holds. If* $A_{24} = 0$ *and* $A_{23} - A_{14} \neq 0$, *then* (2.7) *and* (2.10) *imply the existence of a set* E *outside of which* (3.4) *is valid, and again there is a family of circles* C_m *of radii increasing by* 1 *when* m *increases by* 1, *and* C_m *does not intersect* E *for large* m. *Another case in which* (3.4) *holds is when* $A_{13} \neq 0$ *and all other* $A_{ij} = 0$.

Theorem 3.1. *If Green's function* Γ *for the problem* (3.2) *satisfies* (3.4) *for all sufficiently large* $|\rho|$, *there exists a solution* G *of the integral equation* (3.3) *for all sufficiently large* $|\rho|$, ρ *outside* E.

Proof. The method of successive approximations can be used. Let $G_0(t,\tau,l)$ be the zero function, and define G_{p+1}, $p = 0, 1, 2, \ldots,$ by

$$G_{p+1}(t,\tau,l) = \Gamma(t,\tau,l) - \int_0^\pi \Gamma(t,s,l)q(s)G_p(s,\tau,l)\,ds \tag{3.5}$$

for all $|\rho|$ sufficiently large, where $l = \rho^2$. Let

$$\max |G_{p+1}(t,\tau,\rho^2) - G_p(t,\tau,\rho^2)|(h_v(t,\tau))^{-1}|\rho| = k_p \tag{3.6}$$

where the maximum is taken over $0 \leqq t \leqq \pi$ for fixed τ and $|\rho|$ large, ρ outside E. By (3.4) and (3.5) it is clear that (3.6) holds for $p = 0$ with $k_0 = k$. Suppose now that it has been shown that

$$k_j \leqq \frac{k}{2^j} \qquad (j = 0, 1, \ldots, p) \tag{3.7}$$

Then it will be proved that (3.7) holds for $j = p + 1$ as well. Indeed, from (3.5),

$$k_{p+1} \leqq kk_p|\rho|^{-1} \max_{0\leqq t\leqq\pi} \int_0^\pi |q(s)|h_v(t,s)h_v(s,\tau)(h_v(t,\tau))^{-1}\,ds$$

Using $|s - t| + |\tau - s| \geqq |t - \tau|$, $\pi - |t - s| + |\tau - s| \geqq \pi - |t - \tau|$, $\pi - |\tau - s| + |t - s| \geqq \pi - |t - \tau|$, and $2\pi - |t - s| - |\tau - s| \geqq |t - \tau|$, there results

$$h_v(t,s)h_v(s,\tau) \leqq 2h_v(t,\tau)$$

Thus

$$k_{p+1} \leqq 2kk_p|\rho|^{-1} \int_0^\pi |q(s)|\,ds \tag{3.8}$$

and if $|\rho|$ is large enough

$$2k|\rho|^{-1} \int_0^\pi |q(s)|\, ds < \tfrac{1}{2}$$

Used in (3.8), this yields (3.7) for $j = p + 1$, and hence establishes (3.7) for all j by induction.

The uniform convergence of the sequence $\{G_p\}$ to a limit function G now follows readily. This G satisfies (3.3) for all sufficiently large $|l|$, proving the theorem.

From (3.7) it also follows that $|G|$ satisfies the same inequality (3.4) as Γ, with k replaced by $2k$. Using this in (3.3) again, there results

$$|G(t,\tau,\rho^2) - \Gamma(t,\tau,\rho^2)| \leqq 4k^2|\rho|^{-2}h_v(t,\tau) \int_0^\pi |q(s)|\, ds \tag{3.9}$$

Let f be an integrable function on $0 \leqq t \leqq \pi$, and let

$$S_m(t) = -\frac{1}{2\pi i} \int_{C_m} \left(\int_0^\pi G(t,\tau,\rho^2) f(\tau)\, d\tau \right) \rho\, d\rho$$

and recall that σ_m is the corresponding sum for f using Γ,

$$\sigma_m(t) = -\frac{1}{2\pi i} \int_{C_m} \left(\int_0^\pi \Gamma(t,\tau,\rho^2) f(\tau)\, d\tau \right) \rho\, d\rho$$

If Γ satisfies (3.4), it is possible to prove that $\{S_m\}$ is equiconvergent with $\{\sigma_m\}$, which means $S_m - \sigma_m \to 0$ as $m \to \infty$, uniformly in t, $0 \leqq t \leqq \pi$. This reduces the study of the convergence of $\{S_m\}$ to the simple cases considered in Sec. 2.

Theorem 3.2. *If Γ satisfies* (3.4), *then $\{S_m\}$ is equiconvergent with $\{\sigma_m\}$.*

Proof. From (3.9) it follows that for small $\delta > 0$

$$2\pi|S_m(t) - \sigma_m(t)| \leqq 2k_1 \left(\int_\delta^{t-\delta} + \int_{t+\delta}^{\pi-\delta} \right) |f(\tau)|\, d\tau \int_{C_m} e^{-|v|\delta} \left| \frac{d\rho}{\rho} \right| + 4\pi k_1 \left(\int_0^\delta + \int_{t-\delta}^{t+\delta} + \int_{\pi-\delta}^{\pi} \right) |f(\tau)|\, d\tau \tag{3.10}$$

where

$$k_1 = 4k^2 \int_0^\pi |q(s)|\, ds$$

and where obvious changes in the limits are made if $t < \delta$ or $t > \pi - \delta$. Given any $\epsilon > 0$, δ can be chosen small enough so that the last term on the right of (3.10) can be made less than $\epsilon/2$, independent of t on $0 \leqq t \leqq \pi$. Having so chosen δ, then by taking m large enough it is possible to make the first term on the right of (3.10) less than $\epsilon/2$, by making use of (2.14). This proves the equiconvergence of $\{S_m\}$ with $\{\sigma_m\}$.

A particular case where Theorem 3.2 is valid is when $A_{24} \neq 0$. Coupling this with Theorem 2.1, it follows that $\{S_m\}$ is equiconvergent with the Fourier cosine expansion of f on $0 \leqq t \leqq \pi$ in this case.

In case (3.4) does not hold, the construction of G from Γ becomes more complicated. The behavior of $G(t,\tau,l)$ as $|l| \to \infty$ can be found by using the results of Theorem 3.1, Chap. 6. For an nth-order equation, Sec. 5, Chap. 5, is relevant and $Lx = lx$ of (3.1) is considered in the example at the end of Sec. 5. With $l = \rho^2$, $Lx = lx$ has for $\Im\rho \geqq 0$ two solutions φ_1 and φ_2 with

$$\begin{aligned} \varphi_1(t,\rho) &\sim e^{i\rho t}\left[1 + \frac{1}{2i\rho}\int^t q(s)\,ds + 0\left(\frac{1}{|\rho|^2}\right)\right] \\ \varphi_2(t,\rho) &\sim e^{-i\rho t}\left[1 - \frac{1}{2i\rho}\int_0^t q(s)\,ds + 0\left(\frac{1}{|\rho|^2}\right)\right] \end{aligned} \tag{3.11}$$

Moreover,

$$\varphi_1'(t,\rho) \sim e^{i\rho t}\left[i\rho + \frac{1}{2}\int_0^t q(s)\,ds + 0\left(\frac{1}{|\rho|}\right)\right] \tag{3.12}$$

and similarly for φ_2'.

Expressing Δ, $K(t,\tau,l)$ of Chap. 7, and $G(t,\tau,l)$ in terms of φ_1 and φ_2 and using (3.11) and (3.12), the behavior of Δ and G for large $|\rho|$ is found (for $\Im\rho \geqq 0$). Since $l = \rho^2$, this determines the behavior of G for large $|l|$ and thus the convergence of

$$-\frac{1}{2\pi i}\int_{C_m}\left(\int_0^\pi G(t,\tau,l)f(\tau)\,d\tau\right)dl$$

as $m \to \infty$ may be considered.

4. The nth-order Case

The generalization of the method exploited in Secs. 2 and 3 to the case of the nth-order linear differential operator is straightforward. Consider the operator L, where

$$Lx = x^{(n)} + p_1(t)x^{(n-1)} + \cdots + p_n(t)x$$

and the p_j are continuous complex-valued functions on some closed bounded finite interval which, with no real restriction, may be assumed to be $0 \leqq t \leqq \pi$. Let U be a boundary operator with components U_1, . . . , U_n,

$$U_ix = \sum_{j=1}^{n} (a_{ij}x^{(j-1)}(0) + b_{ij}x^{(j-1)}(\pi)) \qquad (i = 1, \ldots, n)$$

the a_{ij}, b_{ij} being constants. The boundary-value problem of interest is

$Lx = lx$, $Ux = 0$. A slightly more general problem

$$Lx = lr(t)x \qquad Ux = 0$$

where $r \neq 0$ on $0 \leqq t \leqq \pi$ and of class C^n there, can be reduced to the case where $r(t) = 1$ on $0 \leqq t \leqq \pi$ by the substitution $ds = (r(t))^{1/n}\,dt$. It is also possible to assume that $p_1(t) = 0$ on $0 \leqq t \leqq \pi$, if p_1 is of class C^{n-1}, because the substitution $x = qy$, where q is a solution of $nq' + p_1 q = 0$, effects this. Thus it will be assumed that the boundary-value problem is of the form

$$Lx = x^{(n)} + p_2(t)x^{(n-2)} + \cdots + p_n(t)x = lx \qquad Ux = 0 \tag{4.1}$$

where $p_2, \ldots, p_n$ are continuous functions on $0 \leqq t \leqq \pi$.

The associated simpler problem

$$\Lambda x = x^{(n)} = lx \qquad Ux = 0 \tag{4.2}$$

is considered. If the nth roots of 1 are $\alpha_1, \ldots, \alpha_n$, then a set of linearly independent solutions of $\Lambda x = lx$ for $l \neq 0$ is $e^{\alpha_1 \rho t}, e^{\alpha_2 \rho t}, \ldots, e^{\alpha_n \rho t}$, where $l = \rho^n$. Using these solutions, the explicit nature of Green's function Γ for (4.2) can be analyzed as in Sec. 2, and analogous expansion results considered.

The relationship between Green's function G for (4.1) and Γ is obtained as follows. Since, as a function of t, G and Γ are of class C^{n-2} and since they have the same discontinuity in the $(n-1)$st derivative at $t = \tau$, it follows that $G - \Gamma$ is of class C^{n-1} in t. From the differential equations in (4.1) and (4.2) it follows that the nth derivative of $G - \Gamma$ is continuous at $t = \tau$ so that, in fact, $G - \Gamma \in C^n$ in t on $0 \leqq t \leqq \pi$. Clearly, except at $t = \tau$,

$$\Lambda(G - \Gamma) - l(G - \Gamma) = -p_2 G^{(n-2)} - \cdots - p_n G \tag{4.3}$$

where $G^{(k)}$ denotes the kth derivative of G with respect to t. Since $U(G - \Gamma) = 0$, (4.3) implies that

$$G(t,\tau,l) - \Gamma(t,\tau,l) = \int_0^\pi \Gamma(t,s,l) f(s,\tau,l)\, ds \tag{4.4}$$

where

$$f(s,\tau,l) = -p_2(s)G^{(n-2)}(s,\tau,l) - \cdots - p_n(s)G(s,\tau,l)$$

From (4.4) follows

$$G^{(k)}(t,\tau,l) - \Gamma^{(k)}(t,\tau,l) = \int_0^\pi \Gamma^{(k)}(t,s,l) f(s,\tau,l)\, ds \tag{4.5}$$

For cases where Γ has reasonably smooth behavior, (4.4) and (4.5) can be used for large $|\rho|$ $(l = \rho^n)$ as *defining* G. As in Sec. 3, (4.4) and (4.5) can be dealt with by a successive-approximation procedure. This also yields a bound on $|G - \Gamma|$ which can be used as in Sec. 3 to get an equiconvergence theorem for the representations of a function by G and by Γ.

Where Γ has complicated behavior, the behavior of G for large $|l|$ can be obtained directly by using the result of Chap. 6, as was indicated at the end of Sec. 3 for the case $n = 2$.

5. The Form of the Expansion

Before proving the expansion theorem, the nature of the expansion will be considered. The orthogonality of the eigenfunctions which holds when π is self-adjoint need not hold when π is no longer self-adjoint. It is replaced by a biorthogonality relationship involving the eigenfunctions of π and those of the adjoint problem π^+ defined in Chap. 11.

Let $l = \lambda$ be an eigenvalue for π, which means that

$$(L - \lambda)x = 0 \qquad Ux = 0 \tag{5.1}$$

has k independent solutions, $k \geqq 1$. By Theorem 3.4, Chap. 11, this implies that

$$(L^+ - \bar{\lambda})x = 0 \qquad U^+x = 0 \tag{5.2}$$

also has k independent solutions. Let λ_p and λ_q be eigenvalues of π and let χ_p be an eigenfunction of π for $l = \lambda_p$. Let ψ_q be an eigenfunction of π^+ for $l = \bar{\lambda}_q$. That

$$(\chi_p,\psi_q) = \int_a^b \chi_p \bar{\psi}_q \, dt = 0 \tag{5.3}$$

follows at once from the adjointness relationship

$$(L\chi_p,\psi_q) - (\chi_p,L^+\psi_q) = 0 \tag{5.4}$$

stated below (3.2), Chap. 11, if $\lambda_p \neq \lambda_q$.

In case $\lambda_p = \lambda_q$, the relationship between χ and ψ is treated, in an important case, in the following theorem.

Theorem 5.1. *If $G = G(t,\tau,l)$, the Green's function for π, has a simple pole at $l = \lambda_p$, then the residue of G at the pole is*

$$-\sum_{j=m_p}^{n_p} \chi_j(t)\bar{\psi}_j(\tau) \tag{5.5}$$

where the χ_j and ψ_j are eigenfunctions of π for $l = \lambda_p$ and of π^+ for $l = \bar{\lambda}_p$, respectively. Moreover,

$$(\chi_i,\psi_j) = \delta_{ij} \qquad (m_p \leqq i, j \leqq n_p)$$

and the χ_j, $m_p \leqq j \leqq n_p$, are a complete set of eigenfunctions of π at $l = \lambda_p$ and similarly the ψ_j for π^+ at $l = \bar{\lambda}_p$.

From this and (5.3) it follows now that χ_i with eigenvalue λ_p is orthogonal to all except one of the eigenfunctions of π^+. The exception is ψ_i with eigenvalue $\bar{\lambda}_p$, and $(\chi_i,\psi_i) = 1$. As a corollary there is the following theorem:

Theorem 5.2. *If all the poles of G are simple and if the eigenfunctions of π are $\{\chi_j\}$, then the eigenfunctions of π^+ can be arranged in a sequence $\{\psi_j\}$ so that*

$$(\chi_i,\psi_j) = \delta_{ij}$$

If χ_i has eigenvalue λ_p, then ψ_i has eigenvalue $\bar{\lambda}_p$.

In this case, the expansion (1.1) becomes, because (5.5) is valid at all poles,

$$f(t) = \sum_{j=1}^{\infty} \chi_j(t) \int_a^b f(\tau)\bar{\psi}_j(\tau)\,d\tau \tag{5.6}$$

which, in the self-adjoint case, of course becomes the familiar orthogonal-function expansion theorem with $\psi_j = \chi_j$. The expansion (1.1) does not take the simple form (5.6) if G has poles of order higher than 1.

Proof of Theorem 5.1. Let the residue of G at the simple pole $l = \lambda_p$ be $G_0(t,\tau)$. Let χ_j be an eigenfunction of π at $l = \lambda_p$. Then

$$(L - l)\chi_j = (\lambda_p - l)\chi_j$$

so that

$$\chi_j(t) = (\lambda_p - l) \int_a^b G(t,\tau,l)\chi_j(\tau)\,d\tau$$

Letting $l \to \lambda_p$, $(l - \lambda_p)G(t,\tau,l) \to G_0(t,\tau)$. Thus

$$\chi_j(t) = -\int_a^b G_0(t,\tau)\chi_j(\tau)\,d\tau \tag{5.7}$$

From (2.6), Chap. 7, it follows that $G(t,\tau,l) - K(t,\tau,l)$ is of class $C^n[a,b]$ as a function of t. Moreover, $K(t,\tau,l)$ is an entire function of l. Thus $G_0(t,\tau)$ is also the residue of $G - K$ at $l = \lambda_p$ and therefore

$$G_0(t,\tau) = \frac{1}{2\pi i} \int (G(t,\tau,l) - K(t,\tau,l))\,dl$$

where the integral is over a small circle with the center at λ_p. Thus $G_0 = G_0(t,\tau)$ is of class C^n as a function of t. From $(L - l)\,G = 0$, $t \neq \tau$, where L operates on G as a function of t, follows

$$(L - \lambda_p)G = (l - \lambda_p)G \qquad UG = 0$$

and, expanding G as a Laurent series in the neighborhood of $l = \lambda_p$, there follows for $t \neq \tau$

$$(L - \lambda_p)G_0 = 0 \qquad UG_0 = 0 \tag{5.8}$$

Since $G_0 \in C^n$ as a function of t, (5.8) is valid at $t = \tau$ also and therefore

$$G_0(t,\tau) = \sum_{j=m_p}^{n_p} C_j(\tau)\chi_j(t) \tag{5.9}$$

where the χ_j, $m_p \leqq j \leqq n_p$, are the independent eigenfunctions of π at $l = \lambda_p$ and the C_j are functions of τ.

Since $G^+(t,\tau,l) = \bar{G}(\tau,t,\bar{l})$, it follows that $\bar{G}_0(\tau,t)$, the residue of G^+ at $l = \bar{\lambda}_p$, is also of class C^n as a function of t, so that for π^+ the analogue of (5.8) is

$$(L^+ - \bar{\lambda}_p)_t \bar{G}_0(\tau,t) = 0 \qquad U_t \bar{G}_0(\tau,t) = 0 \tag{5.10}$$

Thus since the χ_j are independent, each $\bar{C}_j(t)$ in (5.9) must be an eigenfunction of π^+ at $l = \bar{\lambda}_p$. Denoting these eigenfunctions $\bar{C}_j(t)$ by $-\psi_j(t)$, the result (5.5) follows.

Using (5.5) in (5.7)

$$\chi_j(t) = \sum_{i=m_p}^{n_p} \chi_i(t) \int_a^b \bar{\psi}_i(\tau)\chi_j(\tau)\,d\tau$$

Since the χ_j are independent, the relation $(\chi_j,\psi_i) = \delta_{ji}$ follows at once. This shows that no $\psi_i \equiv 0$ and thus all the eigenfunctions χ_j of π at $l = \lambda_p$ occur in (5.5). If the ψ_i were not independent, the relationship $(\chi_i,\psi_j) = \delta_{ij}$ would be impossible. This completes the proof of Theorem 5.1.

The Green's function need not have simple poles. The problem

$$-x'' = lx \qquad x(0) = 0 \qquad x'(0) + x'(\pi) = 0$$

has double poles at all its eigenvalues, $\lambda_k = (2k + 1)^2$. Indeed, here

$$G(t,\tau,l) = \frac{\sin l^{\frac{1}{2}}t \cos l^{\frac{1}{2}}(\pi - \tau)}{2l^{\frac{1}{2}} \cos^2 (\frac{1}{2}l^{\frac{1}{2}}\pi)} - \begin{cases} 0 & (t < \tau) \\ \dfrac{\sin l^{\frac{1}{2}}(t - \tau)}{l^{\frac{1}{2}}} & (t > \tau) \end{cases}$$

PROBLEMS

1. Discuss the nature of the residue of $G(t,\tau,l)$ at a multiple pole of G.

2. Consider the system

$$Lx = x' - A(t)x = lR(t)x$$

on an interval $a \leqq t \leqq b$, where x is a vector and where A and R are continuous n-by-n matrices and the characteristic roots of R are distinct on $a \leqq t \leqq b$. Let M and N be constant matrices and $Ux = Mx(a) + Nx(b) = 0$ be a boundary condition for solutions of $Lx = lR(t)x$. There exists a continuous nonsingular matrix T on $a \leqq t \leqq b$ such that $T^{-1}RT = D$, a diagonal matrix. Setting $x = Ty$, one obtains

$$y' - (T^{-1}AT - T^{-1}T')y = lDy$$

provided T' exists. Thus assume that R is diagonal. Let $A = A_1 + A_2$, where A_1 is the diagonal matrix consisting of the diagonal of A, and let

$$P_0(t) = \exp\left[\int_a^t A_1(s)\,ds\right]$$

so that $P_0' = A_1P_0$. Suppose P_1 is the unique matrix with diagonal terms zero satisfying $P_1R - RP_1 = A_2P_0$. Let

$$\Psi(t,l) = (P_0(t) + P_1(t)l^{-1}) \exp[l\Lambda(t)]$$

where

$$\Lambda(t) = \int_a^t R(s)\,ds$$

Show that Ψ is a fundamental matrix for a system

$$x' - A(t)x = lR(t)x + B(t,l)l^{-1}x$$

where $|B(t,l)|$ is bounded for $a \leq t \leq b$ and for $|l|$ large. Since Ψ is known, Green's matrix function Γ for the problem

$$(*) \qquad Lx = lR(t)x + B(t,l)l^{-1}x \qquad Ux = 0$$

can be explicitly determined; do this. (See Prob. 16, Chap. 7.) Let G be Green's matrix for $Lx = lR(t)x$, $Ux = 0$. Prove that

$$G(t,\tau,l) = \Gamma(t,\tau,l) - l^{-1}\int_a^b \Gamma(t,s,l)B(s,l)G(s,\tau,l)\,ds$$

and develop results for (*) analogous to those given in Secs. 2 and 3.

3. In Prob. 4, Chap. 9, the solution φ_1 can be used to construct a Green's function $G(t,\tau,s^2)$ on $0 \leq t < \infty$ for $\Im s > 0$. Indeed, if $Ae^{i\alpha} = F$, as in Prob. 5, Chap. 9,

$$G(t,\tau,s^2) = \begin{cases} \dfrac{\psi(t,s)\varphi_1(\tau,s)}{F(s)} & (t < \tau) \\[2ex] \dfrac{\psi(\tau,s)\varphi_1(t,s)}{F(s)} & (t > \tau) \end{cases}$$

Now $\int G\,dl$ corresponds to $2\int Gs\,ds$. For a function f of class C^1 and vanishing for t small and for t large let

$$J(t) = \int_C s\,ds \int_0^\infty G(t,\tau,s^2)f(\tau)\,d\tau$$

where C is the path consisting of the line from $-R + i\epsilon$ to $R + i\epsilon$ and the semicircle $i\epsilon + Re^{i\theta}$, $0 \leq \theta \leq \pi$, in the s plane. If q is restricted so that for some $\delta > 0$, $\int_0^\infty e^{\delta t}|q(t)|\,dt < \infty$ (very much less will do), show that F is analytic for $\Im s > -\delta$ and that $F \to 1$ as $|s| \to \infty$. Thus F has a finite number of zeros for $\Im s \geq 0$. Consider J as $\epsilon \to 0$ and $R \to \infty$. That part of J which is in the semicircle can be computed as $R \to \infty$. Note that $\psi(t,s) \sim (\sin ts)/s$ as $|s| \to \infty$. Proceed similarly for $\Im s < 0$. Combine and get the expansion theorem of Prob. 4. Relax the conditions on f. The problem $x'(0) + ax(0) = 0$ for real a can also be solved as above.

The interest of the method here is that, as a little consideration will show, it is valid for complex-valued $q(t)$ and complex a. Carry out the case $x(0) = 0$ with q complex-valued and obtain the expansion theorem for that case. Note that the problem is no longer self-adjoint and thus F can have zeros off the imaginary axis. The eigenfunctions are no longer necessarily orthogonal.

CHAPTER 13

ASYMPTOTIC BEHAVIOR OF NONLINEAR SYSTEMS; STABILITY

1. Asymptotic Stability

The treatment of nonlinear systems presented here will be restricted to local behavior, that is, to the behavior of solutions starting near a known solution of a system.

A solution ψ of a system

$$x' = F(t,x) \qquad \left(' = \frac{d}{dt}\right)$$

which is defined for $t \geqq 0$ is said to be *stable* if, given any $\epsilon > 0$, there exists a $\delta > 0$ such that any solution φ of the system satisfying

$$|\varphi(0) - \psi(0)| < \delta$$

satisfies

$$|\varphi(t) - \psi(t)| < \epsilon \qquad (t \geqq 0)$$

Note that this requires solutions starting nearby $\psi(0)$ to exist for all $t \geqq 0$. The solution ψ is said to be *asymptotically stable* if, in addition to being stable,

$$|\varphi(t) - \psi(t)| \to 0 \qquad (t \to \infty)$$

The following result of Perron is the simplest example of asymptotic stability.

Theorem 1.1. *Let*

$$x' = Ax + f(t,x) \tag{1.1}$$

where A is a real constant matrix with the characteristic roots all having negative real parts. Let f be real, continuous for small $|x|$ and $t \geqq 0$, and

$$f(t,x) = o(|x|) \qquad (|x| \to 0)$$

uniformly in t, $t \geqq 0$. Then the identically zero solution is asymptotically stable.

The conditions that A and f be real or that f be continuous can be replaced by any other conditions which assure the local existence of a solution for (1.1) for small $|x|$ and $t \geqq 0$.

The fact that the characteristic roots of A have negative real parts assures that the linear system $y' = Ay$ has the trivial solution as an asymptotically stable solution.

Proof of Theorem 1.1. The solution φ of (1.1) with $|\varphi(0)|$ small can be continued for increasing t so long as $|\varphi(t)|$ remains small. So long as $\varphi(t)$ exists, it follows from (1.1) that

$$\varphi(t) = e^{tA}\varphi(0) + \int_0^t e^{(t-s)A}f(s,\varphi(s))\,ds \tag{1.2}$$

Because the real parts of the characteristic roots of A are negative, there exist positive constants K and σ such that

$$|e^{tA}| \leqq Ke^{-\sigma t} \qquad (t \geqq 0) \tag{1.3}$$

Using (1.3), (1.2) yields

$$|\varphi(t)| \leqq K|\varphi(0)|e^{-\sigma t} + K\int_0^t e^{-\sigma(t-s)}\,|f(s,\varphi(s))|\,ds$$

Given $\epsilon > 0$, there exists a δ such that $|f(t,x)| \leqq \epsilon|x|/K$ for $|x| \leqq \delta$. Thus, so long as $|\varphi(t)| \leqq \delta$, it follows that

$$e^{\sigma t}|\varphi(t)| \leqq K|\varphi(0)| + \epsilon\int_0^t e^{\sigma s}|\varphi(s)|\,ds$$

This inequality yields, by Prob. 1, Chap. 1,

$$e^{\sigma t}|\varphi(t)| \leqq K|\varphi(0)|e^{\epsilon t}$$

or

$$|\varphi(t)| \leqq K|\varphi(0)|e^{-(\sigma-\epsilon)t} \qquad (t \geqq 0) \tag{1.4}$$

If ϵ is chosen so that $\epsilon < \sigma$, then (1.4) shows that $|\varphi(t)| \leqq K|\varphi(0)|$ so long as $|\varphi(t)| \leqq \delta$. Thus, if $|\varphi(0)| < \delta/K$, it follows that (1.4) is valid for all $t \geqq 0$, which completes the proof of Theorem 1.1.

Let the characteristic roots of A be λ_k, $k = 1, 2, \ldots, n$, and let

$$\max\,(\Re\lambda_k) = -\mu < 0 \tag{1.5}$$

Then *any solution* φ *of* (1.1) *which tends to zero as* $t \to \infty$ *satisfies*

$$\limsup_{t\to\infty} \frac{\log|\varphi(t)|}{t} \leqq -\mu \tag{1.6}$$

Thus, by Theorem 1.1, all solutions with $|\varphi(0)|$ sufficiently small satisfy (1.6).

To prove (1.6), it is noted that σ in (1.3) can be taken as $\mu - \epsilon$ for any given $\epsilon > 0$. This may necessitate taking $K = K_\epsilon$ large. Since $\varphi(t) \to 0$, it is the case that $|\varphi(t_0)|$ can be made as small as is required by taking t_0 large enough. Thus, applying Theorem 1.1 for $t \geqq t_0$, it follows as in (1.4) that

$$e^{(\sigma-\epsilon)(t-t_0)}|\varphi(t)| = 0(1) \qquad (t \to \infty)$$

Since $\sigma = \mu - \epsilon$, it follows that

$$\limsup_{t\to\infty} \frac{\log|\varphi(t)|}{t} \leqq -\mu + 2\epsilon$$

Since $\epsilon > 0$ is arbitrary, (1.6) follows.

A more general statement of Theorem 1.1 weakens the requirement $|f| = o(|x|)$. It is sufficient to assume that, for some $k > 0$,

$$|f(t,x)| \leqq k|x| \qquad (t \geqq 0) \tag{1.7}$$

for all small $|x|$, and that, given any $\epsilon > 0$, there exist δ and T such that

$$|f(t,x)| \leqq \epsilon|x| \qquad (|x| \leqq \delta,\ t \geqq T) \tag{1.8}$$

To show that (1.7) and (1.8) suffice, observe that, with (1.7), (1.1) yields

$$\|\varphi\|' \leqq (\|A\| + kn^{\frac{1}{2}})\|\varphi\|$$

where $\|\varphi\|$ is the Euclidean length of φ. Here use is made of the fact that $n^{-\frac{1}{2}}|x| \leqq \|x\| \leqq |x|$, where x has n components. Thus, so long as $\|\varphi(t)\|$ is small,

$$\|\varphi(t)\| \leqq \|\varphi(0)\|e^{(\|A\|+kn^{\frac{1}{2}})t}$$

or

$$|\varphi(t)| \leqq n^{\frac{1}{2}}|\varphi(0)|e^{(|A|+kn^{\frac{1}{2}})t} \qquad (t \geqq 0) \tag{1.9}$$

so long as $|\varphi(t)|$ is small. In the same way,

$$|\varphi(0)| \leqq n^{\frac{1}{2}}|\varphi(t)|e^{(|A|+kn^{\frac{1}{2}})t} \qquad (t \geqq 0)$$

Having chosen ϵ, (1.8) is used for $t \geqq T$, and Theorem 1.1 is applied to the interval $t \geqq T$, assuming $|\varphi(T)|$ to be small. But by (1.9) it is the case that $|\varphi(T)|$ is small if $|\varphi(0)|$ is small enough. This proves that (1.7) and (1.8) can replace $|f| = o(|x|)$ in Theorem 1.1. It is also the case that (1.6) is valid here.

The inequality (1.9) and that below (1.9) show that stability over $[0, \infty)$ and $[T, \infty)$ are equivalent.

A special case of some interest where (1.7) and (1.8) hold is the case where $f(t,x)$ in (1.1) is replaced by $B(t)x + g(t,x)$, where the matrix $B(t) \to 0$ as $t \to \infty$ and $|g(t,x)| = o(|x|)$ uniformly in $t \geqq 0$ for small $|x|$. In this case, (1.1) would be written as

$$x' = Ax + B(t)x + g(t,x) \tag{1.10}$$

For Theorem 1.1 to be true for $t \geqq T$, it suffices for (1.8) to hold not for arbitrarily small ϵ but merely for $\epsilon < \sigma/K$, where σ and K are from (1.3). With this less restrictive hypothesis, (1.6) need not hold.

In case the matrix A in (1.1) has one or more characteristic roots with positive real parts, then it is not possible for the solution $\varphi = 0$ to be stable. In this sense Theorem 1.1 and the results following it are the best possible.

Theorem 1.2. *Let at least one characteristic root of A in* (1.1) *have its real part positive. Let $f(t,x)$ satisfy* (1.8). *Then the solution $\varphi = 0$ of* (1.1) *is not stable.*

REMARK: With a slightly more restricted hypothesis the result is a consequence of Theorem 4.1 of this chapter.

Proof of Theorem 1.2. To prove the theorem, a transformation $x = Py$, P a constant matrix, is made, resulting in an equation of the form

$$y' = By + g(t,y) \tag{1.11}$$

where $B = P^{-1}AP$. It will be shown that the zero solution of (1.11) is not stable, and this clearly implies that $\varphi = 0$ is not stable for (1.1). By proper choice of P, the matrix B can be put in the form

$$B = \begin{pmatrix} B_1 & 0 \\ 0 & B_2 \end{pmatrix} \tag{1.12}$$

where B_1 is a canonical matrix of k rows and columns with its characteristic roots all having positive real parts, while B_2 is a canonical matrix with characteristic roots all having nonpositive real parts. The characteristic roots are in the main diagonal. Those elements off the main diagonal which are not zero can be assumed to be $\gamma > 0$, where γ can be made as small as any assigned positive quantity by proper choice of P. While y corresponding to real x may be complex, Py will be real. Thus

$$g(t,y) = P^{-1}f(t,Py)$$

is defined.

Let the components of φ be φ_j and let

$$R^2 = \sum_{i=1}^{k} |\varphi_i|^2 \qquad \text{and} \qquad \rho^2 = \sum_{i=k+1}^{n} |\varphi_i|^2$$

Let the real parts of the characteristic roots of B_1 exceed some $\sigma > 0$. Choose $\epsilon < \sigma/10$ and choose η and T so that

$$|g(t,y)| \leqq \epsilon\|y\| \qquad (t \geqq T) \tag{1.13}$$

for $\|y\| \leqq \eta$.

Suppose the zero solution of (1.11) is stable. Thus for η and T as chosen above there exists a $\delta > 0$ such that, if φ is a solution of (1.11) with $\rho(T) + R(T) < \delta$, $\rho(t) + R(t) < \eta$ for $t \geqq T$. Choose such a solution φ with $R(T) = 2\rho(T) > 0$.

With σ defined as above, it follows from the use of (1.11), (1.12), and (1.13) that, for $t \geqq T$

$$\sum_{i=1}^{k} (\varphi_i'\bar{\varphi}_i + \varphi_i\bar{\varphi}_i') = 2RR' \geqq 2\sigma R^2 - 2\gamma R^2 - 2\epsilon(\rho + R)R$$

Or, since γ can be chosen smaller than $\sigma/20$ and since ϵ is chosen less than $\sigma/10$, it follows that

$$R' \geqq \tfrac{1}{2}\sigma R - \epsilon\rho \tag{1.14}$$

In the same way,

$$\rho' \leqq \epsilon(\rho + R) + \frac{\sigma}{20}\rho \tag{1.15}$$

From (1.14) and (1.15) follows

$$(R - \rho)' \geqq \tfrac{1}{4}\sigma(R - \rho)$$

Thus

$$R(t) - \rho(t) \geqq (R(T) - \rho(T))e^{\sigma(t-T)/4}$$

Since $R(T) = 2\rho(T)$, it follows that $R(t) \geqq \rho(T)e^{\sigma(t-T)/4}$. This is impossible, since under the hypotheses of stability $\rho(t) + R(t) < \eta$ for $t \geqq T$, and thus the theorem is proved.

Let $f(t,x)$ consist of a linear term $B(t)x$ and a term that for small $|x|$ is $0(|x|^{1+a})$, $a > 0$. An assumption of this kind about f leads to the possibility that, as a function of t, for fixed x, $f(t,x)$ can grow large as $t \to \infty$ without affecting asymptotic stability. This case is treated in the following theorem.

Theorem 1.3. *In Theorem* 1.1 *let the condition* $|f(t,x)| = o(|x|)$ *be replaced by the conditions that for small* $|x|$ *and all* $t \geqq 0$

$$|f(t,x)| \leqq k|x| + |x|^{1+a}t^b \tag{1.16}$$

where $a > 0$, b, *and* k *are constants, and that, given any* $\epsilon > 0$, *there exist* $\delta > 0$ *and* $T \geqq 0$ *such that for* $|x| \leqq \delta$ *and* $t \geqq T$

$$|f(t,x)| \leqq \epsilon|x| + |x|^{1+a}t^b \tag{1.17}$$

Then the solution $\varphi = 0$ *of* (1.1) *is asymptotically stable.*

Proof. With K and σ determined as in (1.3), choose $\eta < \sigma$. Choose ϵ in (1.17) so that $\epsilon K < \frac{1}{2}\eta$. The choice of ϵ determines δ and T. From (1.16) for $|x| < 1$ and $0 \leqq t \leqq T$,

$$|f(t,x)| \leqq (k + T^b)|x| \tag{1.18}$$

Using (1.18) much as (1.7) is used in deriving (1.9), it follows that $|\varphi(T)|$ can be made arbitrarily small by taking $|\varphi(0)|$ small enough. Let $|\varphi(0)|$ be small enough so that

$$|\varphi(T)| < \frac{\delta}{2K} \tag{1.19}$$

and so that for $t \geqq T$

$$K^{1+a}|\varphi(T)|^a e^{-a(\sigma-\eta)(t-T)} t^b < \tfrac{1}{4}\eta \tag{1.20}$$

The requirement certainly can be fulfilled since $\sigma > \eta$. From (1.3) applied for $t \geqq T$, and (1.17),

$$|\varphi(t)| \leqq K|\varphi(T)|e^{-\sigma(t-T)} + \epsilon K \int_T^t e^{-\sigma(t-s)}|\varphi(s)|\,ds + K \int_T^t e^{-\sigma(t-s)}|\varphi(s)|^{1+a} s^b\,ds \tag{1.21}$$

as long as $|\varphi(t)| \leqq \delta$. From (1.3) for $t \geqq 0$, $K \geqq 1$. So long as

$$|\varphi(t)|^a t^b \leqq \frac{\eta}{2K} \tag{1.22}$$

it follows from (1.16), using $\epsilon K < \frac{1}{2}\eta$, that

$$|\varphi(t)|e^{\sigma t} \leqq K|\varphi(T)|e^{\sigma T} + \eta \int_T^t e^{\sigma s}|\varphi(s)|\,ds$$

where this inequality and the following are valid as long as $|\varphi(t)| \leqq \delta$ and (1.22) holds. Applying Prob. 1, Chap. 1, the above inequality yields

$$e^{\sigma t}|\varphi(t)| \leqq K|\varphi(T)|e^{\sigma T}e^{\eta(t-T)}$$

or

$$|\varphi(t)| \leqq K|\varphi(T)|e^{-(\sigma-\eta)(t-T)} \tag{1.23}$$

Since (1.19) is satisfied, (1.23) implies that $|\varphi(t)| < \delta$ for $t \geqq T$. From (1.20) and (1.23) it follows that (1.22) is also satisfied for all $t \geqq T$ and thus (1.23) holds for all $t \geqq T$, proving the theorem.

It is easy to show that (1.6) is valid, with the hypothesis of Theorem 1.3, for all solutions of (1.1) which tend to zero as $t \to \infty$.

For the result of Theorem 1.3 it is evident from the proof that it is not necessary for ϵ in (1.17) to be arbitrarily small. It is enough if $\epsilon < \sigma/K$.

An important case where Theorem 1.3 applies is the system

$$\frac{dx}{ds} = \left(s^r \sum_{m=0}^{\infty} s^{-m} A_m\right) x + g(s,x) \tag{1.24}$$

where the A_m are constant matrices, r is a constant and $r > -1$, and $g(s,x)$ is a power series in the x_i for small $|x|$, beginning with at least second-degree terms, with coefficients which are $O(s^b)$ for large s with b

constant. The change of variables $t = s^{r+1}/(r+1)$ is made, and (1.24) assumes the required form for Theorem 1.3 with $A_0 = A$. The system (1.24) is a generalization to the nonlinear case of the irregular singular point at infinity. The treatment of (1.24) for complex s, in case g is analytic, requires little modification.

The results obtained for (1.1) can be easily extended to the case

$$x' = Ax + f(t,x,x') \tag{1.25}$$

In Theorem 1.1 it is only necessary to require that $f(t,x,w)$ satisfy

$$f(t,x,w) = o(|x| + |w|) \tag{1.26}$$

uniformly in $t \geqq 0$ for small $|x| + |w|$. The conclusion is that, if $|\varphi(0)|$ and $|\varphi'(0)|$ are small enough, then $\varphi(t) \to 0$ as $t \to \infty$. The various extensions which apply to (1.1) also apply to (1.25).

To prove these statements, α and $\beta < 1$ are determined so that, with K and σ as in (1.3),

$$0 < \gamma = \frac{\alpha + \beta|A|}{1-\beta} K < \sigma \tag{1.27}$$

From (1.26) it follows that there exists a $\delta > 0$ such that for $|x| + |w| \leqq \delta$

$$|f(t,x,w)| \leqq \alpha|x| + \beta|w|$$

Let $|\varphi(0)|$ and $|\varphi'(0)|$ be small. Then so long as $|\varphi(t)|$ and $|\varphi'(t)|$ are small, (1.25) implies

$$|\varphi'| \leqq |A|\,|\varphi| + |f| \leqq (|A| + \alpha)|\varphi| + \beta|\varphi'|$$

or

$$|\varphi'| \leqq \frac{|A| + \alpha}{1-\beta} |\varphi| \tag{1.28}$$

Thus $|\varphi'(t)|$ remains small so long as $|\varphi(t)|$ is small. From (1.3)

$$|\varphi(t)| \leqq K|\varphi(0)|e^{-\sigma t} + K \int_0^t e^{-\sigma(t-s)}(\alpha|\varphi(s)| + \beta|\varphi'(s)|)\,ds$$

Using (1.28) and (1.27),

$$|\varphi(t)| \leqq K|\varphi(0)|e^{-\sigma t} + \gamma \int_0^t e^{-\sigma(t-s)}|\varphi(s)|\,ds$$

But this implies

$$|\varphi(t)| \leqq K|\varphi(0)|e^{-(\sigma-\gamma)t} \qquad (t \geqq 0)$$

Thus, if $|\varphi(0)|$ and $|\varphi'(0)|$ are small enough, $|\varphi(t)|$ and $|\varphi'(t)|$ remain small, and the result is proved.

Theorem 1.4. *Theorems 1.1, 1.2, and 1.3 all apply equally well in case the constant matrix A is replaced by a real periodic matrix $P(t)$ and*

$$y' = P(t)y \tag{1.29}$$

has all n characteristic exponents with negative real parts.

Proof. A fundamental matrix solution Φ of (1.29) is given by (Theorem 5.1, Chap. 3)

$$\Phi(t) = Z(t)e^{tB}$$

where Z is a periodic nonsingular matrix and B is a constant matrix with all its characteristic roots having negative real parts. Let $x = Z(t)w$ in

$$x' = P(t)x + f(t,x) \tag{1.30}$$

Then

$$w' = Bw + Z^{-1}(t)f(t,Z(t)w) \tag{1.31}$$

so that all the theorems apply to (1.31) and hence also to solutions of (1.30). Indeed, a real solution φ of (1.30) gives rise to a solution $\psi = Z^{-1}\varphi$ of (1.31) which can be shown to go to zero as in Theorem 1.1 or 1.3. It then follows that φ itself goes to zero.

The conditions that P and f be real in (1.30) are used only to get the local existence of a solution and can be replaced by any other conditions that assure local existence of solutions, such as f analytic in x, for example.

2. First Variation: Orbital Stability

Let p be a real solution of

$$x' = F(t,x) \qquad \left(' = \frac{d}{dt}\right) \tag{2.1}$$

for $0 \leqq t < \infty$, where F is real, continuous, and has continuous first-order partial derivatives with respect to x_i, $i = 1, \ldots, n$, in a region of (t,x) space which contains the solution curve $(t,p(t))$, $0 \leqq t < \infty$. [The requirement that p and F be real can be replaced by any other condition that assures the local existence of solutions of (2.1). Thus it would suffice for F to be analytic in x for each t.] Let $z = \varphi - p$, where φ satisfies (2.1), and let the matrix with columns $(\partial F/\partial x_i)(t,p(t))$ be denoted by $F_x(t,p(t))$. Then

$$\begin{aligned} z' &= F(t, z + p(t)) - F(t,p(t)) \\ &= F_x(t,p(t))z + f(t,z) \end{aligned} \tag{2.2}$$

where, by the theorem of the mean,

$$f(t,z) = o(|z|) \tag{2.3}$$

for small $|z|$ uniformly in t over any finite t interval. If f is omitted from (2.2), there occurs the linear system

$$y' = F_x(t,p(t))y \tag{2.4}$$

which is called the *first variation* of (2.1) with respect to the solution $p(t)$. The first variation determines in some cases the nature of the stability of the solution p of (2.1).

An important case arises when F is periodic in t. Let the solution p be periodic of least period ω and let F be periodic of period ω in t. [Note that ω need not be the least period of F. If the smallest period of F in t is ω/m, where m is an integer and $m > 1$, then p is called a *subharmonic* solution of (2.1).] The equation (2.4) now has a periodic coefficient matrix of period ω. The appraisal (2.3) holds uniformly in t, $0 \leqq t < \infty$. The equation (2.2) is of the form considered in Theorem 1.4. Indeed, the following result is an immediate consequence of Theorem 1.4:

Theorem 2.1. *If the characteristic exponents associated with the equation of first variation* (2.4) *all have negative real parts, then the periodic solution p of* (2.1) *is asymptotically stable as $t \to \infty$.*

A much more subtle case arises when the right member of (2.1) does not depend on t. In this case,

$$x' = F(x) \tag{2.5}$$

The system (2.5) arises in classical mechanics as Hamilton's equations, for example.

It is assumed that p is a periodic solution of (2.5) of period T and that F is real and of class C^1 in some region of x space which contains the closed curve $x = p(t)$, $0 \leqq t \leqq T$. The case where F is analytic (and thus need not be real) is of special interest and will be mentioned later.

Since p is a solution of (2.5), $p'(t) = F(p(t))$. On differentiating this equation, there results the fact that $p'(t)$ is a solution of the equation of first variation

$$y' = F_x(p(t))y \tag{2.6}$$

Clearly p' has period T and thus the characteristic exponent associated with it as a solution of the linear system (2.6) may be taken as zero. In this case, the equation (2.6) can have at most $n - 1$ characteristic exponents with negative real parts and thus the hypothesis of Theorem 2.1 cannot be satisfied. Indeed, the conclusion of Theorem 2.1 does not hold in this case. To see this, note that p_δ, where $p_\delta(t) = p(t + \delta)$, is a solution of (2.5) for any constant δ. By taking δ small enough, p and p_δ can be made arbitrarily close at $t = 0$. Nevertheless, it is obvious that $|p(t + \delta) - p(t)|$ does not tend to zero as $t \to \infty$, so that asymptotic stability does not prevail.

It is possible, however, for a type of asymptotic stability to prevail which is of great importance. The solution $x = p(t)$ may be regarded as a closed curve or *orbit* in x space with t as a parameter. If $n - 1$ characteristic exponents of (2.6) have negative real parts, then the closed orbit is asymptotically stable in the sense that *any solution of* (2.5) *which comes near a point of the orbit tends to the orbit as* $t \to \infty$. This is called *asymptotic orbital stability.* Indeed, the following theorem is true:

Theorem 2.2. *Let* $n - 1$ *characteristic exponents of* (2.6) *have negative real parts. Then there exists an* $\epsilon > 0$ *such that if a solution* φ *of* (2.5) *satisfies* $|\varphi(t_1) - p(t_0)| < \epsilon$ *for some* t_0 *and* t_1, *there exists a constant* c *such that*

$$\lim_{t\to\infty} |\varphi(t) - p(t + c)| = 0 \tag{2.7}$$

Thus not only is there asymptotic orbital stability but each solution near the orbit possesses an asymptotic phase c.

In the proof that follows it will be shown that there is a surface S in x space which has dimension $n - 1$ such that all solutions of (2.5) which start on S at $t = 0$ tend to the curve $x = p(t)$ as $t \to \infty$. From this the result will follow readily.

Proof of Theorem 2.2. It will be assumed that the coordinates have been translated and rotated so that $p(0) = 0$ and $p'(0)$ is a vector with all components zero except the first. Thus $p'(0)$ is a multiple of the unit vector e_1 with components $(1, 0, 0, \ldots, 0)$.

The equation (2.2) now has the form

$$z' = F_x(p(t))z + f(t,z) \tag{2.8}$$

Since

$$f_z(t,z) = F_x(z + p(t)) - F_x(p(t))$$

it follows from the continuity of F_x that

$$f_z = o(1) \qquad (|z| \to 0) \tag{2.9}$$

uniformly in t.

Any real fundamental solution $\tilde{\Psi}$ of (2.6) satisfies

$$\tilde{\Psi}(t + T) = \tilde{\Psi}(t)C$$

where C is a real nonsingular matrix. Since p' is a solution of (2.6) of period T, one characteristic root of C is one. All other characteristic roots are less than one in magnitude, since it was assumed that $n - 1$ characteristic exponents of (2.6) have negative real parts. Thus there exists a real constant nonsingular matrix M such that

$$M^{-1}CM = \begin{pmatrix} 1 & 0 \\ 0 & C_1 \end{pmatrix}$$

where C_1 is a square matrix with $n - 1$ rows and with all characteristic roots less than one in magnitude. The matrix $\Psi = \tilde{\Psi}M$ is also a fundamental matrix for (2.6) and it satisfies

$$\Psi(t + T) = \Psi(t) \begin{pmatrix} 1 & 0 \\ 0 & C_1 \end{pmatrix} \tag{2.10}$$

Let ψ_1 be the first column of Ψ. Then (2.10) implies $\psi_1(T) = \psi_1(0)$ so that ψ_1 has period T. Because $n - 1$ characteristic exponents have negative real parts, there cannot be two independent solutions of (2.6) of period T. Therefore $\psi_1 = kp'$ for some constant k. Thus, with no restriction, the first column of $\Psi(0)$ can be taken as e_1, where e_1 is the vector with the first component one and all other components zero.

The solution matrix Ψ can be expressed as

$$\Psi(t) = Z(t)e^{tB} \tag{2.11}$$

where Z has period T, $e^{TB} = M^{-1}CM$, so that

$$B = \begin{pmatrix} 0 & 0 \\ 0 & B_1 \end{pmatrix}$$

where $B_1 = \log C_1/T$, and B_1 has all its characteristic roots with negative real parts. Therefore

$$\Psi(t) = Z(t) \begin{pmatrix} 1 & 0 \\ 0 & e^{tB_1} \end{pmatrix}$$

Let

$$U_1(t,s) = Z(t) \begin{pmatrix} 0 & 0 \\ 0 & e^{(t-s)B_1} \end{pmatrix} Z^{-1}(s)$$

and

$$U_2(t,s) = Z(t) \begin{pmatrix} 1 & 0 \\ 0 & 0 \end{pmatrix} Z^{-1}(s)$$

Clearly

$$U_1(t,s) + U_2(t,s) = \Psi(t)\Psi^{-1}(s) \tag{2.12}$$

is real and as a function of t for fixed s is a solution of (2.6). Since the first column of

$$Z(t) \begin{pmatrix} 1 & 0 \\ 0 & 0 \end{pmatrix} \tag{2.13}$$

is the first column of $\Psi(t)$, this matrix is real and is a solution of (2.6). The first row of the matrix

$$\begin{pmatrix} 1 & 0 \\ 0 & 0 \end{pmatrix} Z^{-1}(s) \tag{2.14}$$

is the first row of $\Psi^{-1}(s)$ and hence is real. Since $U_2(t,s)$ is the product of the matrices in (2.13) and (2.14), it follows that $U_2(t,s)$ is a real matrix which is a solution of (2.6) for fixed s. Since by (2.12) $U_1 + U_2$ is a real solution, U_1 is also a solution of (2.6) for fixed s, and is real.

Let the real parts of the characteristic roots of B_1 all be less than $-\sigma$, where $\sigma > 0$. Then there is a constant K such that

$$|U_1(t,s)| \leqq Ke^{-\sigma(t-s)} \qquad (t \geqq s) \tag{2.15}$$

$$|U_2(t,s)| \leqq K \tag{2.16}$$

Consider the integral equation

$$\theta(t) = \Psi(t)a + \int_0^t U_1(t,s)f(s,\theta(s))\,ds - \int_t^\infty U_2(t,s)f(s,\theta(s))\,ds \tag{2.17}$$

where a is a constant vector with the first component zero. For $t \geqq 0$, it follows from (2.11) that there exists a K_1 such that

$$|\Psi(t)a| \leqq K_1|a|e^{-\sigma t} \tag{2.18}$$

It is readily verified, with the use of (2.12) and the fact that the U_j are solutions of (2.6), that if the integral on the right of (2.17) and its derivative converge, then θ is a solution of (2.8). By (2.9) there exists a δ such that

$$|f(\tilde{z}) - f(z)| \leqq \frac{\sigma}{8K}|\tilde{z} - z| \qquad (|\tilde{z}|,\ |z| < \delta) \tag{2.19}$$

It will be shown by successive approximations that if $|a| < \delta/(2K_1)$, then (2.17) has a solution $\theta = \theta(t,a)$ for $t \geqq 0$ and

$$|\theta(t,a)| \leqq 2K_1|a|e^{-\frac{1}{2}\sigma t} \tag{2.20}$$

Let $\theta_{(0)}(t,a) = 0$ and let $\theta_{(k+1)}(t,a)$ be given by replacing $\theta(t)$ on the right of (2.17) by $\theta_{(k)}(t,a)$. Clearly, by (2.18),

$$|\theta_{(1)}(t,a) - \theta_{(0)}(t,a)| \leqq K_1|a|e^{-\frac{1}{2}\sigma t}$$

for $t \geqq 0$. It follows readily that if for $t \geqq 0$

$$|\theta_{(j)}(t,a) - \theta_{(j-1)}(t,a)| \leqq \frac{K_1|a|e^{-\frac{1}{2}\sigma t}}{2^{j-1}} \tag{2.21}$$

for $j \leqq k$, then $|\theta_{(j)}(t,a)| < 2K_1|a| < \delta$ for $j \leqq k$. Using (2.15), (2.16), and (2.19),

$$|\theta_{(k+1)}(t,a) - \theta_{(k)}(t,a)| \leqq K\int_0^t e^{-\sigma(t-s)}\frac{\sigma}{8K}|\theta_{(k)}(s,a) - \theta_{(k-1)}(s,a)|\,ds$$
$$+ K\int_t^\infty \frac{\sigma}{8K}|\theta_{(k)}(s,a) - \theta_{(k-1)}(s,a)|\,ds$$

Using (2.21) with $j = k$,

$$|\theta_{(k+1)}(t,a) - \theta_{(k)}(t,a)| \leqq \frac{K_1|a|}{2^{k-1}}\frac{\sigma}{8}\left[e^{-\sigma t}\int_0^t e^{\frac{1}{2}\sigma s}\,ds + \int_t^\infty e^{-\frac{1}{2}\sigma s}\,ds\right] \leqq \frac{K_1|a|}{2^k}e^{-\frac{1}{2}\sigma t}$$

which proves (2.21) by induction. From (2.21) it follows that the sequence $\{\theta_{(j)}\}$ converges uniformly for $0 \leqq t < \infty$ and $|a| < \delta/2K_1$ to a limit $\theta = \theta(t,a)$ which satisfies (2.20). Because of the uniformity of convergence, θ is a continuous function of (t,a) for $0 \leqq t < \infty$ and $|a|$ small. This and (2.20) in (2.17) show that θ is a solution of (2.8) which tends to zero uniformly in a as $t \to \infty$.

By (2.11) $Z(0) = \Psi(0)$ so that the first column of $Z(0)$ is e_1. Putting $t = 0$ in (2.17) and using the definition of U_2

$$\theta(0,a) = \Psi(0)a - \begin{pmatrix}1 & 0\\ 0 & 0\end{pmatrix}\int_0^\infty Z^{-1}(s)f(s,\theta(s,a))\,ds \tag{2.22}$$

The integral in (2.22) contributes nothing to the last $n - 1$ components of the vector equation. Taking the last $n - 1$ components of the equation (2.22) and observing that the cofactor of the first element of the first column of $\Psi(0)$ must be nonvanishing, it follows that the components $\theta_j(0,a)$, $j = 2, \ldots, n$, are linear combinations of the a_j, $j = 2, \ldots, n$, and conversely. If the initial values $\theta(0,a)$ are represented as points in z space, then $\theta(0,a)$ is taken as z in (2.22). Taking the first component of (2.22), it is seen that the initial values $z_i = \theta_i(0,a)$ satisfy an equation

$$z_1 + \sum_{j=2}^n b_j z_j + \tilde{H}(a_2, \ldots, a_n) = 0 \tag{2.23}$$

where the b_j are constants and $\tilde{H}$ is the first component of the integral on the right of (2.22). By (2.9) and (2.20),

$$\tilde{H}(a_2, \ldots, a_n) = o(|a|) \qquad (|a| \to 0) \tag{2.24}$$

Since the a_j are linear homogeneous in z_j, $j \geqq 2$, it follows that

$$\tilde{H}(a_2, \ldots, a_n) = H(z_2, \ldots, z_n)$$

Thus (2.23) becomes

$$z_1 + \sum_{j=2}^n b_j z_j + H(z_2, \ldots, z_n) = 0 \tag{2.25}$$

where $H = o(|z_2| + \cdots + |z_n|)$. The equation (2.25) is the equation of a surface S in z space from which solutions θ of (2 8) emanate at $t = 0$

which tend to zero as $t \to \infty$. The surface is only defined near $z = 0$. Clearly the tangent plane to S at the origin is given by (2.25) with H replaced by zero. Since $x = z + p(t)$, it follows because $p(0) = 0$ that the initial manifold in x space, which will also be denoted by S, has the same equation as in z space; that is,

$$x_1 + \sum_{j=2}^{n} b_j x_j + H(x_2, \ldots, x_n) = 0$$

Because $p'(0)$ is parallel to the x_1 axis, it follows that the curve $x = p(t)$ crosses S at $x = 0$ and is not tangent to S.

If a solution φ of (2.5) satisfies $|\varphi(t_1) - p(t_0)| < \epsilon$ for some t_1 and t_0, then because $\psi(t) = \varphi(t - t_0 + t_1)$ is also a solution, $|\psi(t_0) - p(t_0)| < \epsilon$. Because p has period T and because ψ and p are both solutions of (2.5), it follows that $|\psi(t) - p(t)|$ remains small for $|t - t_0| < 2T$ if ϵ is small. Thus the solution ψ crosses the surface S for some $\bar{t}$, where $|\bar{t} - t_0| < 2T$. But the solution $\tilde{\psi}$ of (2.5), where $\tilde{\psi}(t) = \psi(t + \bar{t})$, which has $\tilde{\psi}(0)$ on S satisfies $\tilde{\psi}(t) - p(t) \to 0$. Thus $\varphi(t - t_0 + t_1 + \bar{t}) - p(t) \to 0$ so that if $c = t_0 - t_1 - \bar{t}$, Theorem 2.2 is proved.

If F is analytic in x, then it follows readily from the uniform convergence of $\{\theta_{(j)}\}$ that the surface S is an analytic surface because $\theta(0,a)$ is analytic in $(a_2, \ldots, a_n)$, from which it follows that H is analytic in $(x_2, \ldots, x_n)$.

3. Asymptotic Behavior of a System

Theorem 3.1. *In the system*

$$x' = Ax + f(t,x) + g(t,x) \tag{3.1}$$

let f and g be continuous for small $|x|$ and $t \geqq 0$. For small $|x|$ let

$$g(t,x) \to 0 \text{ as } t \to \infty \tag{3.2}$$

uniformly in x. Let the characteristic roots of A have negative real parts, and given any $\epsilon > 0$, let there exist δ and t_ϵ so that $|f| \leqq \epsilon|x|$ for $|x| \leqq \delta$ and $t \geqq t_\epsilon$. Then there exists a T such that any solution $\varphi(t) \to 0$ as $t \to \infty$ if $|\varphi(T)|$ is small enough.

REMARK: Even though $\varphi(t) \to 0$ as $t \to \infty$, it is not the case that $\varphi = 0$ is a solution of (3.1) unless $g(t,0) \equiv 0$. In particular, g may be independent of x.

Proof of Theorem 3.1. To prove the above result, the constants K and σ of (1.3) are required. Let δ and T be chosen so that for $|x| \leqq \delta$

$$|f(t,x)| \leqq \epsilon|x| < \frac{\sigma|x|}{2K} \qquad (t \geqq T)$$

and let T be increased if necessary so that for $t \geqq T$ there is an $a > 0$ such that

$$|g(t,x)| \leqq a < \frac{\sigma - K\epsilon}{K}\,\delta \tag{3.3}$$

Much as in Theorem 1.1, if φ is a solution of (3.1)

$$|\varphi(t)| \leqq K|\varphi(T)|e^{-\sigma(t-T)} + K\epsilon \int_T^t e^{-\sigma(t-s)}|\varphi(s)|\,ds + K\int_T^t e^{-\sigma(t-s)}|g(s,\varphi(s))|\,ds \tag{3.4}$$

so long as $|\varphi(t)| \leqq \delta$. Let max $|\varphi(s)|$ for $T \leqq s \leqq t$ be denoted by $M(t)$. Then

$$M(t) \leqq K|\varphi(T)| + \frac{K\epsilon M(t)}{\sigma} + \frac{Ka}{\sigma}$$

or

$$M(t) \leqq \frac{K\sigma}{\sigma - K\epsilon}|\varphi(T)| + \frac{Ka}{\sigma - K\epsilon} \tag{3.5}$$

Clearly, since $Ka/(\sigma - K\epsilon) < \delta$, it follows that if $|\varphi(T)|$ is small enough $M(t) < \delta$ for all $t \geqq T$.

Let

$$\gamma = \limsup_{t\to\infty} |\varphi(t)|$$

Clearly $0 \leqq \gamma \leqq \delta < \infty$, and there exists a sequence $\{t_j\}, j = 1, 2, \ldots,$ such that as $j \to \infty$, $t_j \to \infty$, and $|\varphi(t_j)| \to \gamma$. From (3.4) it follows that

$$\begin{aligned}|\varphi(t_j)| \leqq K|\varphi(T)|e^{-\sigma(t_j-T)} &+ K\epsilon \int_T^{t_j/2} e^{-\sigma(t_j-s)}|\varphi(s)|\,ds \\ &+ K\epsilon\int_{t_j/2}^{t_j} e^{-\sigma(t_j-s)}|\varphi(s)|\,ds + K\int_T^{t_j/2} e^{-\sigma(t_j-s)}|g(s,\varphi(s))|\,ds \\ &+ K\int_{t_j/2}^{t_j} e^{-\sigma(t_j-s)}|g(s,\varphi(s))|\,ds\end{aligned}$$

Given any $\eta > 0$, there exists an integer J_η such that for all $j \geqq J_\eta$, $|\varphi(t_j)| > \gamma - \eta$, and $|\varphi(t)| < \gamma + \eta$ for $t \geqq t_j/2$. Thus, for $j \geqq J_\eta$,

$$\begin{aligned}\gamma - \eta \leqq K|\varphi(T)|e^{-\sigma(t_j-T)} + \frac{K\epsilon\delta}{\sigma}e^{-\frac{1}{2}\sigma t_j} + \frac{K\epsilon(\gamma+\eta)}{\sigma} + \frac{Ka}{\sigma}e^{-\frac{1}{2}\sigma t_j} \\ + \frac{K}{\sigma}\max_{\frac{1}{2}t_j \leqq s \leqq t_j}|g(s,\varphi(s))|\end{aligned}$$

Letting $j \to \infty$, it follows that $\gamma - \eta \leqq (K\epsilon/\sigma)(\gamma + \eta)$; and since $K\epsilon/\sigma < \frac{1}{2}$, $\gamma < 3\eta$, which implies $\gamma = 0$.

REMARK: If g does not necessarily satisfy (3.2) but does satisfy (3.3), it follows from (3.5) that φ exists and is bounded over (T, ∞) if $|\varphi(T)|$ is small enough.

Theorem 3.1 may be applied to the case of a first-order equation of classical interest. Let y and s be scalars and let

$$\frac{dy}{ds} = s^{-m}[ay + bs + h(s,y)] \tag{3.6}$$

for small $s \geqq 0$ and small y. Let $a > 0$ and b be constants and let $h(s,y)$ be a power series in s and y involving terms of the second and higher powers. (Much less restrictive hypotheses on h suffice, as will be clear from the proof.) Suppose $m > 1$ and let $(m - 1)s^{m-1} = 1/t$. Then (3.6) becomes

$$\frac{dy}{dt} = -ay + g(t) + f(t,y)$$

for large t and small y, where g and f (scalars) satisfy the hypotheses for (3.1) for $t > 0$. Indeed, $g(t)$ comes from $-bs - h(s,0)$ while $f(t,y)$ comes from $-h(s,y) + h(s,0)$. Thus any solution φ of (3.6), with $|\varphi(s_0)|$ sufficiently small for some $s_0 > 0$, satisfies $\varphi(s) \to 0$ as $s \to +0$.

The case $m = 1$ is handled by setting $s = e^{-t}$. (The case $m < 1$ is not singular at the origin, the equation being, in fact, Lip (s), and thus the existence theorem of Prob. 4, Chap. 1, is applicable in this case.)

It is clear that s^m in (3.6) can be replaced by a positive function of $s, p(s)$, defined, for $s > 0$ and with

$$\int_s^1 \frac{ds}{p(s)} < \infty \qquad (s > 0)$$

but with

$$\int_0^1 \frac{ds}{p(s)} = \infty$$

The substitution $\int_s^1 ds/p(s) = t$ can be made here. The term bs in the numerator of (3.6) can also be replaced by more general terms.

4. Conditional Stability

If some but not all of the characteristic roots of A have negative real parts, then some but not all of the solutions φ of

$$x' = Ax + f(t,x) \tag{4.1}$$

with $|\varphi(0)|$ small tend to zero as $t \to \infty$, providing f is suitably restricted.

It will be assumed here that f is continuous in (t,x) for small $|x|$ and $t \geqq 0$; moreover, given any $\epsilon > 0$, there exists a δ and T such that for $t \geqq T$

$$|f(t,\tilde{x}) - f(t,x)| \leqq \epsilon|\tilde{x} - x| \tag{4.2}$$

for $|x| \leqq \delta$, $|\tilde{x}| \leqq \delta$. A sufficient condition for (4.2) is that the matrix f_x exist and that as $|x| \to 0$

$$f_x = o(1)$$

uniformly in $t \geqq 0$. It will be assumed that $f(t,0) = 0$. It will also be assumed that A and f are real-valued but, as in previous results of this chapter, this requirement can be omitted if f is, for example, analytic.

Theorem 4.1. *Let the above assumptions hold and let k characteristic roots of A have negative real parts and $n - k$ have positive real parts. Then for any large t_0 there exists in x space a real k-dimensional manifold S containing the origin such that any solution φ of (4.1) with $\varphi(t_0)$ on the manifold S satisfies $\varphi(t) \to 0$ as $t \to \infty$. Moreover, there exists an η such that any solution φ near the origin but not on S at $t = t_0$ cannot satisfy $|\varphi(t)| \leqq \eta$, $t \geqq t_0$. If f is analytic in x for each $t \geqq 0$ and $|x|$ small, then S is an analytic manifold.*†

More precisely, it will be shown that there exists a real nonsingular constant matrix P such that if $y = Px$ then there are $n - k$ real continuous functions $\psi_j = \psi_j(y_1, \ldots, y_k)$ defined for small $|y_i|$, $i \leqq k$, such that

$$y_j = \psi_j(y_1, \ldots, y_k) \qquad (j = k + 1, \ldots, n) \tag{4.3}$$

define a k-dimensional manifold $\tilde{S}$ in y space. The manifold S in x space is obtained from $\tilde{S}$ by applying P^{-1} to y so that

$$x = P^{-1} \begin{pmatrix} y_1 \\ \vdots \\ y_k \\ \psi_{k+1} \\ \vdots \\ \psi_n \end{pmatrix}$$

defines S in terms of k curvilinear coordinates $y_1, \ldots, y_k$.

If there is a constant c such that for each fixed $t \geqq t_0$ f is analytic in x for $|x| < c$, where x is a vector with complex components, then it will be shown that the ψ_j are analytic in $(y_1, \ldots, y_k)$.

Proof of Theorem 4.1. There exists a real nonsingular constant matrix P such that

$$PAP^{-1} = \begin{pmatrix} B_1 & 0 \\ 0 & B_2 \end{pmatrix} = B$$

where B_1 is a matrix of k rows and columns having all its characteristic roots with negative real parts and B_2 is a matrix of $n - k$ rows and columns having all its characteristic roots with positive real parts. Letting $y = Px$, (4.1) becomes

$$y' = By + g(t,y) \tag{4.4}$$

† See also Prob. 11.

where $g = Pf(t,P^{-1}y)$. Thus from (4.2) it follows that, given any ϵ, there exist a δ and T, not necessarily equal to those of (4.2), such that

$$|g(t,\tilde{y}) - g(t,y)| \leqq \epsilon|\tilde{y} - y| \tag{4.5}$$

for $|\tilde{y}| \leqq \delta$, $|y| \leqq \delta$, $t \geqq T$. Let

$$U_1(t) = \begin{pmatrix} e^{tB_1} & 0 \\ 0 & 0 \end{pmatrix} \tag{4.6}$$

and

$$U_2(t) = \begin{pmatrix} 0 & 0 \\ 0 & e^{tB_2} \end{pmatrix} \tag{4.7}$$

Then $e^{tB} = U_1(t) + U_2(t)$ and

$$U_j' = BU_j \qquad (j = 1, 2) \tag{4.8}$$

Let $\alpha > 0$ be chosen so that the real parts of the characteristic roots of B_1 are less than $-\alpha$. Then there exist positive constants K and σ such that

$$|U_1(t)| \leqq Ke^{-(\alpha+\sigma)t} \qquad (t \geqq 0) \tag{4.9}$$
$$|U_2(t)| \leqq Ke^{\sigma t} \qquad (t \leqq 0) \tag{4.10}$$

Consider the integral equation

$$\theta(t,a) = U_1(t - t_0)a + \int_{t_0}^{t} U_1(t - s)g(s,\theta(s,a))\,ds - \int_{t}^{\infty} U_2(t - s)g(s,\theta(s,a))\,ds \qquad (t_0 \geqq T) \tag{4.11}$$

where a is a constant vector. Let ϵ in (4.5) be chosen so that $2\epsilon K/\sigma < \frac{1}{2}$ and let $|a|$ satisfy $2K|a| < \delta$. Using successive approximations to solve (4.11) with $\theta_{(0)}(t,a) = 0$, it follows readily that

$$|\theta_{(l+1)}(t,a) - \theta_{(l)}(t,a)| \leqq \frac{K|a|e^{-\alpha(t-t_0)}}{2^l}$$

which leads to the existence of a solution θ of (4.11) which satisfies

$$|\theta(t,a)| < 2K|a|e^{-\alpha(t-t_0)} \tag{4.12}$$

From (4.11) it is clear that the last $n - k$ components of the vector a do not enter into the solution and may be taken as zero. That θ is a solution of (4.4) is immediate for small $|a|$, since by (4.10) the integral in (4.11) converges. It is also clear from the uniform convergence of the successive approximations that θ is continuous in (t,a) for $t \geqq t_0$ and $|a|$ small.

From (4.11) it follows that the first k components of $\theta_j(t_0,a)$ are

$$\theta_j(t_0,a) = a_j \qquad (j = 1, \ldots, k)$$

and the later components are given by

$$\theta_j(t_0,a) = -\left(\int_{t_0}^{\infty} U_2(t_0 - s)g(s,\theta(s,a))\,ds\right)_j \qquad (j = k+1, \ldots, n)$$

where $(\ \ )_j$ denotes the jth component. If the functions ψ_j are defined by

$$\psi_j(a_1, \ldots, a_k) = -\left(\int_{t_0}^{\infty} U_2(t_0 - s)g(s,\theta(s,a))\,ds\right)_j$$

for $j = k + 1, \ldots, n$, then clearly the initial values $y_j = \theta_j(t_0,a)$ satisfy the equations

$$y_j = \psi_j(y_1, \ldots, y_k) \qquad (j = k+1, \ldots, n)$$

in y space, which define a manifold $\tilde{S}$ in y space. The condition (4.5) implies the uniqueness of solutions of (4.4) which start near the origin. Therefore, if p is any solution of (4.4) with $|p(t_0)|$ small and $p(t_0)$ on $\tilde{S}$, then $p(t) = \theta(t,a)$ for some a, where θ is the solution of (4.11) satisfying $\theta(t_0,a) = p(t_0)$, and $p(t) \to 0$ as $t \to \infty$.

It will be shown next that no solution p of (4.4) with $|p(t_0)|$ small and $p(t_0)$ not on $\tilde{S}$ can satisfy $|p(t)| \leqq \delta$ for $t \geqq t_0$, where δ is the same as below (4.5). Indeed, suppose $|p(t)| \leqq \delta$ for $t \geqq t_0$. Then it follows readily from (4.4) that

$$p(t) = e^{(t-t_0)B}p(t_0) + \int_{t_0}^{t} e^{(t-s)B}g(s,p(s))\,ds$$

Using (4.6) and (4.7), this can be written as

$$p(t) = U_1(t - t_0)p(t_0) + U_2(t - t_0)c + \int_{t_0}^{t} U_1(t - s)g(s,p(s))\,ds - \int_{t}^{\infty} U_2(t - s)g(s,p(s))\,ds \tag{4.13}$$

where c is the constant vector

$$c = \int_{t_0}^{\infty} U_2(t_0 - s)g(s,p(s))\,ds + p(t_0)$$

and the integral above converges because of (4.10) and the fact that by (4.5) $|g(s,p(s))|$ is bounded for $|p(s)| \leqq \delta$ and $s \geqq t_0$.

Clearly all the terms on the right of (4.13) are bounded as $t \to \infty$ except possibly the term $U_2(t - t_0)c$. Unless all the components c_j, $j > k$, of c vanish, it will be shown that this term is unbounded as $t \to \infty$. Each component of $U_2(t - t_0)c$ is the sum of polynomials multiplied by exponential terms of increasing magnitude. Thus by Prob. 26, Chap. 3, each component is unbounded unless it vanishes identically. By (4.7) all components can vanish only if all c_j are zero for $j > k$. Since the left side of (4.13) is bounded as $t \to \infty$, it follows that the right side must be bounded and thus that all components c_j, $j > k$, of c vanish so that p satisfies (4.11).

It will now be shown that, if θ is a solution of (4.11) satisfying $|\theta(t,a)| \leqq \delta$ for $t \geqq t_0$, then it is unique. This will prove that if p is any solution of (4.4) for which $|p(t)| \leqq \delta$, $t \geqq t_0$, then $p(t)$ is $\theta(t,a)$ for some a, where θ is the solution of (4.11) constructed above by successive approximations. Thus $p(t_0)$ is on $\tilde{S}$, contradicting the fact that $p(t_0)$ is not on $\tilde{S}$.

To prove the uniqueness of solutions of (4.11), let θ and $\tilde{\theta}$ be solutions for the same a and let $|\theta(t,a)| \leqq \delta$ and $|\tilde{\theta}(t,a)| \leqq \delta$. Then (4.11) yields, with (4.5),

$$|\tilde{\theta}(t,a) - \theta(t,a)| \leqq K\epsilon e^{-\sigma t} \int_{t_0}^{t} e^{\sigma s}|\tilde{\theta}(s,a) - \theta(s,a)|\, ds + K\epsilon e^{\sigma t} \int_{t}^{\infty} e^{-\sigma s}|\tilde{\theta}(s,a) - \theta(s,a)|\, ds$$

If $\sup |\tilde{\theta}(t,a) - \theta(t,a)| = M$ $(t \geqq t_0)$, then $M \leqq 2K\epsilon M/\sigma$ so that, since $\epsilon < \sigma/2K$, it follows that $M = 0$, proving the uniqueness.

In case f is analytic in x for each $t \geqq 0$ and $|x|$ small, it follows in the usual way as a consequence of the uniform convergence of the successive approximations procedure for (4.11) that θ is analytic in a for fixed t and therefore that $\tilde{S}$ is an analytic manifold. This completes the proof of Theorem 4.1.

In the case where f has continuous first derivatives with respect to the x_i, the manifold S is of class C^1, as the following theorem shows.

Theorem 4.2. *The manifold S of Theorem 4.1 is differentiable if $\partial f/\partial x_i$ exists and is continuous for $i = 1, \ldots, n$, and t_0 is sufficiently large. To be more precise, the functions ψ_j, $j = k + 1, \ldots, n$, are of class C^1 for $|y_l|$ sufficiently small, $l \leqq k$. Moreover, $\partial\psi_j/\partial y_l = 0$ at $y_1 = \cdots = y_k = 0$*

Proof. The proof is equivalent to proving that $(\partial\theta/\partial a_i)(t_0,a)$, $i = 1, \ldots, k$, exist and are continuous for small $|a|$.

Let h be a scalar and j be fixed and let $a + h$ be used to denote $a + he_j$. Let $p(t) = |\theta(t,a+h) - \theta(t,a)|$. Then, using (4.5) in (4.11), it follows for small $|h|$ that

$$p(t) \leqq K|h| + K\epsilon \int_{t_0}^{t} e^{-\sigma(t-s)}p(s)\, ds + K\epsilon \int_{t}^{\infty} e^{-\sigma(s-t)}p(s)\, ds$$

Let $M = \sup p(t)$ for $t \geqq t_0$. Then the above yields

$$M \leqq K|h| + \frac{2K\epsilon M}{\sigma}$$

Since $2K\epsilon/\sigma < \frac{1}{2}$, it follows that $M \leqq 2K|h|$ or that $|p(t)| \leqq 2K|h|$. Let $q(t,a,h) = [\theta(t,a+h) - \theta(t,a)]/h$. Then the above result shows that

$$|q| \leqq 2K$$

From (4.11) follows

$$q(t,a,h) = U_1(t - t_0)e_j + \int_{t_0}^{t} U_1(t - s)[g_y(s,\theta(s,a))q(s,a,h) + \Delta]\,ds$$
$$- \int_{t}^{\infty} U_2(t - s)[g_y(s,\theta(s,a))q(s,a,h) + \Delta]\,ds \quad (4.14)$$

where g_y is the matrix $(\partial g_i/\partial y_j)$ and

$$\Delta = \frac{1}{h}[g(s,\theta(s,a + h)) - g(s,\theta(s,a))] - g_y(s,\theta(s,a))q(s,a,h)$$

Given any $\eta > 0$, $|h|$ can be taken so small that by the theorem of the mean and the continuity of g_y for small s and by (4.5) and (4.12) for large s it follows that

$$|\Delta| \leqq \eta|q(s,a,h)|$$

and since $|q| \leqq 2K$

$$|\Delta| \leqq 2K\eta \quad (4.15)$$

If $|a|$ is small, it follows from (4.5) and (4.12) that

$$|g_y(s,\theta(s,a))| \leqq \epsilon n \quad (4.16)$$

Let ϵ be small enough so that $2K\epsilon n/\sigma < \frac{1}{2}$. Let

$$\psi(t,a) = U_1(t - t_0)e_j + \int_{t_0}^{t} U_1(t - s)g_y(s,\theta(s,a))\psi(s,a)\,ds$$
$$- \int_{t}^{\infty} U_2(t - s)g_y(s,\theta(s,a))\psi(s,a)\,ds \quad (4.17)$$

That a continuous solution ψ of the linear system (4.17) exists follows from the use of successive approximations. Subtracting (4.17) from (4.14), denoting sup $|q - \psi|$ for $t \geqq t_0$ by $m(h)$, it follows, using (4.16), that

$$m(h) \leqq K\epsilon n m(h)\frac{2}{\sigma} + 2K^2\eta\frac{2}{\sigma}$$

where the last term results from the use of (4.15). Since $2K\epsilon n/\sigma < \frac{1}{2}$, it follows that

$$m(h) \leqq \frac{8K^2\eta}{\sigma}$$

Since $\eta \to 0$ as $h \to 0$, it follows that $m(h) \to 0$ as $h \to 0$. Thus $q \to \psi$ as $h \to 0$. This means that $\partial\theta/\partial a_j$ exists and is the solution ψ of (4.17).

From (4.12) and (4.11) it follows that

$$\sum_{j=k+1}^{n} |\theta_j(t_0,a)| \leqq 2K^2|a|\epsilon \int_{t_0}^{\infty} e^{-\sigma(s-t_0)}\,ds \leqq \frac{2K^2|a|\epsilon}{\sigma}$$

Since ϵ can be made arbitrarily small by taking $|a|$ small enough, it follows that under the hypothesis of Theorem 4.1 $(\partial\psi_j/\partial y_i)(0, \ldots, 0) = 0$, $j > k$. This is, of course, also true under the stricter hypothesis of Theorem 4.2. This completes the proof of Theorem 4.2.

In the next theorem, (4.2) can be replaced by a weaker condition which is (4.2) with $\tilde{x} = 0$. In that case, solutions of (4.1) need no longer be unique.

Theorem 4.3. *Let φ be a solution of (4.1) and let*

$$\limsup_{t\to\infty} \frac{\log|\varphi(t)|}{t} = b < 0 \tag{4.18}$$

Given any $\epsilon > 0$, let there exist a $\delta > 0$ and T such that

$$|f(t,x)| \leqq \epsilon|x|$$

for $|x| < \delta$ and $t \geqq T$. Let k of the characteristic roots λ_i, $i = 1, \ldots, k$, of A have negative real parts and the remaining $n - k$ roots have nonnegative real parts. Let $\Re\lambda_i = \mu_i$ and let $\mu_1 \leqq \mu_2 \leqq \cdots \leqq \mu_k < 0$. Then $b = \mu_j$ for some $j \leqq k$ or else $\varphi(t) \equiv 0$.

Proof. Suppose $\mu_m < b < \mu_{m+1}$ for some $m \leqq k - 1$. Let B_1 have the characteristic roots $\lambda_1, \ldots, \lambda_m$ and let B_2 have the remaining $n - m$ characteristic roots with real parts all exceeding μ_m. Much as in the proof of Theorem 4.1, there is a P such that

$$PAP^{-1} = B = \begin{pmatrix} B_1 & 0 \\ 0 & B_2 \end{pmatrix}$$

Let $\alpha > 0$ and $\sigma > 0$ be chosen so that

$$\mu_m < -\alpha - \sigma < -\alpha + \sigma < \mu_{m+1} \tag{4.19}$$

and also

$$-\alpha - \tfrac{1}{2}\sigma < b < -\alpha \tag{4.20}$$

With B_1 and B_2 defined as above, let U_1 and U_2 be given by (4.6) and (4.7). Then (4.8) holds but, instead of (4.9) and (4.10),

$$|U_1(t)| \leqq Ke^{-(\alpha+\sigma)t} \qquad (t \geqq 0) \tag{4.21}$$
$$|U_2(t)| \leqq Ke^{-(\alpha-\sigma)t} \qquad (t \leqq 0) \tag{4.22}$$

Setting $y = Px$, the analogue of (4.4) is obtained and, corresponding to the solution φ of (4.1), there is $P\varphi = \tilde{\varphi}$ of the analogue of (4.4). Choose ϵ so that $2K\epsilon/\sigma < \frac{1}{2}$. This fixes δ. The variation-of-constants formula shows that for any fixed $t_1 \geqq T$ there are vectors $c^{(1)}$ and $c^{(2)}$, $c^{(1)}$ having its last $n - m$ components all zero and $c^{(2)}$ its first m, such that

$$\bar{\varphi}(t) = U_1(t - t_1)c^{(1)} + U_2(t - t_1)c^{(2)} + \int_{t_1}^{t} U_1(t - s)g(s,\bar{\varphi}(s))\, ds - \int_{t}^{\infty} U_2(t - s)g(s,\bar{\varphi}(s))\, ds \quad (4.23)$$

An easy appraisal of the terms in (4.23) shows that all terms except $U_2(t - t_1)c^{(2)}$ are of exponential growth at most $e^{-\alpha t}$ while the term $U_2(t - t_1)$ is of growth at least $e^{-(\alpha-\sigma)t}$. Thus the equation (4.23) cannot hold unless $c^{(2)} = 0$.

Using (4.23) and recalling that $c^{(2)} = 0$ and denoting $|c^{(1)}|$ by c,

$$|\bar{\varphi}(t)| \leqq cKe^{-(\alpha+\sigma)(t-t_1)} + K\epsilon \int_{t_1}^{t} e^{-(\alpha+\sigma)(t-s)}|\bar{\varphi}(s)|\, ds + K\epsilon \int_{t}^{\infty} e^{(\alpha-\sigma)(s-t)}|\bar{\varphi}(s)|\, ds \quad (4.24)$$

Let

$$\max_{s \geqq t} e^{\alpha(s-t_1)}|\bar{\varphi}(s)| = M(t)$$

Then by (4.20) $M(t)$ exists and is monotone nonincreasing. For each t there exists a $\bar{t} \geqq t$ such that

$$M(t) = e^{\alpha(\bar{t}-t_1)}|\bar{\varphi}(\bar{t})| = M(\bar{t})$$

Thus (4.24) yields, with $t = \bar{t}$,

$$M(\bar{t}) \leqq cKe^{-\sigma(\bar{t}-t_1)} + K\epsilon \int_{t_1}^{\bar{t}} e^{-\sigma(\bar{t}-s)}M(s)\, ds + M(\bar{t})K\epsilon \int_{\bar{t}}^{\infty} e^{-\sigma(s-\bar{t})}\, ds$$

Since $M(s) = M(t) = M(\bar{t})$ for $t \leqq s \leqq \bar{t}$, this gives

$$M(t) \leqq cKe^{-\sigma(t-t_1)} + K\epsilon \int_{t_1}^{t} e^{-\sigma(t-s)}M(s)\, ds + \frac{2K\epsilon M(t)}{\sigma}$$

Since $2K\epsilon/\sigma < \frac{1}{2}$,

$$M(t)e^{\sigma t} \leqq 2cKe^{\sigma t_1} + 2K\epsilon \int_{t_1}^{t} e^{\sigma s}M(s)\, ds$$

Using the inequality in Prob. 1, Chap. 1,

$$M(t)e^{\sigma t} \leqq 2cKe^{\sigma t_1}e^{2K\epsilon(t-t_1)}$$

Since $2K\epsilon < \frac{1}{2}\sigma$, this gives

$$M(t) \leqq 2cKe^{-\frac{1}{2}\sigma(t-t_1)}$$

or

$$|\bar{\varphi}(t)| \leqq 2cKe^{-(\alpha+\frac{1}{2}\sigma)(t-t_1)}$$

which by (4.18) implies $b \leqq -(\alpha + \frac{1}{2}\sigma)$ contrary to (4.20). This proves then that $b \leqq \mu_m$, and thus contradicts the assumption $\mu_m < b < \mu_{m+1}$. Therefore, if $\mu_1 \leqq b \leqq \mu_k$, then $b = \mu_j$ for some $j \leqq k$.

In case $b < 0$, it follows easily that

$$b \leqq \mu_k < 0 \tag{4.25}$$

and in case $b < \mu_1$ it follows easily that $\varphi = 0$.

The following generalization of Theorem 4.1 is proved with only minor changes.

Theorem 4.4. *Let the real parts of the characteristic roots λ_i of A be denoted by μ_i and let $\mu_i \leqq \mu_{i+1}$. Suppose m is such that*

$$\mu_m < \mu_{m+1} < 0 \tag{4.26}$$

and otherwise let the hypothesis of Theorem 4.1 hold. Then for any large t_0 there exists a real m-dimensional open manifold S_m containing the origin such that any solution φ of (4.1) with $\varphi(t_0)$ on the manifold S_m satisfies

$$\limsup_{t\to\infty} \frac{\log |\varphi(t)|}{t} \leqq \mu_m < 0 \tag{4.27}$$

Moreover, there exists an $\eta > 0$ such that any solution satisfying $|\varphi(t)| < \eta$ for $t \geqq t_0$ but not on S_m at $t = t_0$ satisfies

$$\limsup_{t\to\infty} \frac{\log |\varphi(t)|}{t} \geqq \mu_{m+1} > \mu_m \tag{4.28}$$

If for each i, $\partial f/\partial x_i$ exists and is continuous in (t,x) for $t \geqq t_0$ and $|x|$ small, then the analogue of Theorem 4.2 is true for S_m. If for each t, f is analytic in x for small $|x|$, then S_m is an analytic surface.

Proof. The proof of Theorem 4.1 is modified by defining B_1 and B_2 as in the proof of Theorem 4.3. It is also assumed that (4.19) holds. With U_1 and U_2 defined as before, (4.21) and (4.22) also hold. The equation (4.11) is now considered with U_1 and U_2 defined as above and with the vector a having its last $n - m$ components all zero. As before, successive approximations lead easily to a solution $\theta = \theta(t,a)$. The analogue of (4.12) is

$$|\theta(t,a)| \leqq 2K|a|e^{-\alpha(t-t_0)} \tag{4.29}$$

where $|a|$ is small enough so that $2K|a| < \delta$. The existence of $\tilde{S}_m$ and consequently S_m follows as before.

Since $-\alpha < \mu_{m+1}$, it follows from Theorem 4.3 that (4.29) implies (4.27) for θ. A solution of (4.4) starting from $\tilde{S}_m$ at $t = t_0$ coincides with $\theta(t,a)$ for some choice of a by the uniqueness theorem for (4.4). This completes the proof of (4.27).

Suppose a solution φ of (4.1) is not on S_m at $t = t_0$ and that (4.28) does not hold. Then by Theorem 4.3, (4.27) must hold and therefore the solution $\tilde{\varphi} = P\varphi$ of (4.4) must satisfy (4.23) with $c^{(2)} = 0$ as proved below (4.23). Let t_1 in (4.23) be taken as t_0. Thus $\tilde{\varphi}$ is a solution of the integral equation (4.11), (with U_1, U_2 and a modified as above). Since $|\varphi(t)| < \eta$, η can be chosen small enough so that

$$2K|\tilde{\varphi}(t_0)| < \delta \qquad \text{and} \qquad |\tilde{\varphi}(t)| < \delta \qquad (t \geqq t_0)$$

Let the first components of a in (4.11) be those of $\tilde{\varphi}(t_0)$. Suppose now the integral equation (4.11) has two solutions θ and $\tilde{\theta}$ for the same a with both $|\theta|$ and $|\tilde{\theta}|$ less than δ for $t \geqq t_0$. Then subtracting the two equations and using (4.21) and (4.22),

$$|\theta - \tilde{\theta}|e^{\alpha t} \leqq K\epsilon \int_{t_0}^{t} e^{-\sigma(t-s)}|\theta - \tilde{\theta}|e^{\alpha s}\,ds + K\epsilon \int_{t}^{\infty} e^{-\sigma(s-t)}|\theta - \tilde{\theta}|e^{\alpha s}\,ds$$

Letting sup $|\theta - \tilde{\theta}|e^{\alpha s}$, $s \geqq t_0$, be denoted by M, there follows $M \leqq 2K\epsilon M/\sigma$ which implies $M = 0$ because $2K\epsilon/\sigma < 1$. Thus $\tilde{\varphi}(t)$ coincides with $\theta(t,a)$ and therefore lies on $\tilde{S}_m$ at $t = t_0$. This proves (4.28).

The proof of the analogue of Theorem 4.2 is immediate. The remarks on analyticity follow in the same way as at the end of the proof of Theorem 4.1.

Theorem 4.5. *Let the hypothesis of Theorem* 4.3 *be satisfied and let there be a* $\Delta > 0$ *such that for small* $|x|$

$$f(t,x) = 0(|x|^{1+\Delta}) \tag{4.30}$$

uniformly in $t \geqq 0$. *By Theorem* 4.3 *there exist integers* p *and* q, $1 \leqq p \leqq q \leqq k$ *such that*

$$\Re\lambda_{p-1} < \Re\lambda_p = \Re\lambda_{p+1} = \cdots = \Re\lambda_q = b < \Re\lambda_{q+1}$$

There exists a $\delta > 0$ *and a solution* ψ *of* $x' = Ax$,

$$\psi(t) = \sum_{j=p}^{q} Q_j(t)e^{\lambda_j t} \tag{4.31}$$

where the $Q_j(t)$ *are column vectors not all zero which are polynomials in* t, *such that*

$$\varphi(t) = \psi(t) + 0(e^{(b-\delta)t}) \tag{4.32}$$

as $t \to \infty$. *Conversely, if* (4.2) *holds, then corresponding to any solution* ψ *of* $x' = Ax$ *which is of the form* (4.31) *there is a solution* φ *of* (4.1) *which satisfies* (4.32). *Furthermore, if* $p = 1$, *then* φ *is uniquely determined by* ψ.

Thus (4.32) shows that solutions of (4.1) which go to zero as $t \to \infty$ are equal to solutions of $x' = Ax$ to within an error term of exponentially smaller order.

Proof of Theorem 4.5. There exists a real nonsingular matrix P such that

$$PAP^{-1} = \begin{pmatrix} B_1 & 0 & 0 \\ 0 & B_2 & 0 \\ 0 & 0 & B_3 \end{pmatrix} = B$$

where B_1 has characteristic roots with real parts less than b, B_2 has characteristic roots with real parts equal to b, and B_3 has characteristic roots with real parts greater than b.

Let $z = Px$. Then (4.1) becomes

$$z' = Bz + g(t,z) \tag{4.33}$$

where $g(t,z) = Pf(t,P^{-1}z)$. By (4.30)

$$g(t,z) = 0(|z|^{1+\Delta}) \tag{4.34}$$

Let $\tilde{\varphi} = P\varphi$. Then $\tilde{\varphi}$ is a solution of (4.33) and

$$\limsup_{t\to\infty} \frac{\log|\tilde{\varphi}(t)|}{t} = b = \mu_p \tag{4.35}$$

where $\mu_p = \Re\lambda_p$. By (4.34) and (4.35) there exists an $\eta > 0$ such that for large t

$$g(t,\tilde{\varphi}(t)) = 0(e^{(b-\eta)t}) \tag{4.36}$$

Let

$$U_1(t) = \begin{pmatrix} e^{tB_1} & 0 & 0 \\ 0 & 0 & 0 \\ 0 & 0 & 0 \end{pmatrix}$$

$$U_2(t) = \begin{pmatrix} 0 & 0 & 0 \\ 0 & e^{tB_2} & 0 \\ 0 & 0 & 0 \end{pmatrix}$$

and similarly for $U_3(t)$ so that $e^{tB} = U_1 + U_2 + U_3$, and moreover

$$U_j' = BU_j \qquad (j = 1, 2, 3)$$

It follows from the definitions of B_1, B_2, and B_3 that there exists a δ, $0 < \delta < \eta$, such that

$$U_1(t) = 0(e^{(b-\delta)t}) \qquad (t > 0) \tag{4.37}$$
$$U_j(t) = 0(e^{(b-\delta)t}) \qquad (t < 0,\ j = 2, 3) \tag{4.38}$$

Because $\tilde{\varphi}$ is a solution of (4.33), it follows readily that

$$\tilde{\varphi}(t) = e^{tB}c_0 + \int_{t_0}^{t} e^{(t-s)B}g(s,\tilde{\varphi}(s))\,ds$$

where c_0 is the constant $e^{-t_0 B}\bar{\varphi}(t_0)$. This can be written as

$$\bar{\varphi}(t) = U_1(t)c_0 + U_2(t)c + J_1 + J_2 + J_3 \tag{4.39}$$

where c is a constant vector and

$$J_1 = U_3(t)c$$
$$J_2 = \int_{t_0}^{t} U_1(t-s)g(s,\bar{\varphi}(s))\,ds$$
$$J_3 = -\int_{t}^{\infty} [U_2(t-s) + U_3(t-s)]\,g(s,\bar{\varphi}(s))\,ds$$

From (4.36) and (4.37) it follows easily that

$$J_2 = 0(e^{(b-\delta)t})$$

as $t \to \infty$, and from (4.36) and (4.38)

$$J_3 = 0(e^{(b-\eta)t})$$

The term J_1 on the right of (4.39) is either identically zero or else

$$\limsup_{t\to\infty} \frac{\log |J_1|}{t} > b$$

Because of the size of all other terms in (4.39), the above is impossible so that J_1 is identically zero. Thus by (4.37) and the appraisals for J_2 and J_3

$$\bar{\varphi}(t) = U_2(t)c + 0(e^{(b-\delta)t}) \tag{4.40}$$

as $t \to \infty$. Since $\varphi = P^{-1}\bar{\varphi}$, the result (4.32) follows. That not all the $Q_j(t)$ can be zero is a consequence of Theorem 4.3.

Given any ψ, then $P\psi(t)$ is $U_2(t)c$ for some choice of the vector c. There is a solution of (4.39), with $J_1 \equiv 0$, just as in the proof of the existence of a solution of (4.11), and (4.40) follows much as before. This proves that there is at least one φ corresponding to a given ψ.

Finally, in case $p = 1$, then $U_1 \equiv 0$. Any solution of (4.1) satisfying (4.32) must satisfy the integral equation (4.39) with $J_1 \equiv 0$. The impossibility of the integral equation having two distinct solutions now follows, much as in the uniqueness proof at the end of Theorem 4.4.

5. Behavior of Solutions off the Stable Manifold

In this section it is necessary to introduce a real canonical form of the real matrix A. A real nonsingular matrix P exists so that

$$PAP^{-1} = B$$

where

$$B = \begin{pmatrix} D_1 & 0 & \cdots & 0 \\ 0 & D_2 & \cdots & 0 \\ \cdot & \cdot & \cdots & \cdot \\ 0 & 0 & \cdots & D_m \end{pmatrix} \tag{5.1}$$

the D_j being real square matrices and all other elements of B being zero. Each D_j is either of the form

$$\begin{pmatrix} \lambda_j & 0 & 0 & \cdots & 0 \\ \gamma & \lambda_j & 0 & \cdots & 0 \\ 0 & \gamma & \lambda_j & \cdots & 0 \\ \cdot & \cdot & \cdot & \cdots & \cdot \\ 0 & 0 & 0 & \cdot\,\cdot\ \gamma & \lambda_j \end{pmatrix} \tag{5.2}$$

where γ may be taken as any real number not zero, or else D_j is of the form

$$\begin{pmatrix} S_j & 0 & 0 & \cdots & 0 \\ \gamma E_2 & S_j & 0 & \cdots & 0 \\ 0 & \gamma E_2 & S_j & \cdots & 0 \\ \cdot & \cdot & \cdot & \cdots & \cdot \\ 0 & 0 & 0 & \cdot\,\cdot\ \gamma E_2 & S_j \end{pmatrix} \tag{5.3}$$

where S_j is real and

$$S_j = \begin{pmatrix} \alpha_j & -\beta_j \\ \beta_j & \alpha_j \end{pmatrix}$$

and E_2 is the unit matrix of two rows and columns.

The matrix (5.2) may contain a single row and column. It is associated, of course, with the characteristic root λ_j, while (5.3) is associated with the conjugate characteristic roots $\alpha_j \pm i\beta_j$. In the simplest case, (5.3) is S_j itself.

Clearly

$$e^{tB} = \begin{pmatrix} e^{tD_1} & 0 & \cdots & 0 \\ 0 & e^{tD_2} & \cdots & 0 \\ \cdot & \cdot & \cdots & \cdot \\ 0 & 0 & \cdots & e^{tD_m} \end{pmatrix} \tag{5.4}$$

For the form (5.2)

$$e^{tD_j} = e^{\lambda_j t} \begin{pmatrix} 1 & 0 & 0 & \cdots & 0 \\ \gamma t & 1 & 0 & \cdots & 0 \\ \dfrac{(\gamma t)^2}{2!} & \gamma t & 1 & \cdots & 0 \\ \cdot & \cdot & \cdot & \cdots & \cdot \\ \cdot & \cdot & \cdot & \cdot\,\cdot\ \gamma t & 1 \end{pmatrix} \tag{5.5}$$

while for (5.3)

$$e^{tD_j} = \begin{pmatrix} e^{tS_j} & 0 & 0 & \cdots & 0 \\ \gamma t e^{tS_j} & e^{tS_j} & 0 & \cdots & 0 \\ \dfrac{(\gamma t)^2}{2!} e^{tS_j} & \gamma t e^{tS_j} & e^{tS_j} & \cdots & 0 \\ \cdot & \cdot & \cdot & \cdots & \cdot \\ \cdot & \cdot & \cdot & \cdots & e^{tS_j} \end{pmatrix} \tag{5.6}$$

where

$$e^{tS_j} = e^{\alpha_j t}\begin{pmatrix}\cos \beta_j t & -\sin \beta_j t\\ \sin \beta_j t & \cos \beta_j t\end{pmatrix} \tag{5.7}$$

The proof of (5.6) is an easy consequence of

$$D_j = \begin{pmatrix} S_j & 0 & \cdots & 0\\ 0 & S_j & \cdots & 0\\ \cdot & \cdot & \cdots & \cdot\\ 0 & 0 & \cdots & S_j\end{pmatrix} + \gamma\begin{pmatrix} 0 & 0 & \cdots & 0 & 0\\ E_2 & 0 & \cdots & 0 & 0\\ 0 & E_2 & \cdots & 0 & 0\\ \cdot & \cdot & \cdots & \cdot & \cdot\\ 0 & 0 & \cdots & E_2 & 0\end{pmatrix}$$

and the fact that the two matrices above commute, together with the fact that

$$\begin{pmatrix} 0 & 0 & \cdots & 0 & 0\\ E_2 & 0 & \cdots & 0 & 0\\ 0 & E_2 & \cdots & 0 & 0\\ \cdot & \cdot & \cdots & \cdot & \cdot\\ 0 & 0 & \cdots & E_2 & 0\end{pmatrix}^2 = \begin{pmatrix} 0 & 0 & \cdots & 0\\ 0 & 0 & \cdots & 0\\ E_2 & 0 & \cdots & 0\\ 0 & E_2 & \cdots & 0\\ \cdot & \cdot & \cdots & \cdot\\ 0 & 0 & \cdots & 0\end{pmatrix}$$

etc.

A case of great interest occurs when $f(t,x)$ of (4.1) becomes a function of x only. In this case the k-dimensional manifold of initial values shown to exist in Theorem 4.1 is clearly invariant under t and, moreover, any point on the manifold and close enough to the origin remains on the manifold as t increases. In case f is a function of x only, the system is called *autonomous* and the solutions may be regarded as curves in x space with t as a parameter. Through each point of x space there passes a unique solution curve. In the autonomous case, by changing the sign of t, it follows under the hypothesis of Theorem 4.1 that there exists an $(n - k)$-dimensional manifold containing the origin such that any point on the manifold near the origin tends to the origin as $t \to -\infty$ and any solution starting off this manifold cannot remain arbitrarily close to the origin as $t \to -\infty$. This result can be made more precise. This will be shown for the autonomous case but can be generalized also to the case where f depends on t as well as x.

It will be assumed that the transformation $y = Px$ has been made and that the system is in the form

$$y' = By + g(y) \tag{5.8}$$

where

$$B = \begin{pmatrix} B_1 & 0\\ 0 & B_2\end{pmatrix}$$

and B is in the form (5.1). The k characteristic roots of B_1 have negative real parts and the $n - k$ roots of B_2 have positive real parts. The constant γ will be specified later.

It is convenient to set

$$y = y_{(1)} + y_{(2)}$$

where $y_{(1)}$ is a vector with its first k components equal to those of y and the last $n - k$ all zero and $y_{(2)} = y - y_{(1)}$. The stable manifold introduced in the proof of Theorem 4.1 may be given by the vector equation

$$y_{(2)} = \psi(y_{(1)}) \tag{5.9}$$

since it does not depend on t_0. Note that the first k components of ψ are zero. Consider now any solution φ of (5.8) with $|\varphi(0)|$ small. Let $\varphi = \varphi_{(1)} + \varphi_{(2)}$ and let

$$\xi(t) = \varphi_{(2)}(t) - \psi(\varphi_{(1)}(t)) \tag{5.10}$$

Theorem 5.1. *Let f of (4.1) be a function of x only and let the hypotheses of Theorems 4.1 and 4.2 be satisfied. Let $|\varphi(0)|$ be small, and suppose φ does not start on the manifold (5.9) at $t = 0$. Then so long as $|\varphi(t)|$ remains small, the Euclidean length of $\xi(t)$, as given by (5.10), is an exponentially increasing function of t.*

This theorem shows that the distance of the solution $\varphi(t)$ from the stable manifold, if taken normal to $y_{(2)} = 0$, is an exponentially increasing function of t.

Proof of Theorem 5.1. Clearly

$$\xi'(t) = \varphi'_{(2)}(t) - \psi_y(\varphi_{(1)}(t))\varphi'_{(1)}(t)$$

where ψ_y is the n-by-n matrix with columns $\partial\psi/\partial y_j$. Note that the last $n - k$ columns of ψ_y are zero. If $g = g_{(1)} + g_{(2)}$, it follows from (5.8) that

$$\varphi'_{(j)} = B\varphi_{(j)} + g_{(j)}(\varphi) \qquad (j = 1, 2)$$

Thus

$$\xi' = B\varphi_{(2)} + g_{(2)}(\varphi) - \psi_y(\varphi_{(1)})(B\varphi_{(1)} + g_{(1)}(\varphi)) \tag{5.11}$$

For solutions on the stable manifold, ξ is identically zero. For such a solution θ, (5.11) holds with $\xi' = 0$ and gives

$$0 = B\psi(\theta_{(1)}) + g_{(2)}(\theta_{(1)} + \psi(\theta_{(1)})) \\ - \psi_y(\theta_{(1)})[B\theta_{(1)} + g_{(1)}(\theta_{(1)} + \psi(\theta_{(1)}))] \tag{5.12}$$

Because the above holds for all solutions θ on the manifold and since there is a solution through each point of the manifold, it follows that it is an identity and thus holds if $\theta_{(1)}$ is replaced by $y_{(1)}$, with $|y_{(1)}|$ small.

Using $\varphi = \varphi_{(1)} + \varphi_{(2)}$ in (5.11) and subtracting (5.12), in which $\theta_{(1)}$ is replaced by $\varphi_{(1)}$, from (5.11) it follows that

$$\xi' = B\xi + g_{(2)}(\varphi_{(1)} + \varphi_{(2)}) - g_{(2)}(\varphi_{(1)} + \psi(\varphi_{(1)})) \\ - \psi_y(\varphi_{(1)})(g_{(1)}(\varphi_{(1)} + \varphi_{(2)}) - g_{(1)}(\varphi_{(1)} + \psi(\varphi_{(1)}))) \quad (5.13)$$

If ξ^* is the transpose of ξ, then $J = \xi^*\xi$ is the square of the Euclidean length of ξ, and clearly $J' = \xi^*\xi' + \xi^{*\prime}\xi$. Since $\varphi_{(2)} = \xi + \psi(\varphi_{(1)})$, it follows that

$$|g_{(j)}(\varphi_{(1)} + \varphi_{(2)}) - g_{(j)}(\varphi_{(1)} + \psi(\varphi_{(1)}))| \leqq \epsilon|\xi| \qquad (j = 1, 2)$$

for $|\varphi| \leqq \delta$, much as in (4.5). Using this in (5.13),

$$J' = \xi^*(B + B^*)\xi + J_1 \quad (5.14)$$

where

$$|J_1| \leqq K\epsilon J \quad (5.15)$$

for some constant K. Because the first k components of ξ and ξ^* are zero, only the matrices B_2 and B_2^* enter in the evaluation of $\xi^*(B + B^*)\xi$. The elements of $B_2 + B_2^*$ on the main diagonal are all positive and real and exceed some number $2d$. The other nonvanishing elements of $B_2 + B_2^*$ are at most $2(n - k - 1)$ constants γ. Thus from (5.14) and (5.15)

$$J' \geqq (2d - 2(n - k - 1)\gamma - K\epsilon)J$$

Thus, if γ is taken small by proper choice of P, and ϵ is taken small by a proper choice of δ, then

$$J' \geqq dJ$$

which proves the result so long as $|\varphi(t)| \leqq \delta$.

Corollary to Theorem 5.1. *Under the hypothesis of Theorem* 5.1 *there exists an $(n - k)$-dimensional manifold*

$$y_{(1)} = \chi(y_{(2)})$$

of solutions stable for $t \to -\infty$. By reversing t, Theorem 5.1 *leads to the fact that, for φ a solution as before, the vector*

$$\varphi_{(1)}(t) - \chi(\varphi_{(2)}(t))$$

has a Euclidean length which is decreasing as t increases so long as $|\varphi(t)|$ remains small.

PROBLEMS

1. Let all solutions of the linear system with constant coefficients $y' = Ay$ be bounded for $t \geqq 0$, that is, let $|e^{tA}| \leqq M$, $t \geqq 0$, for some constant M. Let f be of class C and let there exist a constant a and a $g(t)$ such that $|f(t,x)| \leqq g(t)|x|$ for

$|x| \leqq a$ and $t \geqq 0$ and let $\int_0^\infty g(t)\,dt < \infty$. Show that there exists a constant M_1 such that any solution φ of the system $x' = Ax + f(t,x)$ satisfies $|\varphi(t)| < M_1|\varphi(0)|$ if $|\varphi(0)| \leqq a/M_1$.

HINT: Show that $|\varphi(t)| \leqq M|\varphi(0)| \exp\left[M \int_0^t g(s)\,ds\right]$.

2. In Prob. 1 it is clear that $e^{tA} = U_1(t) + U_2(t)$, where $U_1(t)$ contains elements which are sums of exponential terms $e^{i\lambda_j t}$ for real λ_j and $|U_1(t)| \leqq K_1$, $-\infty < t < \infty$, and $|U_2(t)| \leqq K_2 e^{-\sigma t}$, $0 \leqq t < \infty$ for some $\sigma > 0$, where K_1 and K_2 are constants. Show that corresponding to the solution φ there is a constant vector p such that $\varphi(t) - U_1(t)p \to 0$ as $t \to \infty$.

HINT: Show that there is a p such that

$$\varphi(t) = e^{tA}p + \int_0^t U_2(t-s)f(s,\varphi(s))\,ds - \int_t^\infty U_1(t-s)f(s,\varphi(s))\,ds$$

3. Another method of dealing with (3.6) and, indeed, for getting further results is based on the variation-of-constants formula

$$\varphi(s) = cE(s) - bE(s)\int_s^1 \frac{\sigma^{1-m}}{E(\sigma)}\,d\sigma - E(s)\int_s^1 \frac{h(\sigma,\varphi(\sigma))\sigma^{-m}}{E(\sigma)}\,d\sigma$$

where $E(s) = e^{-as^{-m+1}/(m-1)}$ for $m > 1$ and a similar $E(s) = s^a$ for $m = 1$. The above is used with c a small constant, and a successive-approximation procedure is developed based on setting $\varphi_{n-1}(\sigma)$ in place of $\varphi(\sigma)$ on the right and $\varphi_n(s)$ in place of $\varphi(s)$ on the left. The term $\varphi_0(s) \equiv 0$. Show that the process converges for small c and $s > 0$ and leads to the existence of $\varphi(s)$ and to the fact that $\varphi(s) \to 0$ as $s \to +0$.

4. Show that the results of Secs. 3 and 4 apply to the case where the constant matrix A is replaced by a real periodic matrix P of period ω.

HINT: Set $x = Z(t)y$, where $Z(t)e^{tB}$ is a solution of $x' = P(t)x$ and where B and the periodic matrix $Z(t)$ can be taken as real for real $P(t)$. An alternative approach is to take $Z(t)e^{tB}$ which follows by assigning it the initial value E and, if necessary, break e^{tB} up into several parts, for example, $U_1(t) + U_2(t)$, where each part is determined by the rate of growth of the exponential terms assigned it. Since Ze^{tB} is real, it follows that ZU_1 and ZU_2 are real. Thus in the variation-of-constants formula appears

$$V_j(t,\tau) = Z(t)U_j(t-\tau)Z^{-1}(\tau) \qquad (j = 1, 2)$$

5. Consider the real system

$$x_i' = a_i x_i + f_i(x_1,x_2) \qquad (i = 1, 2)$$

with $a_1 < a_2 < 0$. Let f_i be of class C^1 and let f_i and its first-order partial derivatives vanish at (0,0). Using Theorem 4.4, show that, except for translations of t, the system has exactly one solution $\varphi = (\varphi_1,\varphi_2)$ such that

$$\limsup_{t\to\infty} \log \frac{(|\varphi_1(t)| + |\varphi_2(t)|)}{t} = a_1$$

Moreover, show by use of the extension of Theorem 4.2 for the case of Theorem 4.4 that this solution lies on a curve of class C^1 $x_2 = \psi(x_1)$ with $\psi'(0) = 0$, and that ψ is an analytic function if the f_i are analytic.

6. Show by Theorem 4.5 that if $f_i = 0[(|x_1| + |x_2|)^{1+\Delta}]$, $\Delta > 0$, then for each choice of the constant c the differential equation of Prob. 5 has a solution $\varphi = (\varphi_1,\varphi_2)$ satisfying $\varphi_1(t) = 0(e^{(a_2-\delta)t})$ and $\varphi_2(t) = ce^{a_2t} + 0(e^{(a_2-\delta)t})$ for some $\delta > 0$ as $t \to \infty$. Since the case $c = 0$ must be that discussed in Prob. 5, show that all other solutions must satisfy the equations above. Show therefore that for all solutions other than that of Prob. 5

$$\lim_{t \to \infty} \frac{\varphi_1(t)}{\varphi_2(t)} = 0$$

Express the results of Probs. 5 and 6 in terms of the geometric configuration of the solutions as curves in the (x_1,x_2) plane in the neighborhood of the origin.

7. Using Theorems 4.1 and 4.2, show that the real system considered in Prob. 5 but with $a_2 > 0 > a_1$ has, except for translations in t, exactly one solution tending to zero as $t \to \infty$ and that this solution lies on a curve of class C^1 $x_2 = \psi(x_1)$, $\psi'(0) = 0$. Show that for $t \to -\infty$ a similar situation prevails with x_1 and x_2 interchanging their roles. This is the case of a *saddle point*. State the significance of Sec. 5 for this case. Show that ψ is an analytic function if the f_i are analytic.

8. Consider the real system

$$\begin{aligned} x_1' &= -\alpha x_1 + \beta x_2 + f_1(x_1,x_2) \\ x_2' &= -\beta x_1 - \alpha x_2 + f_2(x_1,x_2) \end{aligned}$$

where $\alpha > 0$ and the f_i are of class C^1 and vanish with their first-order partial derivatives at the origin. Let $f_i = 0[(|x_1| + |x_2|)^{1+\Delta}]$ for some $\Delta > 0$. Using Theorem 4.5, show that there is a $\delta > 0$ such that corresponding to any choice of the constants c and γ there is a unique solution $\varphi = (\varphi_1,\varphi_2)$, where

$$\begin{aligned} \varphi_1(t) &= ce^{-\alpha t}\cos(\beta t + \gamma) + 0(e^{-(\alpha+\delta)t}) \\ \varphi_2(t) &= ce^{-\alpha t}\sin(\beta t + \gamma) + 0(e^{-(\alpha+\delta)t}) \end{aligned}$$

as $t \to \infty$. For $\beta \neq 0$ this is the case of *spirals*, and for $\beta = 0$, a case of a *node* at the origin.

9. In Prob. 8 show that if $\omega = \tan^{-1}(\varphi_2/\varphi_1)$ and if $\rho = (\varphi_1^2 + \varphi_2^2)^{\frac{1}{2}}$, then

$$\lim_{t \to \infty} \left[\omega(t) + \frac{\beta}{\alpha}\log\rho(t)\right] = \gamma + \frac{\beta}{\alpha}\log c$$

Show that, except for translations in t, the constant on the right determines the solution of Prob. 8 uniquely. Note that the case $\beta = 0$ is also valid. Show that lim can be omitted if $f_1 \equiv f_2 \equiv 0$.

10. Consider the real system

$$\begin{aligned} x_1' &= -\lambda x_1 + f_1(x_1,x_2) \\ x_2' &= x_1 - \lambda x_2 + f_2(x_1,x_2) \end{aligned}$$

where $\lambda > 0$ and the f_i are as in Prob. 8. Show that for any choice of the constants c_1 and c_2 there is a unique solution $\varphi = (\varphi_1,\varphi_2)$, where

$$\begin{aligned} \varphi_1(t) &= c_1e^{-\lambda t} + 0(e^{-(\lambda+\delta)t}) \\ \varphi_2(t) &= c_1te^{-\lambda t} + c_2e^{-\lambda t} + 0(e^{-(\lambda+\delta)t}) \end{aligned}$$

with $\delta > 0$. Conversely, for any solution show there are c_1 and c_2 for which the above holds. Show that

$$\lim_{t \to \infty} \left(\frac{\varphi_2}{\varphi_1} + \frac{\log \varphi_1}{\lambda}\right) = \frac{c_2}{c_1} + \frac{\log c_1}{\lambda}$$

and that, except for translations in t, the constants on the right side above determines the solution uniquely. Show that lim can be omitted if $f_1 \equiv f_2 \equiv 0$.

11. Let the hypothesis of Theorem 4.1 be modified so that instead of $n - k$ characteristic roots of A having positive real parts they now have nonnegative real parts. Show that the conclusion of Theorem 4.1 remains valid with the following change: There is no solution φ not on S at $t = t_0$ satisfying $|\varphi(t)| < \eta$ and

$$\lim_{t\to\infty} \sup \frac{\log |\varphi(t)|}{t} < 0$$

HINT: $|U_1| \leqq Ke^{-(\alpha+\sigma)t}$, $t > 0$, $|U_2| \leqq Ke^{-\sigma t}$, $t < 0$.

12. Let $F = F(t,x)$ be of period T in t. Let $x' = F(t,x)$, (2.1), have a periodic solution p of period T and let F and F_x be of class C in a region of (t,x) space which contains $(t,p(t))$, $0 \leqq t < \infty$. Let $\varphi = \varphi(t,a)$ be the solution of (2.1) with $\varphi(0,a) = p(0) + a$, where a is a vector with n components. Show that the variational equation (2.4) has the matrix $\varphi_a(t,0)$ as a fundamental solution.

13. Assume F and p are as in Prob. 12. Let $\xi = h(x)$ be a one-to-one transformation with nonvanishing Jacobian from x to ξ space of a region of x space containing the closed curve $x = p(t)$. Let $x = g(\xi)$ and let h and g be of class C^2. Then, corresponding to (2.1), there is the equation

$$(*) \qquad \xi' = h_x(g(\xi))F(t,g(\xi))$$

which has the periodic solution $h(p(t))$. Show that $h(\varphi(t,a))$ is a solution of (*) for small $|a|$. Show that $h_x(\varphi(t,a))\varphi_a(t,a)$ for $a = 0$ is a fundamental solution of the variational equation of (*) with respect to the solution $h(p(t))$. Show that the characteristic exponents are the same for the transformed case as for the original case.

HINT: Since $h_x(p(t))\varphi_a(t,0)$ is a fundamental matrix, the characteristic exponents are obtained from the logarithm of the matrix

$$\varphi_a^{-1}(0,0)h_x^{-1}(p(0))h_x(p(T))\varphi_a(T,0)$$

which is the matrix $\varphi_a^{-1}(0,0)\varphi_a(T,0)$.

14. Let the function V be of class C^1 in x space for small $|x|$. Let $V(x) > 0$ for $x \neq 0$ and $V(0) = 0$. Show that, if $|\xi|$ is small, solutions of

$$x' = f(t,x) \qquad x(0) = \xi$$

tend to zero as $t \to \infty$ if there is a constant $k > 0$ such that

$$\sum_{i=1}^{n} \frac{\partial V}{\partial x_i}(x)\, f_i(t,x) \leqq -kV(x)$$

for $t \geqq 0$ and small $|x|$.

HINT: Let φ be a solution of $x' = f(t,x)$. Let $F(t) = V(\varphi(t))$. Then $dF/dt \leqq -kF$.

CHAPTER 14

PERTURBATION OF SYSTEMS HAVING A PERIODIC SOLUTION

1. Nonautonomous Systems

The behavior of the system

$$x' = g(t,x) + \mu h(t,x,\mu) \tag{1.1}$$

for small $|\mu|$, based on the behavior for $\mu = 0$, is of considerable importance. The system (1.1) is a special case of

$$x' = f(t,x,\mu) \tag{1.2}$$

In (1.2) μ may be a real vector. The general dependence of solutions on a parameter μ has been considered in Sec. 7, Chap. 1. Here the case of special importance is considered where (1.2) with $\mu = 0$ has a real periodic solution p of period T. It is assumed that f is periodic in t of period T. (Note that T need not be the least period of either p or of f. Also note that f need not depend on t at all.)

In Theorem 7.5, Chap. 1, where the existence of the partial derivatives $\partial\varphi/\partial\xi_i$ of a solution φ with respect to initial values ξ_i is proved, it is assumed that f_μ is continuous, as well as f_x. However, by using the method of Theorem 7.2, Chap. 1, the existence of φ_ξ follows if only f_x is assumed to be continuous. This will be done here, although in practice f_μ usually exists, and indeed f is usually analytic in (x, μ).

It is assumed that f is real and continuous in (t,x,μ) when (t,x) is in some domain V of (t,x) space containing the curve $(t,p(t))$ and when $|\mu|$ is small. It is also assumed that f has first-order partial derivatives with respect to the components x_i of x which are continuous in (t,x,μ).

The first variation as defined in Sec. 2, Chap. 13, will occur in this chapter.

Theorem 1.1. *If f satisfies the conditions above and if the first variation of* (1.2) *for $\mu = 0$ with respect to the solution p has no solution of period T, then for small $|\mu|$ the equation* (1.2) *has a solution $q = q(t,\mu)$, periodic in t of period T, continuous in (t,μ), and with $q(t,0) = p(t)$. There is only one such solution for each μ.*

The first variation is the linear system with periodic coefficients

$$y' = \sum_{j=1}^{n} \frac{\partial f}{\partial x_j}(t,p(t),0)y_j = f_x(t,p(t),0)y \tag{1.3}$$

where the matrix $f_x(t,p(t),0)$ has period T. The condition that (1.3) have no solution of period T is equivalent to there being no characteristic exponent of the linear system (1.3) which is an integral multiple of $2\pi i/T$.

Proof of Theorem 1.1. The solution of (1.2) which assumes at $t = 0$ the initial value $p(0) + \alpha$, where $|\alpha|$ is small, will be denoted by

$$\varphi = \varphi(t,\alpha,\mu)$$

It follows from the uniqueness of φ that for this solution to be periodic of period T it is necessary and sufficient that $\varphi(T,\alpha,\mu) = \varphi(0,\alpha,\mu)$ or that

$$\varphi(T,\alpha,\mu) - p(0) - \alpha = 0$$

For $\mu = 0$, this system has as a solution $\alpha = 0$. If the Jacobian of this system taken with respect to the components of α is nonvanishing at $\mu = 0$, $\alpha = 0$, then it follows that the system has a unique solution $\alpha = \alpha(\mu)$ in the neighborhood of $\mu = 0$, $\alpha = 0$, where α is continuous in μ and $\alpha(0) = 0$. The Jacobian is the determinant of the matrix

$$\varphi_\alpha(T,0,0) - E \tag{1.4}$$

If the Jacobian does not vanish, then the existence of a periodic solution q of (1.2) is established for small $|\mu|$ by setting $\varphi(t,\alpha(\mu),\mu) = q(t,\mu)$. Moreover, in the neighborhood of $x = p(t)$, this solution is uniquely determined since $\alpha(\mu)$ is unique. It is an important fact that the Jacobian depends only on $f(t,x,0)$ since $\mu = 0$ in (1.4). Thus in the case (1.1) the Jacobian does not depend on h.

The Jacobian is closely related to the first variation (1.3) of (1.2) with respect to the solution p. In fact, if the equation

$$\varphi'(t,\alpha,\mu) = f(t,\varphi(t,\alpha,\mu),\mu)$$

is differentiated with respect to the components α_i of α, there results at $\mu = 0$, $\alpha = 0$

$$\varphi'_\alpha(t,0,0) = f_x(t,\varphi(t,0,0),0)\varphi_\alpha(t,0,0)$$

or, since $\varphi(t,0,0) = p(t)$,

$$\varphi'_\alpha(t,0,0) = f_x(t,p(t),0)\varphi_\alpha(t,0,0)$$

Thus the matrix Ψ given by

$$\Psi(t) = \varphi_\alpha(t,0,0) \tag{1.5}$$

is a solution matrix of (1.3), and since $\varphi_\alpha(0,0,0) = E$ it is a fundamental matrix. Hence the multipliers associated with (1.3) are the roots of

$$\det(\Psi(T) - \lambda E) = 0 \tag{1.6}$$

But by (1.5), $(\Psi(T) - E)$ is precisely the matrix (1.4), the determinant of which is the Jacobian. Thus the Jacobian vanishes if and only if (1.6) has $\lambda = 1$ as a root. A necessary and sufficient condition for (1.3) to have a solution of period T is precisely that $\lambda = 1$ be a root of (1.6), so that the theorem is established.

If the characteristic exponents of (1.3) all have negative real parts, it is the case that the solution p of (1.2) for $\mu = 0$ is asymptotically stable. The following result holds.

Theorem 1.2. *If the real parts of the characteristic exponents of* (1.3), *the first variation of* (1.2) *for* $\mu = 0$ *with respect to* p, *are all negative, then* (1.3) *can have no periodic solution, so that the conclusion of Theorem* 1.1 *is valid. Moreover, in this case the periodic solution* $q = q(t,\mu)$ *of* (1.2) *is asymptotically stable providing* $|\mu|$ *is small.*

Proof. The first part of the theorem is obvious. To prove the stability of q, we observe that the first variation in this case is the system with periodic coefficients

$$y' = f_x(t,q(t,\mu),\mu)y \tag{1.7}$$

and where (1.7) becomes (1.3) for $\mu = 0$. Thus, if $\Psi = \Psi(t,\mu)$ is a fundamental solution of (1.7) with $\Psi(0,\mu) = E$, then the multipliers of (1.7) are the characteristic roots of the matrix $\Psi(T,\mu)$. Since Ψ is continuous in μ for small $|\mu|$, it follows that, since the characteristic roots of $\Psi(T,0)$ are less than one in magnitude, the same is true for $\Psi(T,\mu)$ for small $|\mu|$. The proof of Theorem 1.2 now follows by using Theorem 2.1, Chap. 13.

For each t let f satisfy the further condition of being analytic in (x,μ) for (t,x) in V and $|\mu|$ small. Then, from the existence Theorem 8.4, Chap. 1, it follows that $\varphi = \varphi(t,\alpha,\mu)$ is analytic in α and μ for small $|\alpha|$ and $|\mu|$ over $0 \leqq t \leqq T$. The nonvanishing of the Jacobian of (1.4) now assures not only the existence and uniqueness of $\alpha = \alpha(\mu)$ but also that α is analytic in μ. Thus q is analytic in μ for small $|\mu|$ and for any t.

For simplicity it will now be assumed that μ has only one component and thus need not be treated as a vector. Because of analyticity, q has for each t a power-series representation in μ with coefficients which are continuous in t since q is continuous in (t,μ). The power series is of the form

$$q(t,\mu) = p^{(0)}(t) + \mu p^{(1)}(t) + \mu^2 p^{(2)}(t) + \cdots \tag{1.8}$$

where $p^{(0)}(t) = p(t)$ and because of the periodicity of q it follows that $p^{(j)}$ is also periodic of period T for all j.

Using (1.8) in the differential equation (1.2), there results

$$\sum_{j=0}^{\infty} \frac{dp^{(j)}}{dt}(t)\mu^j = f(t,q(t,\mu),\mu)$$

Since the right side is analytic in μ for small $|\mu|$, it can be expanded in powers of μ. Equating powers of μ, there results, with $p^{(0)}(t) = p(t)$, the sequence

$$\begin{aligned} \frac{dp^{(0)}}{dt}(t) &= f(t,p^{(0)}(t),0) \\ \frac{dp^{(1)}}{dt}(t) &= f_x(t,p^{(0)}(t),0)p^{(1)}(t) + \frac{\partial f}{\partial \mu}(t,p^{(0)}(t),0) \\ \frac{dp^{(2)}}{dt}(t) &= f_x(t,p^{(0)}(t),0)p^{(2)}(t) + F^{(2)}(t) \\ &\cdots\cdots\cdots\cdots\cdots\cdots \end{aligned} \tag{1.9}$$

where $F^{(2)}$ is determined by $p^{(0)}$ and $p^{(1)}$, and in general $F^{(m)}$ is determined by $p^{(j)}$, $j = 0, 1, \ldots, m-1$. Thus the procedure leads to a formal process for obtaining each $p^{(k)}$ in terms of the $p^{(j)}$, $j < k$, by solving a linear system of differential equations.

Since the existence of q as an analytic function of μ has been demonstrated, it follows that the system (1.9) has the desired periodic $p^{(j)}$, $j \geqq 1$, as a solution. It is also the case that there is no other solution of period T possible for (1.9); this will now be proved by induction. Since $p^{(0)}(t) = p$, which is given as part of the hypothesis, there is no question about uniqueness here. Suppose then that the next equation of (1.9) has two different solutions $p^{(1)}$ and $\tilde{p}^{(1)}$ of period T. Then the difference $p^{(1)} - \tilde{p}^{(1)}$ is clearly a solution of the first-variation equation (1.3). But by hypothesis this equation is assumed to have no periodic solution of period T. Thus $p^{(1)} = \tilde{p}^{(1)}$.

Suppose now that $p^{(j)}$ is determined for all $j < m$ for some m. Then

$$\frac{dp^{(m)}}{dt}(t) = f_x(t,p^{(0)}(t),0)p^{(m)}(t) + F^{(m)}(t) \tag{1.10}$$

where $F^{(m)}$ is uniquely determined by $p^{(j)}$, $j < m$. Thus for the same reason as in $m = 1$ it follows that $p^{(m)}$ is uniquely determined by (1.10) and the fact that it is periodic of period T. This proves the statement that (1.9) has one and only one system of periodic solutions $p^{(j)}$, $j \geqq 1$, of period T and that these may be obtained by solving the sequence of linear differential equations (1.9) for $j = 1, 2, \ldots$.

2. Autonomous Systems

An example of the system to be treated here is

$$x' = f(x) + \mu g(x,\mu) \tag{2.1}$$

where $|\mu|$ is small. A more general formulation in which μ may be a vector is

$$x' = f(x,\mu) \tag{2.2}$$

The system (2.2) is real for real x and μ. Here it is assumed that, for $\mu = 0$, (2.2) has a real periodic solution p of period T_0. This solution, $x = p(t)$, may be regarded as a closed curve in x space. Let V be a domain in x space containing the curve $x = p(t)$. Let f be of class C^1 in (x,μ) for x in V and $|\mu|$ small. In fact, it suffices for f_x to exist and be continuous for x in V and $|\mu|$ small.

The first variation of (2.2) for $\mu = 0$ with respect to the solution p is

$$y' = f_x(p(t),0)y \tag{2.3}$$

The system (2.3) has a periodic solution of period T_0, namely p'. Thus in the case (2.2) the hypothesis of Theorem 1.1 cannot be satisfied and a different result is required.

Let Ψ be the fundamental matrix for (2.3) which satisfies $\Psi(0) = E$. The fact that (2.3) has a solution of period T_0 is equivalent to the matrix $\Psi(T_0)$ having one as a characteristic root.

Theorem 2.1. *If f satisfies the conditions stated above and if* (2.3) *has one as a simple characteristic root of the matrix $\Psi(T_0)$, then for small $|\mu|$ the equation* (2.2) *has a solution $q = q(t,\mu)$ of period $T(\mu)$. Both q and T are continuous in μ for small $|\mu|$, $q(t,0) = p(t)$ and $T(0) = T_0$. The determination of $q(t,\mu)$ and $T(\mu)$ is unique.*

In order to motivate the proof presented below the following geometric discussion is given. By rotating and translating the x space the tangent vector $p'(0)$ can be made parallel to the x_1 axis and the first component of $p(0)$, $p_1(0)$, can be made zero. Thus $p_1(0) = 0$ and $p_j'(0) = 0, j = 2, \ldots, n$. A solution of (2.2) for $|\mu|$ small which is near some point of $x = p(t)$ will, because of the continuity of the solution with respect to initial value and μ, stay near $x = p(t)$ over a range of t of $2T_0$. Since p must cross the plane $x_1 = 0$, in this range of t, so does the solution of (2.2). Because (2.2) does not involve t explicitly, the solution curve in x space is uniquely determined by the coordinates of the point where it crosses $x_1 = 0$. Thus for any fixed μ it takes only $n - 1$ parameters to determine a solution of (2.2) as a curve in x space.

Proof of Theorem 2.1. As above, it is assumed that $p'(0)$ is parallel to the x_1 axis. Consider, for small $|\mu|$, the solution of (2.2) which assumes at $t = 0$ the value $p(0) + \alpha$, where $|\alpha|$ is small and $\alpha_1 = 0$. This solution is designated by $\varphi = \varphi(t,\alpha,\mu)$. For this solution to be periodic of period

$T = T_0 + \tau$, where $|\tau|$ is small, it is necessary and sufficient that

$$\varphi(T_0 + \tau, \alpha, \mu) - p(0) - \alpha = 0 \tag{2.4}$$

The equation (2.4) has at $\mu = 0$ the solution $\alpha = 0$, $\tau = 0$. If the Jacobian with respect to $(\tau, \alpha_2, \ldots, \alpha_n)$ does not vanish at $(\alpha = 0, \tau = 0, \mu = 0)$, then (2.4) has a unique solution $\alpha = \alpha(\mu)$, $\tau = \tau(\mu)$ for small $|\mu|$, $|\alpha|$, and $|\tau|$. Because $(\partial\varphi/\partial\tau)(T_0,0,0) = p'(T_0) = p'(0)$, the Jacobian is the determinant of the matrix

$$\begin{pmatrix} \dfrac{dp_1}{dt}(0) & \dfrac{\partial\varphi_1}{\partial\alpha_2} & \cdots & \dfrac{\partial\varphi_1}{\partial\alpha_n} \\ 0 & \dfrac{\partial\varphi_2}{\partial\alpha_2} - 1 & \cdots & \dfrac{\partial\varphi_2}{\partial\alpha_n} \\ \cdot & \cdot & \cdots & \cdot \\ 0 & \dfrac{\partial\varphi_n}{\partial\alpha_2} & \cdots & \dfrac{\partial\varphi_n}{\partial\alpha_n} - 1 \end{pmatrix} \tag{2.5}$$

where $\partial\varphi_i/\partial\alpha_j$ is evaluated at $(T_0,0,0)$. If the Jacobian does not vanish, then α and τ are uniquely determined as continuous functions of μ for small $|\mu|$. Also $\alpha(0) = 0$, $\tau(0) = 0$. Thus $\varphi(t,\alpha(\mu),\mu) = q(t,\mu)$ is periodic of period $T_0 + \tau(\mu)$. Because $p_1'(0) \neq 0$, the nonvanishing of the Jacobian is equivalent to the nonvanishing of the cofactor of dp_1/dt in (2.5).

If α_1 is not held fixed, then the fundamental solution Ψ is given by the matrix $\varphi_\alpha(t,0,0)$ with elements $\partial\varphi_i/\partial\alpha_j$. Thus the equation for the characteristic roots of the matrix $\Psi(T_0)$ is given by

$$\det\,(\varphi_\alpha(T_0,0,0) - \lambda E) = 0 \tag{2.6}$$

Since $(\partial\varphi_j/\partial\alpha_1)(0,0,0) = 0$ for $j \geqq 2$ and since $(\partial\varphi/\partial\alpha_1)(t,0,0)$ is a solution of (2.3), it follows that $(\partial\varphi/\partial\alpha_1)(t,0,0)$ is some multiple of p' and is therefore periodic of period T_0 so that $(\partial\varphi_j/\partial\alpha_1)(T_0,0,0) = 0$, $j \geqq 2$. Thus only the first element in the first column of the matrix on the left side of (2.6) is not zero. Since $(\partial\varphi_1/\partial\alpha_1)(0,0,0) = 1 = (\partial\varphi_1/\partial\alpha_1)(T_0,0,0)$, it follows that the cofactor of dp_1/dt in (2.5) can vanish only if $\lambda = 1$ is a multiple root of (2.6). This proves the theorem.

Because p' is a solution of the first-variation equation (2.3), it follows that one is a characteristic root of $\Psi(T_0)$, as already stated. If the remaining $n - 1$ characteristic roots are all less than one in magnitude, then the solution p has asymptotic orbital stability and the following result holds.

Theorem 2.2. *If $n - 1$ of the characteristic roots associated with the first variation, (2.3), of (2.2) with $\mu = 0$ are less than one in magnitude, then clearly Theorem 2.1 must hold. Moreover, the periodic solution of (2.2), $q = q(t,\mu)$, has asymptotic orbital stability for small $|\mu|$.*

Proof. The proof is very similar to that of Theorem 1.2. Briefly, if $\Psi = \Psi(t,\mu)$ is the solution of $y' = f_x(q(t,\mu),\mu)y$ with $\Psi(0,\mu) = E$, then the matrix $\Psi(T(\mu),\mu)$ is continuous for small μ and for $\mu = 0$, $\Psi(T_0,0) = \Psi(T_0)$. Thus, because $\Psi(T_0)$ has $n - 1$ characteristic roots less than one in magnitude, the same must be true of $\Psi(T(\mu),\mu)$ by continuity considerations, and this proves the theorem.

In case f in (2.2) is analytic in (x,μ) for x in V and $|\mu|$ small, it follows from Theorem 8.1, Chap. 1, that p is analytic in t; and then an easy modification of the existence theorem 8.4, Chap. 1, implies that

$$\varphi = \varphi(t,\alpha,\mu)$$

is analytic in (t,α,μ) for small $|\alpha|$ and $|\mu|$ over any fixed finite range of t. The nonvanishing of the Jacobian [the determinant of (2.5)] implies that α and τ are analytic in μ. Thus q is analytic in (t,μ) and T is analytic in μ for small $|\mu|$.

For simplicity it will be assumed from here on that μ has only one component, that is, μ need no longer be regarded as a vector. Because of the analyticity of q and T, it follows that if $t = sT(\mu)$ then

$$q(sT(\mu),\mu) = r(s,\mu)$$

is analytic in μ for small $|\mu|$ and periodic in s of period 1. Thus

$$r(s,\mu) = r^{(0)}(s) + \mu r^{(1)}(s) + \mu^2 r^{(2)}(s) + \cdots \tag{2.7}$$

$$T(\mu) = T_0 + \mu T_1 + \mu^2 T_2 + \cdots \tag{2.8}$$

where $r^{(i)}(s)$ has period 1 since r does. Letting $\mu \to 0$, $r^{(0)}(s) = p(sT_0)$. The system (2.2) becomes

$$\frac{dx}{ds} = T(\mu)f(x,\mu)$$

Since r is a solution, it follows that for small $|\mu|$

$$\frac{dr^{(0)}}{ds}(s) + \mu\frac{dr^{(1)}}{ds}(s) + \mu^2\frac{dr^{(2)}}{ds}(s) + \cdots$$
$$= (T_0 + \mu T_1 + \cdots)f(r(s,\mu),\mu)$$

Expanding the right member in powers of μ yields

$$\begin{aligned}
\frac{dr^{(0)}}{ds} &= T_0 f(r^{(0)},0)\\
\frac{dr^{(1)}}{ds} &= T_0 f_x(r^{(0)},0)r^{(1)} + T_1 f(r^{(0)},0) + T_0\frac{\partial f}{\partial \mu}(r^{(0)},0)\\
&\;\;\vdots\\
\frac{dr^{(m)}}{ds} &= T_0 f_x(r^{(0)},0)r^{(m)} + T_m f(r^{(0)},0) + F^{(m)}\\
&\cdots\cdots\cdots\cdots\cdots
\end{aligned} \tag{2.9}$$

where $F^{(m)}$ depends only on $r^{(j)}, j < m$, and $T_j, j < m$. Aside from (2.9), it is also the case that $r_1(0,\mu) = p_1(0)$. Thus $r_1^{(j)}(0) = 0, j = 1, 2, \ldots$.

The expressions for $r(s,\mu)$ and $T(\mu)$ given by (2.7) and (2.8) are then solutions of the system (2.9). It will be shown that the system (2.9) determines $r^{(j)}$ and T_j uniquely. Thus if the system (2.9) is solved for $r^{(j)}$ and $T_j, j \geqq 1$, subject to the conditions $r_1^{(j)}(0) = 0$ and $r^{(j)}$ periodic of period 1, then the result is r and T. The system provides then a convenient means for obtaining r and T.

Since $r^{(0)}$ and T_0 are given, the proof begins with $r^{(1)}$ and T_1. Let there be two solutions to (2.9) for $m = 1$, $(r^{(1)},T_1)$, and $(\tilde{r}^{(1)},\tilde{T}_1)$. Denote the difference $r^{(1)} - \tilde{r}^{(1)}$ by h and $T_1 - \tilde{T}_1$ by aT_0. Then h is periodic of period 1 and $h_1(0) = 0$. Subtracting the equations for the two cases shows that h is a solution of

$$\frac{dw}{ds} = P(s)w + aT_0 f(r^{(0)}(s),0) \tag{2.10}$$

where $P(s) = T_0 f_x(r^{(0)}(s),0)$ is the matrix coefficient of the variational equation for (2.2) with t replaced by sT_0. Indeed, the variational equation is

$$\frac{dy}{ds} = P(s)y \tag{2.11}$$

and a periodic solution of (2.11) is $\psi^{(1)} = dr^{(0)}/ds$. Moreover, by hypothesis the characteristic root 1 associated with the periodic solution is a simple root. The other $n - 1$ roots are all different from one. Using the first equation of (2.9), (2.10) becomes

$$\frac{dw}{ds} = P(s)w + a\psi^{(1)}(s) \tag{2.12}$$

with solution $w = h(s)$. Since $\psi^{(1)}$ is a solution of (2.11), it is readily verified that if

$$\psi(s) = h(s) - as\psi^{(1)}(s) \tag{2.13}$$

then ψ is a solution of (2.11).

If the solutions of (2.11) are in canonical form $\psi^{(i)}$, where $\psi^{(1)}$ is already defined and

$$\psi^{(i)}(1) = \lambda_i\psi^{(i)}(0) + d_i\psi^{(i-1)}(0) \qquad (i = 2, \ldots n)$$

then $d_2 = 0$, since $\lambda_1 = 1$ is a simple characteristic root. Using

$$\psi = \sum_{i=1}^{n} k_i\psi^{(i)}$$

the periodicity of h yields

$$\sum_{i=2}^{n} k_i(\lambda_i - 1)\psi^{(i)}(0) + \sum_{i=3}^{n} k_i d_i \psi^{(i-1)}(0) + a\psi^{(1)}(0) = 0 \qquad (2.14)$$

Since the solutions are linearly independent, this implies that $a = 0$. Thus (2.13) yields $h = \psi$. By the periodicity of h this means ψ must be $k\psi^{(1)}$ for some constant k. Since $h_1(0) = 0$ and $\psi_1^{(1)}(0) \neq 0$, it follows that $k = 0$. Thus $h(s) \equiv 0$, and the uniqueness of $r^{(1)}$ and T_1 is proved. The uniqueness of $r^{(2)}$ and T_2, etc., proceeds in identical fashion, and the proof follows by induction.

3. Perturbation of a Linear System with a Periodic Solution in the Nonautonomous Case

Consider the second-order differential equation

$$u'' + u = \mu f(t,u,u',\mu) \qquad (3.1)$$

where f is periodic in t of period 2π. The first variation for the case $\mu = 0$ is simply $u'' + u = 0$ which has two independent solutions of period 2π. Thus, if (3.1) is rewritten as a system, neither Theorem 1.1 nor 2.1 (in case f does not contain t explicitly) can apply, since the relevant Jacobians vanish. Nevertheless, the equation (3.1) is of considerable importance.

A sketch of a procedure for treating (3.1) will be given and then a general system will be treated. A minor simplification can be made by translating t in (3.1) by γ, where the constant γ will be specified later. Thus (3.1) becomes

$$u'' + u = \mu f(t + \gamma,\ u,\ u',\ \mu) \qquad (3.2)$$

Let a be a constant and consider the solution $\varphi(t) = \varphi(t,a,\gamma,\mu)$ of (3.2) which satisfies $\varphi(0) = a$, $\varphi'(0) = 0$. Clearly (3.2) implies

$$\varphi(t) = a \cos t + \mu \int_0^t \sin (t - s)\, f(s + \gamma,\ \varphi(s),\ \varphi'(s),\ \mu)\, ds \qquad (3.3)$$

In order that φ be periodic of period 2π in t, it is necessary and sufficient for $\varphi(2\pi) = a$, $\varphi'(2\pi) = 0$, or for

$$H_j(a,\gamma,\mu) = 0 \qquad (j = 1, 2) \qquad (3.4)$$

where

$$H_1(a,\gamma,\mu) = \int_0^{2\pi} \sin s\, f(s + \gamma,\ \varphi(s,a,\gamma,\mu),\ \varphi'(s,a,\gamma,\mu),\ \mu)\, ds$$

and similarly for H_2 but with $\sin s$ replaced by $\cos s$.

Suppose that the equations (3.4) have for $\mu = 0$ a solution $a = a_0$ and $\gamma = \gamma_0$ so that $H_j(a_0,\gamma_0,0) = 0$, $j = 1, 2$. Then if the Jacobian

$$\frac{\partial(H_1,H_2)}{\partial(a,\gamma)}(a_0,\gamma_0,0) \neq 0 \tag{3.5}$$

it follows that a and γ are uniquely determined as continuous functions of μ near $\mu = 0$. Because the Jacobian is evaluated at $\mu = 0$, the explicit form of H_1 and H_2 for use in (3.5) is found from the use of (3.3) with $\mu = 0$. Thus

$$H_1(a,\gamma,0) = \int_0^{2\pi} \sin s\, f(s + \gamma,\ a \cos s,\ -a \sin s,\ 0)\, ds \tag{3.6}$$

and similarly for $H_2(a,\gamma,0)$.

If $f(t,u,u',\mu) = \alpha u + \beta u^3 + c \cos t$, where α, β, and c are constants, then a and γ become functions of α, β, and c as well as of μ. It is easily seen that $H_1 = 0$ with the present choice of f yields $\gamma_0 = 0$, and the equation $H_2 = 0$ becomes

$$\alpha a + \frac{3\beta}{4} a^3 + c = 0 \tag{3.7}$$

which is then the equation that determines a_0. It is easily verified that (3.5) is now equivalent to the condition that a_0 be a simple root of (3.7).

The equation (3.1) is a special case of the system

$$x' = Ax + \mu f(t,x,\mu) \tag{3.8}$$

where A is a constant real matrix with Ni as a characteristic root for some integer N, and f is real and has period 2π in t. (Note that 2π need not be the least period of f.) The unperturbed system with $\mu = 0$,

$$x' = Ax \tag{3.9}$$

has then a solution of period 2π so that Theorem 1.1 does not apply. Let $\varphi^{(j)}$, $j = 1, \ldots, k$, be solutions of (3.9) of period 2π. Then for any constants c_j and γ_j

$$\sum_{j=1}^{k} c_j \varphi^{(j)}(t + \gamma_j) \tag{3.10}$$

is also a periodic solution of (3.9). It is not obvious for what c_j and γ_j, if any, (3.8) may have a periodic solution which tends to (3.10) as $\mu \to 0$. Here sufficient conditions will be given for (3.8) to have a periodic solution.

Setting $x = Py$, where P is a real nonsingular constant matrix, the system (3.8) can be replaced by a system for y where the coefficient matrix $B = P^{-1}AP$, when $\mu = 0$, is in real canonical form. Moreover, this new system satisfies the same assumption as (3.8). It will therefore be assumed that A already has the following real canonical form

$$A = \begin{pmatrix} A_1 & & & & & & & & \\ & A_2 & & & & & & & \\ & & \ddots & & & & & & \\ & & & A_k & & & & & \\ & & & & B_1 & & & & \\ & & & & & B_2 & & & \\ & & & & & & \ddots & & \\ & & & & & & & B_m & \\ & & & & & & & & C \end{pmatrix} \tag{3.11}$$

where the elements not shown are zeros. Each A_j, $j = 1, \ldots, k$, is a matrix of α_j (α_j even) rows and columns of the form

$$A_j = \begin{pmatrix} S_j & 0_2 & \cdots & 0_2 & 0_2 \\ E_2 & S_j & \cdots & 0_2 & 0_2 \\ 0_2 & E_2 & \cdots & 0_2 & 0_2 \\ \cdot & \cdot & \cdots & \cdot & \cdot \\ 0_2 & 0_2 & \cdots & E_2 & S_j \end{pmatrix} \tag{3.12}$$

where 0_2 is the 2-by-2 zero matrix and

$$S_j = \begin{pmatrix} 0 & -N_j \\ N_j & 0 \end{pmatrix} \qquad E_2 = \begin{pmatrix} 1 & 0 \\ 0 & 1 \end{pmatrix}$$

N_j being a positive integer. (In the following, E_k will always denote the k-dimensional unit matrix, $0 < k < n$, and $E = E_n$.) A matrix A_j may have only two rows and columns, in which case it is S_j. Each matrix B_j has β_j rows and columns, $j = 1, \ldots, m$, and is of the form

$$B_j = \begin{pmatrix} 0 & 0 & \cdots & 0 & 0 \\ 1 & 0 & \cdots & 0 & 0 \\ 0 & 1 & \cdots & 0 & 0 \\ \cdot & \cdot & \cdots & \cdot & \cdot \\ 0 & 0 & \cdots & 1 & 0 \end{pmatrix} \tag{3.13}$$

where B_j may have only one row and column, in which case B_j consists of the single element 0. The matrix C has $\gamma = n - \sum_{j=1}^{k} \alpha_j - \sum_{j=1}^{m} \beta_j$ rows and columns, and has no characteristic roots of the form iN for any integer N, including $N = 0$. It is useful to notice that C need not be in canonical form.

A fundamental matrix for (3.9) is given by

$$e^{tA} = \begin{pmatrix} e^{tA_1} & & & & & & & & & & \\ & e^{tA_2} & & & & & & & & & \\ & & \cdot & & & & & & & & \\ & & & \cdot & & & & & & & \\ & & & & \cdot & & & & & & \\ & & & & & e^{tA_k} & & & & & \\ & & & & & & e^{tB_1} & & & & \\ & & & & & & & e^{tB_2} & & & \\ & & & & & & & & \cdot & & \\ & & & & & & & & & \cdot & \\ & & & & & & & & & & \cdot \\ & & & & & & & & & & & e^{tB_m} \\ & & & & & & & & & & & & e^{tC} \end{pmatrix} \tag{3.14}$$

Here

$$e^{tA_j} = \begin{pmatrix} e^{tS_j} & 0_2 & \cdots & 0_2 \\ te^{tS_j} & e^{tS_j} & \cdots & 0_2 \\ \cdot & \cdot & \cdots & \cdot \\ \dfrac{t^{p_j-1}e^{tS_j}}{(p_j-1)!} & \dfrac{t^{p_j-2}e^{tS_j}}{(p_j-2)!} & \cdots & e^{tS_j} \end{pmatrix} \tag{3.15}$$

where $\alpha_j = 2p_j$, and

$$e^{tS_j} = \begin{pmatrix} \cos N_jt & -\sin N_jt \\ \sin N_jt & \cos N_jt \end{pmatrix} \tag{3.16}$$

while

$$e^{tB_j} = \begin{pmatrix} 1 & 0 & \cdots & 0 \\ t & 1 & \cdots & 0 \\ \cdot & \cdot & \cdots & \cdot \\ \dfrac{t^{\beta_j-1}}{(\beta_j-1)!} & \dfrac{t^{\beta_j-2}}{(\beta_j-2)!} & \cdots & 1 \end{pmatrix} \tag{3.17}$$

Concerning the matrix e^{tC}, the fact that the characteristic roots of C are not of the form iN, for N an integer, implies that

$$\det\,(e^{2\pi C} - E_\gamma) \neq 0 \tag{3.18}$$

Suppose now that (3.8) has a unique solution $\varphi = \varphi(t,\mu,c)$, where $\varphi(0,\mu,c) = c$, which exists for t in some interval containing $0 \leqq t \leqq 2\pi$, and is continuous in μ for μ sufficiently near $\mu = 0$. From (3.8), using the variation-of-constants formula,

$$\varphi(t,\mu,c) = e^{tA}c + \mu \int_0^t e^{(t-s)A} f(s,\varphi(s,\mu,c),\mu)\,ds \tag{3.19}$$

It follows directly from (3.19), and uniqueness, that a necessary and sufficient condition for φ to be periodic of period 2π is that

$$(e^{2\pi A} - E)c + \mu \int_0^{2\pi} e^{(2\pi - s)A} f(s,\varphi(s,\mu,c),\mu)\, ds = 0 \tag{3.20}$$

This represents a system of n equations for the n unknowns consisting of the components of c. In order to state sufficient conditions for the existence of c, the structure of (3.20) is analyzed in more detail. If there exists a function $c = c_\mu$, continuous in μ for small $|\mu|$, such that $\varphi(t,\mu,c_\mu)$ has period 2π, then from (3.20), letting $\mu \to 0$,

$$(e^{2\pi A} - E)c_0 = 0 \tag{3.21}$$

Thus (3.21) is a necessary condition for the existence of such a periodic solution $\varphi = \varphi(t,\mu,c_\mu)$. Note that from (3.19) we have $\varphi(t,0,c_0) = e^{tA}c_0$, and (3.21) just expresses the necessary and sufficient condition that $\varphi(t,0,c_0)$ be a periodic solution of (3.9) with period 2π.

From (3.15) and (3.16) we have

$$e^{2\pi A_j} - E_{\alpha_j} = \begin{pmatrix} 0_2 & 0_2 & \cdots & 0_2 \\ 2\pi E_2 & 0_2 & \cdots & 0_2 \\ \cdot & \cdot & \cdots & \cdot \\ \dfrac{(2\pi)^{p_j-1}E_2}{(p_j-1)!} & \dfrac{(2\pi)^{p_j-2}E_2}{(p_j-2)!} & \cdots & 0_2 \end{pmatrix} \tag{3.22}$$

and from (3.17)

$$e^{2\pi B_j} - E_{\beta_j} = \begin{pmatrix} 0 & 0 & \cdots & 0 \\ 2\pi & 0 & \cdots & 0 \\ \cdot & \cdot & \cdots & \cdot \\ \dfrac{(2\pi)^{\beta_j-1}}{(\beta_j-1)!} & \dfrac{(2\pi)^{\beta_j-2}}{(\beta_j-2)!} & \cdots & 0 \end{pmatrix} \tag{3.23}$$

From (3.22), (3.23), and (3.18) it is now clear that (3.21) implies that all components c_{0i} of c_0 are zero, except possibly those with index i corresponding to the last two rows of any A_j or to the last row of any B_j. These indices are all those with the following forms

$$\begin{aligned} i &= \alpha_1 - 1,\ \alpha_1 \\ i &= \alpha_1 + \alpha_2 - 1,\ \alpha_1 + \alpha_2 \\ &\cdot \\ &\cdot \\ &\cdot \\ i &= \alpha_1 + \cdots + \alpha_k - 1,\ \alpha_1 + \cdots + \alpha_k \\ i &= \alpha_1 + \cdots + \alpha_k + \beta_1 \\ i &= \alpha_1 + \cdots + \alpha_k + \beta_1 + \beta_2 \\ &\cdot \\ &\cdot \\ &\cdot \\ i &= \alpha_1 + \cdots + \alpha_k + \beta_1 + \cdots + \beta_m \end{aligned} \tag{3.24}$$

The indices i having the forms given in (3.24) will be called *exceptional indices,* and the corresponding components of any vector will be called *exceptional components.* They are $2k + m$ in number, and the exceptional components c_{0i} of c_0 are not determined by (3.21).

To proceed further, consider the components of the vector on the left in (3.20) with indices

$$
\begin{aligned}
j &= 1, 2 \\
j &= \alpha_1 + 1, \alpha_1 + 2 \\
&\cdots\cdots\cdots\cdots\cdots\cdots \\
j &= \alpha_1 + \cdots + \alpha_{k-1} + 1, \alpha_1 + \cdots + \alpha_{k-1} + 2 \\
j &= \alpha_1 + \cdots + \alpha_k + 1 \\
j &= \alpha_1 + \cdots + \alpha_k + \beta_1 + 1 \\
&\cdots\cdots\cdots\cdots\cdots\cdots \\
j &= \alpha_1 + \cdots + \alpha_k + \beta_1 + \cdots + \beta_{m-1} + 1
\end{aligned}
\tag{3.25}
$$

These $2k + m$ indices will be called *singular indices,* and the corresponding components of a vector *singular components.* From (3.22) and (3.23) it is evident that the singular components of $(e^{2\pi A} - E)c_\mu$ are all zero. Thus the singular components of the integral in (3.20) must vanish for all μ sufficiently near $\mu = 0$. This gives

$$
\left[\int_0^{2\pi} e^{(2\pi - s)A} f(s, \varphi(s, \mu, c), \mu) \, ds \right]_j = 0 \tag{3.26}
$$

for any j in the set (3.25), where $[\quad]_j$ represents the jth component. In particular, if $\mu = 0$, then $\varphi(s,0,c_0) = e^{sA}c_0$, and thus for the singular indices

$$
H_j(c_0) = \left[\int_0^{2\pi} e^{(2\pi - s)A} f(s, e^{sA}c_0, 0) \, ds \right]_j = 0 \tag{3.27}
$$

If the periodicity of $e^{sA}c_0$ and f are used, then (3.27) can be replaced by

$$
H_j(c_0) = \left[\int_0^{2\pi} e^{sA} f(-s, e^{-sA}c_0, 0) \, ds \right]_j = 0
$$

For $\mu = 0$, all components of c_0 other than the exceptional ones are zero. Thus (3.27) is a system of $2k + m$ equations in $2k + m$ unknowns, namely, the components c_{0i} of c_0 with i exceptional. If $\varphi = \varphi(t,\mu,c_\mu)$ is to exist as a periodic solution, it is *necessary* for (3.27) to have a solution. *Note that* (3.27) *can be written out explicitly without solving the nonlinear system,* (3.8).

Suppose that the system (3.27) has a solution for the exceptional components of c_0, say $c_{0i} = a_i$. Denoting by a the vector with components a_i for the exceptional indices and zero otherwise, it follows that

$$
p(t) = \varphi(t,0,a)
$$

is a periodic solution of (3.9) with period 2π.

To prove the existence of a periodic solution of (3.8) for $|\mu|$ small, the following assumptions will be needed:

(i) A is a real constant matrix with the canonical form (3.11) through (3.13), with at least one characteristic root of the form iN, where N is an integer;
(ii) μ,f are real, and f has period 2π in t;
(iii) a vector a exists which satisfies (3.21) and (3.27) for $c_0 = a$;
(iv) f, f_x are continuous in (t,x,μ) for (t,x) in a vicinity V containing the periodic solution $p(t) = e^{tA}a$ of (3.9), $0 \leqq t \leqq 2\pi$, in its interior, and for $|\mu| < \delta$, for some $\delta > 0$;†
(v) the Jacobian of the $(2k + m)$ H_j of (3.27), j singular, with respect to the $(2k + m)$ c_{0i}, i exceptional, does not vanish for $c_{0i} = a_i$.

Theorem 3.1. *Under the assumptions* (i) *through* (v) *above, there exists a unique periodic solution* $q = q(t,\mu)$ *of* (3.8), *of period* 2π *in* t, *which is continuous in* (t,μ) *for all* t, *and* $|\mu|$ *sufficiently small, and which for* $\mu = 0$ *reduces to* $q(t,0) = p(t)$.

REMARK: If assumption (v) does not hold, then a more detailed analysis is required which will not be undertaken here.

Proof of Theorem 3.1. It will be shown that, for sufficiently small $|\mu|$, (3.20) has a unique solution $c = c_\mu$, continuous in μ, and with $c_0 = a$. From this it follows at once that $q(t,\mu) = \varphi(t,\mu,c_\mu)$ is the desired solution. The existence and uniqueness of a solution $\varphi(t,\mu,c)$ of (3.8) for $|\mu|$ and $|c - c_0|$ sufficiently small and with $\varphi(0,\mu,c) = c$ follow directly from the assumption (iv).

To show the existence of $c = c_\mu$, the system of equations (3.20) is replaced by the systems $S_1(\mu,c)$ and $S_2(\mu,c)$, where $S_1(\mu,c)$ consists of the components of (3.20) with nonsingular indices, while $S_2(\mu,c)$ consists of the equations (3.26) with singular indices. As discussed after (3.23), $S_1(0,c)$ is a homogeneous linear system for the nonexceptional components c_i of c, and since the determinant Δ of the coefficients is nonvanishing, the only solution is $c_{0i} = 0$, i nonexceptional. $S_2(0,c)$ is just the system of Eqs. (3.27) with $c = c_0$, and by assumption this has a solution $c_{0i} = a_i$, i exceptional. The first partial derivatives of the left side of $S_1(\mu,c)$ with respect to the exceptional components of c are all zero at $\mu = 0$, $c = a$. Thus the over-all Jacobian $D(\mu,c)$ of the left sides of $S_1(\mu,c)$ and $S_2(\mu,c)$ with respect to the components of c, when evaluated at $\mu = 0$, $c = a$, is the Jacobian $D_1(\mu,c)$ of the left side of $S_1(\mu,c)$ with respect to the nonexceptional components of c, evaluated at $\mu = 0$, $c = a$, multiplied by the Jacobian $D_2(\mu,c)$ of the left side of (3.26) with respect to the exceptional components of c, evaluated at $\mu = 0$, $c = a$. But $D_1(0,a)$ is just the determinant Δ which is not zero, and $D_2(0,a)$ is just the Jacobian in

† See the second paragraph of Sec. 1.

assumption (v), which is not zero. Therefore, by the implicit-function theorem, the combined system $S_1(\mu,c)$, $S_2(\mu,c)$ has a unique solution $c = c_\mu$ for sufficiently small $|\mu|$, which is continuous in μ, and such that $c_0 = a$. This completes the proof.

The situation as regards analytic perturbations f is as follows:

Theorem 3.2. *Suppose the assumptions* (i) *through* (v) *are satisfied, and in addition let f be an analytic function of (x,μ) for (t,x) in V and $|\mu| < \delta$. Then q is analytic in μ for sufficiently small $|\mu|$.*

Proof. For (μ,c) sufficiently close to $(0,c_0)$, the solution $\varphi = \varphi(t,\mu,c)$ of (3.8) is analytic in μ and c by the existence theorem for such systems. Also, c_μ is analytic in μ by the implicit-function theorem for analytic systems. Thus $q(t,\mu) = \varphi(t,\mu,c_\mu)$ is analytic in μ.

From the practical point of view, it is important to know (in the analytic case) whether the periodic coefficients (with period 2π) $q^{(i)}(t)$ in the convergent power-series expansion

$$q(t,\mu) = \sum_{i=0}^{\infty} \mu^i q^{(i)}(t) \tag{3.28}$$

can be calculated recursively. As is to be expected, this is indeed true. Let the jth component of $q^{(i)}$ be denoted by $q_j^{(i)}$. Placing (3.28) into (3.8) and comparing coefficients of powers of μ, there results

$$\begin{aligned} q^{(0)}(t) &= e^{tA}a, \\ \frac{dq^{(1)}}{dt}(t) &= Aq^{(1)}(t) + f(t,q^{(0)}(t),0), \\ \frac{dq^{(2)}}{dt}(t) &= Aq^{(2)}(t) + f_x(t,q^{(0)}(t),0)q^{(1)}(t) + \frac{\partial f}{\partial \mu}(t,q^{(0)}(t),0), \ldots, \\ \frac{dq^{(j)}}{dt}(t) &= Aq^{(j)}(t) + f_x(t,q^{(0)}(t),0)q^{(j-1)}(t) + F^{(j)}(t), \ldots \end{aligned} \tag{3.29}$$

where $F^{(j)}$ depends only on $q^{(l)}$ for $0 \leqq l \leqq j - 2$. That there exist solutions $q^{(i)}$ to the system of differential equations (3.29) follows from the existence of q. It will be shown that each equation in (3.29) has at most one solution of period 2π and thus the formal process of solving for the $q^{(i)}$ recursively yields q. Clearly $q^{(1)}$ is determined by the second equation in (3.29) only to within a periodic solution of the homogeneous equation (3.9). However, the requirement that the next equation of (3.29) have a periodic solution $q^{(2)}$ determines $q^{(1)}$ uniquely. For suppose this is not the case. Then there are two distinct solutions for $q^{(1)}$ each of which in the next equation allows for the solution of a periodic $q^{(2)}$. Denoting the differences between the two $q^{(1)}$'s by $\tilde{q}^{(1)}$ and the two $q^{(2)}$'s by $\tilde{q}^{(2)}$, it follows, by subtracting the two equations for the $q^{(1)}$'s from each other and the two for the $q^{(2)}$'s, that

$$\frac{d\tilde{q}^{(1)}}{dt}(t) = A\tilde{q}^{(1)}(t) \tag{3.30}$$

$$\frac{d\tilde{q}^{(2)}}{dt}(t) = A\tilde{q}^{(2)}(t) + f_x(t,e^{tA}a,\, 0)\tilde{q}^{(1)}(t) \tag{3.31}$$

If $\tilde{q}^{(2)}(0) = \tilde{a}^{(2)}$, then from (3.31),

$$\tilde{q}^{(2)}(t) = e^{tA}\tilde{a}^{(2)} + \int_0^t e^{(t-s)A}f_x(s,e^{sA}a,0)\tilde{q}^{(1)}(s)\, ds$$

Since $\tilde{q}^{(2)}$ has period 2π, this yields

$$(e^{2\pi A} - E)\tilde{a}^{(2)} + \int_0^{2\pi} e^{(2\pi-s)A}f_x(s,e^{sA}a,0)\tilde{q}^{(1)}(s)\, ds = 0$$

Taking the components of the above with singular indices, j, it follows that

$$\left[\int_0^{2\pi} e^{(2\pi-s)A}f_x(s,e^{sA}a,0)\tilde{q}^{(1)}(s)\, ds\right]_j = 0 \tag{3.32}$$

Clearly $\tilde{q}^{(1)}$, being a periodic solution of (3.30), is of the form

$$\tilde{q}^{(1)}(s) = e^{sA}\tilde{a}^{(1)}$$

where the only nonvanishing components $\tilde{a}_j^{(1)}$ of $\tilde{a}^{(1)}$ are those with exceptional indices. Now (3.32) is linear homogeneous in these $\tilde{a}_j^{(1)}$, and the determinant of the coefficients of the left side of (3.32) with respect to the $\tilde{a}_j^{(1)}$ is precisely the Jacobian of the $(2k + m)$ H_j of Theorem 3.1 which is assumed not to vanish. Hence $\tilde{a}^{(1)} = 0$, and thus $\tilde{q}^{(1)}(t) \equiv 0$. Precisely the same argument shows that if the $q^{(l)}$ are uniquely determined for $l \leqq j$, where $j > 1$, then $q^{(j+1)}$ is also uniquely determined, thus yielding the result by induction.

Theorem 3.3. *If the assumptions of Theorem* 3.2 *hold, then the analytic solution* q *of* (3.8) *can be obtained recursively by solving the system* (3.29) *for the periodic coefficients* $q^{(i)}$, *of period* 2π, *in the convergent power-series expansion* (3.28). *Each* $q^{(i)}, i \geqq 1$, *is determined uniquely by the ith equation in* (3.29) *and the fact that the* $(i + 1)$*st equation has a periodic solution.*

4. Perturbation of an Autonomous System with a Vanishing Jacobian

Here the real system

$$x' = Ax + \mu f(x,\mu) \tag{4.1}$$

will be considered, where A is a real constant matrix, $|\mu|$ is small, and f is real continuous in (x,μ) for small $|\mu|$ and x in a region V to be described later. In fact, it will be assumed that f_x is continuous in (x,μ) for x in V and $|\mu|$ small. With $\mu = 0$, the system (4.1) becomes

$$x' = Ax \tag{4.2}$$

It is assumed that (4.2) has a periodic solution of period 2π. Note that 2π need not be its least period. It is assumed to be the case that $e^{2\pi A}$ does not have one as a simple characteristic root. This last statement is equivalent to the vanishing of the relevant Jacobian (2.5).

As in Sec. 3, it can be assumed with no restriction that A is in real canonical form given by formulas (3.11) through (3.13). Thus (3.14) through (3.18) also hold. As before, C need not be in canonical form.

Let (4.1) have a unique solution $\varphi = \varphi(t,\mu,c)$, where $\varphi(0,\mu,c) = c$, which exists for t in some finite interval and is continuous in μ for μ near $\mu = 0$. Then, as before,

$$\varphi(t,\mu,c) = e^{tA}c + \mu \int_0^t e^{(t-s)A}f(\varphi(s,\mu,c),\mu)\,ds \tag{4.3}$$

Necessary and sufficient for φ to be periodic of period $2\pi + \tau$ is that

$$(e^{(2\pi+\tau)A} - E)c + \mu \int_0^{2\pi+\tau} e^{(2\pi+\tau-s)A}f(\varphi(s,\mu,c),\mu)\,ds = 0$$

or

$$(e^{2\pi A} - E)c + e^{2\pi A}(e^{\tau A} - E)c + \mu \int_0^{2\pi+\tau} e^{(2\pi+\tau-s)A}f(\varphi(s,\mu,c),\mu)\,ds = 0 \tag{4.4}$$

If $\varphi = \varphi(t,\mu,c)$ is periodic of period $2\pi + \tau$ and if $c = c_\mu$ and $\tau = \tau(\mu)$ are continuous for small enough $|\mu|$, and $\tau(0) = 0$, then it follows, since $\varphi(t,0,c_0) = e^{tA}c_0$, that

$$(e^{2\pi A} - E)c_0 = 0 \tag{4.5}$$

As in Sec. 3, this implies that only the components of c_0 with exceptional indices can be different from zero. (Clearly τ and c need, in fact, only exist for small $\mu > 0$ with limiting values as $\mu \to 0+$.)

It is assumed here that in the canonical form for A there appears at least one matrix of the type A_j. The exceptional indices associated with A_1 are $\alpha_1 - 1$ and α_1. The component of $e^{tA}c_0$ with index α_1 is $(c_0)_{\alpha_1-1}\sin N_1t + (c_0)_{\alpha_1}\cos N_1t$, and hence, for any specific choice of $(c_0)_{\alpha_1-1}$ and $(c_0)_{\alpha_1}$, this sinusoid vanishes for some value of t and has there a nonvanishing first derivative; if not, both $(c_0)_{\alpha_1-1}$ and $(c_0)_{\alpha_1}$ vanish. By continuity, the component φ_{α_1} of $\varphi(t,\mu,c_\mu)$ must cross the t axis at some t also. The system (4.1) is invariant under translations in t. In what follows it will be assumed that φ_{α_1} vanishes at $t = 0$. This means that

$$\varphi_{\alpha_1}(0,\mu,c_\mu) = (c_\mu)_{\alpha_1} = 0$$

for sufficiently small $|\mu|$, including $\mu = 0$. Thus the problem becomes one in obtaining sufficient conditions for the existence of $c = c_\mu$ and $\tau = \tau(\mu)$ with $(c_\mu)_{\alpha_1} = 0$ satisfying (4.4).

As in Sec. 3, the components of $(e^{2\pi A} - E)c_\mu$ with singular indices are all zero. Thus (4.4) shows for the components with these indices

$$\left[e^{2\pi A}\left(\frac{e^{\tau A} - E}{\tau}\right) c_\mu \left(\frac{\tau}{\mu}\right) + \int_0^{2\pi+\tau} e^{(2\pi+\tau-s)A} f(\varphi(s,\mu,c_\mu),\mu)\, ds\right]_j = 0 \quad (4.6)$$

Letting $\mu \to 0$ in (4.6), the term involving the integral tends to a limit, and hence the other term approaches a limit. Since $\tau \to 0$ as $\mu \to 0$, (4.6) gives for the components with singular indices

$$\left[e^{2\pi A} A c_0 \lim_{\mu\to 0}\left(\frac{\tau}{\mu}\right) + \int_0^{2\pi} e^{(2\pi-s)A} f(e^{sA}c_0,0)\, ds\right]_j = 0 \qquad (4.7)$$

Changing the variables from s to $2\pi - s$ and using the periodicity of $e^{sA}c_0$, there follows

$$\left[e^{2\pi A} A c_0 \lim_{\mu\to 0}\left(\frac{\tau}{\mu}\right) + \int_0^{2\pi} e^{sA} f(e^{-sA}c_0,0)\, ds\right]_j = 0$$

which can be used instead of (4.7). If at least one singular component of $e^{2\pi A}Ac_0$ is different from zero, then (4.7) mplies the existence of the limit of τ/μ as $\mu \to 0$, whereas the existence of this limit is not implied by (4.7) if all the components of $e^{2\pi A}Ac_0$ with singular indices are zero. The system (4.7), which holds for singular indices only, can be regarded as a system of $2k + m$ equations for the $2k + m$ unknowns consisting of the $2k + m - 1$ components of c_0 with exceptional indices $((c_0)_{\alpha_1} = 0)$ and the unknown limit of (τ/μ), $\mu \to 0$.

Let $\tau = \mu\nu$, and let

$$H_j(c_0,\nu) = \left[\nu e^{2\pi A} A c_0 + \int_0^{2\pi} e^{(2\pi-s)A} f(e^{sA}c_0,0)\, ds\right]_j \qquad (4.8)$$

where j goes through the singular indices. [It is to be observed that H_j is given explicitly in (4.8) as a function of c_0 and ν, and does not require that (4.1) be solved.]

For the following existence theorem it is *assumed* that

(i) A is a real constant matrix with canonical form (3.11) through (3.13);

(ii) μ, $f = f(x,\mu)$ are real;

(iii) a vector a, with $a_{\alpha_1} = 0$ and with nonexceptional components zero, and a number ν_0 exist which satisfy $H_j(a,\nu_0) = 0$, j singular;

(iv) f, f_x are continuous in (x,μ) for x in a vicinity V containing the periodic solution $e^{tA}a$, of period 2π, of (4.2), and for $|\mu| < \delta$, for some $\delta > 0$;†

†See the second paragraph of Sec. 1.

(v) the Jacobian determinant of the $(2k + m)$ H_j of (4.8) with respect to the $(2k + m - 1)$ $(c_0)_i$, i exceptional ($i \neq \alpha_1$), and with respect to ν, does not vanish for $(c_0)_i = a_i$ ($(c_0)_{\alpha_1} = 0$), and $\nu = \nu_0$.

Theorem 4.1. *Under the assumptions* (i) *through* (v) *above there exists a periodic solution $q = q(t,\mu)$ of* (4.1) *with period $2\pi + \tau(\mu)$, where q is continuous in (t,μ) for all t and $|\mu|$ sufficiently small, $\tau = \tau(\mu)$ is continuous in μ, $q(t,0) = e^{tA}a$, $\tau(\mu)/\mu \to \nu_0$ as $\mu \to 0$, and $q_{\alpha_1}(0,\mu) = 0$. There is no other periodic solution of* (4.1) *which, when $\mu \to 0$, becomes $e^{tA}a$.*

REMARK: The role of α_1 can be taken by $\alpha_1 + \cdots + \alpha_j$, for any $j \leq k$.

Proof of Theorem 4.1. Since ν enters H_j linearly, the Jacobian will certainly vanish if the coefficients $(e^{2\pi A}Aa)_j = 0$, j singular. Note that, as in Sec. 3, the a_j with nonexceptional indices all must vanish. From this it follows that the terms $(e^{2\pi A}Aa)_j$, j singular, can be different from zero only for those j associated with an A_i which has exactly two rows and columns, and for no B_i. *Thus* (v) *really implies that there is at least one A_i which must have two rows and columns.*

The proof follows that of Theorem 3.1 closely and hence will be omitted.

Theorem 4.2. *Let f be analytic in (x,μ) for x in V and $|\mu| < \delta$, and suppose the assumptions of Theorem 4.1 hold. Then the periodic solution q is analytic in (t,μ) for all t and for $|\mu|$ sufficiently small, and τ is analytic in μ for $|\mu|$ sufficiently small.*

Proof. The proof is very much like that of Theorem 3.2.

In the analytic case, the coefficients in the power-series expansions for q and τ can be calculated recursively. Here it is convenient to replace t by s, where $t = s(1 + \tau/2\pi)$, and let $q(s(1 + \tau/2\pi),\mu) = p(s,\mu)$. Clearly p is analytic in μ for small $|\mu|$, and periodic in s of period 2π. Therefore there exist expansions

$$p(s,\mu) = \sum_{i=0}^{\infty} \mu^i p^{(i)}(s) \tag{4.9}$$

$$\frac{\tau}{2\pi} = \sum_{i=1}^{\infty} \mu^i b_i \tag{4.10}$$

where both series converge for small enough $|\mu|$. Clearly, $2\pi b_1 = \nu_0$, and $p^{(0)}(s) = e^{sA}a$. Since the component of $q(0,\mu)$ with index α_1 vanishes, it follows that this component of $p^{(i)}(0)$ must vanish for all $i \geq 0$.

The differential equation (4.1) becomes

$$\frac{dx}{ds} = \left(1 + \frac{\tau}{2\pi}\right)(Ax + \mu f(x,\mu)) \tag{4.11}$$

If (4.9) and (4.10) are substituted into (4.11) and coefficients of the powers of μ are compared, there results the following system of equations:

$$\begin{aligned}
\frac{dp^{(0)}}{ds} &= Ap^{(0)} \\
\frac{dp^{(1)}}{ds} &= Ap^{(1)} + b_1Ap^{(0)} + f(p^{(0)},0) \\
\frac{dp^{(2)}}{ds} &= Ap^{(2)} + b_1Ap^{(1)} + b_2Ap^{(0)} + b_1f(p^{(0)},0) + f_x(p^{(0)},0)p^{(1)} \\
&\qquad + \frac{\partial f}{\partial \mu}(p^{(0)},0) \\
&\cdots\cdots\cdots\cdots\cdots\cdots\cdots\cdots \\
\frac{dp^{(j)}}{ds} &= Ap^{(j)} + b_1Ap^{(j-1)} + b_jAp^{(0)} + f_x(p^{(0)},0)p^{(j-1)} + F^{(j)} \\
&\cdots\cdots\cdots\cdots\cdots\cdots\cdots\cdots
\end{aligned} \tag{4.12}$$

where $F^{(j)}$ depends on $p^{(0)}, p^{(1)}, \ldots, p^{(j-2)}$, and $b_1, b_2, \ldots, b_{j-1}$. That the system (4.12) has a solution for $p^{(i)}$ and b_i is clear.

Theorem 4.3. *Under the assumptions of Theorem 4.2, the analytic solution q of (4.1) can be obtained by solving Eqs. (4.12) in succession for the periodic coefficients $p^{(i)}$, of period 2π, of the power series (4.9) for*

$$p(s,\mu) = q\left(s\left(1 + \frac{\tau}{2\pi}\right),\mu\right)$$

and the constants b_i in the expansion (4.10) for $\tau/2\pi$. The $p^{(i)}$ and b_i are uniquely determined in (4.12) by the requirements that $p^{(0)}(s) = e^{sA}a$, $2\pi b_1 = \nu_0$, $p^{(i)}(s + 2\pi) = p^{(i)}(s)$, and $p^{(i)}_{\alpha_1}(0) = 0$.

Proof. Suppose there are two functions $p^{(1)}, \hat{p}^{(1)}$, satisfying the second equation in (4.12), and two constants $b_2, \hat{b}_2$, such that to the pairs $(p^{(1)},b_2)$, $(\hat{p}^{(1)},\hat{b}_2)$ there correspond $p^{(2)}, \hat{p}^{(2)}$, respectively, satisfying the third equation of (4.12). Subtracting the third equation for one case from that for the other case, and denoting $\tilde{p}^{(1)} = p^{(1)} - \hat{p}^{(1)}$, $\tilde{p}^{(2)} = p^{(2)} - \hat{p}^{(2)}$, $\tilde{b}_2 = b_2 - \hat{b}_2$,

$$\frac{d\tilde{p}^{(2)}}{ds} = A\tilde{p}^{(2)} + b_1A\tilde{p}^{(1)} + \tilde{b}_2Ap^{(0)} + f_x(p^{(0)},0)\tilde{p}^{(1)}$$

From the second equation for each case follows

$$\frac{d\tilde{p}^{(1)}}{ds} = A\tilde{p}^{(1)}$$

Let $\tilde{p}^{(1)}(0) = \tilde{a}^{(1)}$, and $\tilde{p}^{(2)}(0) = \tilde{a}^{(2)}$. The α_1 component of each vector is zero. Since $\tilde{p}^{(1)}$ is periodic, it follows that

$$\tilde{p}^{(1)}(s) = e^{sA}\tilde{a}^{(1)}$$

where only the exceptional components of $\tilde{a}^{(1)}$ can be different from zero. Thus $\tilde{a}^{(1)}$ has at most $2k + m - 1$ components that are not known to be zero. From the differential equation for $\tilde{p}^{(2)}$, and the fact that

$$\tilde{p}^{(2)}(0) = \tilde{p}^{(2)}(2\pi)$$

it follows that

$$(E - e^{2\pi A})\tilde{a}^{(2)} = b_1 A \int_0^{2\pi} e^{(2\pi-\sigma)A}\tilde{p}^{(1)}(\sigma)\,d\sigma + \tilde{b}_2 A \int_0^{2\pi} e^{(2\pi-\sigma)A} p^{(0)}(\sigma)\,d\sigma + \int_0^{2\pi} e^{(2\pi-\sigma)A} f_x(p^{(0)}(\sigma),0)\tilde{p}^{(1)}(\sigma)\,d\sigma$$

Since the singular components of the left side vanish, the same is true for the right side. Setting $p^{(0)}(\sigma) = e^{\sigma A}a$, and $2\pi b_1 = \nu_0$, it follows that

$$\left[\nu_0 e^{2\pi A} A\tilde{a}^{(1)} + 2\pi\tilde{b}_2 e^{2\pi A} Aa + \int_0^{2\pi} e^{(2\pi-\sigma)A} f_x(e^{\sigma A}a,0)e^{\sigma A}\tilde{a}^{(1)}\,d\sigma\right]_j = 0$$

for all singular j. These $2k + m$ equations are linear homogeneous in the $2k + m$ terms consisting of $2\pi\tilde{b}_2$ and the $2k + m - 1$ components of $\tilde{a}^{(1)}$ not fixed at zero. However, the determinant made up of the coefficients of these $2k + m$ terms is precisely the Jacobian of the $(2k + m)$ $H_j(c_0,\nu)$ with respect to (c_0,ν) evaluated at $\nu = \nu_0$, $c_0 = a$. This Jacobian is not zero. Thus $\tilde{b}_2 = 0$ and $\tilde{a}^{(1)} = 0$, which proves that b_2 and $p^{(1)}$ are uniquely determined by (4.12). In the same way, it follows that if $p^{(0)}, \ldots, p^{(j-1)}$ and $b_1, \ldots, b_j$ $(j > 2)$ are determined uniquely, then (4.12) determines $p^{(j)}$ and b_{j+1} uniquely, thus yielding the result by induction.

Examples analogous to those in (3.1) can be considered. The case

$$u'' + u = \mu f(u,u',\mu),$$

where u,μ are scalars, yields for the periodicity equations (4.6)

$$c_\mu \frac{(\cos\tau - 1)}{\tau}\cdot\frac{\tau}{\mu} + \int_0^{2\pi+\tau} \sin(\tau - s)\, f(\varphi(s,\mu),\varphi'(s,\mu),\mu)\,ds = 0$$

$$-c_\mu \frac{\sin\tau}{\tau}\cdot\frac{\tau}{\mu} + \int_0^{2\pi+\tau} \cos(\tau - s)\, f(\varphi(s,\mu),\varphi'(s,\mu),\mu)\,ds = 0$$

where c_μ here represents a scalar. Letting $\mu \to 0$ and recalling that $\tau(0) = 0$, there results

$$\int_0^{2\pi} \sin s\, f(c_0\cos s, -c_0\sin s, 0)\,ds = 0$$

$$-c_0\nu + \int_0^{2\pi} \cos s\, f(c_0\cos s, -c_0\sin s, 0)\,ds = 0$$

PROBLEMS

1. Show that the consideration of

$$u'' + u = \mu f(\omega t, u, u', \mu)$$

where ω is near 1, can be transformed to the case (3.1) by setting $\tau = \omega t$ and

$$k = \frac{\omega^2 - 1}{\omega^2 \mu}$$

giving

$$\frac{d^2u}{d\tau^2} + u = \mu g\left(\tau, u, \frac{du}{d\tau}, \mu\right)$$

where $g = ku + (1 - k\mu)f(\tau, u, \omega\, du/d\tau, \mu)$. In particular, if

$$f = \alpha u + \beta u^3 + c \cos \omega t$$

show that the effect on a_0 of varying ω is the same as that of varying α in (3.7). Show that by varying α continuously, $|a_0|$ may be discontinuous.

2. Consider the system

$$x' = Ax + f(t,x) + \mu F(t)$$

where A is a constant matrix, f is continuous in (t,x) for small $|x|$ and all t, and periodic in t of period T, F is continuous and periodic of period T, and μ is a constant. Let f_x exist for small $|x|$ and all t and be continuous in (t,x). Let $y' = Ay$ have no solution of period T. Let $f_x(t,0) = 0$, and let $f(t,0) = 0$. Prove that for small μ there exists a unique solution $\varphi = \varphi(t,\mu)$ of period T which is continuous in (t,μ) for all $|t|$ and small μ. Show that $\varphi(t,0) = 0$.

HINT: Use Theorem 1.1.

3. In Prob. 2 let f be independent of t and thus of the form $f(x)$ and let F be an almost periodic function. Let A have no characteristic root with real part zero. Show that for small μ the differential equation has a unique almost periodic solution $\varphi = \varphi(t,\mu)$.

HINT: $e^{tA} = \Psi_1(t) + \Psi_2(t)$, where $|\Psi_1(t)| \leqq Ke^{-\sigma t}$, $t \geqq 0$, and $|\Psi_2(t)| \leqq Ke^{\sigma t}$, $t \leqq 0$. By successive approximations, show that if ϵ is taken as $\sigma/(4K)$ and $|\mu| < \delta\sigma/(4KM)$, where $M = \max |F(t)|$, then the system

$$\varphi(t) = \int_{-\infty}^{\infty} H(t-s) f(\varphi(s))\, ds + \mu g(t) \tag{*}$$

has a solution. Here $H(t) = \Psi_1(t)$ for $t > 0$, and $H(t) = -\Psi_2(t)$ for $t < 0$ and

$$g(t) = \int_{-\infty}^{\infty} H(t-s) F(s)\, ds$$

Show that $|\varphi(t,\mu)| \leqq 4KM|\mu|/\sigma$. In (*), g is almost periodic. Given $\eta > 0$, let L be a translation number of g such that $|g(t+L) - g(t)| < \eta$. Show from (*) that $|\varphi(t+L) - \varphi(t)| < 4K\eta/\sigma$ and thus that φ is almost periodic. That φ is unique follows from the fact that any small uniformly bounded solution of the differential equation on $(-\infty, \infty)$ is a solution of (*).

4. State and prove the analogues of Theorems 3.1, 3.2, and 3.3 for the case where A in (3.8) is replaced by a periodic matrix $A(t)$ of period T.

CHAPTER 15

PERTURBATION THEORY OF TWO-DIMENSIONAL REAL AUTONOMOUS SYSTEMS

1. Two-dimensional Linear Systems

Consider the real linear system

$$\begin{aligned} x_1' &= ax_1 + bx_2 \\ x_2' &= cx_1 + dx_2 \end{aligned} \qquad \left(' = \frac{d}{dt}\right) \tag{1.1}$$

where a, b, c, d are real constants such that the determinant $ad - bc$ does not vanish. Clearly $(x_1,x_2) = (0,0)$ is then the *only* critical point of this system, that is, the only point where the right member of (1.1) vanishes. Let the coefficient matrix of (1.1) be denoted by

$$A = \begin{pmatrix} a & b \\ c & d \end{pmatrix}$$

Then (1.1) can be written as $x' = Ax$, where $x = (x_1,x_2)$. Let A have the characteristic roots λ, μ. These roots can be real or complex, but if one is complex, say $\lambda = \alpha + i\beta$ (α, β real, $\beta \neq 0$), then $\mu = \alpha - i\beta$ is the other root, for the coefficients of the characteristic equation for A are real. It is known that there exists a *real* nonsingular constant matrix T such that, if $y = Tx$, then the transformed system $y' = (TAT^{-1})y$ has a real coefficient matrix $J = (TAT^{-1})$ which has one of the following real canonical forms:

(I) $\begin{pmatrix} \lambda & 0 \\ 0 & \lambda \end{pmatrix}$ $(\lambda \neq 0)$ (II) $\begin{pmatrix} \lambda & 0 \\ 0 & \mu \end{pmatrix}$ $(\mu < \lambda < 0,$ or $0 < \mu < \lambda)$

(III) $\begin{pmatrix} \lambda & 0 \\ \gamma & \lambda \end{pmatrix}$ $(\lambda \neq 0, \gamma > 0)$ (IV) $\begin{pmatrix} \lambda & 0 \\ 0 & \mu \end{pmatrix}$ $(\lambda < 0 < \mu)$

(V) $\begin{pmatrix} \alpha & \beta \\ -\beta & \alpha \end{pmatrix}$ $(\alpha \neq 0, \beta \neq 0)$ (VI) $\begin{pmatrix} 0 & \beta \\ -\beta & 0 \end{pmatrix}$ $(\beta \neq 0)$

Thus, in order to discuss the nature of the orbits of (1.1) near (0,0), one may assume A has one of the forms (I) through (VI).

Before taking up each of these cases individually, a matter of notation will be settled. In general, a solution of a two-dimensional system

$$x_1' = g_1(x_1,x_2) \qquad x_2' = g_2(x_1,x_2) \tag{1.2}$$

will be denoted by $\varphi = (\varphi_1,\varphi_2)$, and it will often be convenient to consider the *polar functions* ρ, ω, associated with the solution φ, defined by

$$\rho(t) = (\varphi_1^2(t) + \varphi_2^2(t))^{\frac{1}{2}} \qquad \omega(t) = \tan^{-1}\frac{\varphi_2(t)}{\varphi_1(t)}$$

It is stressed that ρ, ω are defined with respect to a *particular solution* φ

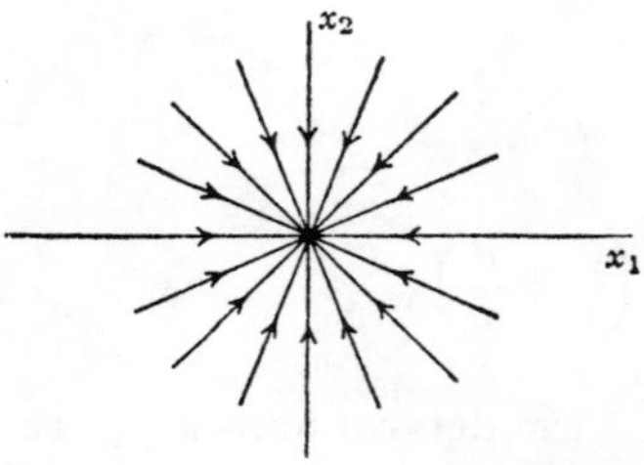

FIG. 3. (I) Proper node, $\lambda < 0$.

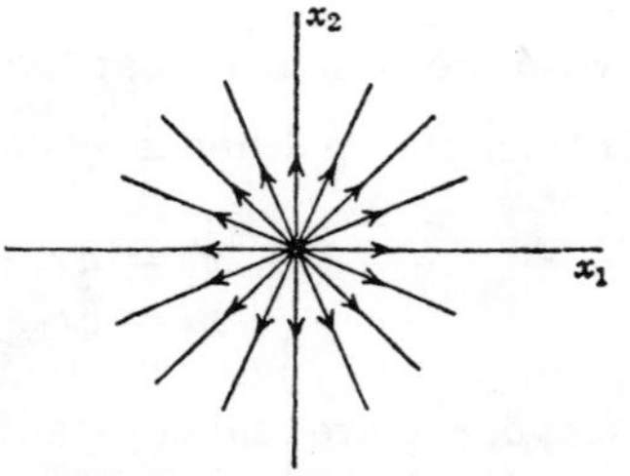

FIG. 4. (I) Proper node, $\lambda > 0$.

of (1.2), and are consequently *functions of t*. Thus ρ, ω are to be distinguished from the *polar coordinates* r,θ in the (x_1,x_2) plane defined by

$$r = (x_1^2 + x_2^2)^{\frac{1}{2}} \qquad \theta = \tan^{-1}\frac{x_2}{x_1}$$

just as the solution coordinate functions φ_1,φ_2 are to be distinguished from the Cartesian coordinates x_1,x_2 in the plane.

(I) Here the system is given by

$$x_1' = \lambda x_1 \qquad x_2' = \lambda x_2$$

and therefore, if (c_1,c_2) is any initial point except (0,0), a solution through this point is given by $\varphi_1(t) = c_1e^{\lambda t}$, $\varphi_2(t) = c_2e^{\lambda t}$. If $\lambda < 0$, then $\rho(t) \to 0$ as $t \to +\infty$, and if $\lambda > 0$, $\rho(t) \to 0$ as $t \to -\infty$. The orbit through (c_1,c_2) is an open half line passing through this point and with an end point at (0,0). See Figs. 3 and 4, where the arrows indicate the direction of increasing t. This type of critical point is called a *proper node*. Its distinguishing feature is that *every orbit tends to the origin in a definite direction* as $t \to +\infty$ (for $\lambda < 0$), or as $t \to -\infty$ (for $\lambda > 0$), and, *given any direction, there exists an orbit which tends to the origin in this direction.* Thus the origin is (asymptotically) stable in case $\lambda < 0$, and unstable when $\lambda > 0$.

(II) The system for Case (II) is

$$x_1' = \lambda x_1 \qquad x_2' = \mu x_2$$

and the solution passing through $(c_1,c_2) \neq (0,0)$ at $t = 0$ is given by $\varphi_1(t) = c_1e^{\lambda t}$, $\varphi_2(t) = c_2e^{\mu t}$. Assume $\mu < \lambda < 0$, for example. Then as $t \to +\infty$, $(\varphi_1(t),\varphi_2(t)) \to (0,0)$, and if $c_1 \neq 0$, $\varphi_2(t)/\varphi_1(t) = (c_2/c_1)e^{(\mu-\lambda)t} \to 0$, as $t \to +\infty$. If $c_1 = 0$, $c_2 \neq 0$, $(\varphi_1(t),\varphi_2(t)) = (0,c_2e^{\mu t})$, which is just the open positive or negative x_2 axis, according as $c_2 > 0$ or $c_2 < 0$. In this

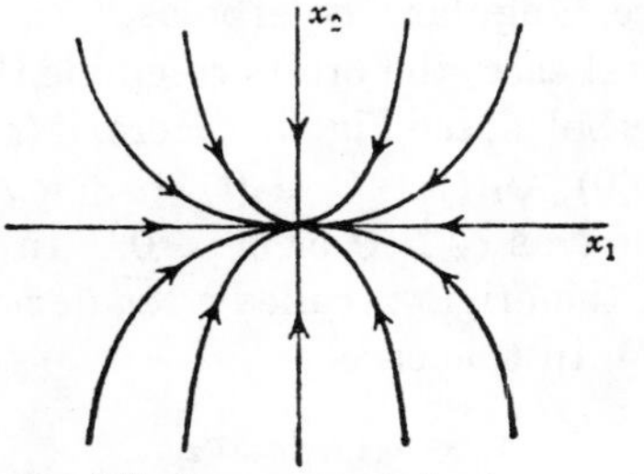

FIG. 5. (II) Improper node, $\mu < \lambda < 0$.

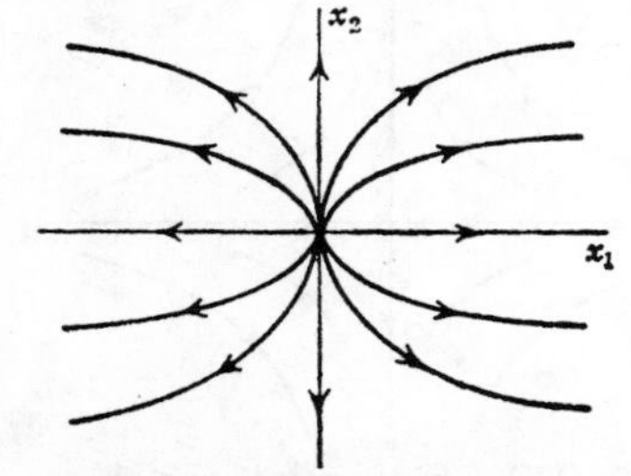

FIG. 6. (II) Improper node, $0 < \mu < \lambda$.

case, the origin is called an *improper node.* A qualitative picture of the orbits is shown in Figs. 5 and 6. Here, *every orbit, except one, has the same limiting direction at the origin.* The origin is (asymptotically) stable in case $\mu < \lambda < 0$, and unstable when $0 < \mu < \lambda$.

(III) The equations in this case are

$$x_1' = \lambda x_1 \qquad x_2' = \gamma x_1 + \lambda x_2$$

and it is easy to see that $\varphi_1(t) = c_1e^{\lambda t}$, $\varphi_2(t) = (c_2 + c_1\gamma t)e^{\lambda t}$, is the solution passing through (c_1,c_2) at $t = 0$. Suppose $\lambda < 0$, for example. Then as $t \to +\infty$, φ_1 and φ_2 tend to 0. If $c_1 \neq 0$, $\varphi_2(t)/\varphi_1(t) = c_2/c_1 + \gamma t \to \pm\infty$,

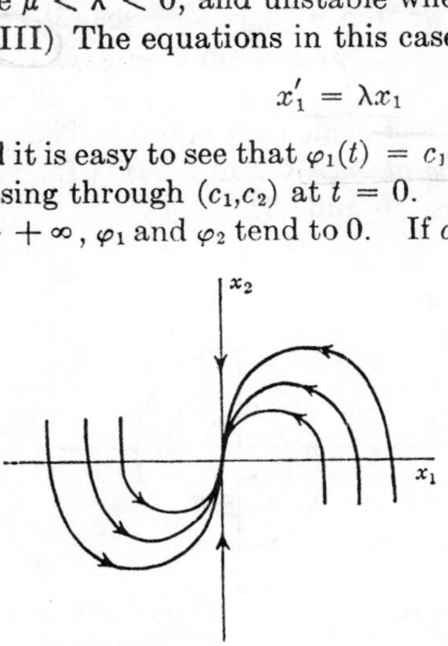

FIG. 7. (III) Improper node, $\lambda < 0$.

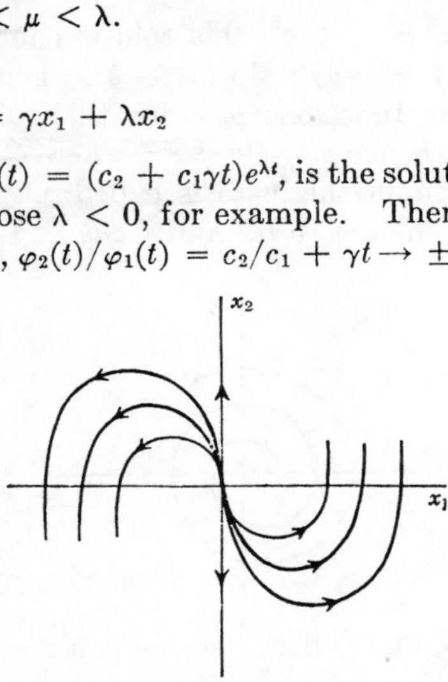

FIG. 8. (III) Improper node, $\lambda > 0$.

as $t \to \pm\infty$. If $c_1 \gtrless 0$, then $\varphi_2(t) \gtrless 0$, for t positive and large enough, and if $c_1 = 0$, then $(\varphi_1(t),\varphi_2(t)) = (0,c_2e^{\lambda t})$, which gives an orbit which is a half x_2 axis. Also, if $c_1 \neq 0$, $\varphi_2'(t)/\varphi_1'(t) = (\gamma/\lambda) + \varphi_2(t)/\varphi_1(t) \to \pm\infty$, as $t \to \pm\infty$. Thus *every orbit has the same limiting direction at* (0,0). The origin in this case is also called an *improper node.* The nature of the orbits is sketched in Figs. 7 and 8.

(IV) Here the equations are

$$x_1' = \lambda x_1 \qquad x_2' = \mu x_2$$

and a solution is given by $\varphi_1(t) = c_1e^{\lambda t}$, $\varphi_2(t) = c_2e^{\mu t}$, where now $\lambda < 0$, $\mu > 0$. If $|\lambda| = |\mu|$, the orbits would be rectangular hyperbolas. In the general case, the orbits resemble these hyperbolas; see Fig. 9. Here, if $(c_1,c_2) \neq (0,0)$, $\varphi_1(t) \to 0$, $\varphi_2(t) \to \pm\infty$, according as $c_2 > 0$ or $c_2 < 0$. In this case, the origin is called a *saddle point*.

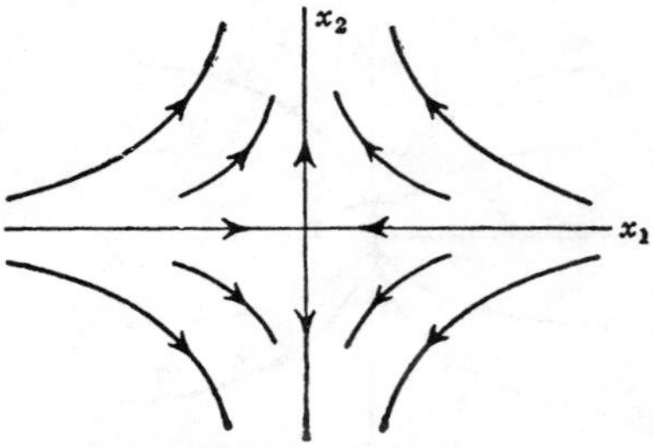

FIG. 9. (IV) Saddle point, $\lambda < 0 < \mu$.

(V) In this case

$$x_1' = \alpha x_1 + \beta x_2$$
$$x_2' = -\beta x_1 + \alpha x_2$$

and the solution which passes through (c_1,c_2) at $t = 0$ is given by

$$\varphi_1(t) = e^{\alpha t}(c_1 \cos \beta t + c_2 \sin \beta t) \qquad \varphi_2(t) = e^{\alpha t}(-c_1 \sin \beta t + c_2 \cos \beta t)$$

If $\rho_0^2 = c_1^2 + c_2^2$, this solution may be written $\varphi_1(t) = \rho_0e^{\alpha t} \cos (\beta t - \delta)$, $\varphi_2(t) = -\rho_0e^{\alpha t} \sin (\beta t - \delta)$, where $\cos \delta = c_1/\rho_0$ and $\sin \delta = c_2/\rho_0$. The polar functions ρ, ω for this solution are $\rho(t) = \rho_0e^{\alpha t}$, $\omega(t) = -\beta t + \delta$, and hence $\rho = Ce^{-(\alpha/\beta)\omega}$, where $C = \rho_0e^{(\alpha/\beta)\delta}$, which is a spiral. Thus the origin in this case is called a *spiral point.* (Alternate terms for such a point are *vortex* and *focus.*) See Figs. 10 and 11.

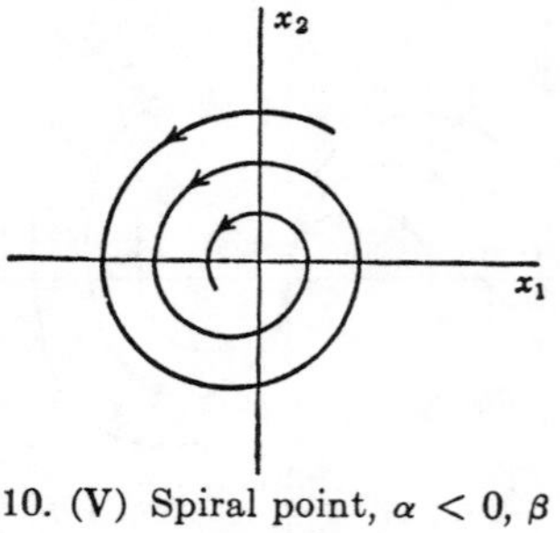

FIG. 10. (V) Spiral point, $\alpha < 0$, $\beta < 0$.

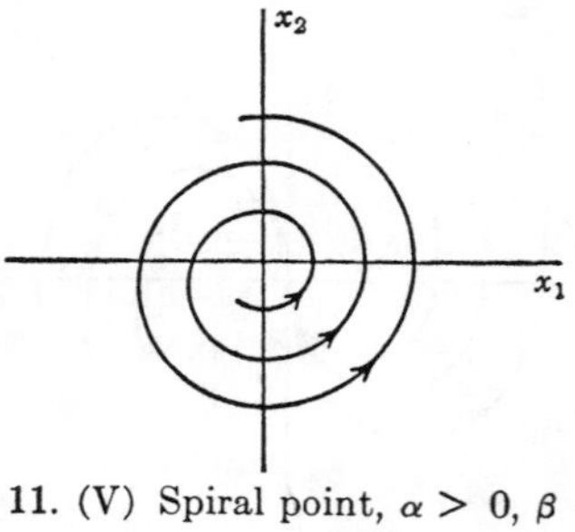

FIG. 11. (V) Spiral point, $\alpha > 0$, $\beta < 0$.

(VI) This is just a special case of (V) where $\alpha = 0$. In this situation a solution through (c_1,c_2) at $t = 0$ is $\varphi_1(t) = c_1 \cos \beta t + c_2 \sin \beta t$, $\varphi_2(t) = -c_1 \sin \beta t + c_2 \cos \beta t$, or, as in (V), $\rho(t) = \rho_0$, which is a circle of radius ρ_0 with (0,0) as the center. The origin is called a *center* in this case; see Figs. 12 and 13.

From the definition of stability, it is easy to see by considering the six cases (I) through (VI) above that the following theorem holds. The pictures in Figs. 3 through 13 give a nice qualitative idea of the notion of stability in each of the cases.

Theorem 1.1. *Necessary and sufficient for the origin to be stable for the system* (1.1) *is that the characteristic roots of the real nonsingular coefficient matrix A should have negative or zero real parts.*

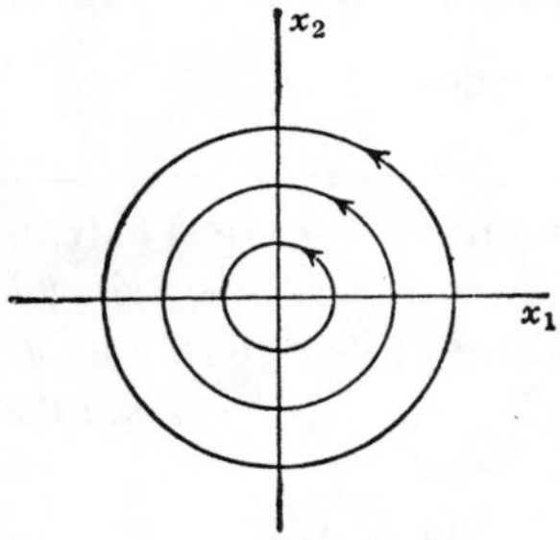

FIG. 12. (VI) Center, $\beta < 0$.

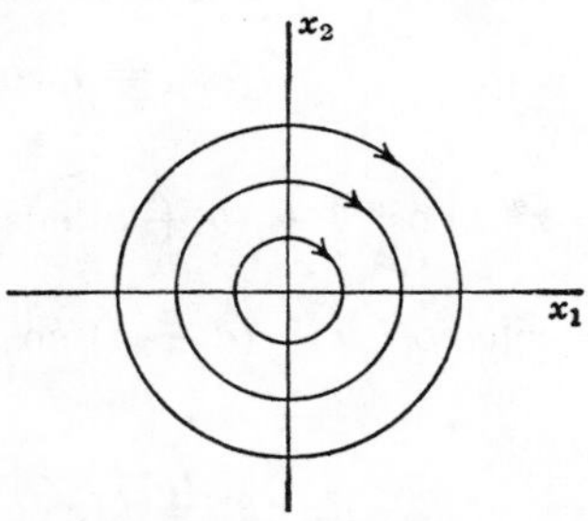

FIG. 13. (VI) Center, $\beta > 0$.

2. Perturbations of Two-dimensional Linear Systems

Consider now the *nonlinear* two-dimensional real autonomous system

$$\text{(NL)} \qquad \begin{aligned} x_1' &= ax_1 + bx_2 + f_1(x_1,x_2) \\ x_2' &= cx_1 + dx_2 + f_2(x_1,x_2) \end{aligned}$$

where a, b, c, d are real constants, $ad - bc \neq 0$, and f_1,f_2 are real continuous functions defined in some circle about the origin $(x_1,x_2) = (0,0)$ with radius $r_0 > 0$. The functions f_1 and f_2 are called *perturbations*, and the system (NL) will be referred to as the *perturbed system* corresponding to the *linear* system

$$\text{(L)} \qquad x_1' = ax_1 + bx_2 \qquad x_2' = cx_1 + dx_2$$

Intuitively, if the perturbations f_1 and f_2 are "small" in some sense, one would expect that the behavior of the orbits of (NL) near the origin in the (x_1,x_2) plane would be very similar to the behavior of the orbits of (L). It will be shown that this is in general true, provided that f_1,f_2 satisfy certain minimum assumptions.

In addition to the given assumptions on f_1,f_2, it will be assumed that

$$f_1 = o(r) \qquad f_2 = o(r) \qquad (\text{as } r \to 0+) \tag{2.1}$$

This guarantees that the perturbations tend to zero faster than the linear terms in (NL). Also, it is easily seen that this condition, and the fact that $ad - bc \neq 0$, imply that the origin is an *isolated* critical point for (NL); that is, there exists a circle about the origin in which the origin is the only point where the right member of (NL) vanishes. An isolated critical point such as the origin in (NL), with $ad - bc \neq 0$, is called a *simple critical point.*

It is to be remarked that the assumptions made on f_1 and f_2 do not imply the uniqueness of solutions of (NL).

One of the most important methods for analyzing the orbits of (NL) is to use the *polar equations* obtained from (NL) via the substitution

$$x_1 = r \cos \theta \qquad x_2 = r \sin \theta$$

namely,

$$\begin{aligned} rr' &= r^2[a \cos^2 \theta + (b + c) \cos \theta \sin \theta + d \sin^2 \theta] + r \cos \theta \, F_1(r,\theta) \\ &\qquad + r \sin \theta \, F_2(r,\theta) \\ r^2\theta' &= r^2[c \cos^2 \theta + (d - a) \cos \theta \sin \theta - b \sin^2 \theta] + r \cos \theta \, F_2(r,\theta) \\ &\qquad - r \sin \theta \, F_1(r,\theta) \end{aligned}$$

where

$$F_j(r,\theta) = f_j(r \cos \theta, r \sin \theta) \qquad (j = 1, 2)$$

Clearly, if $\varphi = (\varphi_1,\varphi_2)$ is a solution of (NL), then the polar functions (ρ,ω) constitute a solution of the polar equations.

Before proceeding to the detailed statements and proofs of the results, the definitions of the various types of critical points will be made precise. If there exists a δ, $0 < \delta \leqq r_0$, such that, for any solution path $(\varphi_1(t),\varphi_2(t))$ of (NL) which has at least one point in $0 < r < \delta$, the solution exists over a t half line, and if $(\varphi_1(t),\varphi_2(t)) \to (0,0)$ as $t \to +\infty$ or $-\infty$, then the origin is called an *attractor* for (NL). In the case $f_1 = f_2 = 0$, then the nodes and spiral points are attractors, whereas saddle points and centers are not. The origin is said to be a *node* for (NL) if it is an attractor for which all orbits arrive at the origin in a definite direction, and it is said to be a *proper node* if it is a node and every half line through the origin is tangent to some orbit there. The origin is called a *spiral point* for (NL) if it is an attractor such that $|\omega(t)| \to +\infty$ as $t \to +\infty$ or $-\infty$, where $\omega(t) = \tan^{-1} (\varphi_2(t)/\varphi_1(t))$ and (φ_1,φ_2) is any solution of (NL) which enters the region $0 \leqq r < \delta$. If there exists a sequence of periodic orbits $\{C_n\}$ of (NL) each of which contains all later orbits and the origin in its interior and such that C_n tends to the origin as $n \to \infty$, then the origin is called a *center* for (NL).

The following theorem is a special case of Theorem 1.1, Chap. 13.

Theorem 2.1. *If the origin is an attractor for the linear system* (L), *it is one for the nonlinear system* (NL).

Theorem 2.2. *If the origin is a spiral point for the linear system* (L), *it is one for the nonlinear system* (NL).

Proof. By Theorem 2.1, the origin is an attractor for (NL). The polar equation for θ is given by

$$r^2\theta' = x_1x_2' - x_1'x_2 = -\beta r^2 + o(r^2) \qquad (r \to 0)$$

But $r \to 0$, as $t \to +\infty$ (in case $\alpha < 0$). Thus as $t \to +\infty$

$$\theta' = -\beta + o(1)$$

and therefore, for any solution φ of (NL) starting sufficiently near the origin,

$$\omega(t) = -\beta t + o(t)$$

Thus $\omega(t)/t \to -\beta$ as $t \to +\infty$, which proves $\omega(t) \to \pm\infty$, as $t \to +\infty$, according as $\beta < 0$ or $\beta > 0$, and hence proves the theorem.

3. Proper Nodes and Proper Spiral Points

Although attractors of (L) go into attractors of (NL), it is not in general true that a node goes into a node. This is illustrated by the following example, where a proper node for (L) goes over into a spiral point for (NL). Consider the system

$$x_1' = -x_1 - \frac{x_2}{\log (x_1^2 + x_2^2)^{\frac{1}{2}}} \qquad x_2' = -x_2 + \frac{x_1}{\log (x_1^2 + x_2^2)^{\frac{1}{2}}} \tag{3.1}$$

Clearly (3.1) satisfies the same hypothesis as the system (NL). The polar equations corresponding to (3.1) are

$$\theta' = \frac{1}{\log r} \qquad r' = -r$$

Thus $r = \rho(t) = ce^{-t}$, for some constant $c > 0$, and hence

$$\theta' = \omega'(t) = \frac{1}{(\log c - t)}$$

Therefore $\omega(t) = -\log (t - \log c) + k$, where

$$k = \omega(t_0) + \log (t_0 - \log c)$$

This implies $\omega(t) \to -\infty$, as $t \to +\infty$, and the origin is a spiral point for (3.1), although the origin is a proper node for the corresponding linear system $x_1' = -x_1$, $x_2' = -x_2$.

The above example shows that in order for a node for (L) to become a node for (NL) it is necessary to restrict the functions f_1, f_2 still further. It can be shown, in fact, that if f_1, f_2 satisfy the conditions for (NL), and if, further, $f_1 = 0(r^{1+\epsilon})$, $f_2 = 0(r^{1+\epsilon})$, $r \to 0$, for some $\epsilon > 0$, then if the origin is a proper node for (L) it is one for (NL) also.

This result is a special case of one which holds for a spiral point (see Corollary to Theorem 3.1 below). Consider the canonical form of (NL) when the corresponding linear system has a spiral point at the origin

$$\begin{aligned} x_1' &= \alpha x_1 + \beta x_2 + f_1(x_1,x_2) \\ x_2' &= -\beta x_1 + \alpha x_2 + f_2(x_1,x_2) \end{aligned} \qquad (\alpha \neq 0,\ \beta \neq 0) \tag{3.2}$$

For the linear case,

$$x_1' = \alpha x_1 + \beta x_2 \qquad x_2' = -\beta x_1 + \alpha x_2 \tag{3.3}$$

the polar equations are

$$r' = \alpha r \qquad \theta' = -\beta$$

If $\alpha < 0$, $\beta < 0$, for example, then for every solution path of (3.3), $r = \rho(t) \to 0$, and $\theta = \omega(t) \to +\infty$, as $t \to +\infty$. Further, $\rho e^{\alpha\omega/\beta}$ is a positive constant or, what is the same, $\omega + (\beta/\alpha) \log \rho = c$, for some constant c. Also, given any constant c, there exists a solution of (3.3) such that $\omega + (\beta/\alpha) \log \rho = c$. This prompts the following definition. If $\alpha + i\beta$, $\alpha - i\beta$ $(\alpha \neq 0)$ are the characteristic roots of the coefficient matrix

$$\begin{pmatrix} a & b \\ c & d \end{pmatrix}$$

in (NL), the origin is said to be a *proper spiral point* for (NL) if it is an attractor such that for every solution tending to the origin as $t \to +\infty$ (or $t \to -\infty$), $\omega + (\beta/\alpha) \log \rho$ tends to some constant c as $t \to +\infty$ (or $t \to -\infty$), and given *any* constant c, there exists a solution of (NL) such that $\omega + (\beta/\alpha) \log \rho \to c$ as $t \to +\infty$ (or $t \to -\infty$). If $\beta = 0$, then this reduces to the definition of a *proper node.*

The following example shows that a spiral point for (L) (which is, of course, a proper spiral point) may fail to go into a proper spiral point for (NL). Consider

$$\begin{aligned} x_1' &= -x_1 + x_2 + \frac{x_1}{\log (x_1^2 + x_2^2)^{\frac{1}{2}}} \\ x_2' &= -x_1 - x_2 + \frac{x_2}{\log (x_1^2 + x_2^2)^{\frac{1}{2}}} \end{aligned} \tag{3.4}$$

The polar equation involving r' is

$$\frac{r'}{r} = -1 + \frac{1}{\log r}$$

and this implies $r = \rho(t)$ satisfies $\rho e^t = \rho_0/(\log \rho - 1)$, for some constant ρ_0. Thus $\rho e^t \to 0$, as $t \to +\infty$; for $\rho \to 0$, as $t \to +\infty$.

By virtue of the example (3.4), one is led to inquire under what conditions on f_1, f_2 in (NL) a proper spiral point for (L) becomes one for (NL). The following general result gives a sufficient condition.

Theorem 3.1. *Let f_1, f_2 in* (NL) *satisfy the inequalities*

$$|f_i(x_1,x_2)| \leqq \psi((x_1^2 + x_2^2)^{\frac{1}{2}}) \qquad (i = 1, 2) \tag{3.5}$$

where $\psi = \psi(r)$ is a continuous function defined on $0 \leqq r \leqq r_0$ such that as $r \to 0$

$$\psi(r) = o(r) \tag{3.6}$$

and

$$\int_0^{r_0} \frac{\psi(r)}{r^2}\, dr < \infty \tag{3.7}$$

Then, if the origin is a spiral point (or a proper node) for (L), *it is a proper spiral point (or a proper node) for* (NL).

Using a more restrictive hypothesis, stronger results are obtained in Probs. 8 and 9, Chap. 13.

Proof of Theorem 3.1. It can be assumed that the equations (NL) and (L) are in the canonical forms (3.2) and (3.3), and that $\alpha < 0$, $\beta \leqq 0$. Letting $x_1 = r \cos \theta$, $x_2 = r \sin \theta$ in (3.2), one obtains

$$\begin{aligned} rr' &= \alpha r^2 + r \cos \theta\, f_1(r \cos \theta, r \sin \theta) + r \sin \theta\, f_2(r \cos \theta, r \sin \theta) \\ r^2\theta' &= -\beta r^2 + r \cos \theta\, f_2(r \cos \theta, r \sin \theta) \\ &\qquad - r \sin \theta\, f_1(r \cos \theta, r \sin \theta) \end{aligned} \tag{3.8}$$

From the first equation in (3.8)

$$rr' = \alpha r^2 + o(r^2) \qquad (r \to 0)$$

and this implies that not only does any solution $r = \rho(t)$, which starts sufficiently near the origin, tend to the origin, but that for any such solution $\rho' < 0$, for t sufficiently large. Therefore, if t is sufficiently large, $r = \rho(t)$ is a monotone function of t, and thus determines the inverse function $t = g(r)$ [by $t = g(\rho(t))$], which is monotone in a vicinity of the origin $r = 0$, say for $0 < r \leqq r_1$. If $\theta = \omega(t)$ is a solution of the second equation of (3.8), define $\tilde{\omega}$ by $\tilde{\omega}(r) = \omega(g(r))$, $0 < r \leqq r_1$. Then clearly $\theta = \tilde{\omega}(r)$ is a solution of the equation obtained from (3.8) by formally dividing, that is,

$$\frac{d\theta}{dr} = F(r,\theta) \tag{3.9}$$

where

$$F(r,\theta) = \frac{-\beta + F_1(r,\theta)}{\alpha(r + F_2(r,\theta))} \tag{3.10}$$

and

$$\begin{aligned} F_1(r,\theta) &= \frac{\cos \theta}{r} f_2(r \cos \theta, r \sin \theta) - \frac{\sin \theta}{r} f_1(r \cos \theta, r \sin \theta) \\ F_2(r,\theta) &= \frac{\cos \theta}{\alpha} f_1(r \cos \theta, r \sin \theta) + \frac{\sin \theta}{\alpha} f_2(r \cos \theta, r \sin \theta) \end{aligned} \tag{3.11}$$

Equation (3.9) can be rewritten in the following form

$$\frac{d\theta}{dr} + \frac{\beta}{\alpha}\frac{1}{r} = \tilde{F}(r,\theta) \tag{3.12}$$

where

$$\tilde{F}(r,\theta) = \frac{\beta F_2(r,\theta) + rF_1(r,\theta)}{\alpha r(r + F_2(r,\theta))} \tag{3.13}$$

From (3.5), (3.6), and (3.11), it follows that if r is sufficiently small, say $0 \leqq r \leqq r_2$, then

$$|\tilde{F}(r,\theta)| \leqq 4\,\frac{|\alpha| + |\beta|}{\alpha^2}\,\frac{\psi(r)}{r^2} \tag{3.14}$$

By virtue of (3.14) and the assumption (3.7), the integral

$$\int_0^{\tilde{r}} \tilde{F}(r,\tilde{\omega}(r))\,dr \qquad (\tilde{r} = \min\,(r_1,r_2))$$

is convergent. Therefore, from (3.12) it follows that

$$\tilde{\omega}(r) + \frac{\beta}{\alpha}\log r \to \tilde{\omega}(\tilde{r}) + \frac{\beta}{\alpha}\log \tilde{r} - \int_0^{\tilde{r}} \tilde{F}(r,\tilde{\omega}(r))\,dr \qquad (r \to 0)$$

From the definition of $\tilde{\omega}$, one has $\omega(t) + (\beta/\alpha)\log\rho(t) \to c$, $t \to +\infty$, where c is a constant.

Conversely let c be a real constant and consider the integral equation

$$\Phi(r) = c + \int_0^r \tilde{F}(s,\Phi(s) - \frac{\beta}{\alpha}\log s)\,ds \tag{3.15}$$

Because $\tilde{F}$ satisfies (3.14) and ψ satisfies (3.7), an equicontinuous sequence can be constructed for (3.15) over some interval $0 \leqq r \leqq r_3$, just as it was in the proof of the Carathéodory existence theorem. From this sequence a convergent subsequence can be extracted which leads to the existence of a solution Φ of (3.15). Let $\tilde{\omega}(r) = \Phi(r) - (\beta/\alpha)\log r$. Then by (3.15), $\theta = \tilde{\omega}(r)$ is a solution of (3.12) and clearly

$$\tilde{\omega}(r) + \frac{\beta}{\alpha}\log r \to c \text{ as } r \to 0$$

Corresponding to this solution $\theta = \tilde{\omega}(r)$ there exists a solution $r = \rho(t)$ of the first equation of (3.8) for which $\rho(t) \to 0$, as $t \to +\infty$, and if $\omega(t) = \tilde{\omega}(\rho(t))$, the pair $(\rho(t), \omega(t))$ determines a solution of (3.2), for which $\omega(t) + (\beta/\alpha)\log\rho(t) \to c$, as $t \to \infty$. This completes the proof of the theorem.

Corollary. *The conclusion of Theorem* 3.1 *remains valid for* (NL) *if* (3.5), (3.6), (3.7) *are replaced by*

$$f_1 = O(r^{1+\epsilon}) \qquad f_2 = O(r^{1+\epsilon}) \qquad (r \to 0)$$

for some $\epsilon > 0$.

Proof. Choose, in Theorem 3.1, $\psi(r) = Cr^{1+\epsilon}$, where C is a constant such that $|f_1| \leqq Cr^{1+\epsilon}$, $|f_2| \leqq Cr^{1+\epsilon}$, for all sufficiently small r. Clearly (3.6) and (3.7) hold, and hence the conclusion of Theorem 3.1 is valid.

4. Centers

The case where the origin is a center for (L) will now be investigated. In order to illustrate what can happen in passing to the perturbed system (NL) in this case, consider the example

$$x_1' = -x_2 - x_1\sqrt{x_1^2 + x_2^2} \qquad x_2' = x_1 - x_2\sqrt{x_1^2 + x_2^2} \tag{4.1}$$

This system satisfies the assumptions for (NL), and the polar equations corresponding to (4.1) are $r' = -r^2$, and $\theta' = 1$. The solution of this system passing through (r_0,θ_0), where $r_0 \neq 0$, at $t = 0$ is given by

$$\rho(t) = \left(t + \frac{1}{r_0}\right)^{-1} \qquad \omega(t) = t + \theta_0$$

and therefore $\rho(t) \to 0$ and $\omega(t) \to +\infty$, as $t \to +\infty$. Hence the origin is a *spiral point* for the system (4.1), although the origin is a *center* for the corresponding linear system.

Actually, the perturbed system (NL) can be much more complicated than (L) in this case, while still remaining a center. As an example consider the system

$$\begin{aligned} x_1' &= -x_2 + x_1(x_1^2 + x_2^2)\sin\frac{\pi}{(x_1^2 + x_2^2)^{\frac{1}{2}}} \\ x_2' &= x_1 + x_2(x_1^2 + x_2^2)\sin\frac{\pi}{(x_1^2 + x_2^2)^{\frac{1}{2}}} \end{aligned} \tag{4.2}$$

The nonlinear perturbations have continuous first derivatives everywhere, and therefore there exists a *unique* solution through any given point $(c_1,c_2) \neq (0,0)$ at $t = 0$. The polar equations for (4.2) are

$$r' = r^3 \sin\frac{\pi}{r} \qquad \theta' = 1$$

The circles $r = 1/n$, $n = 1, 2, \ldots$, are periodic orbits, represented by the solutions $\rho(t) = 1/n$, $\theta(t) = t + \theta_0$, where θ_0 is a constant. Also

$$\begin{aligned} &r' > 0, \qquad r > 1 \\ &r' < 0, \qquad \frac{1}{2m} < r < \frac{1}{2m-1} \qquad (m = 1, 2, \ldots) \\ &r' > 0, \qquad \frac{1}{2m+1} < r < \frac{1}{2m} \end{aligned}$$

Therefore no orbits but $r = 1/n$ can be periodic, and every nonperiodic orbit must remain completely within one of the regions $r > 1$, $1/2m < r < 1/(2m-1)$, $1/(2m+1) < r < 1/2m$, $m = 1, 2, \ldots$. Since ρ and ω are monotonic as $t \to +\infty$ $(\theta \to +\infty)$, these nonperiodic orbits must tend to the circles $r = 1/n$ as $t \to +\infty$ or $t \to -\infty$, or in the case $r > 1$, $\rho \to +\infty$. Thus the origin is a *center* for (4.2).

The examples (4.1), (4.2) actually exhaust the possibilities for (NL) when the origin is a center for the linear system (L). In fact, the following theorem holds.

Theorem 4.1. *If the origin is a center for* (L), *then it is either a center or a spiral point for* (NL).

Proof. The canonical form of the equations under consideration is

$$x_1' = +\beta x_2 + f_1(x_1,x_2) \qquad x_2' = -\beta x_1 + f_2(x_1,x_2) \tag{4.3}$$

and

$$x_1' = \beta x_2 \qquad x_2' = -\beta x_1$$

Assume $\beta < 0$; otherwise the roles of t and $-t$ are interchanged.

The polar equations for (4.3) give

$$r' = o(r) \qquad \theta' = -\beta + o(1) \qquad (\text{as } r \to 0) \tag{4.4}$$

From (4.4) it follows that if φ is a solution of (4.3) starting near enough the origin at $t = 0$, then its polar functions $r = \rho(t)$, $\theta = \omega(t)$ satisfy, for any $\epsilon > 0$, and $\rho_0 > 0$ small enough,

$$\rho_0 e^{-\epsilon t} < \rho(t) \qquad \omega' > 0$$

for $t > 0$. Therefore $\rho > 0$ for all finite $t > 0$ for which it exists, and ω is a monotone function of t. Let the inverse of ω be h, that is, $t = h(\theta)$, and define $\tilde{\rho}$ by $\tilde{\rho}(\theta) = \rho(h(\theta))$. Then $r = \tilde{\rho}(\theta)$ satisfies the differential equation

$$\frac{dr}{d\theta} = F(r,\theta) \tag{4.5}$$

where

$$F(r,\theta) = \frac{\cos\theta\, f_1(r\cos\theta, r\sin\theta) + \sin\theta\, f_2(r\cos\theta, r\sin\theta)}{-\beta + (\cos\theta/r) f_2(r\cos\theta, r\sin\theta) - (\sin\theta/r) f_1(r\cos\theta, r\sin\theta)}$$

Conversely, if $r = \tilde{\rho}(\theta)$ is a solution of (4.5) starting sufficiently near the origin, the polar equation for θ will give a solution $\theta = \omega(t)$ which is monotone in t. Then, if $\rho(t) = \tilde{\rho}(\omega(t))$, the pair $(\rho(t),\omega(t))$ gives rise to a solution of (4.3) starting near the origin.

From this discussion it follows that the behavior of the solutions of (4.3) near the origin can be studied by investigating the behavior of the solutions of (4.5) near the origin.

The function F is continuous in (r,θ) in some circle $0 \leqq r \leqq r_1$, $(r_1 > 0)$, $F(r,\theta + 2\pi) = F(r,\theta)$, and $F(r,\theta) = o(r)$, $r \to 0$, uniformly in θ. These facts do not guarantee the uniqueness of solutions of (4.5).

Let r_2, $0 < r_2 \leqq r_1$, and $\eta > 0$ be given, and set $M = \max |F|$ on $0 \leqq r \leqq r_2$. Then, by the local-existence theorem, there exists a circle $0 \leqq r \leqq r_2/2$ such that, if (ρ_0,θ_0) is inside this circle, (4.5) has a solution $r = \tilde{\rho}(\theta)$, $\tilde{\rho}(\theta_0) = \rho_0$, which exists for $0 \leqq |\theta - \theta_0| \leqq \min (2\pi + \eta, r_2/2M)$ and stays within the circle $0 \leqq r \leqq r_2$. However, $F = o(r)$, $r \to 0$, implies that if r_2 is chosen small enough $r_2/2M > 2\pi + \eta$. In this case $\tilde{\rho}$ exists on $0 \leqq |\theta - \theta_0| \leqq 2\pi + \eta$ and remains within $0 \leqq r \leqq r_2$.

Suppose the origin is not a center. Then by decreasing r_2 if necessary there exist no periodic orbits in $r < r_2$. Consider again the solution $r = \tilde{\rho}(\theta)$ through (ρ_0,θ_0). Then either $\tilde{\rho}(\theta_0 + 2\pi) < \tilde{\rho}(\theta_0)$ or else $\tilde{\rho}(\theta_0 + 2\pi) > \tilde{\rho}(\theta_0)$. With no loss in generality, only the first case need be considered for $\tilde{\rho}$. If $\tilde{\rho}(\theta) - \tilde{\rho}(\theta + 2\pi)$ vanishes as θ increases, there is a periodic orbit in $r < r_2$. Thus $\tilde{\rho}(\theta) > \tilde{\rho}(\theta + 2\pi)$ for $\theta \geqq \theta_0$. Since the sequence $\{\tilde{\rho}(\theta_0 + 2\pi k)\}$, $k = 0, 1, \ldots$, is monotone decreasing, positive, its limit $\hat{r}$ exists. If $\hat{r} = 0$, then $\tilde{\rho}(\theta) \to 0$ as $\theta \to \infty$. If $\hat{r} > 0$, let $\tilde{\rho}(\theta_0 + \theta + 2\pi k) = \tilde{\rho}_k(\theta)$. As $|d\tilde{\rho}/d\theta| \leqq M$, the sequence $\{\tilde{\rho}_k\}$ is equicontinuous on $[0,2\pi]$. Clearly $\tilde{\rho}_k(0) \to \hat{r}$ and $\tilde{\rho}_k(2\pi) \to \hat{r}$ as $k \to \infty$. Thus there is a subsequence of $\{\tilde{\rho}_k\}$ converging to a solution $\hat{\rho}$ of (4.5), and $\hat{\rho}(0) = \hat{\rho}(2\pi) = \hat{r}$. Thus the solution is periodic, contrary to the assumption of no periodic orbits in $r < r_2$. Thus $\hat{r} = 0$ and $\tilde{\rho}(\theta) \to 0$ as $\theta \to \infty$.

In case there is a unique solution through each point, this completes the proof. In the general case, consider the minimum and maximum solutions of (4.5) $\tilde{\rho}_m$ and $\tilde{\rho}_M$ through (ρ_0,θ_0). Clearly $\tilde{\rho}_m$ spirals to the origin as θ increases because $\tilde{\rho}$ does. Suppose $\tilde{\rho}_M$ does not do so. Then

$$\tilde{\rho}_M(\theta_0 + 2\pi) > \tilde{\rho}_M(\theta_0)$$

and $$\tilde{\rho}_m(\theta_0 + 2\pi) < \tilde{\rho}_m(\theta_0) = \rho_0 = \tilde{\rho}_M(\theta_0) < \tilde{\rho}_M(\theta_0 + 2\pi)$$

Thus, by the corollary to Theorem 1.3, Chap. 2, there must be a periodic solution through (ρ_0,θ_0) contrary to assumption. Thus all solutions through (ρ_0,θ_0) spiral toward the origin as θ increases.

The solutions through any point near the origin must spiral inward as θ increases. For if not, there is a solution $\hat{\rho}$ which spirals outward as θ increases. The maximum solution $\tilde{\rho}_M$ considered above must intersect this solution, that is, there must exist a $\theta_1 > \theta_0$ such that

$$\tilde{\rho}_M(\theta_1) = \hat{\rho}(\theta_1 + 2k\pi)$$

for some integer k. Consider now the solution $\hat{\rho}_0$, where

$$\begin{aligned} \hat{\rho}_0(\theta) &= \tilde{\rho}_M(\theta) \qquad (\theta_0 \leqq \theta \leqq \theta_1) \\ \hat{\rho}_0(\theta) &= \hat{\rho}(\theta + 2k\pi) \qquad (\theta_1 \leqq \theta) \end{aligned}$$

This solution through (ρ_0,θ_0) clearly exceeds $\tilde{\rho}_M(\theta)$ for $\theta > \theta_1$, contrary to the definition of maximum solution. Thus $\hat{\rho}$ cannot exist and all solutions spiral inward as θ increases. This proves the theorem.

5. Improper Nodes

Consider the case where the origin is an improper node of Type (II) for the linear system (L), and suppose for simplicity that the system is in the canonical form

$$x_1' = \lambda x_1 \qquad x_2' = \mu x_2 \qquad (\mu < \lambda < 0) \tag{5.1}$$

The nonlinear system (NL) corresponding to (5.1) is then

$$x_1' = \lambda x_1 + f_1(x_1,x_2) \qquad x_2' = \mu x_2 + f_2(x_1,x_2) \tag{5.2}$$

and the following theorem illustrates how the geometry of the orbits of (5.2) near the origin compares with the geometry of those of (5.1).

The results of Probs. 5 and 6, Chap. 13, are closely related to the following theorem and include (c).

Theorem 5.1. (a) *Every orbit of* (5.2) *near the origin tends to the origin and has a limiting direction which makes an angle of* 0, $\pi/2$, π, *or* $3\pi/2$ *with the positive* x_1 *axis. Moreover, there are an infinite number of orbits tending to the origin at angles* 0 *and* π.

(b) *There exists at least one orbit tending to the origin at an angle* $\pi/2$, *and at least one at* $3\pi/2$.

(c) *If* $\partial f_1/\partial x_1$, $\partial f_2/\partial x_1$ *exist and are continuous on* $0 \leqq r \leqq r_0$, *then there exists exactly one orbit tending to the origin in the directions* $\pi/2$ *and* $3\pi/2$.

Proof of (a). From Theorem 2.1 the origin is an attractor for (5.2). Therefore there exists a δ, $0 < \delta \leqq r_0$, such that any solution path starting in $0 \leqq r < \delta$ exists for $t \geqq t_0$ for some t_0, and tends to the origin as $t \to +\infty$. From (5.2) it follows that, for any solution starting in $0 < r < \delta$,

$$r^2\theta' = (\mu - \lambda)r^2 \cos\theta \sin\theta + o(r^2) \qquad (r \to 0)$$

or

$$\theta' = \frac{(\mu - \lambda)}{2} \sin 2\theta + o(1) \qquad (r \to 0) \tag{5.3}$$

Consider any ϵ, $0 < \epsilon < \pi/4$, and the regions

$$T_1: \quad |\theta| \leqq \epsilon \qquad\qquad T_2: \quad |\theta - \pi| \leqq \epsilon$$

$$T_3: \quad \left|\theta - \frac{\pi}{2}\right| \leqq \epsilon \qquad\qquad T_4: \quad \left|\theta - \frac{3\pi}{2}\right| \leqq \epsilon$$

On the line $\theta = \epsilon$, $\sin 2\epsilon > 0$, and so by (5.3) $\theta' < 0$ there if r is small enough. Similarly, $\theta' > 0$ on $\theta = -\epsilon$ if r is small enough. Thus, if r is small enough, any orbit starting within T_1 cannot leave T_1. A similar argument shows this is true of T_2. On the other hand, if r is small enough, the directions of orbits on the boundaries of T_3 and T_4 are toward the exterior of these regions. Thus any orbit starting outside of T_3 and

T_4 cannot get into T_3 or T_4. Let $\delta_1 \leqq \delta$ be so small that orbits starting inside $0 < r \leqq \delta_1$ behave as outlined above.

It is clear that a necessary and sufficient condition for an orbit to approach the origin at an angle of π is that for any ϵ, $0 < \epsilon < \pi/4$, there exists a t_ϵ such that the orbit lies in T_1 (corresponding to this ϵ) for all $t \geqq t_\epsilon$. Note that an orbit approaches at an angle of π if it does so along the positive x_1 axis.

It will now be shown that if an orbit C starts inside $0 < r \leqq \delta_1$, it tends to the origin at an angle of 0, $\pi/2$, π, or $3\pi/2$. Suppose the assertion is not true. Then for some ϵ_0, $0 < \epsilon_0 < \pi/4$, C does not lie in T_1, T_2, T_3, or T_4 (corresponding to ϵ_0). Suppose, for example, C is in the region S: $\epsilon_0 < \theta < \pi/2 - \epsilon_0$. Then it eventually enters T_1. For suppose this were not true. Then C stays in S for all sufficiently large t. But in S, by (5.3), $\theta' < [(\mu - \lambda)/4] \sin 2\epsilon_0 < 0$. Thus if $\zeta = -[(\mu - \lambda)/4] \sin 2\epsilon_0$, C must leave S and enter T_1 in a t interval of less than $\pi/(2\zeta)$. This is a contradiction, and hence C enters T_1 for every ϵ and thus tends to the origin at an angle of π. A similar argument holds if C is in any of the regions other than T_1, T_2, T_3, or T_4. This completes the proof of (a).

Proof of (b). Let $\epsilon > 0$ and let the sector OAB be bounded by the radii OB and OA emanating from the origin O at angle $\frac{1}{2}\pi - \epsilon$ and $\frac{1}{2}\pi + \epsilon$, respectively, and let the radius of the sector be small enough so that in the sector r is a decreasing function of t. Because r is monotone decreasing in t, the system (5.2) can be replaced by a first-order equation $d\theta/dr = F(r,\theta)$ in the sector. Consider the points S on the arc AB with the property that all solutions of $d\theta/dr = F$ emanating from points of AB to the left of any point of S pass out of the sector OAB by intersecting the open radial segment OA. The points S form an interval AQ which does not include points near B. It will now be shown that S does not contain the end point Q of the interval. That is, AQ is open on the right. Indeed, suppose all solutions from Q cross the open segment OA. Then, in particular, the minimum solution does so. Thus by Theorem 1.4, Chap. 2, the minimum solution for nearby points to the right of Q will cross OA. The maximum solutions have θ at least as large as do the minimum solutions and thus will surely cross OA. Thus all solutions starting at Q and all points near it will cross OA, which is impossible by the definition of Q.

Since the maximum solutions are continuous from above, the maximum solution from Q crosses OA or stays in the sector OAB. The minimum solution does not cross the open segment OA. If the minimum solution does not tend to O in OAB, then the minimum solution crosses the open segment OB. Let the maximum solution from Q cross OA at C and the minimum solution cross OB at D. Let A_1 lie on OA closer to O than C is, let B_1 be on OB closer to O than D is, and let $OA_1 = OB_1$.

Consider the sector OA_1B_1. Proceed on the arc A_1B_1 as with AB above. Let the point that corresponds to Q on this arc be denoted by Q_1. Consider the solutions of $d\theta/dr = F$ from Q_1 for *increasing* r. They cannot cross OA or OB. Thus they must leave OAB by first meeting the solutions CQ or DQ. But a solution which meets CQ at a point K, other than Q, can be continued as a solution along CQ from K to Q, and similarly for solutions meeting DQ. Thus there is at least one solution of $d\theta/dr = F$ which goes from Q to Q_1. In the same way, there is a point Q_2 on an arc A_2B_2 closer to O and a solution from Q_1 to Q_2 and thus from Q to Q_2. This solution can clearly be continued indefinitely toward O. Thus $QQ_1Q_2 \cdots$ is a solution that tends to O. Since θ is close to $\pi/2$ in OAB, it is clear from Part (a) that the solution tends to O at an angle of $3\pi/2$ with the positive x_1 direction as $t \to \infty$.

Proof of (c). The proof will be given for the case $3\pi/2$. For any fixed solution path (φ_1,φ_2) tending to the origin at an angle of $3\pi/2$, $\varphi_1/\varphi_2 \to 0$, and thus from (5.2) $\varphi_2'/\varphi_2 = \mu + o(1)$, as $t \to +\infty$. Thus for $\varphi_2 > 0$, $\varphi_2' < 0$ for all sufficiently large t, and hence $x_2 = \varphi_2(t)$ may be introduced as a new variable. Let $\dot{} = d/dx_2$.

Suppose there were two distinct orbits tending to the origin at $3\pi/2$ as $t \to +\infty$. Let the corresponding orbits be represented for all sufficiently large t by $x_1 = \psi_1(x_2)$, $x_1 = \psi_2(x_2)$. Clearly $\psi_j(x_2)/x_2 \to 0$ as $x_2 \to 0+$. From (5.2)

$$\dot{\psi}_i(x_2) = \frac{\lambda\psi_i(x_2) + f_1(\psi_i(x_2),x_2)}{\mu x_2 + f_2(\psi_i(x_2),x_2)} \qquad (i = 1, 2)$$

and by subtraction, if $\psi = \psi_1 - \psi_2$,

$$\dot{\psi}(x_2) = \frac{\lambda\psi(x_2) + [f_1(\psi_1(x_2),x_2) - f_1(\psi_2(x_2),x_2)]}{\mu x_2 + f_2(\psi_1(x_2),x_2)} + \frac{[\lambda\psi_2(x_2) + f_1(\psi_2(x_2),x_2)][f_2(\psi_2(x_2),x_2) - f_2(\psi_1(x_2),x_2)]}{[\mu x_2 + f_2(\psi_1(x_2),x_2)][\mu x_2 + f_2(\psi_2(x_2),x_2)]} \qquad (5.4)$$

By the uniqueness theorem $\psi \neq 0$, and thus, with no real restriction, it can be assumed that $\psi > 0$. Because of (2.1), $\partial f_i/\partial x_j$ is zero at the origin. Clearly

$$f_i(\psi_1(x_2),x_2) - f_i(\psi_2(x_2),x_2) = \psi(x_2)\,\frac{\partial f_i}{\partial x_1}\,(\zeta_i,x_2)$$

where $\psi_2(x_2) < \zeta_i < \psi_1(x_2)$, and since $\partial f_i/\partial x_1 \to 0$ as $r \to 0$, (5.4) implies that

$$\dot{\psi}(x_2) = \frac{\lambda\psi(x_2)}{\mu x_2}\,(1 + o(1)) \qquad (x_2 \to 0) \qquad (5.5)$$

Therefore, if x_2 is small enough, $x_2\dot{\psi}/\psi < \gamma < 1$ for $\gamma > \lambda/\mu$, and thus

$$c_1 x_2^{\gamma - 1} < \frac{\psi(x_2)}{x_2} \tag{5.6}$$

where $c_1 > 0$ is a constant, so that (5.6) implies $\psi(x_2)/x_2 \to +\infty$ as $x_2 \to 0$, and this contradicts the fact that

$$\frac{\psi(x_2)}{x_2} = \left(\frac{\psi_1(x_2) - \psi_2(x_2)}{x_2}\right) \to 0$$

as $x_2 \to 0$. This proves that there exists exactly one orbit tending to the origin at $3\pi/2$.

The case when the linear system (L) has an improper node of Type (III) can be discussed in an analogous fashion. This case will be left to the reader. This case is also treated in Prob. 10, Chap. 13.

6. Saddle Points

For the case of the saddle point at the origin, let the equations (NL) and (L) assume the canonical forms

$$x_1' = \lambda x_1 + f_1(x_1,x_2) \qquad x_2' = \mu x_2 + f_2(x_1,x_2) \tag{6.1}$$

and

$$x_1' = \lambda x_1 \qquad x_2' = \mu x_2 \tag{6.2}$$

respectively, where $\lambda < 0 < \mu$. Then the geometry of the orbits of (6.1) near the origin is described in the following theorem. The problem is also treated in Prob. 7, Chap. 13.

Theorem 6.1. (a) *There exists at least one orbit tending to the origin at each of the angles* 0 *and* π.

(b) *If, further,* $\partial f_1/\partial x_2$, $\partial f_2/\partial x_2$ *exist and are continuous on* $0 \leqq r \leqq r_0$, *there exists exactly one orbit tending to the origin at each of the angles* 0 *and* π. *Any orbit starting sufficiently near either of these orbits in the neighborhood of the origin tends away from them as* $t \to +\infty$.

The proof that there exists an orbit tending to the origin in a sector $|\theta| \leqq \epsilon$ is very much like the proof of Theorem 5.1(b). This orbit must tend to the origin at a limiting tangent angle of π, for from (6.1)

$$\theta' = \frac{(\mu - \lambda)}{2} \sin 2\theta + o(1) \qquad (r \to 0)$$

and so $\theta = \omega(t)$ can remain in the sector $|\theta| \leqq \epsilon$ only if $\omega(t) \to 0$ as $t \to +\infty$.

The method of proof of Theorem 5.1(c) may be used with only minor changes to prove (b).

PROBLEMS

1. Locate and classify the singular points of

$$x_1' = x_2 \qquad x_2' = -ax_2 - b \sin x_1 \tag{1}$$

and

$$x_1' = x_2 \qquad x_2' = a(1 - x_1^2)x_2 - bx_1 \tag{2}$$

where the constants a and b satisfy $a \geqq 0$ and $b > 0$.

2. Let

$$x_j' = P_j(x_1,x_2) + f_j(x_1,x_2) \qquad (j = 1, 2) \tag{*}$$

where the P_j are homogeneous polynomials of degree $m > 1$ and $f_j = o(r^m)$ as $r \to 0$, where r and θ are polar coordinates. Suppose all solutions of (*) starting near the origin tend to the origin as $t \to \infty$. Let $Q(\theta) = (x_1P_2 - x_2P_1)/r^{m+1}$ and suppose Q is not identically zero. If $\theta = \omega(t)$ is a solution, show that either $\omega(t)$ tends to a finite limit as $t \to \infty$ or else that $\omega(t) \to +\infty$ (or $-\infty$) as $t \to \infty$.

HINT: $d\theta/dt = r^{m-1}Q(\theta) + o(r^{m-1})$.

CHAPTER 16

THE POINCARÉ-BENDIXSON THEORY OF TWO-DIMENSIONAL AUTONOMOUS SYSTEMS

Let $f = (f_1, f_2)$ be a real continuous vector function defined on a bounded open subset D of the real (x_1, x_2) plane, and consider the two-dimensional autonomous system

$$\text{(E)} \qquad x_1' = f_1(x_1, x_2) \qquad x_2' = f_2(x_1, x_2) \qquad \left(' = \frac{d}{dt}\right)$$

Throughout this chapter it will be assumed that f satisfies the above assumption, and that, further, for each real t_0 and each point $(\xi, \eta) \in D$ there exists a *unique* vector solution $\varphi = \varphi(t, \xi, \eta)$ of (E), with components φ_1, φ_2, such that $\varphi_1(t_0, \xi, \eta) = \xi$, $\varphi_2(t_0, \xi, \eta) = \eta$. Actually, the notation $\varphi(t, \xi, \eta)$, which does not contain t_0 explicitly, is used because the solution of (E) through (ξ, η) considered as a curve in the (x_1, x_2) plane is independent of t_0. If $\varphi(t, \xi, \eta)$ is associated with $t_0 = 0$, then for the same φ, $\varphi(t - t_0, \xi, \eta)$ is the solution through (ξ, η) at $t = t_0$.

By Theorem 4.3, Chap. 2, the assumption that φ is unique is enough to guarantee that φ is a continuous function of (t, ξ, η) for all t for which φ is defined and for all $(\xi, \eta) \in D$. This remark is essential in many of the arguments to follow. Sufficient conditions for the existence and continuity of φ are given in Theorem 7.1, Chap. 1.

A point of D at which both f_1 and f_2 vanish is called a *critical point*. A point of D which is not a critical point will be called a *regular point*.

1. Limit Sets of an Orbit

Suppose C^+ (or C^-) is a semiorbit of (E) with representing solution φ defined for all $t \geq t_0$ (or $t \leq t_0$) for some t_0. That is, C^+ (or C^-) is the set of all points $P(t)$ of D with coordinates $(\varphi_1(t), \varphi_2(t))$, where $t_0 \leq t < +\infty$ (or $-\infty < t \leq t_0$). A point Q in the (x_1, x_2) plane is said to be a *limit point* of C^+ (or C^-) if there exists a sequence of real numbers $\{t_n\}$, $n = 1, 2, \ldots$, where $t_n \to \infty$ (or $t_n \to -\infty$), as $n \to \infty$, such that $P(t_n) \to Q$, as $n \to \infty$. The set of all limit points of a semiorbit C^+ (or C^-) is denoted by $L(C^+)$ [or $L(C^-)$], and these sets are called *limit sets*. If C is a full orbit it can be considered as the sum of a positive semiorbit C^+, and a negative semiorbit C^-. The corresponding limit sets $L(C^+)$ and $L(C^-)$ will be denoted by $L^+(C)$ and $L^-(C)$, respectively, and the set of all the limit points of C, namely, $L^+(C) \cup L^-(C)$, will be denoted by $L(C)$.

The qualitative geometric results concerning the solutions of (E) which will be presented here follow directly from a thorough investigation of these limit sets for the case of a semiorbit which remains inside a compact subset K of D. These investigations lead to one of the few very general theorems (the Poincaré-Bendixson theorem) which asserts the existence of periodic solutions of a (nonlinear) system of differential equations. In what follows, an orbit will always mean an orbit of (E).

Theorem 1.1. *If C^+ is a positive semiorbit contained in a closed subset K of D, then $L(C^+)$ is a nonempty, closed, and connected set.*

Proof. Let C^+ be represented by $\varphi = (\varphi_1,\varphi_2)$ for $t \geqq t_0$. Then the infinite set of points P_n: $(\varphi_1(t_0 + n), \varphi_2(t_0 + n))$, $n = 1, 2, \ldots$, is contained in the bounded set K, and hence this sequence has a subsequence which is convergent to a point which must be in K, for K is closed. Therefore $L(C^+)$ is not empty, and $L(C^+) \subseteq K$.

To prove that $L(C^+)$ is closed, let Q be a cluster point of $L(C^+)$. Then there exists a sequence $Q_n \in L(C^+)$, $n = 1, 2, \ldots$, such that $d(Q_n,Q) \to 0$, as $n \to \infty$, where $d(Q_n,Q)$ represents the distance between Q_n and Q. For each Q_n there exists a $t_n > n$ such that $d((\varphi_1(t_n),\varphi_2(t_n)),Q_n) < 1/n$. Therefore, given any $\epsilon > 0$, there exists an integer N_ϵ such that $d((\varphi_1(t_n),\varphi_2(t_n)),Q_n) < \epsilon/2$, and $d(Q_n,Q) < \epsilon/2$, for $n > N_\epsilon$. This implies $d((\varphi_1(t_n),\varphi_2(t_n)),Q) < \epsilon$, for $n > N_\epsilon$, or $Q \in L(C^+)$, for $t_n \to \infty$, as $n \to \infty$.

Suppose $L(C^+)$ is not connected. Then there exist two nonempty, disjoint, closed sets M,N such that $L(C^+)$ is the sum of M and N, that is, $L(C^+) = M \cup N$. Since M and N are both bounded, they are a finite distance δ apart.† Since the points of M and N are limit points of C^+, there exist arbitrarily large t such that $P(t)$ is within $\delta/2$ of M, and arbitrarily large t such that the distance of $P(t)$ from M is more than $\delta/2$. Since the distance $d(P,M)$ from any point P to M is a continuous function of P, and since the coordinates of $P(t)$ are continuous functions of t, it follows that there must exist a sequence $\{t_n\}$, $t_n \to \infty$, as $n \to \infty$, such that $d(P(t_n),M) = \delta/2$. The sequence of points $\{P(t_n)\}$, being bounded, must contain a subsequence converging to a point Q, which must be a limit point of C^+. Hence $Q \in L(C^+)$, and clearly $d(Q,M) = \delta/2$. But this implies Q is in neither M nor N, for

$$d(Q,N) \geqq d(N,M) - d(Q,M) = \frac{\delta}{2}$$

and this results in a contradiction, since by hypothesis $L(C^+) = M \cup N$. Thus the theorem is proved.

† If S_1 and S_2 are two sets of points, then $d(S_1,S_2) = \inf d(P_1,P_2)$ taken over all points $P_1 \in S_1$, and $P_2 \in S_2$.

Theorem 1.2. *Let C^+ be a positive semiorbit contained in a closed subset K of D, and assume $L(C^+)$ contains a regular point Q. Then the orbit C_Q through Q exists as a full orbit, and $C_Q \subseteq L(C^+)$.*

Proof. Reference to Fig. 14 should prove helpful. Let Q have coordinates $(\hat{\xi},\hat{\eta})$, and let C^+ be represented by the solution $\varphi = (\varphi_1,\varphi_2)$ for $t \geqq t_0$. Now φ is a function of the initial position, and thus if

$$(\varphi_1,\varphi_2) = (\xi,\eta)$$

at t_0, then $\varphi = \varphi(t,\xi,\eta)$. From the definition of Q it follows that there exists a sequence $\{t_n\}$, $t_n \to \infty$, as $n \to \infty$, such that the points P_n, whose coordinates are the components of $\varphi(t_n,\xi,\eta)$, have the property $P_n \to Q$, as $n \to \infty$. The curve through P_n can be reparametrized so that P_n is given by the components of $\varphi(0,\xi_n,\eta_n)$, where P_n: (ξ_n,η_n). Thus $\varphi(t,\xi_n,\eta_n) = \varphi(t + t_n,\ \xi,\ \eta)$.

The orbit C_Q is given by $\varphi(t,\hat{\xi},\hat{\eta})$, where Q is given by $\varphi(0,\hat{\xi},\hat{\eta})$. Thus, if

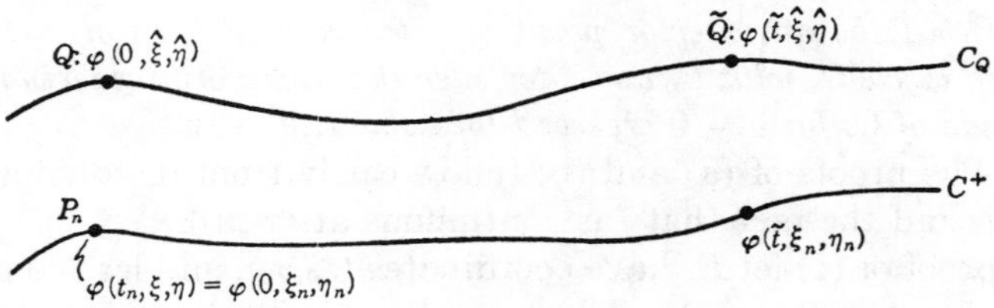

FIG. 14

$\tilde{Q}$ is a point on C_Q, its coordinates are the components of $\varphi(\tilde{t},\hat{\xi},\hat{\eta})$, for some $\tilde{t}$. From the fact that the solution φ is continuous as a function of the initial conditions, $\varphi(\tilde{t},\xi_n,\eta_n) \to \varphi(\tilde{t},\hat{\xi},\hat{\eta})$ as $n \to \infty$, for $P_n \to Q$. But this is the same as $\varphi(\tilde{t} + t_n,\ \xi,\ \eta) \to \varphi(\tilde{t},\hat{\xi},\hat{\eta})$, and this implies $\tilde{Q} \in L(C^+)$, for $\tilde{t} + t_n \to \infty$, as $n \to \infty$. Thus $C_Q \subseteq L(C^+) \subseteq K$, and by a familiar continuation argument this implies C_Q must be a full orbit.

If Q is any regular point in the set $L(C^+)$, where C^+ is any semiorbit satisfying the assumptions of Theorem 1.2, then the orbit C_Q through Q is called a *limit orbit* of C^+. Thus Theorem 1.2 says that $L(C^+)$ is composed of critical points and limit orbits.

2. The Poincaré-Bendixson Theorem

In this section it will always be assumed that C^+ is contained in a closed subset K of D.

If $L(C^+)$ has only regular points, then a description of $L(C^+)$ is given by the Poincaré-Bendixson theorem which asserts that $L(C^+)$ is itself a *periodic orbit* in this case. More precisely, the following is true.

Theorem 2.1 (Poincaré-Bendixson Theorem). *Let C^+ be a positive semiorbit contained in a closed subset K of D. If $L(C^+)$ consists of regular points only, then either*

(i) $C^+(= L(C^+))$ *is a periodic orbit, or*
(ii) $L(C^+)$ *is a periodic orbit.*

If the case (ii) prevails, the limit orbit $L(C^+)$ is called a *limit cycle.* In this case, C^+ actually "spirals" around $L(C^+)$ in a certain sense. This will be shown in Sec. 3.

In order to prove this important theorem, an auxiliary concept will be introduced. A finite closed segment l of a straight line in the (x_1,x_2) plane is called a *transversal* with respect to f if every point of l is regular and if the direction determined by f at every point of l is different from that of l. The properties of a transversal which are required for the proof of Theorem 2.1 are summarized in Lemmas 2.1 and 2.2 below.

Lemma 2.1. (a) *Every regular point* (x_1,x_2) *of* D *is an interior point of some transversal, which may have any direction except that of* $f(x_1,x_2)$.

(b) *Every orbit which meets a transversal must cross it, and all such orbits cross it in the same direction.*

(c) *Let* $P_0 \in D$ *be an interior point of a transversal* l. *For every* $\epsilon > 0$ *there exists a circle* C_ϵ *with* P_0 *as center such that every orbit passing through a point inside of* C_ϵ *for* $t = 0$ *crosses* l *for some* t, $|t| < \epsilon$.

Proof. The proofs of (a) and (b) follow easily from the definition of a transversal, and the fact that f is continuous at (x_1,x_2).

For the proof of (c) let P_0 have coordinates (ξ_0,η_0), and let the points of l satisfy $ax_1 + bx_2 + c = 0$. There exists a circle about P_0 which contains only regular points of f. The solution φ passing through any regular point (ξ,η) near P_0 at $t = 0$ is continuous in (t,ξ,η) together in an open set about $(0,\xi_0,\eta_0)$. Let $L(t,\xi,\eta) = a\varphi_1(t,\xi,\eta) + b\varphi_2(t,\xi,\eta) + c$, where $\varphi = (\varphi_1,\varphi_2)$. Then $L(0,\xi_0,\eta_0) = 0$, and $(\partial L/\partial t)(0,\xi_0,\eta_0) \neq 0$. Therefore, by the implicit-function theorem, there exists a continuous function $t = t(\xi,\eta)$ defined in some circle C about (ξ_0,η_0), satisfying $t(\xi_0,\eta_0) = 0$, and $L(t(\xi,\eta),\xi,\eta) = 0$. Moreover, since the function $t = t(\xi,\eta)$ is continuous at (ξ_0,η_0), for any $\epsilon > 0$, a circle C_ϵ exists about (ξ_0,η_0) such that $|t(\xi,\eta)| < \epsilon$ inside C_ϵ. Therefore the orbit passing through any (ξ,η) inside C_ϵ at $t = 0$ will cross l at $t(\xi,\eta)$, and $|t(\xi,\eta)| < \epsilon$.

Lemma 2.2. *If a finite closed arc* A *of an orbit* C *meets a transversal* l, *it does so in a finite number of points, whose order on* A *is the same as the order on* l. *If* C *is a periodic orbit, it meets* l *in only one point.*

Proof. If φ is a solution representing C, the points of A are of the form $P(t)$: $(\varphi_1(t),\varphi_2(t))$, $\tilde{t} \leqq t \leqq \tilde{\tilde{t}}$, for some finite $\tilde{t}$ and $\tilde{\tilde{t}}$. If A meets l in infinitely many distinct points $P_n = P(t_n)$, then the distinct t_n will have a cluster point $\hat{t}$ on $\tilde{t} \leqq t \leqq \tilde{\tilde{t}}$. Thus there exists a subsequence of $\{t_n\}$, which can again be denoted by $\{t_n\}$, such that $t_n \to \hat{t}$, $n \to \infty$. Then $P_n \to Q = P(\hat{t})$, $n \to \infty$. But $(\varphi(t_n) - \varphi(\hat{t}))/(t_n - \hat{t}) \to f(\varphi_1(\hat{t}),\varphi_2(\hat{t}))$ as $n \to \infty$, and since $(\varphi_2(t_n) - \varphi_2(\hat{t}))/(\varphi_1(t_n) - \varphi_1(\hat{t}))$ is the constant slope of

l it follows that f has the same direction as l at Q, which is a contradiction. Thus A must meet l in a finite number of points.

Now let $P_1 = P(t_1)$ and $P_2 = P(t_2)$ be two successive points of intersection of A with l, where $t_1 < t_2$ (see Figs. 15 and 16). Suppose P_1 is distinct from P_2. Then the curve J consisting of the open arc from P_1 to P_2 on A, denoted by $\widehat{P_1P_2}$, and the closed line segment on l from P_2 to P_1, denoted by $\overline{P_2P_1}$, is a Jordan curve, and thus separates the plane into two regions.† Therefore points Q on C for $t < t_1$, and sufficiently near t_1, will be on the opposite side of J from points R on C for $t > t_2$, and sufficiently near t_2 (see Figs. 15 and 16). There are two cases, according as the points R are inside J or outside J. Suppose the former (Fig. 15):

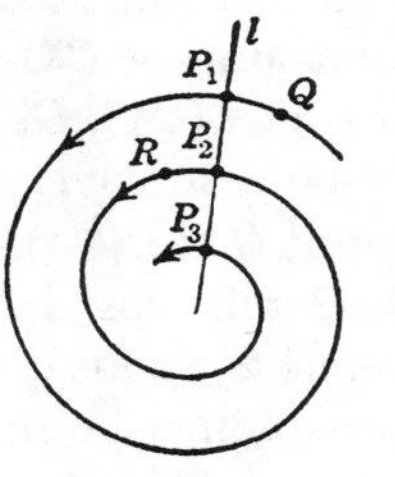

FIG. 15

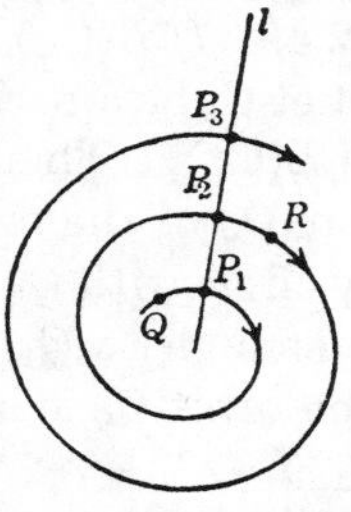

FIG. 16

the other case can be treated similarly. Then, in order for C to get outside J for $t > t_2$, C must cross J. But it cannot cross $\widehat{P_1P_2}$, by uniqueness, and it cannot cross $\overline{P_2P_1}$ in the wrong direction. Hence C remains inside J for $t > t_2$. It is thus clear that the next intersection P_3 (after P_2) of C with l is inside J and is distinct from P_2. Thus P_2 is between P_1 and P_3 on l.

If P_1 is the same as P_2, clearly C is periodic. Suppose P_1 is distinct from P_2, and C is periodic. Then the arc from R on C must return to Q and thus the arc RQ on C must cross J. But, as above, it cannot cross $\widehat{P_1P_2}$ by uniqueness, and cannot cross $\overline{P_2P_1}$ in the wrong direction. Thus P_1 is the same as P_2, and C is periodic. This completes the proof of Lemma 2.2.

The proof of the Poincaré-Bendixson theorem now proceeds via the following two additional lemmas.

Lemma 2.3. *If C^+ and $L(C^+)$ have a point in common, then C^+ is a periodic orbit.*

† A Jordan curve is a topological image of a circle. The separation property of such a curve is based on the *Jordan curve theorem.* This states that if J is a Jordan curve in the (x_1,x_2) plane π, then the complement of J, $\pi - J$, is the union of two disjoint open sets S_i and S_e, each of which has J as boundary. S_i is bounded and is called the interior of J, whereas S_e is unbounded and is called the exterior of J.

Proof. Let $P_1 = P(t_1)$ be a point of C^+ which is also a point of $L(C^+)$. It is a regular point and hence can be made an interior point of a transversal l. Since $P_1 \in L(C^+)$, any circle Γ with P_1 as center must contain in its interior a point $Q = P(\hat{t})$, $\hat{t} > t_1 + 2$. If Γ is the circle for $\epsilon = 1$ in Lemma 2.1(c), then there exists $\tilde{P} = P(\tilde{t})$ of C^+, where $|\tilde{t} - \hat{t}| < 1$ and $\tilde{P}$ is on l. Let $\tilde{P}$ be distinct from P_1. Then the arc $P_1\tilde{P}$ of C^+ intersects l in a finite number of points by Lemma 2.2. Also, the successive intersections of C^+ with l form a monotone sequence which tends away from P_1. Thus P_1 cannot be a limit point of C^+ and so is not in $L(C^+)$. Therefore C^+ meets l only in P_1, and so C^+ is a periodic orbit.

REMARK: This same argument shows that *a transversal cannot meet* $L(C^+)$ *in more than one point.*

Lemma 2.4. *If $L(C^+)$ contains a periodic orbit, it is identical with it.*

Proof. Let C_0 be a periodic orbit in $L(C^+)$, and suppose C_0 is contained properly in $L(C^+)$. Then, by the connectedness of $L(C^+)$, C_0 contains a cluster point Q_0 of the set $L(C^+) - C_0$. Let l be a transversal through Q_0. Every circle with Q_0 as center contains a point Q of $L(C^+) - C_0$, and, for Q close enough to Q_0, the orbit C_Q through Q will cross l, by Lemma 2.1(c). The orbit C_Q is a limit orbit by Theorem 1.2, and is distinct from C_0, for $C_Q \subseteq L(C^+) - C_0$. Hence l contains two distinct points of $L(C^+)$. This contradicts the remark following Lemma 2.3. Thus $C_0 = L(C^+)$.

Proof of the Poincaré-Bendixson Theorem. Clearly, if C^+ is a periodic orbit, then $C^+ = L(C^+)$. Therefore assume C^+ is not periodic. Since $L(C^+)$ is not empty and consists of regular points only, there exists by Theorem 1.2 a limit orbit C_0 in $L(C^+)$. Now $C_0 \subseteq K$ implies that the semiorbit C_0^+ has a limit point P_0, and $P_0 \in L(C^+)$, for $L(C^+)$ is closed. If l is a transversal through P_0, then, since P_0 and C_0^+ are both in $L(C^+)$, l can meet $L(C^+)$ in no point but P_0, by the remark following Lemma 2.3. Since P_0 is a limit point of C_0^+, l must meet C_0^+ in some point, which must be P_0, and hence C_0^+ and $L(C_0^+)$ have the point P_0 in common. By Lemma 2.3, C_0^+ and thus also C_0 are periodic orbits, and this implies, by Lemma 2.4, that $C_0 = L(C^+)$.

Corollary. *If C^+ is a semiorbit contained in a compact set K in which f has no critical points, then K contains a periodic orbit.*

3. Limit Sets with Critical Points

The following result classifies the behavior of $L(C^+)$ when this set contains only a finite number of critical points of f.

Theorem 3.1. *Let C^+ be a semiorbit contained in a closed subset K of D, and suppose D has only a finite number of critical points. Then either*

(i) *$L(C^+)$ consists of a single point, a critical point of f, which C^+ approaches as $t \to \infty$, or*

(ii) *$L(C^+)$ is a periodic orbit, or*

(iii) *$L(C^+)$ consists of a finite number of critical points of f, and a set of orbits, each of which tends to one of these critical points as $t \to \pm\infty$.*

Proof. The set $L(C^+)$ can contain at most a finite number of critical points of f. If $L(C^+)$ contains no regular points of f, then $L(C^+)$ is just one critical point, for $L(C^+)$ is connected. Clearly C^+ must tend to this point as $t \to +\infty$.

If $L(C^+)$ has regular points, it consists of critical points and a set of limit orbits. Let C_0 be a limit orbit. It cannot have a regular limit point unless C_0 is a periodic orbit, by the argument used in the proof of the Poincaré-Bendixson theorem. Thus either C_0 is periodic, in which

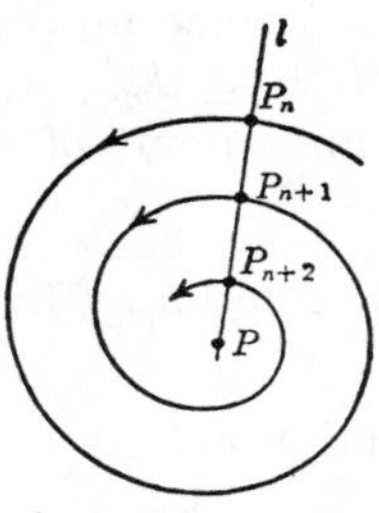

FIG. 17

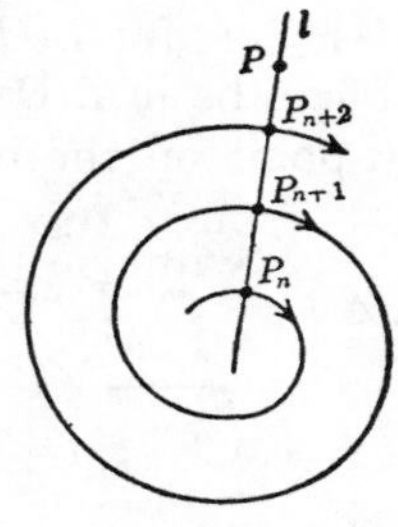

FIG. 18

case $L(C^+) = C_0$, by Lemma 2.4, or C_0 has no regular limit points. Hence either $L(C^+)$ is a periodic orbit, or all orbits in $L(C^+)$ are not periodic and have only critical points of f as limit points. Suppose the latter case, and let C_0 be an orbit in $L(C^+)$. From the proof of (i) it follows that $L^+(C_0)$ and $L^-(C_0)$ each consists of just one critical point of f [$L^+(C_0) = L^-(C_0)$ is not ruled out].

Corollary. *If C^+ is a semiorbit contained in a closed set $K \subset D$, and $L(C^+)$ contains only one critical point $\tilde{P}$ (and regular points), then a limit orbit tends to $\tilde{P}$ as $t \to +\infty$ and $t \to -\infty$.*

Now suppose $L(C^+)$ contains a regular point P of f. If C^+ is a periodic orbit, then $L(C^+) = C^+$, and $L(C^+)$ is completely known. Suppose C^+ is not periodic, and l is a transversal through P. Then, as in Lemma 2.3, C^+ must meet l in an infinite number of points which converge monotonically on l to P. If $P_1 = P(t_1)$ is any such point on l, let P_n, $n = 2, 3, \ldots$, be the successive intersections of C^+ with l for $t > t_1$. Then the curve J_n consisting of the arc $\widehat{P_nP_{n+1}}$ on C^+ and the line segment $\overline{P_{n+1}P_n}$ on l is a Jordan curve which has an interior I_n and an exterior E_n. See Figs. 17 and 18.

There are two cases according as $P_3 \in I_1$ or $P_3 \in E_1$. In the former case, $I_{n+1} \subset I_n$, $n = 1, 2, \ldots$, and in the latter $I_{n+1} \supset I_n$, $n = 1, 2, \ldots$.

For $P_3 \varepsilon I_1$, $P \varepsilon I_1$, and since no point of $L(C^+)$ can be in E_1, $L(C^+) \subset I_1$, and similarly $L(C^+) \subset I_n$ for all n. If $P_3 \varepsilon E_1$, $E_n \supset L(C^+)$, for all n. In the first case, let I denote the nonempty closed set consisting of those points which are in the closure of I_n for all $n > 0$. Thus, if $\bar{I}_n$ is the closure of I_n, I is given by $\bigcap_{n=1}^{\infty} \bar{I}_n$. In case $P_3 \varepsilon E_1$, let I denote the closure of the set of those points which are interior to I_n for some n. Thus I is the closure of $\bigcup_{n=1}^{\infty} I_n$. Let J denote the boundary of I.

Lemma. $L(C^+) = J$.

Proof. Suppose $I = \bigcap_{n=1}^{\infty} \bar{I}_n$; if $I = \overline{\bigcup_{n=1}^{\infty} I_n}$, the reasoning is similar. Clearly $L(C^+) \subseteq I$, for $L(C^+) \subset I_n$ for all n, and since no point of the interior of I can be in $L(C^+)$, it follows that $L(C^+) \subseteq J$. On the other hand, every point on the boundary of I is a limit point of C^+. Thus $L(C^+) = J$.

In the case $I = \bigcap_{n=1}^{\infty} \bar{I}_n$, C^+ is contained in the exterior of the set I, and for the case $I = \overline{\bigcup_{n=1}^{\infty} I_n}$, C^+ is located in the interior of I.

Let C^+ satisfy the same conditions as in Theorem 3.1.

Theorem 3.2. *Suppose $L(C^+)$ has a regular point, and C^+ and $L(C^+)$ have no point in common, i.e., $C^+ \neq L(C^+)$. If C^+ is in the exterior (interior) of I, the semiorbit C_P^+ through any regular point P, sufficiently near $L(C^+)$ and in the exterior (interior) of I, has $L(C^+)$ as its limit set. Moreover, C_P^+ spirals around $L(C^+)$ in the sense that a transversal through any regular point of $L(C^+)$ meets C_P^+ an infinite number of times.*†

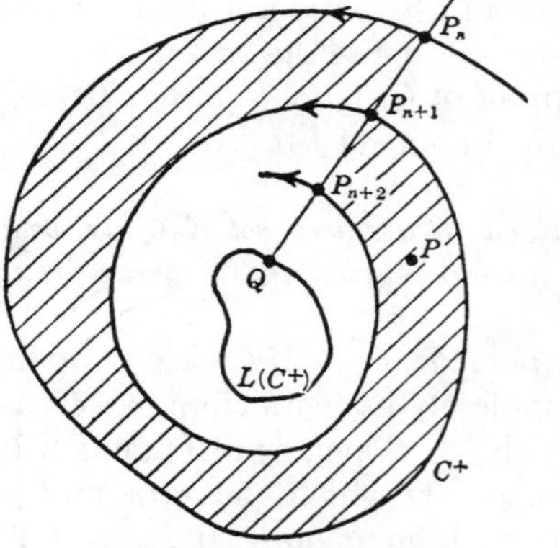

FIG. 19. (C^+ exterior to I.)

REMARK: $L(C^+)$ may contain critical points and thus need not be a periodic orbit.

Proof of Theorem 3.2. Suppose C^+ is in the exterior of I; refer to Fig. 19. If Q is a regular point of $L(C^+)$ and l is a transversal through Q, then by the discussion above there exists a nested sequence of closed sets $\{\bar{I}_n\}$ such that the boundary of $I = \bigcap_{n=1}^{\infty} \bar{I}_n$ is $L(C^+)$. If P is near enough

† In particular, C^+ has this property.

to $L(C^+)$, then P is inside some I_n and exterior to I_{n+1}. Clearly C_P^+ cannot cross the boundary of I_n, and thus C_P^+ either remains in $\overline{I_n - I_{n+1}}$, or else crosses l on the segment $\overline{P_{n+2}P_{n+1}}$ and passes into I_{n+1}. The first possibility implies that $L(C_P^+)$ is also in $\overline{I_n - I_{n+1}}$. If N is large enough, i.e., if P is near enough to $L(C^+)$, there are no critical points of f in $\bar{I}_n - I$, $n > N$, for the critical points are isolated. Thus $L(C_P^+)$ has no critical points, and by the Poincaré-Bendixson theorem $L(C_P^+)$ is a periodic orbit. However, it will be shown in the next section (Theorem 4.4, Corollary 1) that any periodic orbit contains at least one critical point in its interior. This gives a contradiction. Thus C_P^+ must enter I_{n+1} along $\overline{P_{n+2}P_{n+1}}$. The result follows now by induction.

Theorem 3.2 shows that the limit set $L(C^+)$ possesses a type of stability property. This will be defined precisely in the case of a periodic orbit. The periodic orbit C is said to be positively *stable from the outside* (*inside*) if for every $\epsilon > 0$ there exists a $\delta_\epsilon > 0$ such that every positive semiorbit starting at a distance less than δ_ϵ from C and outside (inside) C at $t = 0$ is defined for all $t > 0$ and remains at a distance less than ϵ from C. The periodic orbit C is said to be positively *stable* if it is positively stable from the outside and inside. Negative stability can be defined with $-t$ replacing t in the above. The type of stability defined here is often called *orbital stability*.

Theorem 3.3. *Necessary and sufficient for a periodic orbit C to be positively stable is that for both the interior and exterior of C, either*

(i) *an orbit approaches C as a limit cycle as $t \to +\infty$, or*
(ii) *there exist periodic orbits in any ϵ neighborhood of C.*

Proof. The sufficiency follows directly from Theorem 3.2. Let C be stable, and suppose there are no periodic orbits or critical points at a distance of less than ϵ from C, for some $\epsilon > 0$. Then a positive semiorbit C^+ starting at a point less than δ_ϵ in distance from C must be such that $L(C^+)$ is a periodic orbit, by the Poincaré-Bendixson theorem. Hence $L(C^+) = C$, that is, C is a limit cycle, and the theorem is proved.

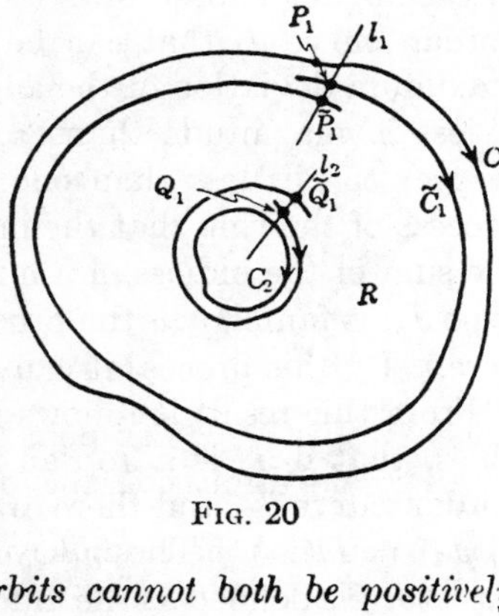

FIG. 20

Now suppose that C_1 and C_2 are periodic orbits, with C_2 contained in the interior of C_1, and there are no critical points or periodic orbits between C_1 and C_2. Then C_1 and C_2 are said to be *adjacent*.

Theorem 3.4. *Two adjacent periodic orbits cannot both be positively stable on the sides facing one another.*

Proof. If C_2 is in the interior of C_1, this means that C_2 cannot be positively stable from the outside and C_1 positively stable from the inside. Suppose C_1 and C_2 are positively stable facing one another. Replace t by $-t$ and let $\tilde{C}_1$ and $\tilde{C}_2$ be two orbits approaching C_1 and C_2, respectively, as $t \to -\infty$, and suppose l_1 and l_2 are two transversals, as shown in Fig. 20. Let R be the closed region bounded by the two Jordan curves

$$J_1: \widehat{P_1\tilde{P}_1} \text{ on } \tilde{C}_1 \text{ and } \overline{\tilde{P}_1P_1} \text{ on } l_1$$

$$J_2: \widehat{Q_1\tilde{Q}_1} \text{ on } \tilde{C}_2 \text{ and } \overline{\tilde{Q}_1Q_1} \text{ on } l_2$$

Any orbit C starting on the boundary of R must remain in R, and since R is free of critical points, $L(C)$ must be a periodic orbit in R, by the Poincaré-Bendixson theorem. This contradicts the assumption that C_1 and C_2 are adjacent.

4. The Index of an Isolated Critical Point

Let $f = (f_1,f_2)$ be a continuous real-valued vector function defined on a bounded open set D in the (x_1,x_2) plane, and suppose f has only isolated critical points on D. Let J be a Jordan curve in D passing through no critical points of f. If $\Delta\theta$ is the total change in the angle θ that $f(x_1,x_2)$ makes with some fixed direction as (x_1,x_2) traverses J once in the positive direction, then the *index of J with respect to f* is defined to be $\Delta\theta/2\pi$, and will be denoted by $I_f(J)$. Clearly this number is an integer. The important property of $I_f(J)$ which is required here is one which is usually proved in topology or complex-variables books.

Theorem 4.1. *If J is a Jordan curve in D containing no critical points of f on it or in its interior, then $I_f(J) = 0$.*

Outline of Proof. Since f is continuous, it is uniformly continuous on any compact subset of D. Thus there is a $\delta > 0$ such that, on any Jordan curve J_δ that can be contained in a square of side δ in D, the maximum deviation of the angle of f from its value at a fixed point on J_δ is less in magnitude than 2π. Thus the index of any such curve with respect to f is less than one in magnitude and is therefore zero. The process of showing that the index of the given Jordan curve J is equal to the sum of the indices of a number of smaller Jordan curves, each of the type J_δ, is familiar in the proof of Cauchy's theorem and will be omitted here. By this process it can be shown that $I_f(J) = 0$.

From this result it follows readily, by usual methods in complex variables, that if J_1 is a Jordan curve contained in the interior of another Jordan curve J_2, and there are no critical points between J_1 and J_2, then $I_f(J_1) = I_f(J_2)$. The *index of an isolated critical point P* with respect to a vector f is defined as the index of any Jordan curve containing P, and no other critical point of f, in its interior. This will be denoted by $I_f(P)$.

Theorem 4.2. *If J is a Jordan curve surrounding a finite number $P_1, \ldots, P_n$ of critical points of f, then* $I_f(J) = \sum_{i=1}^{n} I_f(P_i)$.

Proof. Surround each of the points P_i by a sufficiently small circle C_i containing P_i as the only critical point in its interior, and make the cuts as illustrated in Fig. 21.

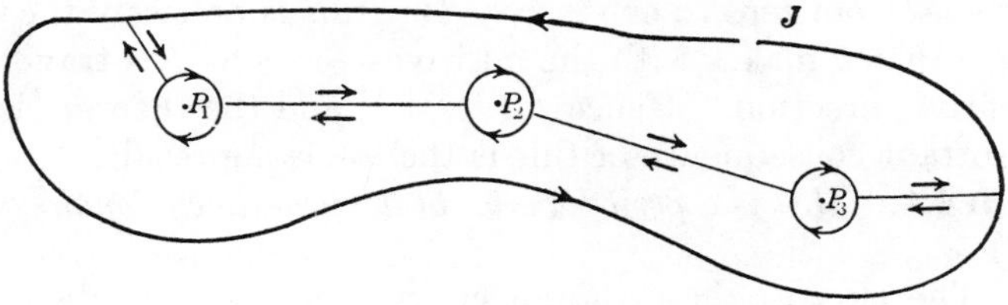

FIG. 21

Theorem 4.3. *If J is a Jordan curve in the (x_1,x_2) plane with a continuously turning tangent vector v, which is nowhere zero on J, then*

$$I_v(J) = 1$$

Proof. If $u(P)$ is the unit tangent vector to J at P, then clearly $I_u(J) = I_v(J)$, and so it suffices to prove the theorem for u. Without loss of generality, assume that J lies entirely in the region $x_2 \geqq 0$, and that the points P of J are given by $P(t)\colon (\alpha_1(t),\alpha_2(t))$, $0 \leqq t \leqq 1$. Thus

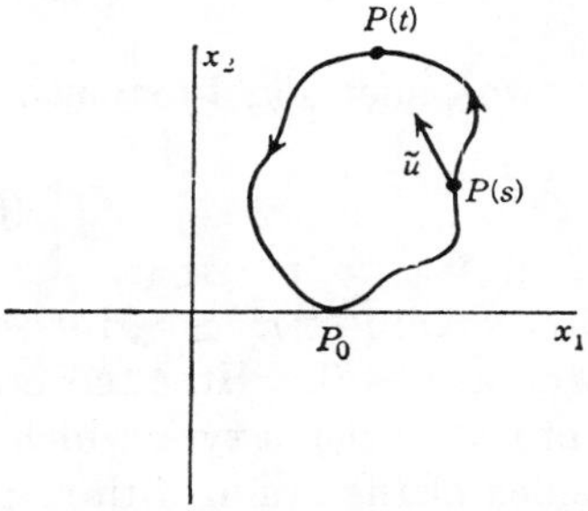

FIG. 22

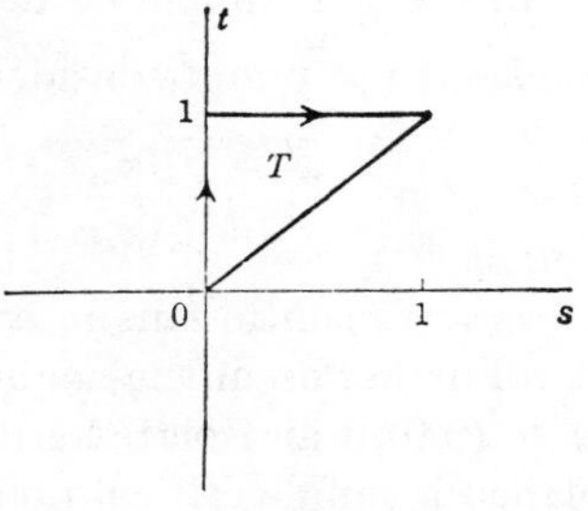

FIG. 23

$v(t) = (\alpha_1'(t),\alpha_2'(t))$, and it can be further assumed that the positive x_1 axis is tangent to J at $P_0 = P(0)$, that is, $v(0)$ has the same direction as the positive x_1 axis (see Fig. 22).

The theorem will be proved by constructing an auxiliary vector $\tilde{u}$ on the closed triangular region

$$T\colon \quad 0 \leqq s \leqq 1 \quad s \leqq t \leqq 1$$

in the (s,t) plane (see Fig. 23) as follows: Define $\tilde{u}(s,s) = u(s)$ for $0 \leqq s \leqq 1$, $\tilde{u}(0,1) = -u(0)$, and for all other (s,t) in T define $\tilde{u}(s,t)$

to be the unit vector in the direction from $P(s)$ to $P(t)$ on J (see Fig. 22). Let $\theta(s,t)$ be the angle that $\tilde{u}(s,t)$ makes with the positive x_1 axis. Clearly $\theta(0,0) = 0$, and since J remains in the region $x_2 \geqq 0$, $\theta(0,t)$ varies from zero to π as t runs from zero to one. Similarly, $\theta(s,1)$ varies from π to 2π as s varies from zero to one. From the definition of $\tilde{u}$ it is clear that $\tilde{u}$ is continuous on T, and $\tilde{u} \neq 0$ there. Hence by Theorem 4.1, applied to the boundary B of T, $I_{\tilde{u}}(B) = 0$. This means that the variation of $\theta(s,s)$ as s goes from zero to one is 2π. But this is precisely the variation in the angle that u makes with the positive x_1 axis as J is traversed once in the positive direction. Hence $I_u(J) = 1$, and the theorem is proved.

An important consequence of this is the following result.

Theorem 4.4. *If C is a periodic orbit of the two-dimensional system* (E), *then $I_f(C) = 1$.*

Proof. The curve C is a Jordan curve, and from its definition f is tangent to C, and $f \neq 0$ on C.

Corollary 1. *A periodic orbit C contains at least one critical point of f in its interior.*

Proof. Otherwise, by Theorem 4.1, $I_f(C) = 0$, and this would be a contradiction.

Corollary 2. *If C is a periodic orbit and the critical points of f are isolated, then the interior of C contains a finite number* ($\geqq 1$) *of critical points of f, the sum of whose indices is one.*

Proof. Apply Theorem 4.2.

5. The Index of a Simple Critical Point

Let $ad - bc \neq 0$ and consider again the two-dimensional real system

$$\begin{aligned} x_1' &= f_1(x_1,x_2) = ax_1 + bx_2 + g_1(x_1,x_2) \\ x_2' &= f_2(x_1,x_2) = cx_1 + dx_2 + g_2(x_1,x_2) \end{aligned} \tag{5.1}$$

where g_1,g_2 are continuous in a circle $0 \leqq r = \sqrt{x_1^2 + x_2^2} \leqq \gamma$, for some $\gamma > 0$. Further assume $g_1 = o(r)$, $g_2 = o(r)$, as $r \to 0$. Here the origin $(x_1,x_2) = (0,0)$ is an isolated critical point of $f = (f_1,f_2)$, a type which has been called a simple critical point. The index of the origin with respect to f is easy to calculate, and indeed this index depends only on the *linear* terms in (5.1). This latter fact depends on the observation embodied in the following lemma.

Lemma. *If v, $\tilde{v}$ are any two continuous vector functions on a Jordan curve J which never have opposite directions, or are zero there, then*

$$I_v(J) = I_{\tilde{v}}(J)$$

Proof. For $0 \leqq s \leqq 1$ define the vector v_s on J by $v_s = (1 - s)v + s\tilde{v}$. Now $v_s \neq 0$ on $0 \leqq s \leqq 1$, for $v_0 = v$, $v_1 = \tilde{v}$, and if $v_s = 0$ for $s \neq 0,1$, then $v = -(s/(1 - s))\tilde{v}$ at such an s, which implies v has a direction opposite to that of $\tilde{v}$. Therefore $I_{v_s}(J)$ exists and is clearly continuous in s. But this index is an integer, and thus a constant. This implies $I_{v_0}(J) = I_{v_1}(J)$, or $I_v(J) = I_{\tilde{v}}(J)$.

Theorem 5.1. *If v is the vector function with components $(ax_1 + bx_2, cx_1 + dx_2)$, and $f = (f_1,f_2)$ is the vector defined in* (5.1), *then $I_f(0) = I_v(0)$, where zero is the origin* (0,0).

Proof. It will be shown that on a sufficiently small circle with center at 0, f and v are never in opposition. Suppose the contrary at some point (x_1,x_2). Then at this point $f + sv = 0$ for some $s \geqq 0$. But $f = v + g$, where $g = (g_1,g_2)$, and hence $(1 + s)v = -g$, or $(1 + s)^2\|v\|^2 = \|g\|^2$ there. Now $\|v\|^2 = (ax_1 + bx_2)^2 + (cx_1 + dx_2)^2$, and if $x_1 = r \cos \theta$, $x_2 = r \sin \theta$, $\|v\|^2 = r^2[(a \cos \theta + b \sin \theta)^2 + (c \cos \theta + d \sin \theta)^2]$. Since $ad - bc \neq 0$, $v = 0$ only at (0,0). Thus v is continuous and not zero on $r = 1$. Therefore

$$m = \inf_{r=1} \|v\| > 0$$

and since $\|v\|$ is homogeneous in r, $\|v\| \geqq mr$ for all $r > 0$. This implies $m^2(1 + s)^2r^2 \leqq \|g\|^2$ at any point (x_1,x_2), where f is in opposition to v. From this it follows that such points cannot be arbitrarily close to (0,0), for such an hypothesis would imply $0 < m^2 \leqq m^2(1 + s)^2 \leqq \|g\|^2/r^2$, and since $\|g\| = o(r)$, as $r \to 0$, this gives a contradiction.

Therefore, for some sufficiently small $\alpha > 0$, f and v are not in opposition in $0 < r < \alpha$. The case $s = 0$ above shows that f does not vanish in $0 < r < \alpha$. Thus, by the previous lemma, $I_f(J) = I_v(J)$ holds for any Jordan curve J in $0 < r < \alpha$ surrounding (0,0), and this proves $I_f(0) = I_v(0)$.

Theorem 5.2. *If $f = (f_1,f_2)$ is the vector defined in* (5.1), *then $I_f(0) = -1$ or $+1$, according as the origin is or is not a saddle point for the linear system $x_1' = ax_1 + bx_2$, $x_2' = cx_1 + dx_2$.*

Proof. By Theorem 5.1 it is sufficient to calculate $I_v(0)$. The unit circle J: $x_1 = \cos \theta$, $x_2 = \sin \theta$, $0 \leqq \theta \leqq 2\pi$, can be used to compute $I_v(0)$. The total change in the angle that v makes with the positive x_1 axis as J is traversed once in the positive direction is clearly

$$2\pi I_v(J) = \int_J d \tan^{-1} \left(\frac{cx_1 + dx_2}{ax_1 + bx_2}\right)$$

or

$$I_v(J) = \frac{(ad - bc)}{2\pi} \int_0^{2\pi} \frac{d\theta}{(a \cos \theta + b \sin \theta)^2 + (c \cos \theta + d \sin \theta)^2} \tag{5.2}$$

The right side of (5.2) is continuous in (a,b,c,d) for $ad - bc \neq 0$. If $ad - bc > 0$, there are two cases according as $ad > 0$ or $ad \leqq 0$. If $ad > 0$, let $b,c \to 0$, and $d \to a$ in (5.2). Then the right side of (5.2), being an integer, remains constant. Thus

$$I_v(J) = \frac{1}{2\pi}\int_0^{2\pi} d\theta = +1$$

in this case. If $ad \leqq 0$, then $bc < 0$, and if ad is increased so that $ad > 0$, the preceding can be applied to give the same result. If $ad - bc < 0$, the same reasoning shows that $I_v(J) = -1$. But $ad - bc < 0$ is just the condition which distinguishes the saddle point from the other types of critical points, and this proves the theorem.

PROBLEMS

1. Show that the system

$$x_1' = x_2 \qquad x_2' = -x_1 + (1 - x_1^2 - x_2^2)x_2 \qquad \left(' = \frac{d}{dt}\right)$$

has $\varphi = (\varphi_1,\varphi_2)$, where $\varphi_1(t) = \sin(t + c)$, $\varphi_2(t) = \cos(t + c)$, and c is an arbitrary constant, as its only periodic solutions (except for the trivial solution $\varphi = 0$).

HINT: Show that on a closed orbit

$$\int (1 - \varphi_1^2 - \varphi_2^2)\varphi_2^2\, dt = 0$$

so that unless $1 - \varphi_1^2 - \varphi_2^2 = 0$ it is necessary for $1 - \varphi_1^2 - \varphi_2^2$ to change sign on the orbit.

2. Show that orbital stability prevails for the solution in Prob. 1. Show that any solution φ, not the trivial one, has an asymptotic phase γ, that is, $\varphi_1(t) \to \sin(t + \gamma)$, $\varphi_2(t) \to \cos(t + \gamma)$ as $t \to \infty$.

HINT: Show that $\varphi_1^2 + \varphi_2^2$ is a decreasing function of t outside of the unit circle.

3. Consider the system (for the damped pendulum) $x_1' = x_2$, $x_2' = -b \sin x_1 - ax_2$, where a and b are positive constants. Show that for any solution $\varphi = (\varphi_1,\varphi_2)$ there is an integer k such that $\varphi_1(t) \to k\pi$, $\varphi_2(t) \to 0$ as $t \to \infty$. Distinguish between the nature of the orbits in the vicinity of $(k\pi,0)$ for the cases k even and k odd. Sketch the orbits in the (x_1,x_2) plane.

HINT: Show that $\lambda = \frac{1}{2}\varphi_2^2 + 2b \sin^2 \frac{1}{2}\varphi_1$ is a monotone decreasing function of t unless $\varphi_2 \equiv 0$ and $\varphi_1 \equiv n\pi$ for some integer n. Show that as $t \to \infty$, $\lambda \to c \geqq 0$, where c is a constant, and that $\varphi_2 \to 0$ so that $2b \sin^2 \frac{1}{2}\varphi_1 \to c$. Thus $\varphi_1 \to c_1$, a constant. Since $\varphi_2' = -b \sin \varphi_1 - a\varphi_2 \to -b \sin c_1$ and $\varphi_2 \to 0$, it follows that $\sin c_1 = 0$.

4. Let $x' = f(t,x)$, where x is a scalar and f and $\partial f/\partial x$ are continuous in (t,x). Let f be real and of period ω in t. If a solution φ satisfies

$$\limsup_{t\to\infty} |\varphi(t)| < \infty$$

then show the equation has a periodic solution.

HINT: Either $\varphi(\omega) = \varphi(0)$ or else $\varphi(k\omega)$, $k = 0, 1, 2, \ldots$, is a monotone sequence.

5. Let f be real and even, g be real and odd, and let $g(x) > 0$, $x > 0$. Let $g \in C^1$ and f be piecewise continuous. Let $F(x) = \int_0^x f(t)\, dt$ and $G(x) = \int_0^x g(t)\, dt$. Let there exist an $a > 0$ such that $F(x) < 0$ for $0 < x < a$ and $F(x) > 0$ and monotone increasing for $x > a$. Let $G(x) \to \infty$ and $F(x) \to \infty$ as $x \to \infty$. Show that, aside from the identically zero solution, the differential equation

$$x'' + f(x)x' + g(x) = 0$$

has (aside from translations in t) a unique periodic solution which is stable. An important example is $x'' + \mu(x^2 - 1)x' + x = 0$.

HINT: Consider the system $x' = y - F(x)$, $y' = -g(x)$ and consider the change in $U = \frac{1}{2}y^2 + G(x)$ along solutions in the right half plane $x > 0$. Let $A'B'C'D'$ (Fig. 24) and $ABEFCD$ each represent a solution in $x > 0$. Using $dU/dx = -gF/(y - F)$, show that $U_B - U_A < U_{B'} - U_{A'}$. Using $dU/dy = F$, show that $U_F - U_E < U_{C'} - U_{B'}$. Show $U_E < U_B$. Show finally that $U_D - U_A < U_{D'} - U_{A'}$. For small A' show that $U_{D'} - U_{A'} > 0$ and for large A show that $U_D < U_A$. From the monotone character of $U_D - U_A$, prove the result.

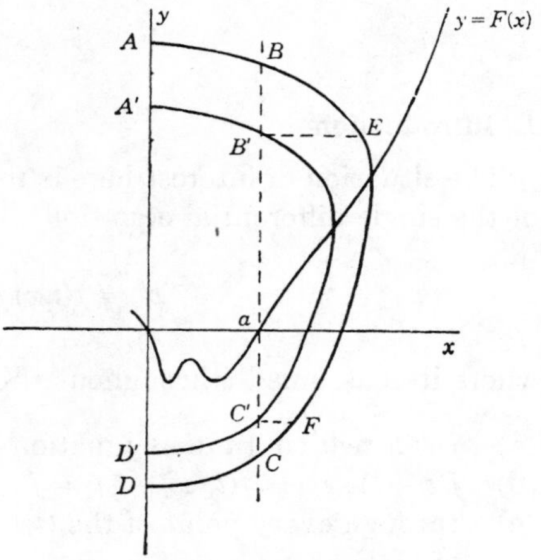

FIG. 24

6. Let F be as in Prob. 5. Show that $y'' + F(y') + y = 0$ has, aside from the trivial solution, a unique periodic solution.

HINT: Let $y' = x$.

7. If the system $x_j' = f_j(x_1,x_2)$, $j = 1, 2$, where $f_j \in C^1$, is considered in the three-dimensional (t,x_1,x_2) space, the solutions may be regarded as determining a $1 - 1$ transformation of the (x_1,x_2) plane into itself, the transformation being continuous in t. The more complicated problem

$$(*) \qquad x_j' = f_j(t,x_1,x_2) \qquad (j = 1, 2)$$

where f_j is periodic in t of period ω and where f_j and $\partial f_j/\partial x_l$ are continuous in (t,x_1,x_2) may be studied by means of the transformation T of the (x_1,x_2) plane into itself defined as follows: Let ξ be a point in the (x_1,x_2) plane. Let φ be the solution of (*) with $\varphi(0) = \xi$. Then $T\xi$ is defined to be $\varphi(\omega)$. Prove that if T exists on the whole plane then $T^{m+n}\xi = T^mT^n\xi$. Prove that if an open simply connected domain D, bounded by a Jordan curve J, satisfies $T\bar{D} \subset \bar{D}$, where $\bar{D} = D \cup J$, then (*) has a periodic solution of period ω.

HINT: Use the Brouwer fixed-point theorem which states that if $T\bar{D} \subset \bar{D}$ then there is a point $P \in \bar{D}$ with $TP = P$, provided that T is continuous.

CHAPTER 17

DIFFERENTIAL EQUATIONS ON A TORUS

1. Introduction

The situation of interest here is the study of the solutions, in the large. of the single differential equation

$$x' = f(t,x) \qquad \left(' = \frac{d}{dt}\right) \tag{1.1}$$

where it is assumed (throughout this chapter) that

(a) f is a real continuous function defined for all real (t,x),
(b) $f(t + 1, x) = f(t, x + 1) = f(t,x)$, and
(c) through every point of the (t,x) plane there passes a unique solution of (1.1).

Because of (a) and (b), f is bounded and hence every solution of (1.1) exists for all t. The periodicity assumption (b) implies that f may be considered as a function on the surface of a torus $\mathfrak{J}$, the points of which can be described by Cartesian coordinates (t,x), where two points $P_1 = (t_1,x_1)$ and $P_2 = (t_2,x_2)$ are regarded as identical if $t_1 - t_2$ and $x_1 - x_2$ are integers. Similarly the solution paths $(t,\varphi(t))$ may be represented on $\mathfrak{J}$. In three-dimensional space with rectilinear coordinates (u,v,w) the torus $\mathfrak{J}$ is given by

$$\begin{aligned} u &= (a + b \cos 2\pi x) \cos 2\pi t \\ v &= (a + b \cos 2\pi x) \sin 2\pi t \\ w &= b \sin 2\pi x \end{aligned}$$

where a and b are constants and $0 < b < a$.

By (c), through every point P of $\mathfrak{J}$ there exists a unique solution path of (1.1). The study, in the large, of the solutions of (1.1) is therefore reduced to the study of the paths $(t,\varphi(t))$, $-\infty < t < +\infty$, on the compact surface $\mathfrak{J}$.

Let $\varphi = \varphi(t,\eta)$ be the solution of (1.1) such that $\varphi(0,\eta) = \eta$. Then consider the transformation ψ of the real line onto itself defined by

$$\psi(\eta) = \varphi(1,\eta) \tag{1.2}$$

From the assumptions (a) and (c) it follows, by applying Theorem 4.3, Chap. 2, that ψ is a *homeomorphism* of the real line onto itself, that is, a continuous mapping whose inverse is continuous.† Let $\mathfrak{C}$ be the circle on $\mathfrak{I}$ defined as the set of all (t,x) on $\mathfrak{I}$ such that $t = 0$. Then ψ induces a homeomorphism T of $\mathfrak{C}$ onto $\mathfrak{C}$ defined by

$$TP = P_1 \tag{1.3}$$

where $P = (0,\eta)$, $P_1 = (1,\varphi(1,\eta)) = (0,\psi(\eta))$. It follows directly from the uniqueness assumption (c) that no two solution paths may cross on $\mathfrak{I}$, and hence ψ must be a *monotone increasing function.* This implies that T *preserves the orientation of the circle* $\mathfrak{C}$.

Consider the solutions $\tilde{\varphi},\hat{\varphi}$ of (1.1) defined by

$$\tilde{\varphi}(t) = \varphi(t, \eta + 1) \qquad \hat{\varphi}(t) = \varphi(t,\eta) + 1$$

It follows easily, from the assumptions (b) and (c), that since

$$\tilde{\varphi}(0) = \varphi(0, \eta + 1) = \eta + 1 = \hat{\varphi}(0)$$

$\tilde{\varphi}(t) \equiv \hat{\varphi}(t)$. Using the definition of ψ [(1.2)], one then obtains

$$\psi(\eta + 1) = \psi(\eta) + 1 \tag{1.4}$$

for all real η. The *continuous monotone increasing function* ψ is said to *represent* T. Actually, from (1.3) it is clear that T can also be represented by $\psi + n$, where n is any integer, for if $P = (0,\eta)$, then

$$P_1 = (1,\psi(\eta)) = (1,\psi(\eta) + n) = (0,\psi(\eta))$$

Also if ψ satisfies (1.4), then so does $\psi + n$.

The investigation of the nature of the solution paths of (1.1) on $\mathfrak{I}$ can now be carried out by studying the homeomorphism T and its representing real function ψ.

2. The Rotation Number

Let ψ^2 be the function defined by $\psi^2(\eta) = \psi(\psi(\eta))$, and, in general, $\psi^n(\eta) = \psi(\psi^{n-1}(\eta))$ for any integer n, where it is understood that $\psi^0(\eta) = \eta$. Similarly, define T^2 by $T^2P = T(TP)$, and the iterates T^n by

$$T^nP = T(T^{n-1}P)$$

$(n = 0, \pm 1, \pm 2, \ldots)$, where $T^0P = P$. The set of iterates of T $\{T^n, n = 0, \pm 1, \pm 2, \ldots\}$ form a group, for clearly $T^kT^m = T^{k+m} = T^mT^k$.

The function ψ^n is of the same type as ψ, that is, it is continuous, monotone increasing, and satisfies (1.4). Also ψ^n represents T^n, if ψ represents

† If f_x is continuous, this follows from Sec. 7, Chap. 1.

T. Often it will be convenient to abbreviate as follows:

$$\eta_n = \psi^n(\eta) \qquad P_n = T^n P \qquad (n = 0, \pm 1, \pm 2, \ldots)$$

Theorem 2.1. *The limit*

$$\rho = \lim_{|n| \to \infty} \frac{\eta_n}{n} \tag{2.1}$$

exists and is independent of $\eta = \eta_0$. *The number* ρ *is rational if and only if some power of* T *has a fixed point.*

Proof. First, suppose ρ exists for some $\tilde{\eta}$, and consider any other η. Then there exists an integer m such that

$$m \leqq \eta - \tilde{\eta} < m + 1 \tag{2.2}$$

Using the monotone character of ψ^n, it follows from (2.2) that $\psi^n(\tilde{\eta} + m) \leqq \psi^n(\eta) < \psi^n(\tilde{\eta} + m + 1)$, or by (1.4) applied to ψ^n, $\tilde{\eta}_n + m \leqq \eta_n < \tilde{\eta}_n + m + 1$. This is readily seen to imply that $(\eta_n - \tilde{\eta}_n)/n \to 0$, as $|n| \to +\infty$. Thus if ρ exists for $\tilde{\eta}$, it exists for η and is the same number.

Now suppose some power of T, say T^m, has a fixed point P (that is, $T^m P = P$), with coordinate η. If $T^m P = P$, then $P = T^{-m}P$, and so T^{-m} has the same fixed point. Thus it can be assumed that $m > 0$. Now $T^m P = P$ implies $\eta_m = \eta + r$, for some integer r. Thus

$$\eta_{2m} = \psi^m(\eta_m) = \psi^m(\eta + r) = \eta_m + r = \eta + 2r$$

and by induction

$$\eta_{mn} = \eta + rn \qquad (n = 0, \pm 1, \pm 2, \ldots) \tag{2.3}$$

Every integer k can be written as $k = mn + s$, where n,s are integers and $0 \leqq s < m$. Thus

$$\eta_k = \eta_{mn+s} = \psi^s(\eta_{mn}) = \psi^s(\eta + rn) = \psi^s(\eta) + rn = \eta_s + rn$$

by (2.3), and

$$\frac{\eta_k}{k} = \frac{\eta_s}{k} + \frac{rn}{k}$$

Since η_s is one of $\eta_0, \eta_1, \ldots, \eta_{m-1}$, $\eta_s/k \to 0$ as $|k| \to \infty$ and

$$\lim_{|k| \to \infty} \frac{\eta_k}{k} = \lim_{|n| \to \infty} \frac{rn}{mn + s} = \frac{r}{m}$$

Hence ρ exists, and is the rational number r/m.

Next it will be shown that if no power of T has a fixed point, then ρ still exists. To say that no power of T has a fixed point means that no integers m and r and real number η exist such that

$$\eta_m = \eta + r$$

Thus corresponding to any integer m and a particular η there exists an integer r such that

$$\eta + r < \eta_m < \eta + r + 1 \tag{2.4}$$

But by continuity considerations (2.4) must then hold for *all real* η, for the corresponding equalities can never take place. If (2.4) is applied to $\eta = 0, 0_m, 0_{2m}, \ldots, 0_{m(n-1)}$, and the inequalities added, one obtains

$$nr < 0_{mn} < n(r + 1)$$

for $n \geqq 1$ and a similar result for $n \leqq -1$. However,

$$r < 0_m < r + 1$$

and hence

$$\left| \frac{0_{mn}}{mn} - \frac{0_m}{m} \right| < \frac{2}{|m|} \tag{2.5}$$

Interchanging m and n in (2.5) and adding to (2.5), there results

$$\left| \frac{0_m}{m} - \frac{0_n}{n} \right| < \frac{2}{|m|} + \frac{2}{|n|}$$

and thus $\rho = \lim (0_m/m)$, $|m| \to \infty$, exists.

Finally it has to be proved that, if ρ is rational, some power of T has a fixed point. Let ρ be rational, that is, there exist integers k and m such that $m\rho + k = 0$. It will be shown T^m has a fixed point. Define χ by $\chi(\eta) = \psi^m(\eta) + k$. Then

$$\chi^n(\eta) = \psi^{mn}(\eta) + kn$$

and thus

$$\frac{\chi^n(\eta)}{n} = m \cdot \frac{\psi^{mn}(\eta)}{mn} + k \to m\rho + k \qquad (|n| \to \infty)$$

Clearly χ represents T^m, and $m\rho + k = 0$ stands in the same relation to χ as ρ does to ψ.

Suppose T^m has no fixed points. Then $\chi(\eta) - \eta \neq 0$ for any η, and hence it can be assumed that $\chi(\eta) > \eta$ for all real η. In particular, $\chi(0) > 0$, and by the monotone nature of χ, $\chi^n(0) > \chi^{n-1}(0) > \cdots > \chi(0) > 0$. Thus $\chi^n(0)$ is increasing in n. Moreover, $\chi^n(0) < 1$, for all n. For suppose not; then from some n onward $\chi^n(0) > 1$. Pick one such n. Then

$$\chi^{2n}(0) > \chi^n(1) = \psi^{mn}(1) + kn = \psi^{mn}(0) + kn + 1 = \chi^n(0) + 1 > 2$$

Thus $\chi^{nl}(0) > l$, and

$$\frac{\chi^{nl}(0)}{nl} > \frac{1}{n} \tag{2.6}$$

Letting $l \to \infty$ in (2.6), one obtains $m\rho + k \geqq (1/n)$, which contradicts the fact $m\rho + k = 0$. Therefore the sequence of numbers $\{\chi^n(0)\}$ is monotone increasing and bounded, and thus has a limit, say $\hat{\eta}$. Now $\chi(\hat{\eta}) = \lim \chi(\chi^n(0)) = \lim \chi^{n+1}(0) = \hat{\eta}$. Hence $\hat{\eta}$ determines a fixed point $\hat{P} = (0,\hat{\eta})$ of T^m, which is a contradiction. This proves ρ rational, $m\rho + k = 0$, implies T^m has a fixed point.

The number ρ is called the *rotation number* of T for the equation (1.1). It measures the average advance of a solution φ of (1.1) starting at $(0,\eta)$ as t changes by a unit. It is independent of the solution used to define it, and is rational if and only if there exists some solution which is periodic with an integer period.

3. The Cluster Set

Let S be the set consisting of all iterates of a point $P \in \mathfrak{C}$ under T, that is,

$$S = \{T^nP;\ n = 0,\ \pm 1,\ \pm 2,\ .\ .\ .\}$$

Denote by S' the set of cluster points of S. The set S' will be called the *cluster set* associated with T and (1.1). When ρ is rational, $m\rho + k = 0$, it follows readily from consideration of T^m that S' consists of a finite number of points. In what follows it will be assumed that ρ is irrational. Superficially S' depends on the P chosen to define S, but the following theorem shows that actually S' is independent of P, and thus the term *the cluster set* is meaningful.

Theorem 3.1. *The cluster set S' is invariant under T, that is, $TS' = S'$, and it is independent of P.*

Proof. Let $Q \in S'$. Then $Q = \lim P_{k_n}$, $k_n \to \infty$, where $P_{k_n} \in S$. Hence $TQ = \lim TP_{k_n} = \lim P_{k_n+1} = \tilde{Q} \in S'$, and therefore $TS' \subseteq S'$. Also $T^{-1}Q = \lim T^{-1}P_{k_n} = \lim P_{k_n-1} = \hat{Q} \in S'$, and so $T^{-1}S' \subseteq S'$. Hence $T(T^{-1}S') \subseteq TS'$, or $S' \subseteq TS'$, proving $TS' = S'$.

In order to prove that S' is independent of P, the following lemma will be required.

Lemma. *Let P_m, $P_n \in S$, and let α and $\tilde{\alpha}$ be the two closed arcs on $\mathfrak{C}$ with P_m and P_n as end points. Then α, $\tilde{\alpha}$ both contain at least one of the transforms Q_l of any point $Q \in \mathfrak{C}$.*

Proof. The proof will be deduced for the arc α. It is easy to see that the arcs α, $T^{m-n}\alpha$, $T^{2(m-n)}\alpha$, . . . , $T^{k(m-n)}\alpha$ are adjacent on $\mathfrak{C}$. Since ρ is irrational, the arcs which are sums of the form

$$\alpha \cup T^{m-n}\alpha \cup T^{2(m-n)}\alpha \cup \cdots \cup T^{k(m-n)}\alpha$$

must cover $\mathfrak{C}$ if k is sufficiently large. For if this were not true the set of points $\{T^{k(m-n)}P_n\}$, $(k = 0, 1, 2, . . .)$, would be monotone and bounded on $\mathfrak{C}$, and thus possess a limit point $\tilde{P}$. But then

$$T^{m-n}\tilde{P} = \lim T^{m-n}(T^{k(m-n)}P_n) = \lim T^{(k+1)(m-n)}P_n = \tilde{P}$$

$k \to \infty$, and therefore $\tilde{P}$ would be a fixed point for T^{m-n}, which is a contradiction. Therefore, for some l, $Q \in T^{l(m-n)}\alpha$, or $T^{l(n-m)}Q \in \alpha$, or $Q_{l(n-m)} \in \alpha$, which proves the lemma.

To continue the proof of Theorem 3.1, let S'_P denote the cluster set associated with T and defined using the point P, whereas let S'_Q denote the corresponding cluster set defined by using any other point Q on $\mathfrak{C}$. If $R \in S'_P$, then there exists a sequence of transforms $P_\nu \to R$, where ν runs through some sequence of integers. By the lemma, there exists a transform Q_μ of Q on the short arc joining any two successive P_ν. Thus there exists a sequence Q_μ of transforms of Q such that $Q_\mu \to R$, where μ runs through some sequence of integers. Hence $R \in S'_Q$, which proves $S'_P \subseteq S'_Q$, and by symmetry this shows S' is independent of P.

Theorem 3.2. *The cluster set S' is perfect, and either*

(a) $S' = \mathfrak{C}$, *or*
(b) S' *is nowhere dense on* $\mathfrak{C}$.

REMARK. In the case (a) T is called *ergodic;* the case (b) is called the *singular case.*

Proof of Theorem 3.2. Since S' is closed, $S'' = (S')' \subseteq S'$. To show $S' \subseteq S''$, choose Q in the last paragraph of the proof of Theorem 3.1 to be a point in S'. Then any point $R \in S'$ is a limit of transforms Q_μ of $Q \in S'$, that is, $R \in S''$. Hence $S'' = S'$, and thus S' is perfect.

The set S is either everywhere dense on $\mathfrak{C}$ or nowhere dense on $\mathfrak{C}$, for if S is dense on an arc of $\mathfrak{C}$, this arc can be assumed to be the α of the previous lemma. Then the sum of a finite number of transforms of α covers $\mathfrak{C}$. Further, S is everywhere dense on $\mathfrak{C}$ if and only if S' is, and this completes the proof of the theorem.

4. The Ergodic Case

From a practical point of view it is desirable to know when the singular case cannot occur. The following theorem, originally due to Denjoy, gives a sufficient condition in order that T be ergodic. The theorem itself was conjectured by Poincaré in the case where f [of (1.1)] is analytic. The variation of the Denjoy proof given here is due to van Kampen.

Theorem 4.1. *Suppose ψ possesses a continuous first derivative $\psi' > 0$ on $0 \leqq \eta \leqq 1$, and ψ' is of bounded variation there. If no power of T has a fixed point, then T is ergodic.*

Proof. Let P be a point on $\mathfrak{C}$, and α an arc with P as end point. Let n be the positive integer such that either P_n or P_{-n} is the only point P_k, $|k| \leqq n$, in the interior of α. Given any integer $N > 0$, then α can be taken small enough so that there is such an $n \geqq N$. Suppose $P_{-n} \in \alpha$; the case $P_n \in \alpha$ can be dealt with similarly.

Lemma. *The two finite sequences*

$$P_0,\ P_1,\ \ldots,\ P_{n-1} \text{ and } P_{-n},\ P_{1-n},\ \ldots,\ P_{-1}$$

alternate on $\mathfrak{C}$.

Proof. Choose the arc $\widehat{P_0P_{-n}}$ which lies in α. It has to be shown that if $0 \leqq k < n$, no point of either of the two sequences is in the *interior* of the arc $\widehat{P_kP_{k-n}}$ on $\mathfrak{C}$ that has the same orientation as $\widehat{P_0P_{-n}}$; that is, the arcs $\widehat{P_kP_{k-n}}$ $(k = 0, 1, \ldots, n-1)$ are disjoint arcs. Suppose the contrary; then for some $l = -n, \ldots, n-1$, $P_l \in \widehat{P_kP_{k-n}}$. There are two cases according as

(i) $k - n \leqq l < n$, or
(ii) $-n \leqq l < k - n < 0$.

Consider Case (i). Then $P_l \in \widehat{P_kP_{k-n}}$ implies, using the orientation-preserving nature of T^{-k}, that $T^{-k}P_l \in \widehat{P_0P_{-n}}$, or $P_{l-k} \in \widehat{P_0P_{-n}}$. But $-n \leqq l - k \leqq l < n$, and hence $P_{l-k} \in \widehat{P_0P_{-n}}$ has been excluded by the choice of n. For Case (ii), $0 \leqq l + n < k$, and hence by Case (i) P_{l+n} is not in $\widehat{P_kP_{k-n}}$. Since P_l is and P_{l+n} is not in $\widehat{P_kP_{k-n}}$, it follows that $\widehat{P_{l+n}P_l}$ and $\widehat{P_kP_{k-n}}$ overlap and thus $P_k \in \widehat{P_{l+n}P_l}$. Applying $T^{-(l+n)}$ to this relation, one obtains $P_{k-l-n} \in \widehat{P_0P_{-n}}$. But this is impossible since, by (ii), $0 < k - l - n < -l \leqq n$. Thus the lemma is proved.

Since ψ' is of bounded variation on $0 \leqq \eta \leqq 1$, and $\psi' > 0$, the function Ψ defined by $\Psi(\eta) = \log \psi'(\eta)$ is defined on $0 \leqq \eta \leqq 1$, and is of bounded variation there. Further, Ψ has period 1, for ψ satisfies (1.4). Let the point P (chosen prior to the above lemma) have coordinate η. Then from the lemma it follows that, since the arcs $\widehat{P_kP_{k-n}}$ $(k = 0, 1, \ldots, n-1)$ are disjoint,

$$\left| \sum_{k=0}^{n-1} (\Psi(\eta_k) - \Psi(\eta_{k-n})) \right| \leqq V \tag{4.1}$$

where V is the total variation of Ψ over $[0,1]$. But

$$\sum_{k=0}^{n-1} \Psi(\eta_k) = \log \left(\prod_{k=0}^{n-1} \psi'(\eta_k) \right) = \log \frac{d\psi^n}{d\eta}(\eta)$$

Similarly

$$\sum_{k=0}^{n-1} \Psi(\eta_{k-n}) = -\log \frac{d\psi^{-n}}{d\eta}(\eta)$$

and therefore, from (4.1),

$$\left| \log \left(\frac{d\psi^n}{d\eta}(\eta) \cdot \frac{d\psi^{-n}}{d\eta}(\eta) \right) \right| \leqq V$$

or

$$e^{-V} \leqq \frac{d\psi^n}{d\eta}(\eta) \cdot \frac{d\psi^{-n}}{d\eta}(\eta) \leqq e^V \tag{4.2}$$

Now η is any number on $0 \leqq \eta \leqq 1$, and hence (4.2) holds for *any* η, $0 \leqq \eta \leqq 1$, and for *arbitrarily large* integral n.

Now let β be any arc on $\mathcal{C}$ with length s. Suppose $T^k\beta$ has length s_k. Then, if we make the nonessential assumption that the radius of $\mathcal{C}$ is $1/2\pi$.

$$s_k = \int_\beta \frac{d\psi^k}{d\eta}\, d\eta \qquad s_{-k} = \int_\beta \frac{d\psi^{-k}}{d\eta}\, d\eta$$

and therefore

$$s_k + s_{-k} = \int_\beta \left(\frac{d\psi^k}{d\eta} + \frac{d\psi^{-k}}{d\eta} \right) d\eta \geqq 2 \int_\beta \left(\frac{d\psi^k}{d\eta} \cdot \frac{d\psi^{-k}}{d\eta} \right)^{\frac{1}{2}} d\eta \geqq 2se^{-V/2}$$

for those k satisfying (4.2). Hence $s_k + s_{-k} \not\to 0$ as $k \to \infty$.

Suppose the open set $\mathcal{C} - S'$ is not empty. Then consider an open arc $\beta \subseteq \mathcal{C} - S'$, with end points in S'. Since $TS' = S'$, and since T is orientation-preserving, all the transforms $T^k\beta$ $(k = 0, \pm 1, \pm 2, \ldots)$ are in $\mathcal{C} - S'$. Clearly no two arcs of $\{T^k\beta\}$ can overlap, for the end points must be in S'. Moreover, no arc can be mapped onto another arc, for an end point of one would be mapped onto an end point of the other and thus ρ would be rational. Thus the arcs $\{T^k\beta\}$ are disjoint and so, as $k \to \infty$, $s_k + s_{-k} \to 0$, which contradicts the result of the previous paragraph. Therefore $\mathcal{C} - S'$ must be empty, that is, T is ergodic.

It remains to give sufficient conditions on f in (1.1) in order that ψ satisfy the conditions of Theorem 4.1.

Theorem 4.2. *Let f in the differential equation* (1.1) $x' = f(t,x)$ *satisfy, besides the assumptions* (a) *through* (c) *below* (1.1), *the hypothesis that $\partial f/\partial x$ exists, is continuous, and is of bounded variation with respect to x for $0 \leqq x \leqq 1$, uniformly in t. Then if no power of T has a fixed point, T is ergodic.*

REMARK: The assumptions on f are satisfied if (a) through (c) below (1.1) hold, and if $\partial^2 f/\partial x^2$ exists and is continuous.

Proof of Theorem 4.2. It will be shown ψ' is positive, continuous, and of bounded variation.

First, since $\partial f/\partial x$ exists and is continuous in (t,x), it follows from Theorem 7.2, Chap. 1, that the solution $\varphi = \varphi(t,\eta)$ of (1.1) such that $\varphi(0,\eta) = \eta$, is a continuously differentiable function of η, and in fact

$$\frac{\partial \varphi}{\partial \eta}(t,\eta) = \exp\left[\int_0^t \frac{\partial f}{\partial x}(\tau,\varphi(\tau,\eta))\, d\tau\right] \tag{4.3}$$

In particular, at $t = 1$, (4.3) gives, since $(\partial\varphi/\partial\eta)(1,\eta) = \psi'(\eta)$,

$$\psi'(\eta) = \exp\left[\int_0^1 \frac{\partial f}{\partial x}(\tau,\varphi(\tau,\eta))\, d\tau\right] \tag{4.4}$$

Therefore $\psi'(\eta)$ exists, and from (4.4) $\psi'(\eta) > 0$.

If

$$g(\eta) = \int_0^1 \frac{\partial f}{\partial x}(\tau,\varphi(\tau,\eta))\, d\tau$$

then clearly, if g is of bounded variation on $0 \leqq \eta \leqq 1$, $\psi'(\eta) = \exp g(\eta)$ will have the same property. To show g is of bounded variation, let $0 = \eta_0 < \eta_1 < \cdots < \eta_n = 1$ be a subdivision of $0 \leqq \eta \leqq 1$. Then

$$\sum_{k=0}^{n-1} |g(\eta_{k+1}) - g(\eta_k)| \leqq \int_0^1 \sum_{k=0}^{n-1} \left| \frac{\partial f}{\partial x}(\tau,\varphi(\tau,\eta_{k+1})) - \frac{\partial f}{\partial x}(\tau,\varphi(\tau,\eta_k)) \right| d\tau \tag{4.5}$$

For fixed t, $0 < t \leqq 1$, the transformation ψ_t defined by

$$\psi_t(\eta) = \varphi(t,\eta) \qquad (0 \leqq \eta \leqq 1)$$

is clearly a homeomorphism of the interval $0 \leqq \eta \leqq 1$ onto a real interval. Since ψ is orientation-preserving, so is ψ_t. Thus if

$$0 = \eta_0 < \eta_1 < \cdots < \eta_n = 1$$

is a subdivision of $0 \leqq \eta \leqq 1$, then

$$\psi_t(0) < \psi_t(\eta_1) < \cdots < \psi_t(\eta_n)$$

or

$$\varphi(t,0) < \varphi(t,\eta_1) < \cdots < \varphi(t,1)$$

is a subdivision of the interval $\varphi(t,0) \leqq x \leqq \varphi(t,1)$. Since $\partial f/\partial x$ is of bounded variation in x *uniformly in* t,

$$\sum_{k=0}^{n-1} \left| \frac{\partial f}{\partial x}(t,\varphi(t,\eta_{k+1})) - \frac{\partial f}{\partial x}(t,\varphi(t,\eta_k)) \right| \leqq V$$

for some constant V, independent of t, $0 \leqq t \leqq 1$. From this it follows from (4.5) that g is of bounded variation on $0 \leqq \eta \leqq 1$, completing the proof of the theorem.

5. Characterization of Solutions in the Ergodic Case

In the ergodic case the solutions of (1.1) have a particular structure. In order to see this, the relationship between the rotation number ρ of T and the set of iterates $\psi^n(\eta)$ of η will be investigated in more detail.

Theorem 5.1. *Suppose ρ is irrational, and η is a fixed real number. Then the number systems A: $\{n\rho + m\}$, and B: $\{\eta_n + m\}$, where m, n are arbitrary integers, are in a one-to-one monotone correspondence.*

Proof. If $n\rho + m$ is associated with $\eta_n + m$, $(\eta_n = \psi^n(\eta))$, then clearly the correspondence is one-to-one because of the irrationality of ρ and the consequent absence of fixed points of T.

Secondly, the order of the elements in B does not depend on the η chosen, that is, if $\eta_n + m < \eta_k + l$, then $\zeta_n + m < \zeta_k + l$, for any real ζ. Equivalently, $\eta_n - \eta_k < l - m$ implies $\zeta_n - \zeta_k < l - m$. The latter follows since the continuous function χ of η, defined by

$$\chi(\eta) = \psi^n(\eta) - \psi^k(\eta) = \eta_n - \eta_k$$

can never be an integer, for ρ is irrational.

Hence it suffices to prove the result for $\eta = 0$. If p,q,r are integers such that

$$p \leqq 0_q \leqq r \qquad (0_q = \psi^q(0)) \tag{5.1}$$

then it will be shown that

$$p < q\rho < r \tag{5.2}$$

To prove (5.2), apply ψ^q to (5.1), obtaining $\psi^q(p) \leqq 0_{2q} \leqq \psi^q(r)$, or $0_q + p \leqq 0_{2q} \leqq 0_q + r$, using (1.4). Thus by (5.1) $2p \leqq 0_{2q} \leqq 2r$, and by repeating it follows for any integer $l > 0$ that $lp \leqq 0_{lq} \leqq lr$, or

$$p \leqq q\frac{0_{lq}}{lq} \leqq r \tag{5.3}$$

Letting $l \to \infty$ in (5.3) one obtains $p \leqq q\rho \leqq r$. But since ρ is irrational, neither equality can occur. Hence (5.2) is proved.

Now if $0_n + m < 0_k + l$, it is to be shown that $n\rho + m < k\rho + l$, and conversely. Applying ψ^{-k} to the first inequality, one sees that what has to be proved is that $0_{n-k} < l - m$ is equivalent to $(n - k)\rho < l - m$. From (5.1) and (5.2) $0_{n-k} < l - m$ implies $(n - k)\rho < l - m$. If $(n - k)\rho < l - m$, and $0_{n-k} \geqq l - m$, then by (5.1) and (5.2) $(n - k)\rho > l - m$, which is a contradiction. This completes the proof of the theorem.

The result of Theorem 5.1 leads immediately to a geometric justification of the term "rotation number" for ρ. Indeed, the following theorem shows that if T is ergodic then T is topologically equivalent to a rotation of the circle $\mathfrak{C}$ by an angle $2\pi\rho$. By this is meant that there exists a homeomorphism H of $\mathfrak{C}$ onto $\mathfrak{C}$ such that $HT = RH$, where R is the rotation of $\mathcal{C}$ by an angle $2\pi\rho$. In terms of representing real functions, this means that there exists a continuous monotone increasing real function h defined for all real y possessing the properties

$$h(y + 1) = h(y) + 1 \tag{5.4}$$
$$h(\psi(y)) = h(y) + \rho \tag{5.5}$$

The continuous increasing nature of h combined with (5.4) implies h represents a homeomorphism H of $\mathcal{C}$ onto $\mathcal{C}$, and (5.5) just says that $HT = RH$, where R, which is represented by the real function

$$r(y) = y + \rho$$

is the rotation of $\mathcal{C}$ by an angle $2\pi\rho$.

Theorem 5.2. *If T is ergodic, then T is topologically equivalent to a rotation of the circle $\mathcal{C}$ by an angle $2\pi\rho$.*

Proof. Let η be a fixed real number and consider the sets A and B defined in Theorem 5.1. If $y \in B$ and $y = \eta_n + m$, define $h_0(y) = n\rho + m$. This defines a real function h_0 on B with values in A, which is monotone increasing. Also A is dense on the real line because ρ is irrational. From this it follows that h_0 is continuous on B. Since T is ergodic, B is dense on the real numbers. Therefore h_0 can be extended in a unique way to a continuous increasing real function $h = h(y)$ defined for *all real* y.

Now suppose $y \in B$, and $y = \eta_n + m$. Then $h(y) = n\rho + m$, and it is clear that $h(y) + 1 = n\rho + (m + 1)$ must correspond to

$$y + 1 = \eta_n + (m + 1)$$

that is, for $y \in B$, (5.4) must hold. But, by continuity, (5.4) must hold for all real y. Similarly, if $y \in B$, $y = \eta_n + m$, then

$$h(y) + \rho = (n + 1)\rho + m$$

must correspond to $\eta_{n+1} + m = \psi(\eta_n) + m = \psi(\eta_n + m) = \psi(y)$. Thus, for $y \in B$, (5.5) must be valid, and by continuity (5.5) holds for all real y. From the remarks just preceding the theorem, it is clear that h represents a homeomorphism H of $\mathcal{C}$ onto $\mathcal{C}$ which establishes the desired topological equivalence.

The importance of Theorems 5.1 and 5.2 from the point of view of differential equations is that they prepare the way for a simple characterization of the solutions of (1.1).

Theorem 5.3 (Bohl). *If T is ergodic, there exists a function $\omega = \omega(t,z)$ which is continuous in (t,z), and periodic*

$$\omega(t + 1, z) = \omega(t, z + 1) = \omega(t,z) \tag{5.6}$$

such that every solution φ *of* (1.1) *can be written in the form*

$$\varphi(t) = t\rho + c + \omega(t, t\rho + c) \tag{5.7}$$

where c is a constant, and ρ *is the rotation number. Conversely, for any constant c,* (5.7) *is a solution of* (1.1) *and to each c,* $0 \leq c < 1$, *there corresponds a unique* $\varphi(0)$ (mod 1). *Indeed,* $c = h(\varphi(0))$.

Proof. Let η be any real number, and let $\varphi = \varphi(t,\eta)$ be the solution of (1.1) with $\varphi(0,\eta) = \eta$. Because $f(t, x + 1) = f(t,x)$

$$\varphi(t, \eta + 1) = \varphi(t,\eta) + 1 \tag{5.8}$$

With $\psi(\eta) = \varphi(1,\eta)$ and because $f(t + 1, x) = f(t,x)$

$$\varphi(t + 1, \eta) = \varphi(t,\psi(\eta)) \tag{5.9}$$

Let $h(\eta) = c$. Then, because h is monotone increasing and continuous, $\eta = g(c)$, where g is monotone increasing and continuous. Moreover, by (5.4) and (5.5),

$$g(c + 1) = g(c) + 1 \tag{5.10}$$
$$\psi(g(c)) = g(c + \rho) \tag{5.11}$$

Let $\tilde{\varphi}(t,c) = \varphi(t,g(c))$. Then, by (5.8) and (5.10),

$$\tilde{\varphi}(t, c + 1) = \tilde{\varphi}(t,c) + 1 \tag{5.12}$$

By (5.9) and (5.11)

$$\tilde{\varphi}(t + 1, c) = \varphi(t + 1, g(c)) = \varphi(t,\psi(g(c))) = \varphi(t,g(c + \rho)) = \tilde{\varphi}(t, c + \rho)$$

so that

$$\tilde{\varphi}(t + 1, c) = \tilde{\varphi}(t, c + \rho) \tag{5.13}$$

Let $\omega(t,z) = \tilde{\varphi}(t, z - t\rho) - z$ for all real t and z. Then, by (5.12) and (5.13),

$$\omega(t + 1, z) = \omega(t, z + 1) = \omega(t,z)$$

Clearly, with $z = t\rho + c$,

$$\tilde{\varphi}(t,c) = t\rho + c + \omega(t, t\rho + c)$$

which proves the theorem.

6. A System of Two Equations

With little change in argument, the pair of equations

$$x_1' = f_1(x_1,x_2) \qquad x_2' = f_2(x_1,x_2) \tag{6.1}$$

may be considered, where it is assumed that f_1,f_2 are real continuous functions defined for all real x_1,x_2 satisfying

$$f_j(x_1 + 1, x_2) = f_j(x_1, x_2 + 1) = f_j(x_1,x_2) \qquad (j = 1, 2)$$
$$f_1^2 + f_2^2 \neq 0$$

Suppose that through each point of the (x_1,x_2) plane passes a unique solution of (6.1), except for translations in t, and that (6.1) has no periodic solutions.

The coordinates (x_1,x_2) may be taken on a torus $\mathfrak{J}$. The role of the circle $\mathfrak{C}$ in the previous sections is taken by the curve $\mathcal{K}$ which will now be defined. Consider the orthogonal system of equations

$$x_1' = -f_2(x_1,x_2) \qquad x_2' = f_1(x_1,x_2) \tag{6.2}$$

Let $\psi = (\psi_1,\psi_2)$ be a solution of (6.2). If it is a closed curve on $\mathfrak{J}$, it may be taken as $\mathcal{K}$. If ψ is not a closed curve, then the points $\psi(t)$, $t = 1, 2, \ldots$, have at least one limit point P on $\mathfrak{J}$. Consider a small curvilinear rectangle R with P in its interior and with its edges consisting of two arcs P_1P_2 and P_3P_4 which are solutions of (6.1) and two arcs P_1P_3 and P_2P_4 which are solutions of (6.2). Let the arcs P_1P_2 and P_1P_3 have equal length on $\mathfrak{J}$ and let R be taken small enough so that the change in direction of the vector with components (f_1,f_2) in R is less than $\pi/100$. Let the solution ψ first intersect arc P_1P_2 (or arc P_3P_4) at $t = t_0$ and leave R for increasing $t > t_0$. Let it return to R for the first time for $t = t_1$ when it meets arc P_3P_4 (or P_1P_2). The curve $\psi(t)$, $t_0 \leqq t \leqq t_1$, can now be closed by an arc from $\psi(t_1)$ to $\psi(t_0)$ which will be entirely in R, will make an angle less in magnitude than $\pi/3$ with the solutions of (6.2) in R, and which can have continuous first derivatives at $\psi(t_0)$ and $\psi(t_1)$ (and also continuous second derivatives if f_1 and f_2 are of class C^1). The closed curve so defined is called $\mathcal{K}$.

$\mathcal{K}$ cannot be continuously deformed to a point since this would imply a critical point on $\mathfrak{J}$ because the index of $\mathcal{K}$ with respect to the vector field (f_1,f_2) is one. Any solution of (6.1) must cut $\mathcal{K}$ as t increases (or decreases). For let this not be the case for a semiorbit C^+ of (6.1). The torus cut along $\mathcal{K}$ is an annulus. Thus, since C^+ does not intersect $\mathcal{K}$, it cannot approach $\mathcal{K}$ and thus C^+ and $L(C^+)$ are in the annulus. The Poincaré-Bendixson argument can be applied to show that $L(C^+)$ is a closed orbit. But, by hypothesis, (6.1) has no periodic orbits. Thus all semiorbits of (6.1) cut $\mathcal{K}$. In particular, all solutions starting on $\mathcal{K}$ return to $\mathcal{K}$. This defines a homeomorphism of $\mathcal{K}$ into itself and the previous results proved for $\mathfrak{C}$ now apply also to $\mathcal{K}$.

REFERENCES

General

A. A. Andronow and C. E. Chaikin, *Theory of oscillations*, Princeton, 1949.

R. Bellman, *Stability theory of differential equations*, New York, 1953.

L. Bieberbach, *Theorie der Differentialgleichungen*, Berlin, 1930; reprinted, New York, 1944.

———, *Theorie der gewöhnlichen Differentialgleichungen auf funktiontheoretischer Grundlage dargestellt*, Berlin, 1953.

A. R. Forsyth, *A treatise on differential equations*, 6th ed., London, 1929.

É. Goursat, *Cours d'analyse mathématique*, vols. 2 and 3, Paris, 1915.

J. Horn, *Gewöhnliche Differentialgleichungen*, 2d ed., Berlin and Leipzig, 1927.

W. Hurewicz, *Ordinary differential equations in the real domain with emphasis on geometric methods* (mimeographed), Brown University, Providence, R.I., 1943.

E. L. Ince, *Ordinary differential equations*, London, 1927.

E. Kamke, *Differentialgleichungen reeller Funktionen*, Leipzig, 1930; reprinted, New York, 1947.

———, *Differentialgleichungen, Lösungsmethoden und Lösungen*, Leipzig, 1943.

N. Kryloff and N. Bogoliuboff, *Introduction to non-linear mechanics*, Princeton, 1943.

S. Lefschetz (ed.), *Contributions to the theory of non-linear oscillations*, vol. 1 (1950), vol. 2 (1952), Princeton.

———, *Lectures on differential equations*, Princeton, 1946.

N. Minorsky, *Introduction to non-linear mechanics*, Ann Arbor, 1947.

F. R. Moulton, *Differential equations*, New York, 1930.

V. V. Nemitzky and V. V. Stepanov, *Qualitative theory of differential equations* (Russian), 2d ed., Moscow, 1949.

É. Picard, *Leçons sur quelques problèmes aux limites de la théorie des équations différentielles*, Paris, 1930.

E. G. C. Poole, *Introduction to the theory of linear differential equations*, Oxford, 1936.

G. Sansone, *Equazioni differenziali nel campo reale*, part 1, 2d ed., Bologna, 1948; part 2, 2d ed., Bologna, 1949.

L. Schlesinger, *Vorlesungen über lineare Differentialgleichungen*, Leipzig, 1908.

J. J. Stoker, *Non-linear vibrations*, New York, 1950.

E. C. Titchmarsh, *Eigenfunction expansions associated with second-order differential equations*, Oxford, 1946.

G. Valiron, *Equations fonctionnelles; applications* (*Cours d'analyse* II) Paris, 1945.

Chapter 1

T. H. Gronwall, *Note on the derivatives with respect to a parameter of the solutions of a system of differential equations*, Ann. of Math. ser. 2 vol. 20 (1919) pp. 292–296.

M. Müller, *Ueber die Eindeutigkeit der Integrale eines Systems gewöhnlicher Differentialgleichungen und die Konvergenz einer Gattung von Verfahren zur Approximation dieser Integrale*, Sitzungsberichte Heidelberg (1927) 9 Abhandlung.

Ch. J. de la Vallée Poussin, *Cours d'analyse infinitésimale* II, reprinted, New York, 1946.

G. Peano, *Démonstration de l'intégrabilité des équations différentielles ordinaires,* Math. Ann. vol. 37 (1890) pp. 182–228.

É. Picard, *Mémoire sur la théorie des équations aux dérivées partielles et la méthode des approximations successives,* J. de Math. ser. 4 vol. 6 (1890) pp. 145–210.

E. Lindelöf, *Démonstration de quelques théorèmes sur les équations différentielles,* J. de Math. ser. 5 vol. 6 (1900) pp. 423–441.

———, *Sur l'application des méthodes d'approximations successives à l'étude des intégrales réeles des équations différentielles ordinaires,* J. de Math. ser. 4 vol. 10 (1894) pp. 117–128.

A. Wintner, *On the exact value of the bound of regularity of solutions of ordinary differential equations,* Amer. J. Math. vol. 57 (1935) pp. 539–540.

Chapter 2

C. Carathéodory, *Vorlesungen über reelle Funktionen,* Leipzig, 1927; reprinted, New York, 1948, pp. 665–688.

E. A. Coddington and N. Levinson, *Uniqueness and convergence of successive approximations,* J. Indian Math. Soc. vol. 16 (1952) pp. 75–81.

E. R. van Kampen, *Remarks on systems of ordinary differential equations,* Amer. J. Math. vol. 59 (1937) pp. 144–152.

M. Müller, *Uber das Fundamentaltheorem in der Theorie der gewöhnlichen Differentialgleichungen,* Math. Zeit. vol. 26 (1927) pp. 619–645.

G. Peano, *Sull' integrabilità delle equazioni differenziali di primo ordine,* Atti Accad. Sci. Torino vol. 21 (1885–1886) pp. 677–685.

O. Perron, *Ein neuer Existenzbeweis für die Integrale eines Systems gewöhnlicher Differentialgleichungen,* Math. Ann. vol. 78 (1917) pp. 378–384.

B. Viswanatham, *The general uniqueness theorem and successive approximations,* J. Indian Math. Soc. vol. 16 (1952) pp. 69–74.

A. Wintner, *On the convergence of successive approximations,* Amer. J. Math. vol. 68 (1946) pp. 13–19.

———, *The infinities in the non-local existence problem of ordinary differential equations.* Amer. J. Math vol. 68 (1946) pp. 173–178.

Chapter 3

G. Darboux, *Leçons sur la théorie générale des surfaces,* 2d ed., Paris, 1915, vol. 2 pp. 112–149.

O. Dunkel, *Regular singular points of a system of homogeneous linear differential equations of the first order,* Proc. Amer. Acad. Arts Sci. vol. 38 (1912–1913) pp. 341–370.

G. Floquet, *Sur les équations différentielles linéaires à coefficients périodiques,* Ann. École Norm. ser. 2 vol. 12 (1883) pp. 47–89.

A. R. Forsyth, *Theory of differential equations,* vol. 4, Cambridge, 1902.

L. Fuchs, *Zur Theorie der linearen Differentialgleichungen mit veränderlichen Koeffizienten,* J. für Math. vol. 66 (1866) pp. 121–160; vol. 68 (1868) pp. 354–385.

N. Levinson, *The asymptotic nature of the solutions of linear systems of differential equations,* Duke Math. J. vol. 15 (1948) pp. 111–126.

L. Schlesinger, *Vorlesungen über lineare Differentialgleichungen,* Leipzig, 1908.

Chapter 4

G. D. Birkhoff, *A simplified treatment of the regular singular point,* Trans. Amer. Math. Soc. vol. 11 (1910) pp. 199–202.

A. R. Forsyth, *Theory of differential equations*, vol. 4, Cambridge, 1902.

G. Frobenius, *Ueber die Integration der linearen Differentialgleichungen durch Reihen*, J. für Math. vol. 76 (1873) pp. 214–235.

L. Fuchs, *Zur Theorie der linearen Differentialgleichungen mit veränderlichen Koeffizienten*, J. für Math. vol. 66 (1866) pp. 121–160; vol. 68 (1868) pp. 354–385.

B. Riemann, *Collected Works*, 2 ed., Leipzig, 1892, reprinted, New York, 1953, p. 379.

A. Schmidt, *Neuer Beweis eines Hauptsatzes über Bestimmtheitsstellen linearer Differentialgleichungssysteme*, J. für Math. vol. 179 (1938) pp. 1–4.

Chapter 5

G. D. Birkhoff, *Singular points of ordinary linear differential equations*, Trans. Amer. Math. Soc. vol. 10 (1909) pp. 436–470.

É. Borel, *Mémoire sur les séries divergentes*, Ann. École Norm. ser. 3 vol. 16 (1899) pp. 9–136.

F. T. Cope, *Formal solutions of irregular linear differential equations* I, Amer. J. Math. vol. 56 (1934) pp. 411–437; II, Amer. J. Math. vol. 58 (1936) pp. 130–140.

J. Horn, *Laplacesche Integrale, Binomialkoeffizientenreihen und Gammaquotientenreihen in der Theorie der linearen Differentialgleichungen*, Math. Zeit. vol. 21 (1924) pp. 85–95.

J. A. Lappo-Danilevsky, *Mémoires sur la théorie des systèmes des équations différentielles linéaires*, Trav. Inst. Phys.-Math. Stekloff vol. 1 (1934); vol. 2 (1935); vol. 3 (1936).

H. Poincaré, *Sur les intégrales irregulières des équations linéaires*, Acta Math. vol. 8 (1886) pp. 295–344.

W. Sternberg, *Ueber die asymptotische Integration von Differentialgleichungen*, Math. Ann. vol. 81 (1920) pp. 119–186.

W. J. Trjitzinsky, *Analytic theory of linear differential equations*, Acta Math. vol. 62 (1933) pp. 167–226.

———, *Singular point problems in the theory of linear differential equations*, Bull. Amer. Math. Soc. vol. 44 (1938) pp. 209–223.

Chapter 6

G. D. Birkhoff, *On the asymptotic character of the solutions of certain linear differential equations containing a parameter*, Trans. Amer. Math. Soc. vol. 9 (1908) pp. 219–231.

R. E. Langer, *The asymptotic solutions of ordinary linear differential equations of the second order, with special reference to a turning point*, Trans. Amer. Math. Soc. vol. 67 (1949) pp. 461–490.

P. Noaillon, *Développements asymptotiques dans les équations différentielles linéaires à paramètre variable*, Mém. Soc. Sci. de Liége ser. 3 vol. 11 (1912) 197 pp.

L. Schlesinger, *Über asymptotische Darstellungen der Lösungen linearer Differentialsysteme als Funktionen eines Parameters*, Math. Ann. vol. 63 (1907) pp. 277–300.

W. J. Trjitzinsky, *Theory of linear differential equations containing a parameter*, Acta Math. vol. 67 (1936) pp. 1–50.

———, *Singular point problems in the theory of linear differential equations*, Bull. Amer. Math. Soc. vol. 44 (1938) pp. 209–223.

H. L. Turrittin, *Asymptotic expansions of solutions of systems of ordinary linear differential equations containing a parameter*, Contributions to the theory of non-linear oscillations, vol. 2, Princeton, 1952, pp. 81–116.

W. Wasow, *Asymptotic solution of the differential equation of hydrodynamic stability in a domain containing a transition point*, Ann. of Math. vol. 58 (1953) pp. 222–252

Chapter 7

G. A. Bliss, *A boundary value problem for a system of ordinary linear differential equations of the first order,* Trans. Amer. Math. Soc. vol. 28 (1926) pp. 561–589.

A. Kneser, *Untersuchungen über die Darstellung willkürlicher Funktionen in der mathematischen Physik,* Math. Ann. vol. 58 (1904) pp. 81–147.

———, *Die Theorie der Integralgleichungen und die Darstellung willkürlicher Funktionen in der mathematischen Physik,* Math. Ann. vol. 63 (1907) pp. 477–524.

D. Hilbert, *Grundzüge einer allgemeinen Theorie der linearen Integralgleichungen,* part 2, Nachr. Ges. Wiss. Göttingen (1904) pp. 213–259.

W. T. Reid, *A new class of self-adjoint boundary value problems,* Trans. Amer. Math. Soc. vol. 52 (1942) pp. 381–425.

Chapter 8

G. D. Birkhoff, *Existence and oscillation theorem for a certain boundary value problem,* Trans. Amer. Math. Soc. vol. 10 (1909) pp. 259–270.

M. Bôcher, *Leçons sur les méthodes de Sturm,* Paris, 1917.

G. Hamel, *Über die lineare Differentialgleichung zweiter Ordnung mit periodischen Koeffizienten,* Math. Ann. vol. 73 (1913) pp. 371–412.

O. Haupt, *Über lineare homogene Differentialgleichungen zweiter Ordnung mit periodischen Koeffizienten,* Math. Ann. vol. 79 (1918) pp. 278–285.

E. Kamke, *Neue Herleitung der Oszillationsätze fur die linearen selbstadjungierten Randwertaufgaben zweiter Ordnung,* Math. Zeit. vol. 44 (1938) pp. 635–658.

M. J. O. Strutt, *Lamésche, Mathieusche und verwandte Funktionen in Physik und Technik,* Berlin, 1932.

———, *Reelle Eigenwerte verallgemeinerter Hillscher Eigenwertaufgaben 2. Ordnung,* Math. Zeit. vol. 49 (1943–44) pp. 593–643.

C. Sturm, *Sur les équations différentielles linéares du second ordre,* J. de Math. vol. 1 (1836) pp. 106–186.

Chapter 9

P. Hartman and A. Wintner, various papers in Amer. J. Math. 1945–1951.

K. Kodaira, *The eigenvalue problem for ordinary differential equations of the second order and Heisenberg's theory of S-matrices,* Amer. J. Math. vol. 71 (1949) pp. 921–945.

N. Levinson, *A simplified proof of the expansion theorem for singular second order linear differential equations,* Duke Math. J. vol. 18 (1951) pp. 57–71; addendum to above vol. 18 (1951) pp. 719–722.

M. H. Stone, *Linear transformations in Hilbert space and their applications to analysis,* Amer. Math. Soc. Colloquium Publications, vol. 15 (1932).

E. C. Titchmarsh, *Eigenfunction expansions associated with second-order differential equations,* Oxford, 1946.

H. Weyl, *Ueber gewöhnliche lineare Differentialgleichungen mit singulären Stellen und ihre Eigenfunktionen,* Nachr. Ges. Wiss. Göttingen (1909) pp. 37–64.

———, *Ueber gewöhnliche Differentialgleichungen mit Singularitäten und die zugehörigen Entwicklungen willkurlicher Funktionen,* Math. Ann. vol. 68 (1910) pp. 220–269.

———, *Ueber gewöhnliche Differentialgleichungen mit singulären Stellen und ihre Eigenfunktionen,* Nachr. Ges. Wiss. Göttingen (1910) pp. 442–467.

K. Yosida, *On Titchmarsh-Kodaira's formula concerning Weyl-Stone's eigenfunction expansion,* Nagoya Math. J. vol. 1 (1950) pp. 49–58.

Chapter 10

E. A. Coddington, *The spectral representation of ordinary self-adjoint differential operators,* Ann. of Math. vol. 60 (1954) pp. 192-211.

———, *The spectral matrix and Green's function for singular self-adjoint boundary value problems,* Canadian J. Math. vol. 6 (1954) pp. 169–185.

I. Halperin, *Closures and adjoints of linear differential operators,* Ann. of Math. vol. 38 (1937) pp. 880–919.

K. Kodaira, *On ordinary differential equations of any even order and the corresponding eigenfunction expansions,* Amer. J. Math. vol. 72 (1950) pp. 502–544.

N. Levinson, *The expansion theorem for singular self-adjoint linear differential operators,* Ann. of Math. ser. 2 vol. 59 (1954) pp. 300–315.

B. M. Levitan, *Proof of a theorem on eigenfunction expansions for self-adjoint differential equations,* C. R. (Doklady) Acad. Sci. URSS vol. 73 (1950) pp. 651–654.

CHAPTER 11

M. Bôcher, *Applications and generalizations of the conception of adjoint systems,* Trans. Amer. Math. Soc. vol. 14 (1913) pp. 403–420.

———, *Leçons sur les méthodes de Sturm,* Paris, 1917.

CHAPTER 12

G. D. Birkhoff, *Boundary value and expansion problems of ordinary differential equations,* Trans. Amer. Math. Soc. vol. 9 (1908) pp. 373–395.

——— and R. E. Langer, *The boundary problems and developments associated with a system of ordinary differential equations of the first order,* Proc. Amer. Acad. Arts Sci. no. 2, vol. 58 (1923) pp. 51–128.

J. D. Tamarkin, *Some general problems of the theory of ordinary linear differential equations and expansions of an arbitrary function in a series of fundamental functions,* Math. Zeit. vol. 27 (1927) pp. 1–54.

CHAPTER 13

R. Bellman, *On the boundedness of solutions of non-linear differential and difference equations,* Trans. Amer. Math. Soc. vol. 46 (1948) pp. 354–388.

N. Levinson, *On stability of non-linear systems of differential equations,* Colloq. Math. II (1949) pp. 40–45.

A. Liapounoff, *Problème général de la stabilité du mouvement,* Ann. Fac. Sci. Univ. Toulouse vol. 9 (1907) pp. 203–475; Ann. of Math. Study no. 17, Princeton, 1947.

O. Perron, *Uber Stabilität und asymptotisches verhalten der Integrale von Differentialgleichungssystemen,* Math. Zeit. vol. 29 (1929) pp. 129–160.

I. Petrovsky, *On the behavior of the integral curves of a system of differential equations in the neighborhood of a singular point,* Rec. Math. (Mat. Sbornik) N.S. vol. 41 (1934) pp. 107–155.

H. Poincaré, *Les méthodes nouvelles de la mécanique céleste,* vol. 1, Paris, 1892, chaps. 3 and 4.

CHAPTER 14

E. A. Coddington and N. Levinson, *Perturbations of linear systems with constant coefficients possessing periodic solutions,* Contributions to the theory of non-linear oscillations, vol. 2, Princeton, 1952, pp. 19–35.

K. O. Friedrichs (and others), *Non-linear mechanics* (mimeographed notes), Brown University, Providence, R.I., 1942–1943.

É. Picard, *Traité d'analyse,* vol. 3, 3d ed., Paris, 1928.

H. Poincaré, *Les méthodes nouvelles de la mécanique céleste,* vol. 1, Paris, 1892, chaps. 3 and 4.

CHAPTER 15

I. Bendixson, *Sur les courbes définies par des équations différentielles,* Acta Math. vol 24 (1901) pp. 1–88.

H. Dulac, *Points singuliers des équations différentielles,* Mém. Sci. Math., Fasc. 61, Paris, 1934.

O. Perron, *Ueber die Gestalt der Integralkurven einer Differentialgleichung erster Ordnung in der Umgebung eines singulären Punktes,* Math. Zeit. vol. 15 (1922) pp. 121–146.

H. Poincaré, *Sur les courbes définies par une équation différentielle,* Oeuvres, vol. 1, Paris, 1892.

A. Wintner, *Asymptotic integration constants in the singularity of Briot-Bouquet,* Amer. J. Math. vol. 68 (1946) pp. 293–300.

———, *Vortices and nodes,* Amer. J. Math. vol. 69 (1947) pp. 815–824.

CHAPTER 16

I. Bendixson, *Sur les courbes définies par des équations différentielles,* Acta Math. vol. 24 (1901) pp. 1–88.

H. Poincaré, *Sur les courbes définies par une équation différentielle,* Oeuvres, vol. 1, Paris, 1892.

CHAPTER 17

P. Bohl, *Über die hinsichtlich der unabhängigen und abhängigen Variabeln periodische Differentialgleichung erster Ordnung,* Acta Math. vol. 40 (1916) pp. 321–336.

A. Denjoy, *Sur les courbes définies par les équations différentielles à la surface du tore,* J. de Math. ser. 9 vol. 11 (1932) pp. 333–375.

E. R. van Kampen, *The topological transformations of a simple closed curve into itself,* Amer. J. Math. vol. 57 (1935) pp. 142–152.

H. Kneser, *Reguläre Kurvenscharen auf den Ringflächen,* Math. Ann. vol. 91 (1923) pp. 135–154.

C. L. Siegel, *Note on differential equations on the torus,* Ann. of Math. ser. 2 vol. 46 (1945) pp. 423–428.

INDEX